PRAISE FOR

The Best of Times

"An earnest, angry, poignant, foreboding book."
—*The New York Times Book Review*

"Beautifully written [and] as full of juicy tidbits as a cherry cake. [Johnson shows] how witty, perceptive and morally grown up American political journalism can be at its best."
—*The Economist*

"An informed, balanced and . . . passionate catalog of the national indulgence and an examination of the forces that fed it. . . . Gripping. A vivid and reliable reminder of what we have been through."
—*The New York Times*

"Drawn with insight, care, and an excellent eye for detail . . . Johnson is among the most brilliant chroniclers of our times, and he scores again here."
—*The Boston Globe*

"Johnson reminds us that good times are never so plentiful that we can afford to squander them on O.J. and Monica. A lesson terrorism has brought home."
—*New York Daily News*

"A wonderful social history of the '90s . . . Johnson's clear prose style and sense of organization make this a fascinating patchwork of events."
—*St. Petersburg Times* (Florida)

"Johnson has a sharp eye for telling a vignette, and it was a stroke of near-genius to open his book on the fateful day in the spring of 1997 when Garry Kasparov, the greatest grand master in the history of chess, was defeated by IBM's chess-playing supercomputer Deep Blue. . . ."
—*Book*

"[Johnson] offers scathing portraits of American popular culture, technology, marketing and the cult of celebrity, and the public's fascination with scandal."
—*The Virginian-Pilot*

"No one can best Haynes Johnson when it comes to giving a panoramic view of an era, boiling it down to its essential elements.... Intriguing, enlightening."
　　　　　　　　　　　　—*Anderson Independent-Mail* (South Carolina)

"Johnson recreates the ups and downs of the '90s with fascinating detail and great insight. The story is a compelling one."
　　　　　　　　　　　　—Book Passage

"An absorbing survey of America's second Gilded Age... A useful summary of recent events [and] a strong candidate for use in survey courses."　　　　　—*Kirkus Reviews* (starred review)

"[Johnson] has written a magnetic book... in startling detail... that every thoughtful American will want to read."
　　　　　　　　　　　　—*Publishers Weekly* (starred review)

"[Johnson's] clear writing and thought-provoking investigations should send this book up the bestseller lists. Strongly recommended for all public libraries."　　　　　　—*Library Journal*

"Johnson has developed a 'franchise': he writes some of his profession's most penetrating 'first drafts of history.'... Expect interest."
　　　　　　　　　　　　—*Booklist*

"Haynes Johnson has painted a vivid, unforgettable portrait of the American 1990s.... *The Best of Times* will long stand as the authoritative evocation of a dizzying decade awash with dot-com dreamers, media madness, runaway scandalmongering, and civic squalor."
　　　—David M. Kennedy, author of the Pulitzer Prize–winning
　　　　　Freedom of Fear: The American People in Depression and War

"In this brilliant and deeply probing work Haynes Johnson takes us to the heart of the American crisis.... It is truly a great and prophetic work."　—James MacGregor Burns, author of the Pulitzer
　　　　　　Prize–winning *Roosevelt: The Soldier of Freedom*

THE
BEST
OF
TIMES

HAYNES JOHNSON

. . . .

THE
BEST
OF
TIMES

*The Boom and Bust Years
of America before and after
Everything Changed*

. . . .

A James H. Silberman Book

A Harvest Book · Harcourt, Inc.

San Diego   New York   London

www.HarcourtBooks.com

Library of Congress Cataloging-in-Publication Data
Johnson, Haynes Bonner, 1931–
The best of times: America in the Clinton years/by Haynes Johnson—1st ed.
p.   cm.
Includes bibliographical references (p. 615) and index.
ISBN 0-15-100445-5
ISBN 0-15-602701-1 (pbk.)
1. United States—Politics and government—1993–2001. 2. United States—
History—1969– 3. Clinton, Bill, 1946– —Influence.   I. Title.
E885.J63 2001
973.929—dc21   2001024753

Text set in Bembo
Designed by G. B. D. Smith

Printed in the United States of America

First Harvest edition 2002
K J I H G F E D C B A

For Kathryn

Contents

Foreword

We all have our memories. Mine began minutes before nine o'clock that Tuesday morning. "Haynes, an airliner just crashed into the World Trade Center in New York," my research assistant shouted as she burst into our seminar room where I was preparing to teach my first class of the fall semester. "It's live on CNN."

I claim no gift of foresight, but my first thought on hearing the news was that this couldn't have been an accident. As a native New Yorker and a veteran air traveler, I know that commercial jets don't fly anywhere near the looming towers at the base of Manhattan. Besides, the weather that morning along the East Coast was close to perfection. Only minutes before, while driving through Rock Creek Park in Northwest Washington, D.C., to the University of Maryland campus at College Park, I was thinking that this was the kind of crystalline day—intense sunshine, light breezes, spectacular foliage—that makes it seem as if all's right with the world. I had other reasons to feel a sense of well-being. Within hours I was to begin a round of interviews about this book, on which I had been laboring for the last four and a half years; the first was scheduled for that night on *The NewsHour with Jim Lehrer.* Now I and countless millions of others found all our plans, all our expectations, being ruthlessly altered by a stunning turn in events.

With word of the crash, I immediately headed for my office upstairs, followed by my students. Together, huddled around the TV set, we witnessed the scenes that became engraved on our, and the nation's, consciousness: the second plane exploding into the other tower; the plumes of billowing smoke and roaring flames; the unbelievable sight of both towers collapsing amid a sea of dust, glass, paper, and twisted metal; the horrifying image of bodies hurtling toward the ground; the pandemonium on the streets as people ran for their lives; the startling live photos from Washington, D.C., where another hijacked plane had just crashed into the nation's military nerve center, the Pentagon, filling the skies with yet more black clouds of smoke and flames and leading to the evacuation of the White House, the Capitol, and the shutdown of the United States government; the unnerving reports, coming amidst a welter of wild rumors about a bombing at the State Department, that a fourth hijacked plane had been headed on a direct path to Washington, with its intended target either the White House or the Capitol building, before it crashed in Pennsylvania.

Wherever we were that day, we understood we were watching something historic. We had left one America behind, an America of tragic complacency and grave distraction. It was an America that had paid scant attention to gathering foreign storms or looming economic collapse, an America whose people were buoyed by get-rich-quick myths, shimmering speculative bubbles, diverting scandals, and arrogant belief in our invulnerability and uniqueness, a special, favored condition that we thought set us apart from the rest of the world.

That America ended in the cataclysmic flames of September 11, 2001; the date is now logged on history's pages simply as *9-11*. In a microsecond, America had entered a far more uncertain era— one that will likely define life in the United States and the world for decades to come as we grapple with the realities of an age of terrorism and confront international forces unlike any that have threatened us in the past.

What follows here is my original portrait of the wonderful but woeful Nineties. That narrative ends with a warning about the dangers posed by what I described as a world that faces "rising tensions between its have and have-not components, growing threats of terrorism accelerated by the dispersal of weapons of mass destruction, and grievously divided fundamentalist and Western industrial societies." For an example of "the fanaticism of extremist elements in some of those societies," I cited shocking actions of "the ruling Taliban movement" in Afghanistan.

Those words were written in March 2001, six months before 9-11. Another six months have passed. The Nineties' bubbles have burst. The party is over. The reckoning has begun. Already, the recklessly self-indulgent Nineties seem as remote and antiquated as the world that awoke after the assassination of the Archduke of Austria in Sarajevo in 1914 to find all previous experience obliterated by the outbreak of a global war. That war would forever change lives, redraw national boundaries, lead to the end of the age of colonialism, and unleash global ideological conflicts lasting the rest of the century. Now 9-11 represents a cleavage in time as sharply delineating an old age from a new as June 28, 1914, when the assassination of Archduke Ferdinand started World War I; October 24 ("Black Thursday") and October 29, 1929 (the day "Wall Street Laid an Egg"), when the stock market crash sank the country, and the world, in economic depression; or December 7, 1941, with the attack on Pearl Harbor.

In offering this paperback edition of the bubble years, I hope the stories I tell of that newly old America will illuminate how in a few short years we went from the best of times to the worst of times. In my afterword, I suggest what lessons we must learn from that experience to avoid further disasters and close the circle on some events that typified the period. For example, readers will not find the word *Enron* in the original manuscript; it now symbolizes the entire giddy era of greed and self-aggrandizement, insider favoritism, manipulation of markets, accumulation of great wealth for the few at the expense of the many, and hubris, all of which characterized the dot-com mania of an age of excess. Finally, I address the formidable new challenges we face, both domestic and foreign, as well as what I believe to be new opportunities at hand if we have the wisdom and the will to seize them.

—*H. J.*
March 2002

To the Reader

Once again, as in *Sleepwalking Through History*, my earlier narrative about America in the Reagan years of the Eighties, I follow the path blazed by that great social historian of the Twenties, Frederick Lewis Allen. His *Only Yesterday*, as he told readers then, was "an attempt to tell, and in some measure to interpret, the story of what in the future may be considered a distinct era in American history." Mine is a similar attempt to tell the story of what I believe will be seen as one of the most consequential eras for America and the world, with ultimate costs yet to be reckoned. It is the story of America during the age of Clinton, an era characterized by accumulation of wealth and self-indulgence, growing out of the great economic expansion of the long boom of the Nineties. As Allen said of his work, the passage of time undoubtedly will reveal errors and deficiencies in his approach, and surely, as he put it, expose "the shortsightedness of many of my judgments and interpretations." The same is so of my effort. My hope is that those inevitable failings will not detract from the larger portrait of a period that already has produced dramatic changes and is certain to lead to even more fateful ones.

About my sources and method: Quotations in the book are

from primary and secondary public sources, so identified, and from my tape-recorded interviews around the country. Where people's thoughts are recalled in the narrative, they come from what people said they *remembered* thinking or feeling at the time or shortly thereafter. Whenever possible, people are identified and quoted directly. In some cases, especially involving the Clinton scandals, I granted specific individuals, at their request, the privilege of anonymity.

Over the course of four years of researching and writing this book, a number of the subjects I initially chose to illustrate the scientific and technological revolution experienced significant changes themselves. This was especially so with Bill Gates and the historic antitrust suit against his Microsoft; J. Craig Venter and Celera's winning the race to map the human genetic code; Bob Shapiro and Monsanto's losing the gamble to be the world leader in genetically modified food and crops. My chapters reflect those changes, but they still remain stories without an end, and so does the greater story of America in its best and worst years of the long boom.

—*H.J.*

It was the best of times,

it was the worst of times,

it was the age of wisdom,

it was the age of foolishness . . .

—CHARLES DICKENS, *A Tale of Two Cities*

THE
BEST
OF
TIMES

Fragments from a Golden Age

When Thackeray presented *Vanity Fair* to the English-speaking people, he described it as "a novel without a hero." He promised his readers they would encounter scenes of high life, scenes of low life, lovemaking for the sentimental, light comedy for the distracted, dreadful combats for the jaded. He hoped the characters strutting across his stage would prove as diverse as they were diverting, but warned sensitive souls they were entering "not a moral place certainly: nor a merry one, though very noisy." For all the promise of that world, he confessed to "a feeling of profound melancholy" as his curtain rose to expose the bustling spectacle behind it.

Thackeray's world was England at its nineteenth-century zenith, proud inheritor of western empires past, and so supreme, so confident in its multiple strengths and successes, that it was the envy of the then-modern world, and devil take those who dared suggest harsher realities existed. My story involves America at *its* zenith a century and a half later, a society so favored as it entered a new millennium that its people could be excused for believing they were experiencing their very best of times, no matter how the glow of new riches blinded them to the evidence of much worse around them. Looking to the future, America's prospects appeared unlimited. Despite an unattractive tendency to boast loudly of their multiple advantages—*We're number one!*—Americans could make a powerful case that not since the peak of the Roman Empire had another society occupied so dominant a world position as they did at century's end.

They enjoyed unprecedented peace and prosperity. Their creation of wealth was unsurpassed, and it was a new wealth shared by more members of the society than ever before. Driven by the force of their longest continuous peacetime boom, the expanding American economy lifted what was already the highest standard of living in the world to even greater heights. So rapid was the growth, so sustained the boom, that Americans were experiencing the best of all economic worlds: low inflation *and* low unemployment, high productivity *and* record high profits. These enviable conditions, emblems of a golden age, prompted scholars to search for comparable historic examples. Few, if any, were found. It was a "new economy," the likes of which had never been experienced before.

In science, in technology, in medicine, in military power, Americans had no equals. As the clock ticked away on the twentieth century, their dominance grew stronger. Nor did their advantages end there. After the global terrors and tragedies into which they had been drawn throughout the century, by the closing years of that epoch they faced no crises domestic or foreign. No specter of another Great Depression then haunted them. American markets continued to soar, setting record after record. The seemingly unstoppable boom roared on. So rewarding was it, so driven by supposedly permanent new factors, that even cautious economic observers began suggesting historic up-and-down market forces had been repealed. Now, the movement was only up—and up, and up. There was no downside.

No threat of another global war existed. The Cold War, the defining episode of the post–World War II era, was over. For half a century, Americans had poured $11 trillion of their treasure into deadly competition with the Soviet Union. They committed the

greatest portion of their military, scientific, and technological resources to that struggle, and an incalculable amount of their psychic energy.

The Cold War affected national attitudes and behavior, government and politics, scientific and technological innovation, public
priorities and personal values. The cost came high in other ways.
Nearly one hundred thousand Americans died in the Cold War
conflicts of Korea and Vietnam. The national debt exploded to historic levels. To finance the demands of the newly emerging National Security State, public funds were diverted to defense from
health and education, public lands and parks, the environment and
the infrastructure. With the Cold War won, no longer did Americans confront the threat of a nuclear exchange with the Soviet
Union, whose empire, even in name, had ceased to exist. Neither
did the spiraling public debt now seem to pose a critical problem.
Suddenly, the deficits turned into surpluses, projected officially to
accumulate far into the next century until they reached $6 trillion.

No new enemies challenged America, certainly none remotely
posing the danger of a Hitler, a Stalin, a Tōjō, a Ho Chi Minh.
The "isms" that bedeviled Americans and the world throughout
the century—fascism, communism—receded into history's pages.
When America stumbled into a war in the Balkans, fears were
expressed that the conflict could spread, like its European predecessors, into the kind of global conflagration that took more than
100 million lives throughout the twentieth century, making it by
far the bloodiest of all centuries.* Yet seventy-eight days after
American involvement began, air strikes alone produced a military
victory without the loss of a single life among the American-led
North Atlantic Treaty Organization (NATO) allies. And this triumph was fashioned largely through technologically superior
American "smart" weapons guided by American earth-orbiting

---

*Since the first century, 149 million people have died in major wars; 111
million of those deaths occurred in the twentieth century. War deaths per
population soared from 3.2 deaths per 1,000 in the sixteenth century to 44.4
per 1,000 in the twentieth.

satellites, assisted by American electronic surveillance, radar-jamming aircraft, and unmanned drones, and delivered by American pilots flying the most modern American warplanes.

No wonder the great majority of Americans tuned out the war news long before it was over. The triumph of American technology simply served to strengthen an already powerful belief that what happens elsewhere has little effect on their lives. Like so many other events that briefly flickered across their television screens or over the Internet, the Balkan War was filed away as a momentary diversion as Americans turned back to another aspect of the post–Cold War era of which they were the great beneficiaries—the triumph of capitalism over communism after a nearly century-long contest. It was a robust, risk-taking variety of capitalistic entrepreneurship that flourished at the millennium, one that created new businesses, forged massive consolidations, amassed vast fortunes.

With all these assets, and with confidence in their future, solid grounds existed for Americans to think their good fortune would continue, perhaps even multiply, propelling them into an even more golden period. A disturbing disconnect was present, however. Despite their blessings, Americans increasingly felt something was wrong with their society.

How to describe *what* was wrong was another matter. It's this, it's that: vague, ill-defined undercurrents of discontent; a sense that something is missing. It's random violence: shootings in schools and workplaces; the terror of strolling down a city sidewalk at night. It's increasing incivility: lack of common courtesy, "road rage" and "air rage," outright hostility, disrespect writ large across the society. It's stress: accumulating pressures on two working parents, on single mothers, on single fathers, on aspiring students, on workers in the fiercely competitive job market. It's frustration: the disappointment at not meeting expectations—yours, your parents', your children's. It's distrust: the lack of confidence in lovers, bosses, coworkers, strangers, authority. It's fear: the growing dread of age, illness, early retirement, relocating, being alone, complex new machines, being overwhelmed by new information and the pace of technological change. It's failing to reach perfection: the perfect romance; the perfect body; the perfect workout; the perfect tummy tuck, eye lift, or penile implant promised by plastic surgeons flaunting youth-enhancing wonders in their glossy ads in the most fashionable magazines.

It's all of these, and none of these. But signals were flashing of surprising public unease—surprising in view of the extraordinary good times that graced America. Were these signs like the symptoms F. Scott Fitzgerald described at the peak of the boom of the Twenties before the crash: "A widespread neurosis began to be evident, faintly signaled, like a nervous beating of the foot"? Perhaps. Probably not.

It was undeniable, however, that a great paradox existed. For all their advantages, Americans felt alienated from their leaders and public institutions, and especially from their political system. Voter participation, a vital sign of the health of a democracy, declined steadily and reached record lows. Fueled by what seemed a wave of scandals, Americans turned inward. Distrust and disbelief spread. Personal, not public, concerns became paramount. The benefits of the boom and the spread of public scandals diverted attention from critical unsolved issues: continuing racial tensions and resentments; inner-city isolation and violence; the failure of public education; the growing health and retirement needs of a rapidly aging population; a vastly widening chasm between the ranks of the very rich and the very poor, while those in the middle were struggling merely to maintain their position; sudden spasms of mass murder and the proliferation of ever more powerful guns in a society that already led all others in deaths from gunfire; alienation and discontent among the nation's children, especially those living in affluent areas supposedly free from the problems afflicting less fortunate youth; depletion of natural resources that threatened another energy crisis in carefree, spendthrift America.

The scandals, one after another, were fed by a mass media scrambling for survival in the face of an increasingly fragmented market. New cable TV outlets proliferated. They attracted viewers by offering sensation and spectacle, the more confrontational and accusatory the better. At the same time, all media, whether mainstream or new, faced fierce competitive challenges from an expanding Internet that fed information into homes and offices everywhere, with a speed and range previously unimaginable. If questions about a possible scandal arose, the most intimate acts of a person's private life were likely to receive the widest possible public exposure. Parents found children asking questions about the most graphic details of sexuality. Gossip, accusation, and rumor— however vile, however unproven—were elevated to new heights and spread instantaneously, unfiltered, into homes everywhere. As the climate of scandal enveloped Washington during the Nineties, it became apparent that something more serious than a series of lurid melodramas rivaling the prime-time TV soaps was at work. In time, everyone touched by them was affected, and all for the worse.

It was against the backdrop of these conflicting forces that I began this work.

The story I tell began as a study of the influence of television on our lives, then expanded gradually to embrace the related cultures of celebrity and scandal (long before someone named Monica entered the scene). Finally, it grew into a chronicle of America, dazzled by celebrity and scandal, while simultaneously being transformed by a scientific/technological/medical revolution. That revolution, creating our new technoworld, has continued at such an astonishingly accelerating rate that even those leading it have difficulty keeping up with all the changes, to say nothing of helping the public understand the implications for society.

In large part, this is a story of opportunities missed: How the United States of America, having arrived at a moment of peak power and influence, possessing the resources and talents necessary to resolve long-term problems affecting all its citizens, given a historic chance to provide leadership for a world increasingly divided between haves and have-nots, and facing more, not fewer, complexities, permitted those opportunities to be squandered through self-destructive squabbles, petty political assaults, growing inattention, and the diversions of the electronic entertainment and scandal culture. It is also the story of how promising leaders betrayed their trust and, instead of rekindling faith, produced discord and cynicism about public institutions, public service, and the political system.

Unlike Thackeray's, this tale is not fiction. It is, however, like his, a story without a hero, but one that contains a symbolic central character: William Jefferson Clinton, forty-first president of the United States, the real-life Gatsby of our times.★ Here is a leader of such talents and complexities, such strengths and weaknesses, that his very successes and failures typify America—an America, like him, both more assured and more ambivalent as it faces its many new challenges in the new era dawning.

Though Thackeray's work and mine are quite different, they have this in common: He drew sketches of English society at its most prosperous for the few as opposed to the many; I attempt sketches of American society at the same stage. His fictional rendering contains elements of high farce and low behavior; my nonfictional account contains elements of low farce and high tragedy. In the Nineties these elements formed a running soap opera, exposing American manners and morals. If presented in a different literary form, or set in a different historical period, hardly anyone would believe these sketches. That they are all true only makes them more incredible—and disturbing.

As my curtain rises, America stands poised to make a great leap into a future that promises to produce the most radical change in all of recorded history. All this raises a question for a new millennium. Is the best—or worst—yet to come? Who knows? It could be either . . . or both.

★He is the forty-first man to be president, even though the State Department, in an act of confounding counting, years ago designated Grover Cleveland as the twenty-second and twenty-fourth because his terms were not consecutive. By that standard, it makes as much sense to count FDR four times, and every two-term president twice. In these pages, I count Clinton as America's forty-first president.

Technotimes

1

Deep (RS/6000 SP) Blue

*"I'm a human being. When I see something
that is well beyond my understanding, I'm afraid."*

G arry Kasparov wasn't just another great chess player, the master of all grandmasters. By universal agreement, he was the greatest chess player in history. In the spring of 1997, at the age of thirty-four, at the peak of the long boom, he had held his world championship for twelve years. Never once had he lost a multigame match against an individual opponent. Never once had he displayed anything but absolute assurance in his chess genius. His attitude toward any rival bordered on the contemptuous, a trait he displayed again after winning, as expected, the first of six games in his heralded rematch that May in New York against an opponent he had soundly defeated just a year before.

As the match resumed, chess experts who gathered to watch the great champion crush his foe witnessed something so unexpected they were left speechless. They were not alone. Millions of observers intently following the contest over the Internet and via worldwide television hookups were astonished to see Kasparov show uncharacteristic signs of confusion. First, he displayed growing doubt, followed by dismay, despair, and loss of control. Finally, he seemed to be having an emotional breakdown. He appeared to be terror stricken.

The first sign that the champion was on the verge of a crack-up came during the second game. It was then that Kasparov encountered something unique in his experience. In the past, he was always able to exploit an opponent's weaknesses by understanding the pattern of thought being employed against him. This time he could not.

That second game ended in a draw. Another draw followed. Then his opponent won a game. When the contestants resumed play on a Saturday, the match was dead even. Kasparov began aggressively, brilliantly; he knew he was winning. His opponent fought back with a series of inspired, indeed brutal, moves that left Kasparov visibly shaken. Grandmasters were shocked to see the champion, for the first time, seem pitiful. He was forced to accept another draw. After a day's break in the match, the denouement came on Monday.

Worldwide attention intensified. Television networks assigned correspondents to cover the event for their lead prime-time broadcasts. Newspapers dispatched top writers, not just their chess analysts, and prepared to open their front pages to report the final results. They and millions more watching on TV and the Internet saw the great Garry Kasparov, the consummate champion whose supreme confidence was matched only by his arrogance, replaced by a nervous, hollowed-out player, his eyes darkened, his manner brooding. He appeared beaten before making his first move.

Kasparov grew even more dispirited as his opponent's swift, ruthless moves drove him into a corner. In a riveting moment captured on television screens, and later on newspaper front pages, after having lost his queen and with his king dangerously exposed to checkmate, the champion leaned forward over the chessboard. He placed his hands over his face and eyes, and lowered his head dejectedly. It became an enduring portrait of human despair.

Moments later Kasparov suddenly stood up. He was resigning the game and match, he announced. Only nineteen moves had been played.

Grandmasters were amazed at the way the champion abruptly crumbled. "It had the impact of a Greek tragedy," said the chairman of the chess committee responsible for officiating the match. Kasparov reacted more simply. "I lost my fighting spirit," he said. "I was not in the mood of playing at all."

Asked to explain why, at a tumultuous news conference minutes later, he replied, "I'm a human being. When I see something that is well beyond my understanding, I'm afraid."

· · ·

Kasparov's opponent had no reaction, maintaining the same state as when the battle began: motionless, positioned inside a bare windowless air-conditioned closet, high over the city in a midtown Manhattan skyscraper, its immense size and weight all but unattended by any human beings.

The victor was the IBM RS/6000 SP supercomputer, christened by its creators "Deep Blue." This behemoth, whose twin metallic structures were described by one *New York Times* writer as resembling nothing so much as amplifiers at a rock concert, stood six feet five inches tall. Each of its towers weighed twenty-eight hundred pounds, for a combined weight of over $2\frac{3}{4}$ tons. Internally, its 516 chess microprocessors were capable of examining 200 million chess positions a second, or 50 billion every three minutes, all while operating at a speed 250 times faster than desktop computers.

In the year since Kasparov had first bested Deep Blue, IBM technicians had doubled its capacity. They also conducted near-daily brainstorming sessions with programmers, researchers, and outside chess experts. Their efforts were rewarded by a spectacular success. In the glow of their triumph, Deep Blue's project manager, C. J. Tan, was magnanimous in victory and praised the dejected Kasparov. "Garry has a brilliant mind, and he's a very brave man," Tan said. "He's a man who sees the future, who understands where technology can take us." When asked by reporters why there had been such global interest in the match, Tan replied: "Because it shows what technology can do for man and how far we can take it."

· · ·

As a news story, the match was a natural: Man versus Machine. Machine wins. As a modern fable, it was fulfillment of an age-old dream. For centuries, scientists and charlatans alike had envisioned the day when machines would beat humans at the intellectually demanding game of chess. After countless failures, that day had come.

Though die-hard chess purists disparaged Deep Blue's victory as nothing more than a highly hyped gimmick, a mechanical game without real significance, it represented something far more important. It was a symbol of the times, a herald of the future.

Long before the match, IBM's supercomputer already had proved its immense value by becoming essential to the successful functioning of the end-of-the-century world. The same type of computer as Deep Blue ran the Web site that enabled millions to watch the chess match around the globe. These computers were also used for a wide variety of other purposes: to enable the newly designed Internet to link people everywhere in ways never before possible; to design new drugs for pharmaceutical companies; to process enormous amounts of data, allowing for computations critical for success in such diverse areas as molecular fluid dynamics, market analysis, and even the functioning of world financial markets; to allow genetic engineers to map genomes of various species, promising to bring revolutionary changes in the very structure and function of life; to simulate nuclear explosions; to coordinate airline reservation and air traffic control systems; to guide rockets as they plunge deep into the universe, exploring other planets and perhaps discovering conclusive evidence of extraterrestrial life.

All these and far more were now part of the rapidly expanding province of the computer. Suddenly human technology was transforming life in the most mundane and the most unimaginable ways, directly affecting individuals, businesses, and nations everywhere. Now computers allowed companies to analyze their data in such a way as to discover previously unnoted trends: a convenience store found that premium beers sold better on Friday nights if located next to diapers and Victoria's Secret realized that women's breast sizes vary in different parts of the country and adjusted their inventories in individual stores accordingly. Welcome to Technotimes. It was a brave new world, and the computer was king.

Without it, the long boom could not have happened. The computer provided the catalyst that ignited the enormous burst of economic energy in the Nineties. Fueled by expanding technologies, including the acceleration of the computer industry itself, new businesses and new fortunes were created at an unprecedented pace.

In the pop-culture aftermath of the match between Kasparov and the supercomputer, journalists scrambled to publish features tracing Deep Blue's antecedents to show how they led to a crowning moment of the computer age. In so doing, they perpetuated one of the great myths of how Technotimes, with all its promise and potential problems, came to be.

*　*　*

The most celebrated tale of the boom was of the rise of new entrepreneurs, bold risk-taking venture capitalists who became the richest of the very rich. The fable comes in two parts. The first, an Eighties tale, tells of youthful geniuses tinkering in garages and emerging with new discoveries that created new technologies that formed new businesses that produced new fortunes. The second, a Nineties tale, tells of young computer "geeks" and "nerds" enjoying the most luxuriant lifestyles after capitalizing beyond wildest dreams from overnight start-up ventures and IPOs that left them, still in their twenties, richer than any Americans before them, richer than their Gatsbyesque counterparts in the Roaring Twenties, richer than those favored few of the Gilded Age in the 1880s.

There's truth in the tales, but as always with tales of heroes, their exploits are colored as much by myth as by reality.

The myth, formed in part by arrogant self-promotion and in part by lack of historical perspective, is that it all began with them. Out of their own singular talents, *they* forged the future. They are the real masters of the universe. The reality tells a different story. Long before Bill Gates and other icons of the dazzling new Technotimes were born, long before popular accounts of their fabulous fortunes fired the public imagination, many of the great developments that ignited the boom were already solidly in place. Interactive computing, software engineering, the Internet, e-mail—all these and much more were waiting to be seized and used, thanks to decades-long creative efforts and strong financial support of others. Not least was the indispensable backing of the American people and their government, which together had mobilized all the resources of the society—economic, physical, and intellectual—in a common partnership that made possible the wonders of the age to come and the great boom that transformed it.

That's the story those historical accounts missed when chronicling Deep Blue's triumph over the great chess master. Focusing on the powers of the new machines without examining how they achieved such power or questioning what problems they might be creating, as many of those accounts did, was akin to describing waves crashing on shore without explaining the seismic activities and midoceanic undercurrents that caused them to crash with such force.

Culture of Success

"This is the center of the universe..."

Although the Internet is still in its infancy, its impact is already being compared to that of Gutenberg's invention of moveable type more than five hundred years before, which led to printing, the end of the Dark Ages, and the widest dissemination of knowledge in history. The rise of the Internet in the mid-Nineties, along with that of its twin, the World Wide Web, initiated the most dramatic phase of the technological revolution that created the long boom. Instantly, life around the globe began to change.

Here, it was immediately said, is the *real* revolution, the truly gigantic leap into a different world. Serious people hailed the Internet as "the most transforming invention in human history" with "the capacity to change everything" and create change "at far greater speed than the other great disruptive technologies of the twentieth century, such as electricity, the telephone and the car." The Internet promised a new way of living, of communicating, of doing business, of engaging in sex, of dealing with each other, not only in our communities and country but throughout the world.

The Internet ushered in a new world, one in which people could sit in their homes and transact business and pay bills, buy and shop, trade stocks and make investments, book travel reservations and rent vacation homes, exchange messages and documents, and move from serious to playtime activities by linking everything from the latest offerings in museum exhibits in Paris and Rome to the most explicit pornography, all in vibrant color. It affected attitudes about society, about work, about government, about private and public interests, about the future. It's the perfect tool for the best of times, the linchpin for the "new economy" of the computer-driven, get-rich-quick, out-for-yourself information age.

Through the ability instantaneously to link the world electronically via the Web, the Internet has become the most rapid agent of technological change ever.

Or so *everyone* says, conveniently ignoring the fundamental technological advances of the industrial age that, in the nineteenth century alone, created change on a scale unmatched by any previous era. In only a few decades, the Western world moved from a rural, agrarian base to an urban, industrial one. Since biblical times people had relied exclusively on sources of power such as windmills and draft animals. Suddenly machines were replacing horses and oxen and would soon replace humans.

With the introduction of entirely new forms of power in the nineteenth century—steam, oil, electricity, even atomic possibilities—as well as the creation of the telegraph, the telephone, and the electric light, the rate of technological change accelerated by what later would be termed quantum leaps. As the American historian Henry Adams observed, a child born in 1900 was entering a world never known before, a world of suddenly erupting new forces that would change life as surely as the scientific discoveries of Copernicus, Galileo, and Newton had created the revolutionary change of centuries earlier.★

★Adams, contemplating the discoveries of the invisible Roentgen X rays and radium within five years of each other in the late 1890s, believed they made the world face the revolutionary concept that force was both inexhaustible and unlimited. At the turn of the twentieth century, he observed that ". . . the

The early decades of the twentieth century witnessed even more explosive technological change with the introduction of automobiles, airplanes, radio, rockets, phonographs, records, motion pictures, television, air conditioning, miracle medical innovations, and discoveries leading to atomic power and man's journey into space.

Yet even in an age of excess, when hyperbole has become the coin of the realm, the Internet's effect on business alone can hardly be overstated. It has altered not only how businesses act but how they're managed and even what they are. Computerized trading radically transformed global capital markets: from London to Tokyo, from New York to Singapore, markets were operating around the clock, with trading taking place on a scale and at a pace impossible in the very recent past. In seconds, transactions representing trillions of dollars hurtle around the globe through cyberspace. With the Internet, businesses everywhere can be linked electronically to global currents that are, to use a cliché, shrinking the world around us.

Official government economic data at the end of 1999 credited the Internet with generating more than a third of America's economic growth between 1995 and 1998, the very peak years of the boom. By the year 2006, government experts estimated, nearly half the American workforce would be employed in industries either producing or using information technology. Even before then, the astonishingly rapid spread of Internet availability would result in a billion—yes, *billion*—people around the world adopting it for daily use.

By June 1999, figures like these, combined perhaps with overbearing arrogance, led Andrew Grove, chairman of Intel, headquartered in the Silicon Valley, to say that in five years' time all companies would be Internet companies or they wouldn't be companies at all. Less blunt, but no less arresting, were widely quoted remarks made around that time by Lou Gerstner, IBM's chief

---

man of science must have been sleepy indeed who did not jump like a scared dog when, in 1898, Mme. Curie threw on his desk the metaphysical bomb she called radium." Until then, Adams also noted, "the atom itself had figured only as a fiction of thought," but now "power leaped from every atom, and enough of it to supply the stellar universe" if it could be harnessed.

executive, in a presentation to leading Wall Street analysts. Gerstner characterized the fervor over new Internet-created businesses as "fireflies before the storm—all stirred up, throwing off sparks." Get ready for what's to come, he warned them. "The storm that's arriving—the *real* disturbance in the force—is when the thousands and thousands of institutions that exist today seize the power of this global computing and communications infrastructure and use it to transform themselves. That's the real revolution."

Time may prove Gerstner right about the "real revolution" yet to come, but in the aftermath of the crash of the technology sector, which started in 2000 and accelerated dramatically into 2001, with Internet businesses either failing or not producing their hoped-for profits, the "old economy" proved less a risk to investors and demonstrated it still possessed the power to be profitable.

· · ·

America's fascination with the wonders—and wealth—being created by the Internet and its new dot-com world can be dated from an August day in 1995. That's when a Silicon Valley start-up company made what became a legendary initial public stock offering.

The company was called Netscape. When the stock exchanges opened that day, Netscape was only sixteen months old. It had never shown a profit, and it was giving away its main product free. Its plan was to offer its stock to the public at fourteen dollars a share, a price based upon a modest expectation of generating $13 million in sales by the end of the next year. Yet even before the trading bell rang, Wall Street demand was so great that Netscape's offering price had doubled to twenty-eight dollars. Within an hour, it was trading at seventy-one dollars a share. By day's end, as the writer Gary Rivlin recounts, Netscape "was worth as much as or more than such Wall Street stalwarts as Boise Cascade, Bethlehem Steel, and Owens-Corning." Within four more months "its market capitalization would exceed those of United Airlines, Apple Computer, Marriott International, and Tyson Foods."

And the reason for this astonishing instant success? Netscape possessed a new browser—a kind of software, or computer program, that enables computer users to tour the World Wide Web and the cyberspace world of the Internet—at a time when people were suddenly discovering the world online. Netscape's browser was not the first to be invented, but it had advantages in permitting both businesses and consumers to view pictures and text as they surfed the Web.

When Netscape went public that day, a young man named Marc Andreessen was credited with creating its browser. Then twenty-four, Andreessen had begun work on this version of the browser several years before as a programmer at the University of Illinois's National Center for Supercomputing Applications. As with many others, he went west to Silicon Valley, leaving the publicly funded academia to develop work begun there for commercial use. His move paid off spectacularly; he was soon worth $130 million. Immediately, he became one of Silicon Valley's celebrated young entrepreneurs, popularized by the press as the fabulously wealthy founders of Technotimes. In Andreessen's case, he posed for the cover of a national magazine barefoot and in jeans. Appropriately for the best of times, he was photographed sitting on a golden throne.*

· · ·

Toward the end of the Nineties, when I began gathering material for this book and reconnected with Silicon Valley people I have known over the years, a new boom was sweeping the valley and the nation, this one the greatest of all. It was a boom for which Silicon Valley credited itself, smugly, for creating and leading.

In the daily press, in national magazines, in pop-culture books

*Gary Rivlin adds a sardonic touch to this overnight success story: The head of the Illinois supercomputing center asked Andreessen to stay on after graduation, but the youth left "because a precondition for a full-time job there was that he leave the browser project, then called Mosaic." Rivlin quotes the director as telling Andreessen: "Some forty people had a role in creating Mosaic. Don't you think it's time to give someone else a chance to share the glory?" Andreessen did not—a familiar aspect to the story of Technotimes.

about Silicon Valley success, fevered tales were told—and retold. There was the story about a six-month waiting list for new Porsches at a car dealership in the valley. There was the tale of the young executive who paid an additional million dollars beyond the $2.4 million asking price for a "home in a comfortable Palo Alto neighborhood." There was the survey showing that sixty-four new millionaires were being created *each day* in Silicon Valley, most of them in their twenties and early thirties. There were numerous stories reporting the latest $100 million high-tech fortune amassed by another young valley "techie," eclipsing the "mere" $10 million—benchmark fortune there in the recent past. There were reports of Silicon Valley company stock certificates being bestowed on a Presbyterian Church instead of cash contributions, of a Roman Catholic school that placed part of its endowment in a venture capital fund. There were stories of bigger signing bonuses— bonuses of $30,000 or more for entry-level jobs—paid by high-tech firms to graduating college seniors, and of the $150,000 salary a valley law firm offered a graduating law student, all driven by frenzied competition that forced employers to offer vastly inflated inducements to prevent new talent from joining rival firms.

"Signing bonuses!" a valley executive told me. "We're taking off from the sports world where you sign somebody to a huge bonus to get him to join your team. It's the whole new economy thing, and the new economy is totally different. In the last ten years we've seen the greatest legal creation of wealth in the history of the world, and a good part of it has happened right here. You don't work for wages anymore. You work for stock options to catch the upside in the creation of wealth. This is my advice as a parent: They say, 'Geez, Dad, when I get out of school I'll get a signing bonus. They'll pay me this.' I say: 'Just make sure you think equity. Equity. All the time think equity. I don't give a crap what you make a year. What's a big number to you, a quarter million, a half million dollars? The way you create wealth is through equity. Put yourself in a position to do that.' More and more kids are thinking that way." More and more Silicon Valley kids, for sure.

Still, no matter what I saw before, or have heard since, about the new boom and the bizarre personal behavior it induced—hot-shots sleeping under desks, stripping nude before computer consoles after hours, racing radio-controlled toy cars between cubicles, or engaging in firefights with Ping-Pong–ball guns—nothing prepared me for the extraordinary scenes of new success.

The landscape of Silicon Valley was being transformed. New shopping centers, offering the world's finest luxury goods, covered what had been vacant land only a few years before. New multimillion-dollar homes and office buildings rose in what already was America's highest-priced real estate market, and that market was increasing in value almost daily. In order to afford the historic highs of Silicon Valley housing prices, observed John Markoff of the *New York Times,* only partly facetiously, instead of requiring two high-income breadwinners "each home will instead soon require two instant-millionaire beneficiaries of initial public offerings."

Everywhere construction cranes towered above the palms and shopping centers and suburban developments. Rolls-Royces, Bentleys, Jaguars, BMWs, filled parking lots at lunchtime. Costly new restaurants, equaling the best of San Francisco and New York, if not Paris and Rome, were crowded every night.

And, omnipresent, moving purposefully about their tasks, were clearly confident, casually dressed young people—they *all* seemed young—and preponderantly male. Cell phones sprouted from their ears. They always seemed to be conferring through them, whether walking, driving in their new cars, or dining in their fine restaurants.

These are anecdotal observations, mere impressions, easily open to exaggeration. But the economic statistics of Silicon Valley tell a story that cannot be misinterpreted. In 1996, the property tax roll in Santa Clara County, within whose borders Silicon Valley mainly lies, increased by $5.2 billion. That represented nearly 25 percent of the property tax roll growth of the entire state of California. In 1997, those figures doubled. In 1998, they doubled again.

Silicon Valley's job production was the envy of the nation.

Adding significantly to the influx of people in the valley were substantial colonies of permanent new immigrants from France, Russia, India, Hong Kong, China, Taiwan.★ They represented some of the best scientific and technological talents of those countries and the valley's gain was a loss for their societies. In the first nine years of the Nineties, the valley added 377,000 high-tech workers to its employment rolls; in 1999, the average high-tech wage there was $72,000, highest in the state. Within another decade, valley businesses expected to create more than 180,000 additional jobs.

"That gives you an idea of what's happening here," said my old friend the county assessor Larry Stone, after providing a host of figures he had gathered for me from his records. "Economically, this place is absolutely on fire."

He was right about that, as well as about the influence of the firms whose names have become synonymous with Silicon Valley—the Suns and SGIs, Ciscos and 3Coms, Netscapes and Yahoos—that were then, as analyst Thomas Scoville put it, "a chain of locomotives pulling the entire U.S. economy forward." Yet even all these dramatic kinds of changes didn't capture the most striking aspect of the new Silicon Valley—the attitudes of the people themselves. In the past the entrepreneurs of Silicon Valley possessed a number of familiar American attributes. They exuded confidence in their ability and destiny. They believed they made their own breaks through talent and perseverance; they subscribed to the Darwinian law-of-the-jungle, winner-take-all school of life. They distrusted government, had contempt for bureaucracies, thought Washington and all its works irrelevant. Their rejection of rules and authority at times bordered on anarchy, but they were convinced that, for all its failings, American society was by far the best. In their energy, their acquisitiveness, their arrogance and con-

---

★In 1999 the DBF (for Doing Business in French) association of new technologists, representing new immigrants from France to the valley, had five thousand members, not counting their family members. Ethnic Chinese and Indian immigrants were then running nearly a quarter of valley high-tech firms.

ceit, their blend of selfishness and idealism, their lack of interest in the past, their studied casualness edged with cynicism, their infatuation with—if not worship of—youth, their behavior reflected the larger society. In the new Silicon Valley, all these behavioral traits and attitudes flourished along with another long-familiar one—a habitual tendency to braggadocio, the kind of boast that years before had led one valley entrepreneur to describe the valley to me as "the spearhead of evolution." Even more than in the past, hyperbole was the calling card of the new valley entrepreneurs. As one person told me, entirely seriously, "This is the center of the universe."

Foolish as such a remark is, it captures the self-absorbed sense of so many that *their* talents and *their* evolving technologies were largely, if not entirely, responsible for creating the great Internet-driven dot-com boom that brought with it such riches.

· · ·

The boom seemed unstoppable. Nothing like it had happened, not in the Gilded Age of the 1880s when predatory Robber Barons flaunted the massive industrial fortunes they had accrued, not in the speculative craze of the 1920s when barbers and shoeshine boys scrambled to put all their meager assets into margin accounts and see them multiply as stock prices kept going up and up.

By the late Nineties, ten thousand Web sites were being created daily, billion-dollar Internet-inspired deals were taking place at a rate of thirty *each day,* and new Internet companies continued to blossom, causing more and more people to join in the fevered rush to invest in start-ups—even though those start-ups had yet to produce a profit, and many never would.

Dreams of becoming the latest dot-com billionaire fired imaginations across the land. And why not? Everywhere, it seemed, were new stories about new winners of Technotimes. Never had so many been so rich. In just a two-year period, from 1995 to 1997, the number of Americans reporting adjusted gross incomes of a million dollars or more increased by nearly two-thirds to 142,566. By 1999, in a survey cited by the *Wall Street Journal,* the number of U.S. households with total assets of a million dollars or more had soared to 6.5 million, an increase of 67 percent since 1995. A third of those households were headed by someone between the ages of eighteen and thirty-nine. By the millennium, the number of American households with a net income of more than $10 million had increased fourfold since the start of the Nineties to about 350,000.

No longer is *millionaire* the ultimate mark of material success in America. Now you have to be a *billionaire* to achieve that status—or, as daily scribes scrambling to find new ways to describe instant "Silicon zillionaires" write of the latest Silicon winners, "*gazillionaires.*"

Even allowing for the overblown predictions about the benefits mankind will experience from the latest great technological advance—historically, the end of war and poverty are the most mentioned in such forecasts—perhaps the Internet will change our lives even more than we can imagine. Perhaps it will allow us to live in a paperless world or go to college at home. Perhaps, as some visionaries believe, it will enable humans to break barriers of language and geography, thus radically changing fundamental concepts of nationhood and culture.

As one of my Silicon Valley friends asked: "What if in ten years—and I don't think this is unreasonable—you could send an e-mail in English to Germany and they get it in their language? What happens when *that* happens?"

Or, as Larry Stone said, "You ask what's on the horizon; well, there's the high-tech industry and there's the biotech industry. Somewhere out in the future, biotech is going to be more important to us here in Silicon Valley. When you get into genomics, we find the possibility of the cure for cystic fibrosis, for cancer obviously. The possibility of finding cures for these gene-oriented diseases is fabulous. What does that mean? What does it mean to increase your life span by twenty-five years? Cloning! What the hell does *that* mean to society?" He laughed. "We might be doing this interview sixty-five years from now." Then he became serious again. "That is really a future we're just beginning to touch."

If all these life-altering changes occur, it becomes a matter of more than passing interest for people and society to understand how these forces came to be, what consequences they bring, what lessons can be learned from them, and what part places like Silicon Valley *really* did—and did not—play in creating the technological revolution.

·   ·   ·

By the end of the Nineties, the best was yet to come, or so people believed. But exactly at that point, shadows began appearing over the valley. Two new factors arose simultaneously to threaten the valley's position as world leader of high tech.

One was familiar—a replay of twenty years before when Sunnyvale, the valley's unofficial capital, declared a moratorium on industrial growth because of concerns that its quality of life was being strangled. Once again, at the end of the Nineties, valley business and civic leaders faced the prospect that forces beyond their control threatened to overwhelm them. A survey they themselves commissioned concluded, as the *New York Times* reported, that "the booming Internet economy is decentralizing technology industries and threatening Silicon Valley's domination of all things digital." Internet companies and the prosperity they generated were being lured into other regions of the country. As those areas attracted high-tech businesses, Silicon Valley faced a critical shortage of skilled workers, a development driven by its high housing and living costs, insane commuting problems, and intractable congestion. Silicon Valley was in danger of becoming a victim of the very culture of success it had created. It wasn't unusual to hear even highly successful people in the valley express growing frustration over feeling more disconnected personally from others while being among the most connected people electronically anywhere. As one executive told me, "It's a real emptiness when your best pal is your PC and your latest software and your knowledge about connectivity."

The valley's second problem posed a more direct threat. Science, it seemed, had reached one of its unbreakable barriers. Suddenly, silicon chips, the marvel of the techno-age, the brains of computers, and the building blocks of everything electronic, appeared to have reached their limits. By the year 2012, scarcely a decade away, the technology that permits chips to keep getting smaller as their power grows greater no longer would be able to accomplish that end.

This news, first disclosed in the magazine *Nature* at the end of June 1999, prompted a small *Washington Post* headline atop a one-inch science note at the bottom of an inside page: "Chips Nearing Smallness Limit."

In the following days, a few longer stories and essays appeared, none on page one, chronicling the approaching end of a historic era. The silicon age seemed to be as doomed as the dinosaurs.

Scarcely three weeks passed before other news overtook word of that supposedly untimely demise. Out of Silicon Valley came the announcement that another, greater, leap into the techno-future was possible. Technological developments were being perfected that could lead to a molecule-size computer chip capable of permitting computers to operate 100 billion times as fast as today's personal computers. Not only that, the new technological breakthrough would open the possibility of supercomputing power "so pervasive and inexpensive that it literally becomes an integral part of every man-made object."

These reports—page one this time, at least in the *New York Times*—came out of Palo Alto in the heart of Silicon Valley. They described how researchers at Hewlett-Packard Co. there and the University of California, Los Angeles, had devised a breathtaking new molecular computer technology that promised to bring even more stunning change in the new century than did silicon chips in the last.

In its inside pages, the *Times* noted, without further elaboration, that the Pentagon "was a major underwriter of the work." That same day, at the very end of its report, the *Wall Street Journal* said the Hewlett-Packard/UCLA project "is just one of many projects in molecular electronics around the world at institutions and corporations like International Business Machines Corp., Hitachi Ltd., the Massachusetts Institute of Technology, Rice University, and others funded by the Pentagon's Defense Advanced Research Projects Agency."

The role of the Defense Advanced Research Projects Agency, or DARPA, was not elaborated upon in either article, but without it, and more important, without the remarkable talents of the people who first created and led it into advanced computer development, the history of the dot-com age and America's place in it over the past half century—and the history of the new economy—would be vastly different.

Many people are responsible for that great achievement, but the efforts of two almost unknown Technotimes pioneers—and the society that supported them—deserve special attention. One was a scientist out of the Massachusetts Institute of Technology named Vannevar Bush, the other an unassuming research scientist named J. C. R. Licklider, also out of MIT, who regarded Bush as "the main external influence" for his own profoundly influential ideas.

Only at century's end are these two unsung heroes of the computer age beginning to receive credit for the parts they played in developing the technological innovations and the concepts that decades later led directly to the boom.

Some scholars now rank Bush as the most influential engineer of the twentieth century and as "the most politically powerful inventor in America since Benjamin Franklin." He is credited with being "the true father of hypertext" and even of the personal computer and the World Wide Web.

Only at the millennium is Licklider, too, receiving credit for being the real creator of that supposedly new technological triumph of the Nineties, the Internet, which has been responsible for much of the wealth spurred by the global computer revolution. Fables notwithstanding, the Internet isn't new, nor did the technological advances that made possible the prosperity of the Nineties suddenly spring to life in places like Silicon Valley. Both were created by government-sponsored work carried out at immense cost over many decades. In Licklider's case, that work took place secretly during the coldest period of the Cold War when a crisis over Sputnik, the Russian earth-orbiting satellite, "a tin can in the sky," led to fear of nuclear annihilation. In Bush's, the work grew out of visionary peacetime leadership stemming from his experience in the most horrific and technologically destructive war in human experience, a war in which he played a significant part in winning.

I tell their stories here not to exhume old bones from a vanished wartime era that, at the millennium, seems as remote to most Americans as the powdered wigs and silken breeches worn by the Founding Fathers. I tell these stories because the largely untold one of how Technotimes was born becomes increasingly relevant as America hurls itself into an ever more technology-driven world.

· · · ·

At the end of the Nineties, an internal IBM paper recounted a long-forgotten article published in the summer of 1945 during the closing weeks of World War II. In it, Vannevar Bush described a theoretical machine he called the Memex, never built, that could extend human memory by mechanically organizing information and making it readily accessible through a web of associations. A Memex, he explained to readers of the *Atlantic Monthly* in July 1945, "is a device in which an individual stores all his books, records, and communications, and which is mechanized so that it may be consulted with exceeding speed and flexibility." Through it, people would be able to store, retrieve, consult, and learn from the information it contained and from the cumulative record of human success and failure to that point in history. By providing people greater access to the bewildering store of knowledge that science was accumulating, the Memex would allow mankind to profit from the inherited knowledge of the ages.

Looking ahead, Bush envisioned machines capable of compressing a library of a million volumes into a box placed on the end of a desk before the person operating it from a keyboard. A "slanting translucent screen" extending from the machine would display information the user wanted. "The world," he noted that July of 1945, "has arrived at an age of cheap complex devices of great reliability; and something is bound to come of it." In recalling Bush's Memex more than half a century later, IBM said his article describing it "turned out to be of those time-bomb essays—a piece so far ahead of its time that it takes decades to recognize its genius. For the Memex in essence is a personal computer, and more than that it is a personal computer in which information is bound together by links of association. Every time a Web user fires up her browser and navigates from site to site, following threads of relationship as she roams, she is, in effect, continuing a journey that began with Vannevar Bush more than half a century ago."

Vannevar Bush's prototype of the personal computer introduced into the American, and world, market thirty-six years after his article has become an innovation so familiar that grandparents—and great-grandparents!—use it to communicate with their families, conduct personal business, and access a variety of information daily. They do so without a clue as to who Vannevar Bush was or the great debt they owe him and other Technotimes pioneers.

History mainly remembers Vannevar Bush, if at all, as the man who briefed Harry Truman about what Truman described as "a scientist's version of the atomic bomb" after Franklin Delano Roosevelt's death in April of 1945. Until that moment, Truman knew nothing of the ultrasecret atomic bomb project.

"Admiral Leahy was with me when Dr. Bush told me this astonishing fact," Truman recalled in his memoirs. He was referring to Fleet Admiral William D. Leahy, who had served as Roosevelt's personal chief of staff, presiding over wartime meetings of the Joint Chiefs of Staff.

Truman vividly remembered the admiral's instantaneous reaction to Bush's Oval Office briefing, delivered in Leahy's "sturdy, salty manner."

"This is the biggest fool thing we have ever done," the admiral told the president as Bush listened. "The bomb will never go off, and I speak as an expert in explosives."

If Leahy's prediction is one of the most wrongheaded in history, Bush's subsequent private report to Truman as the war was ending three months later—the same month his article appeared in *Atlantic*—on "how science could best serve the nation and how government might best support science" stands as one of the most farsighted.

The report Bush handed Truman that July day was thirty-four pages long and titled, simply, *Science—The Endless Frontier*. It became one of the most influential documents in the nation's history, with results felt for the rest of the century and beyond. It set in motion the forces that crested so powerfully in the America of the Nineties with the creation of the great technology-driven boom. Bush's report was instrumental in creating the National Science Foundation and, through federal grants, the further expansion of university and private-industry research laboratories, as well as their connection with such places as the National Institutes of Health. His vision transformed the relationship of science and government by advocating government support for basic research in universities and private industry. From it, too, flowed the same support for basic research in peacetime that his White House's Office of Scientific and Research Development had given to universities and industries in wartime. Despite its influence, except for some academics and aging policymakers, more than half a century later the American people remain largely unaware of the contents, or even the existence, of Vannevar Bush's report. History, capricious as ever, has assigned his credit for shaping the future to others.

Without adoption of Bush's ideas as a pillar of national policy, many of the most significant scientific and technological innovations of the next half century, including those in the field of computer science, would not have occurred. In his paper, commissioned eight months earlier by a dying Franklin D. Roosevelt who had asked him to propose the direction for American science and technology policies in a postwar world, Bush strongly urged Truman to break from the past and set an unprecedented peacetime goal for the United States: To make an open-ended government commitment to support long-term scientific and technological research.

In Bush's mind, this was the great lesson of World War II. All elements of the society, working cooperatively, had led to the scientific and technological developments that changed the nature of warfare and set the stage for the betterment of civilian society: the cracking and harnessing of the atom to produce the atomic bomb and the nuclear age; the development of radar, rockets, missiles, jet engines, the proximity fuse, and ever-more-intricate electronic sensing devices; the discovery of miracle drugs such as sulfathiazole, streptomycin, and penicillin; and, far from incidentally for the future, the development of computers to guide the new weapons systems and rocketry that later put Americans on the moon.

For progress to continue in the new post–World War II era, Bush argued, policymakers had to discard old ways of thinking. In the past, despite the myths about American yeoman tinkerers possessing unique talents, Americans did not lead the world in developing new technologies. They believed they could always import knowledge for their scientific and technological progress, largely by building on basic discoveries of European scientists. That belief had guided the nation until World War II. It was no longer relevant, for, as Bush said: "*A nation which depends upon others for its new basic scientific knowledge will be slow in its industrial progress and weak in its competitive position. . . .*"

Only the federal government possessed the resources necessary to finance—and nurture—the kinds of long-term basic research projects that would take years to carry out but eventually would create fundamental changes far into the future.

At the time he made his historic proposal to Truman, Vannevar Bush occupied a singular national position. Then in his fifties, he was a scientist of towering reputation whose own earlier creative work had earned him a footnote in the development of important innovations that eventually led to modern computers. For a quarter of a century Bush was a member of the electrical engineering department at MIT, contributing original research in optical and photocomposition devices, eventually becoming dean and MIT's vice president; later, he became president of the Carnegie Institution in Washington, D.C. Many of his students went on to become leaders of the computer age. Among them was Frederick Terman, who earned his Ph.D. under Bush and became the legendary Stanford dean of engineering hailed in the Nineties as "the father of Silicon Valley." Terman, who like many other scientists called Bush his mentor, adopted a great number of Bush's ideas in developing the area around Stanford into the world center of high technology. In 1940, with war clouds gathering over the United States, President Roosevelt named Bush chairman of the National Defense Research Committee, and the following year, as America entered the war, the president appointed him to direct the White House's Office of Scientific Research and Development.

There, as the president's science adviser, Bush was charged with the immense responsibility of mobilizing the government's entire wartime research efforts for weapons development, including the secret atomic bomb project. He was directly in charge of coordinating the activities of the six thousand leading U.S. scientists and engineers involved in the total war effort.

Under his leadership, the universities were fully mobilized for war. Large campus laboratories, such as the Radiation and Instrumentation Laboratories at MIT, were created with government funds. The military services expanded their own laboratories as the government's contractual relationships with industry and the universities were greatly extended. The public-private partnership formed through government research contracts vastly hastened technical advances, notably in developing computers essential for military strategy and success.

Until Bush made his proposal to Truman, the United States had never come close to adopting a national industrial policy—and still has not—to say nothing of a national policy for science and technology. The very notion flew in the face of deeply engrained public opposition to such an expansion of government's role, especially at a time when every American instinct was to disengage from the enormous effort of total wartime mobilization. Seen from the vantage of half a century later, the wonder is that Bush's report to Truman, advocating a vast expansion of the federal role, prevailed even as America began a period of swift demobilization. Bush's report became the framework that guided the United States for decades, ushering in a period of remarkable scientific and technological change in the areas that most fundamentally affected life in the United States and, indeed, the world. Even more remarkable, perhaps, is that this vast expansion of the federal role occurred essentially without a galvanizing—and perhaps destructive—national debate. It simply seems to have happened.

It didn't "just happen," of course. It grew directly, and naturally, out of the common experience of suffering that shaped an entire generation of Americans. Self-interest dictated their course. To survive the Great Depression and fight and win World War II, they were compelled to join in common purpose. That cast of mind influenced them when they began considering the world beyond the war, and America's place in it. At the end of World War II, a consensus quickly emerged among the nation's political, scientific, industrial, and intellectual leaders that it was in America's best interest to make a long-term commitment to basic research.*

And the nation did—with results that produced the burst of creativity that exploded into an Internet-driven technological revolution in the Nineties. An inevitable irony arises out of this story. Fifty years later, many wunderkinds of the Nineties believe that *they* created those technological forces. In fact, the fastest-growing parts of the economy in the Nineties were dependent upon the publicly funded support that was the legacy of Vannevar Bush and the World War II era. And none was more dependent on government support for its creation than the Internet, whose development for public use in the mid-Nineties made everything that went before it in the electronic revolution seem merely a prelude.

There, the second great forgotten technotimes pioneer, J. C. R. Licklider, enters the story.

In the summer of 1960, Licklider, an obscure MIT research scientist—"Lick" to his friends, gentle, soft-spoken, modest to a fault—published a seven-page paper in an even more obscure technical journal called *IRE Transactions on Human Factors in Electronics*. His title was hardly crafted to stir public interest: "Man-Computer Symbiosis." Yet that scholarly paper became one of the most influential works of the twentieth century. It provided the

---

*This consensus ranks as one of the two great post–World War II U.S. policy achievements. The first is America's postwar commitment to rebuild the devastated economies of Europe and Asia, including America's erstwhile enemies Germany and Japan, through Marshall Plan aid. The Marshall Plan deserves credit for "saving the world" from communism by stabilizing vulnerable Western democracies. The "Bush Plan," though it never received any official imprimatur and remains essentially unknown to the general public, deserves credit for "changing the world" by spurring the technological revolution of the past half century.

intellectual framework for decades of computer research that led ultimately to the creation of the Internet and the explosion of personal and commercial computer usage destined to affect life even more directly in succeeding centuries than in the closing years of the twentieth. "The hope is that in not too many years," Licklider wrote, "human brains and computing machines will be coupled together very tightly, and that the resulting partnership will think as no human brain has ever thought and process data in a way not approached by the information-handling machines we know today."

His single declarative sentence was breathtaking in its implications, but Licklider offered a grander vision. He foresaw the rapidly approaching day when "artificial intelligence" emanating from machines would operate the most complex technological systems. "It seems entirely possible," he predicted, "that, in due course, electronic or chemical 'machines' will outdo the human brain in most of the functions we now consider exclusively within its province." Already, as he observed, a number of "theorem-proving, problem-solving, chess-playing, and pattern-recognizing" computer programs existed that were "capable of rivaling human intellectual performance in restricted areas."

That was written thirty-seven years before Deep Blue awed the world by displaying powers and a form of new intelligence that frightened and humbled the greatest chess player in history.

Because computers then in use were far too costly for one man—or one business—to afford, Licklider said networks of time-sharing systems would have to be established to reduce expense and broaden the range of interests served. He envisioned a network of such centers connected to one another by communication lines and to individual users by leased-wire services. There it was, the entire essence of the Internet, two generations before the Internet was supposedly created in the mid-Nineties.

Two years after writing his remarkably prophetic paper, the visionary Licklider was picked to head the U.S. government's secret research effort to improve the military's use of computer technology. He assumed that key position at a critical moment in the Cold War when the United States was pouring billions of dollars into countering the perceived threat of attack by Soviet satellite-guided missiles within its borders. A new agency was created within the Defense Department—the Advanced Research Projects Agency, later the Defense Advanced Research Projects Agency, known to scientists and engineers with top secret clearances as DARPA. One area of particular concern was the creation of a decentralized system for the distribution of information that would continue to function despite widespread destruction. Among those recruited to the new agency was Licklider. Quietly, and beyond public notice, he began working to put into practice what he was then calling his "Galactic Network" concept—a globally connected net through which everyone could quickly access computer data and programs from any site in the world. Licklider, following in the footsteps of his model and MIT predecessor, Vannevar Bush, created and managed a DARPA program that funded basic research through government grants to universities and industrial laboratories.★

Six years after joining DARPA, Licklider published another prophetic article, this time in collaboration with fellow computer pioneer Robert Taylor. Their title, in *Science and Technology* magazine, didn't stir the blood, either: "The Computer as a Communication Device." Their opening sentences, however, were riveting—if anyone read them.

"In a few years," they began, "men will be able to communicate more effectively through a machine than face to face.

"That is a rather startling thing to say, but it is our conclusion."

They sketched a vision of a transforming age being born. For the first time, they wrote, a new medium existed to create change on a scale never before experienced. That "is the programmed digital computer" which "can change the nature and value of communication even more profoundly than did the printing press and the picture tube. . . ." In their vision of this new world, they employed terms completely unknown to Americans of the time that are now part of the language itself. Computers connected to other computers

---

★The first university project Licklider funded through DARPA was Project MAC, later MIT's famous Laboratory for Computer Science, which Licklider himself headed upon rejoining the university. Years later this laboratory developed the World Wide Web.

will form an "interactive network" of "on-line communities." The new network will be operated by trained people sitting before "monitors" with a "terminal keyboard," using "electronic pointer controllers called 'mice.'" They will "log in," enter their password, and begin communicating. Linking together these online communities will be a computer network "large enough to support extensive general purpose information processing and storage facilities." This electronic network, they predicted, will change basic functions of work and life.

Licklider and Taylor's glimpse into the future was marvelous and, for the most futuristic of peoples, should have been a natural. But the timing of their article guaranteed it would be ignored and remain essentially unknown to the public. The America in which it was published was a nation in turmoil, divided by war and social unrest, whose disquiet would be felt for the rest of the century.

Their article appeared in April of 1968, one of the most fateful months in American history. The assassination of Martin Luther King that month by a white man in Memphis plunged a nation already more dangerously divided than at any time since the Civil War by civil unrest, poverty, and Vietnam into a state approaching anarchy. Only a brief period before Americans had believed they were heading toward a Great Society through the leadership of Lyndon Johnson, the most powerful president since FDR. The King shooting, occurring less than five years after John F. Kennedy's assassination, intensified conspiracy theories. Two months later Robert F. Kennedy was murdered in a Los Angeles hotel kitchen, the victim of another assassin. Politics and public life, government and leaders, were affected for years to come.

No wonder the futuristic musings of a couple of government computer scientists attracted no attention in that terrible spring of 1968. More's the pity, too, for the remainder of their article contained even more specific descriptions of the new world struggling to rise. It also raised important questions about possible negative effects, as well as positive ones, that society would have to address— and still has not—to fulfill the promise of the new electronic age.

Buried in their article that spring was a brief disclosure that, once again, went unnoticed: "A network of 14 such diverse com-

puters, all of which will be capable of sharing one another's resources, is now being planned by the Defense Department's Advanced Research Projects Agency, and its contractors."

A year and a half later, the fruits of that effort led to the formal
birth of the Internet.

· · ·

Time: Labor Day weekend, 1969. Place: UCLA. Setting: Room
3400, Boelter Hall, the Science Building. Cast: Computer sciences
Prof. Leonard Kleinrock★ and a group of graduate students and
government and business engineers are tensely seeing if they can
transmit bits of information from a huge steel machine to a host
computer twenty feet away. So massive is the machine, with its
four metal hooks on top, that when delivered from Boston earlier
that weekend it has to be winched up the side of the building by
crane because it will not fit in the elevator. They turn on the machine; it works. The information moves from machine to computer. The first test run is successful. A month later, the *real* test of
the DARPA-funded computer network occurs: To see if they can
send data from their computer to one at the Stanford Research
Laboratory. In Los Angeles, they establish a telephone connection
with their Stanford colleagues. They begin typing *login* on their
screen to see if it appears on the Stanford monitor, hundreds of
miles to the north in Palo Alto. They type the L.

"Do you see the L?" a member of Kleinrock's team asks over
the phone.

"Yes, we see the L," Stanford replies.

The next letter is typed into the system, and again the telephonic exchange:

"Do you see the O?"

"Yes, we see the O," comes the answer.

Another letter, G, is struck. Then, the system crashes.

Yet the revolution has begun. It's a moment, brief but historic,
that deserves to take its place with Alexander Graham Bell's famous

---

★Kleinrock is yet another of the remarkable computer talents to have come
out of MIT graduate studies.

first words spoken over the telephone in 1876, "Mr. Watson—come here—I want to see you," and with Wilbur Wright in 1903 watching the frail plane carrying his brother Orville lift into the skies from the sands at Kitty Hawk and fly for fifty-nine seconds at a speed of thirty miles an hour.

Three years later, in 1972, the first use of electronic mail occurs; it stems from the need of DARPA developers to read, file, forward, and respond to messages. The Internet is on the way, and with it, Technotimes.

· · ·

Neither Bush nor Licklider believed their government-financed technological work should be used solely for the benefit of the military. Bush, for one, was an ardent advocate of openness and opposed much of the official secrecy that later cloaked weapons development research. He insisted upon the communication of scientific ideas, the widest dissemination of new discoveries. The results of government-funded research, he believed, should be made readily available to private industry.*

Licklider, too, ardently believed that government-developed technological innovation should be used to benefit all of society. He also believed the interconnected electronic network he was developing had the potential to expand education and knowledge and be "a boon to mankind...beyond measure." He even believed it could so spur the economy that "unemployment would disappear from the face of the earth forever."

---

*The rapidity of the revolution, and the computers that drive it, can be seen in comparing the ENIAC, the government's massive World War II computer, with a single microprocessor chip developed for business by Intel in 1971: Intel's minuscule chip was as powerful as the ENIAC. By the end of the 1980s, modern supercomputers could do 250 million multiplications a second—about 172,800 times faster than ENIAC. Since then their speed and capacity have grown greater and greater, confirming Moore's law, the term adopted worldwide to describe the extraordinary rate of growth of computing power. Moore's law, named after Intel's cofounder Gordon Moore, holds that computing power will double every eighteen months. So it has, with almost incalculable effect on society.

But Licklider, like Bush, was predominantly a realist.

He warned that new technology could also be detrimental to society. In his mind, the principal question was how the new interconnected network would be used.

"For society," he said, writing in 1968, "the impact will be good or bad, depending mainly on the question: Will 'to be on line' be a privilege or a right?"

If only a favored segment of the population got a chance to enjoy the advantages of the new network, he feared the unequal access would further deepen the dangerous divisions that already afflicted American society. Those connected to the network would prosper greatly, he predicted. Those disconnected would fall farther behind.

Which is what happened. By the end of the century this gap was being called "the digital divide."

. . .

In the America of the millennium, high-income, better-educated haves are rapidly expanding their use of the new technology. Low-income, less-educated have-nots are rapidly falling farther behind. The gap between them is widening. The result is a societal "digital divide" that's turning into a "racial ravine," as the U.S. Commerce Department put it in making public the dismaying results of its third national survey on how the American people use, and have access to, computers and the Internet. The results of that survey, published at the end of 1999, revealed a disparity that continues today.

One side, predominantly white and prosperous, was increasing its prosperity by making greater use of computers and the Internet. The other side, primarily people of color and poor, was becoming more impoverished as they had even less access to the new technology. Thus, America's poor were falling farther behind America's affluent, and this separation occurred despite declining costs of computers and the software that ran them. As a result, minorities increasingly were facing many greater disadvantages in competing for the most sought-after entry-level jobs—those requiring knowledge of computers and familiarity in navigating the Internet.

That condition underscores the failure of the Internet to fulfill

Licklider's great hope that the online network would expand education and knowledge for all the people, not only for their betterment but for the nation's. It isn't a failure of the Internet, of course. The failure is society's and especially of its leaders. As the official studies show, the already great gap between Americans with high incomes and higher education and those with lower incomes and less education is widening. People with college degrees are eight times more likely to own a computer and sixteen times more likely to have Internet access than those with only an elementary school education. Families with parents who attended some college are three times more likely to have Internet access than those with parents who have only high school diplomas. A child in a low-income white family is three times more likely to have Internet access than a child in a comparable black family, and four times more likely than a comparable Hispanic child.

These dismal findings mirror those of university and corporate studies on public usage of the computer and the Internet, except that a number of those reports show the racial and economic divisions growing even worse.

As Licklider feared years before, Americans at the bottom, already the most likely to fail, are becoming more disconnected from the society. Their worsening situation brings with it all the social dynamite, all the midnight in their hearts, that their condition implies.

· · ·

One of the most surprising results to emerge from the accumulating official data—surprising, given the breathless media accounts of successes of the boom in the closing years of the Nineties—is the almost startling disparity in incomes that has been developing. By the end of 1999, according to data compiled by the Congressional Budget Office, four out of five American households, or about 217 million people, were taking home a thinner slice of the economic pie than in 1977. At the same time, more than 90 percent of the increase in national family income was going to the richest 1 percent of households. Incomes of the richest Americans were rising twice as fast as those of the middle class.

Even more startling are the figures for the rewards gained by business leaders. In 1980, heads of American corporations were earning over forty times more than their workers. By the early Nineties, just as the boom was getting under way, they were earning more than *ninety* times more than their workers. By the end of the Nineties, the gap between top and bottom had widened even more astoundingly. Then, heads of American corporations were earning 419 times as much as industrial workers! This figure prompted the *Economist* to call it the greatest peacetime transfer of wealth in history, a sober assessment given the dimensions of the extraordinary shift in economic wealth and power.

Income disparity is only one of a host of paradoxes Americans are confronting. The rapid pace and complexity of technological change raise difficult new questions about personal values and lifestyles. While the innovations of Technotimes produce satisfaction and pleasure in many areas of life, they also bring new frustrations and problems: A deluge of information, both wanted and unwanted, may prove overwhelming in a time when the average American office worker sends or receives, on average, 201 messages a day, and spam and junk clutter the screen. The user-friendly interface might turn hostile at any moment as anyone knows who has forgotten a password, been frustrated over opening a lengthy e-mail attachment, read the peremptory announcement: *You have performed an illegal operation; this program is being shut down,* or had his computer crash at an inopportune moment. Being online also means being potentially vulnerable to viruses, worms, and the pranks and predations of hackers. The electronic invasion of privacy, whether licit or illicit, exceeds anything known before, and you needn't log on to experience its effects. Telemarketers call at all hours, endlessly trying to sell you something you don't want, don't need, and probably can't afford. On the other side, dial an office or corporation and you are likely to reach only a tinny telephonic voice: *Here's your menu option. Press one for . . . Press two for . . .* It's hard to escape the tyranny of the machine, the feeling of becoming more like an insignificant pawn in the power of great impersonal technological and corporate forces.

Whether it's the best of Technotimes or the worst, the effect on society is far more complicated than in the past. Charlie Chaplin would have a richer role to play in the new century depicting the average person's frustrations with the complexities of the computer age than he did in the Thirties portraying frustrations with machine age life on the assembly lines.

Demands of the job keep increasing. Time to accomplish tasks keeps shrinking. Keeping up with the latest technological advances becomes more burdensome. Merely getting to work, as we've seen in Silicon Valley, often turns into a horrendous daily ordeal of navigating clogged highways or risking malfunctioning public transportation systems—or no public transit.

Here, perhaps, is the greatest paradox of Technotimes: Despite technological advances that are supposed to free people from burdensome tasks and provide greater leisure time, despite the great drop in the official number of hours in the workweek, from about a hundred to thirty-seven and a half, more people are actually working longer than ever. Not only are they working longer, increasingly, they're working more from their homes than their offices. At century's end, nearly 30 million work at home at least part of the time. Those numbers are rising as more people become professional "telecommuters."

All this causes a profound shift in the nature of work. Therein lies another striking paradox. As long as the bulk of the workforce was in office and blue-collar work, national laws requiring that the regular workweek be forty hours or less could be enforced. At the end of the day people could leave their work when they left the workplace. Now, with laptops, modems, faxes, beepers, and cell phones, the workplace is virtually anyplace, with the result that people are working longer than ever.

Even something as basic as getting adequate sleep appears to be a casualty of the increasingly pressured, around-the-clock technological work environment. Far from being an incidental factor, sleep deprivation appears to be a growing national problem.

At Stanford, a nationally recognized professor of psychology and expert on sleep and dreams, William Dement, is developing research showing that Americans in many respects are living in what is increasingly becoming a sleep-deprived society.

Which isn't exactly what the utopian dreamers of the wonders of Technotimes had in mind.

· · ·

It's a golden age of scientific discovery, David Goodstein, distinguished physicist and assistant provost at Caltech, home of so many Nobel Prize winners, was telling me in Pasadena, California. "Week by week we learn things that are astonishing and exciting. But the profession of science, which is quite different from the discovery of science, is going through a very long, extended, and difficult period. We haven't figured out how to rearrange that. Something's gone awry with the country, too. We've allowed this to happen. They are symptoms of the fact that having won the Cold War and being left without any of the traditional problems, we don't know what to do with ourselves now. We're trying to figure out what our national goal is."

Then Goodstein expressed a theme that applies perfectly to the promise—and problems—of Technotimes. "This is the era of *Pax Americana,*" he said. "We have it all in our lap. We can do anything we want, but we're not doing it."

Those were private remarks, made on the West Coast at the end of the Nineties. About the same time, on the East Coast, another national leader in science and technology made similar remarks in a public address.

The setting for them was Capitol Hill; the occasion, the National Summit on High Technology and the testimony of the president of MIT, Charles M. Vest, the only university representative to appear. Vest came to Washington to deliver an urgent message—and a blunt warning—to the senators gathered to hear him.

In words that sounded eerily like those that Vannevar Bush addressed to Truman so long ago, MIT's president reminded the senators that America's future prosperity rests upon developing new knowledge and then educating and training people to apply that knowledge practically and use it for further innovation. "The knowledge driving today's industries," he said, "has been accumulated during the last forty years of federal and industrial support of long-term research." Economists generally agree, he told them, "that more than half of our economic growth since World War II is due to technological innovation, largely through federally sponsored research in our universities."

Then he got to the heart of his message: Are we doing the right things to generate the knowledge that will drive future economic success? he asked, then quickly answered his own question. "No. We are reducing our investments. We are going in the wrong direction."

For more than a decade, he said, federal expenditure for research and development had been decreasing by about 2.6 percent per year. More troubling, between 1993 and 1997, peak years of the boom, funding for basic and applied research dropped precipitously, falling 12 percent as a share of the nation's gross economic product.

The MIT president was scathing in describing other national failings. The nation's public education system from kindergarten through twelfth grade was "a disgrace." American eighth-grade students ranked behind fifteen other countries in having access to computers in their homes. The nation was failing to attract sufficient numbers of bright young men and women into science, engineering, and mathematics.* A joint study by Harvard and MIT showed the United States falling behind other countries in producing technological innovation. While the United States still

---

*He could have made his case even stronger by citing other findings, such as that U.S. students rank eighteenth worldwide in math and physics. In a major international study comparing the knowledge of high school seniors in math and science in twenty-one countries, only two nations ranked lower than the United States—Cyprus and South Africa. In 1998, U.S. colleges and universities awarded only 12,500 bachelor of science degrees in electrical engineering—less than half those awarded a decade before. Congress contributed greatly to the severe shortage of U.S. physicists by cutting the budget for basic research in physics every year since the 1970s and compounded that situation by accelerating those cuts in the 1990s. This contributed to the sharp decline in the number of students—cut in half—in graduate programs from the peaks of the 1960s. Increasingly, those students are foreign-born. Now, half the entering graduate students are foreign compared to 20 percent in the 1960s.

ranked near the top, "the gap with other nations is becoming in-creasingly small" and within six years America's position will likely "drop below several other countries."

Nor was the steady reduction in federal support for research the only problem. At the same time, major U.S. corporations were also cutting back "very substantially on fundamental long-term re-search. Why? Because it is not clear that the benefits of such re-search will likely accrue directly to the performing company."

In other words, they would not generate immediate profits.

None of this bodes well for future American innovation, Vest warned. As America enters the new millennium, it may be "living off historical assets that are not being renewed."

The strong message he hoped to deliver was: "What is missing is a sense of urgency."

If Vest succeeded in conveying that sense of urgency to the senators, it was not passed on to the public. Not a line about his testimony appeared in next morning's *New York Times* or *Washington Post*.

But why should that be surprising? During the boom, Ameri-cans and their media had many other diversions to claim their attention.

· · · Chapter · · ·

3

Nerd Nirvana

"It's going to be a wild ride . . ."

B y the end of the Nineties, the silicon world was spreading in all directions. Outside of the valley itself, nowhere was the electronic boom more evident than in the Pacific Coast city of Seattle, six hundred miles north of San Francisco.

There, numerous high-tech firms were operating, including at least half a dozen Internet companies boasting of having among the highest valuations of new Net businesses. Of these, the most noted success was Amazon.com. Within four years of its creation in 1995, the online bookseller had become the most powerful Internet merchant. Millions of customers shopped there electronically for the ease, the convenience, and the selection—a happy combination that almost overnight generated $3 million daily in sales of books, music, and videos and made its thirty-plus founder, Jeff Bezos, America's latest instant billionaire. His great personal gain contrasted with the jarring fact that Amazon.com, like so many of the hot new dot-coms, had yet to turn an annual profit.

No matter how spectacular these Internet successes were for their founders, they were overshadowed by the giant winner of them all, Microsoft, the signal success of Technotimes, sitting comfortably in its headquarters in Redmond, seventeen miles from Seattle.

. . .

To someone like me, whose technology business point of reference is the densely packed, flat, unattractive Silicon Valley, the first impression of the world of Microsoft is of a place of surpassing peace and beauty. The drive itself from Seattle, heading across Lake Washington via the Evergreen Point Floating Bridge toward Bellevue and beyond to Redmond, creates a welcome sense of escape from crowds and congestion. To the east, silhouetted against the distant skyline, tower the magnificent Cascades. To the west stand the snowcapped Olympic Mountains. Along the highways, amid verdant fields and evergreens, wildflowers bloom. By the time you turn off the highway and head toward the grounds that Microsoft occupies on its 295 acres, the feeling of having escaped intensifies.

Microsoft appears to be an oasis of calm. Its modern buildings, attractively designed with gleaming glass and metal, stand comfortably apart amid well-tended lawns and firs. Driving through the curving roadways that connect them, you pass crowds of young employees playing soccer on one of the many sports fields dotting the compound. If you didn't know otherwise, this could be a college campus. Instead, it is headquarters of the world's most profitable, most controversial, most intensely competitive corporation, headed by the world's wealthiest man, Bill Gates.

In the view of the U.S. government, Microsoft's Redmond campus is more than a mere corporate headquarters. It's the center of what the government charged, and in a notable trial proved, is the most notorious monopoly to dominate business and affect society since John D. Rockefeller's Standard Oil and its interlocking trusts controlled the supply and production of the new industrial age resource that changed the nature of the world's economy then—petroleum. By the early twentieth century, Rockefeller, with a petroleum-industry market share of more than 80 percent, ruled over the greatest of monopolies; it was a monopoly he had forged through skill, daring, and raw power, and it enabled him to crush competitors, set prices, and control markets. At the turn of the next century, Gates dominated the heartblood of the computer age almost as powerfully as Rockefeller had that of the industrial age. His Microsoft controlled software that operated the basic workings of 90 percent of all personal computers sold; he presided over a computer-generated empire that was transforming society no less than oil had a century earlier. The problem was the way Microsoft conducted its business, which is why in 1998 the government charged it, like Standard Oil long before, with violating antitrust laws passed a century earlier to curb the predations of the Robber Barons. In each case, the government charged the respective company with conspiracy to coerce and crush competitors in order to maintain monopoly positions, all to the detriment of fair trade and the interests of American consumers.

*   *   *

None of this is at all evident when you first visit Microsoft. The initial impression of tranquillity is exactly the one Microsoft's headquarters is designed to create. Its employees and top executives refer to it as their "campus." They all speak of its "culture," their "Microsoft culture." Taking the cue from its leader and founder Bill Gates, Microsoft's is a culture based upon youth, ambition, talent, and work and—as the government demonstrated in winning its case—a killer instinct.

Of the 31,000 Microsoft employees when I visited there at the end of 1998, the average age was thirty-four. Nearly three-quarters were men. They worked "Microsoft hours"—as long as it takes. Time was not a factor; performance was. Which may help explain why by 1997 Microsoft's ranks included 21,000 millionaires, or more than two-thirds of its employees. More were on the way.

Another initial impression is also deliberately created: the casual atmosphere that so strongly strikes the first-time visitor. The dress is most informal. For men: no ties, no dress shirts, no suits, no blazers. Just open sports shirts and slacks, thank you. For women: no designer dresses, no heels, no attention-calling jewelry. It's all most functional.

The pace, too, appears leisurely. Perhaps people *do* race down hallways brimming to disclose the latest major technological innovation they've discovered. If so, the stranger doesn't see, or hear, them. What one sees are employees indistinguishable from employers calmly going about their business, including taking breaks during the day for games on the campus greens. Not just at lunch hours, either; competing teams can be seen throughout the day. What one hears is notable more for its silence than its sound. No shouting, no frantic Wall Street–type phone calls intrude upon the quietude. Indeed, there appears to be little conversation, though that initial impression, too, is misleading.

A recent graduate from Duke University describes how he works with his team of four or five partners on a major new technology project. He has what appears to be a blank check and the personal blessings of Bill Gates himself, who met with him and his team and gave the go-ahead. The team members constantly communicate with each other, he explains, during long work hours that include many meetings, many consultations. But they communicate, it turns out, neither by phone nor face-to-face. Even though they all sit near one another in adjoining cubicles, they do their constant communicating electronically. Their e-mail consultations are among the 3.5 million e-mail messages exchanged daily on the Microsoft campus—new wired world, indeed.

When you do have a chance to talk with them, the experience leaves still another strong impression. It's not that people sound or act alike, robotically speaking, but their remarks are characterized by two striking themes: enthusiasm for their work and a competitive sense of being engaged in something approaching a life-and-death battle.

That young team leader working with Gates's personal blessing is typical.

Scott Guthrie, age twenty-three, was first recruited by Microsoft in Durham, North Carolina, while still a junior at Duke. After earning his degree, he moved to Redmond on a three-month Microsoft internship. He "fell in love" with Microsoft; they, obviously, were impressed by him. A year later, he was program manager for an Internet information server product team that Gates and Microsoft hope will produce important innovations and new profits.

"It's going to be a wild ride," Guthrie says, talking about the intense competition they face from companies "out there." "Microsoft," he explains, "is working on types of Internet infrastructure applications that will be killer applications of the next century, or even of the next five years." He goes on to emphasize the challenges they face as they work to develop dramatic new methods of "fundamentally changing the way business is done on the Web."

Young Guthrie represents something else not at first evident from the seemingly laid-back atmosphere of the Microsoft campus. As Microsoft's renown—and, not incidentally, financial rewards—become more celebrated, the competition to work—and succeed—there becomes even greater. Into Redmond's personnel offices pour 15,000 resumes each month. That amounts to 180,000 applicants a year. Of those, Microsoft hires only 3,000. Even by gaining a foothold on the Microsoft ladder, the chosen job seeker has won a highly competitive battle. But that's just the beginning of the struggle.

From the first moment of arrival, intense competition and pressure to perform are a Microsoft employee's constant companions. There seems to be no letup. If anything, the pressure intensifies, and so do the stakes.

Far from the easygoing campus atmosphere that presents itself so attractively, Microsoft operates on what one senior executive describes as a tightly controlled, monitored, and rated "performance culture." It works like this: Every six months, *every* employee must write down a set of between three and five "performance-oriented objectives" describing what the employee thinks he can accomplish over the next six months. After that time is up, the employee meets with his manager for a formal performance evaluation on how well—or not—the job objectives have been achieved. The employee, in writing, assesses how well *he* thinks he did. The manager, also in writing, gives *his* response to the employee's self-evaluation. The manager adds a critical factor: a numerical grade, just as in school, that gives each employee a performance ranking, based on a scale of 1.0 to 5.0 on a forced-distribution curve, so only a certain percentage of really high scores are allowed. That becomes critical because, an executive explained, "those ratings correlate directly to your pay increase, your bonuses, your future job opportunities in the company, because if there's a job open and I have two candidates, I'm gonna look to the one with the highest performance rating. That's most importantly tied to your eligibility for future stock options, which is of huge importance to this company. You can go from getting zero shares to perhaps thousands of shares depending upon your performance history."

Once understood, the seemingly unstructured and leisurely paced Microsoft corporate campus assumes a far different shape and character. Nor is this highly structured, if not intimidating, employee performance rating system left to random selection or the personal views of individual managers. In the end, the ratings are measured and tabulated by the very soul of the Microsoft business—the machine. All the data is entered into personal computer systems. Managers with several hundred people beneath them are provided scorecards to tabulate the performance results and final rankings. As the executive says, "PC-based systems control it all and store it all, so it's very rigorous and surprising."

In describing this aspect of the Microsoft culture, I do not mean to imply that it resembles an Orwellian world where the individual is trapped by authoritarian or mechanistic forces beyond his control. On the contrary, Microsoft employees fully understand the system and seem, on the basis of their testimony, to relish and thrive under it. As they tell it, they love working there and love the intensely competitive atmosphere of working on the cutting edge of technological innovations that affect life.

At the same time, it would be most naive to ignore the clear sense of Darwinian capitalism at work in a new millennium—and equally foolish not to find clues from it about the behavioral forces and values that drive America's new technology business.

To understand Microsoft *and* the high-tech industry, says Mike Murray, vice president of human resources and administration, and one of Microsoft's most influential executives, you have to understand one basic fact: Microsoft is a manifestation of the very practical entrepreneurial ideas that have been emanating from Silicon

Valley for more than twenty years. "We have no history ourselves," Murray says. "We carry no baggage ourselves, so we can start with a blank sheet of paper. We're highly opportunistic, which is one of the key attributes of any young, technology-driven company."

Then he says, "A generic trait of all such companies is that you must be highly attuned to competition. This is guts ball at its highest level in the sense that there are winners and there are losers. It's very much like politics. You get 50.1 percent of the vote and you're in for four years. The other guy who got 49.9, he's toast—he's history."

Fear of becoming history spurs the high-tech companies to greater efforts, all while looking warily over their respective shoulders.

Inside Microsoft, people uniformly speak of being engaged in battles. They also express fear of Microsoft being beaten or displaced or losing its edge. Along with that external fear comes another, more internal, one: the corporation's fear of losing talented people. At Microsoft, that concern speaks volumes about success, Microsoft-style. Because so many Microsoft employees possess great wealth at a young age, many become far too wealthy to want to continue working. So they leave. "We have a very large number of employees, between the ages of thirty-five and forty-five, who have been here eight to twelve years, who, because of their stock option program have created independent wealth for themselves," says Mike Murray. "In some cases, it's amazing. Our stock option program is an overachiever. We didn't expect it to do so well. We expected our stock option program to allow people to get a downpayment for their house, to put away some money for their kids' education, and to build a bit of a safety net. We didn't expect it to be able to *buy* the safety net company with it. But it has. So many of our employees have already hit the financial finish line in their lives—but way too soon." (Of course, those words came before the collapse of many dot-coms, which left thousands of young workers who accepted those options in lieu of higher salaries with the realization that their options were worthless and they were without jobs.)

As Murray says, in a great understatement about these young techies who have accumulated tens of millions of dollars and even more by their late thirties: "It's an interesting dilemma. Why do I work? It used to be really easy to answer that. 'I've got a mortgage, I've got a family, I've got to take care of them. I've got to salute the flag and do what my dad did. I go to work.' You dismember that and suddenly you're asking yourself: Do I work because I love to work? Because I have to work? Why?"

Microsoft's interest in this sociological dilemma created by its own success is far from abstract. Microsoft worries that it stands to lose talented people years, even decades, before they could be expected to produce some of their most creative work. At the time of my visit, Microsoft executives like Murray said the company didn't face a major attrition problem. But personnel statistics already were showing that employees who had been there a longer period of time were leaving at a higher rate.

Other fears drive the Microsoft culture—the fear of becoming, to quote Murray again, just another "traditional large company" riven with inevitable bureaucratic problems that plague large entities whether in business or in government. The fear of becoming too big is discussed constantly along with the fear of creating conditions that produce failure—of becoming too muscle-bound, too cumbersome, too vulnerable to unknown new competitors.

"It's a natural phenomenon," Murray says, spelling out the corporate philosophy. "If you become the winner, you *really* win. If you become the loser, you *really* lose. It's also the same within our culture, and similar cultures at Apple, Sun Microsystems, Oracle. But your success is never guaranteed. We're always very nervous about whether we can continue to maintain our position, continue to make products people want. If they don't want them—boom! It's over. We're out. We're history."

All these attitudes illuminate the motivations behind Microsoft's sense of being constantly engaged in warfare, its determination to win, its kill-or-be-killed tactics. If this fairly captures the Microsoft mantra, its raison d'être—and I believe it portrays the essence—there's no doubt that Microsoft employees embrace it.

There's no doubt, either, that they take their signals from the attitudes and behavior of their leader.

* * *

Bill Gates needs no introduction. Everyone, everywhere, knows he's "the world's richest man"—literally the first $100 billion man, who in 1998 was on his way to becoming the first trillionaire by the year 2004, if his fortune holds. In many respects, he is an unlikely model for an entrepreneurial titan, the kind who follow the Horatio Alger school of success and make it on their own. Bill Gates didn't start at the bottom, nor did he lack financial resources and personal comfort. He was, in fact, the scion of one of the wealthiest old families in the Pacific Northwest and was given every advantage money and connections could provide, including the best private schools and then on to the East Coast and Harvard.

During the boom, no one attracted greater public attention than Gates or was subjected to such endless scrutiny and speculation. No one received more contradictory accounts. Envy and admiration colored them all. Inevitably, they produced a murky, and certainly an unfinished, portrait. He was a buccaneer or a philanthropist, a bully or a genius, a nerd or a visionary. His company, too, received the most conflicting descriptions. Microsoft was a ruthless monopoly or the model of a forward-looking modern corporation where talent and initiative are rewarded equably from top to bottom. On one point, though, there is no disagreement. The story of Technotimes and its imprint on the future cannot be told without examining the influence of Gates and Microsoft.

The rise of Bill Gates was as swift as John D. Rockefeller's a century before. And the power he exercised in his chosen field was comparable. From November 1975, when Gates and his Seattle boyhood friend and fellow computer aficionado, Paul Allen, founded the then Micro-Soft company, its success was one of the most spectacular in financial history.

At the end of the first year of operation, their books showed seven employees and seven thousand dollars in revenue. In 1979, after relocating from Albuquerque to Seattle, Micro-Soft had grown to twenty-eight employees and produced nearly $2.5 million in revenue. Four years later, after continuing to prosper, they introduced their Windows software, which became the industry standard. By adopting Apple's icon-based system and replacing the clumsy MS-DOS interface, Microsoft revolutionized the PC world. Within a year, their sales doubled to nearly $100 million.

On March 13, 1986, the nine-year-old Microsoft offered its stock for public purchase. The price: twenty-one dollars per share. It was one of the greatest stock buys in history.

By 1999, twenty-four years after its founding, the market value of Microsoft's stock had reached $507 billion. It was the first company in the world to pass the half-trillion level. Its profit margins were breathtaking. Microsoft was then generating forty cents of profit on each dollar of revenue.

If Microsoft were a country, at that moment it would have had the ninth-largest economy in the world, ranking just behind Spain; the gross domestic product of the entire United States didn't reach that half-trillion mark until 1960, when Gates was five years old. Less than forty years later, at the age of forty-three, Bill Gates had become the world's wealthiest man and his Microsoft was the most highly valued company on the planet.*

By the time the U.S. Justice Department filed its antitrust suit against Microsoft, in which twenty states originally joined, Gates and his company had become a symbol worldwide of what someone brightly called the rise of the "billionaire babies," or new technological titans, whose fabulous fortunes wielded such power. Adjectives and alliteration flow easily in popular accounts describing these high-tech leaders; but, however ripe the prose, perhaps in their case it isn't even excessive.

The Microsoft suit and subsequent trial, beginning in October 1998, was the most important antitrust action brought by the

---

*Cofounder Paul Allen left in 1983; Gates got 64 percent of the stock to Allen's 36 percent, leaving Allen, at $30 billion, number three among the world's top ten billionaires in 1999. Steven A. Ballmer, another Harvard classmate of Gates, whom Gates named president and second in line in 1998, ranked fourth with a net worth of $19.5 billion. Two years later, Ballmer became Microsoft's chief executive officer.

United States since the 1911 Standard Oil case and the breakup of AT&T in the early 1980s. The case itself, although confoundingly complex in its details, allegations, charges, and countercharges, deserved its commonly employed description of "historic," or at least potentially so. As Ken Auletta, who had observed the trial closely over a two-year period, wrote in a brilliant *New Yorker* article that chronicled the twists and turns of the Microsoft trial, and was published before the trial was over: "In the end, the Microsoft trial may determine whether antitrust laws can fairly be applied to technology companies and the Internet, where classic monopoly characteristics—rising prices, control of finite resources, distribution barriers to entry, and choke holds on innovation—are not as apparent, even if the allegations of coercive tactics are familiar." The trial could help answer the question, as Auletta also put it, of how "Microsoft became more dominant in software than Saudi Arabia is in the production of oil."★

The stakes went even higher for Microsoft, for the computer industry, and for the public, not only in America but around the world. Its final outcome could determine new ground rules establishing the standards, as the *New York Times* rightly commented, "affecting companies and consumers alike for decades." At the very least, the issues raised in the trial highlighted questions over power, control, and access to computers that society will face in the future as the electronics revolution becomes even more dominant in all aspects of life.

Somewhere between the government's portrait of Gates and Microsoft as a rapacious monopoly and the company's version of itself as an admirable twenty-first-century example of free enterprise capitalism is a relatively simple story.

Essentially, Microsoft found itself in trouble because it bet wrong on a new technological device—the browser introduced, as

★Auletta's 23,000-word article, "Hard Core," with a subheading, "*Why does Bill Gates think that the Microsoft antitrust trial has been such a disaster for him and for the company?*"in the Aug. 16, 1999, *New Yorker,* was the longest in the magazine's recent history, and became the basis of the book *World War 3:0.*

we've seen, by Netscape in that legendary Silicon Valley public stock offering in August 1995. When Microsoft recognized the danger the browser presented to its competitive position, it turned its full energies and powers into beating the opposition. The fight was over nothing less than who controls the Internet.

Until the advent of the browser, Microsoft had achieved near total dominance of the computer software world. It was banking on adding to its strength through further aggressive marketing of the Bill Gates mantra: "A computer on every desk and in every home." As Gates ruefully was forced to concede later, he and Microsoft missed the great Internet wave just cresting over American homes and businesses. Netscape's instant success with its browser quickly changed Microsoft's assessment of this threat to its present and—more troubling—future position.

In less than five months, Gates had redirected the energies of Microsoft into beating Netscape's browser with its own, called Explorer. In a famous internal Microsoft address that came to be known as his "Pearl Harbor Day speech" delivered on December 7, 1995, Gates began a dramatic company turnaround. He shifted Microsoft's focus from dominating the desktop personal computer field to dominating the next big technological wave—the Internet. The war was on.

From that point, Microsoft and Netscape engaged in a classic corporate war, one that led directly to the antitrust suit. Through its overwhelming strength in the computer markets, the government charged that Microsoft forced computer manufacturers to install its Internet Explorer on every new model shipped that contained Microsoft's Windows operating system. It was total warfare, kill or be killed. From Netscape came attacks labeling Microsoft "the Beast from Redmond." Netscape's Marc Andreessen was shown in trial documents to have boasted of his browser that it would "reduce Windows to a set of poorly debugged device drivers"—demonstrating at the least that Microsoft was not alone in wanting to crush its competitor. As the trial dragged on, the rhetoric became more inflated, the insults more personal: Which side most represented a "citadel of greed"? Which most displayed gross extravagance in lifestyle by comparing the sizes of executives' private yachts or their West Coast mansions and vacation homes in the south of France?

In the end, Microsoft won its battle with Netscape. It did so by bundling its browser with its Windows operating system that was used by most computer manufacturers and giving it away—free. In Microsoft's view, it merely gave customers what they wanted. A hell of a deal.

Critics, and later the government, maintained that by giving away its browser, Microsoft provided little incentive for computer makers or users to try and pull Microsoft's browser out of Windows in order to replace it with Netscape's. Further, in the view held by the government when it brought its antitrust suit, Microsoft, as the most powerful computer company, bullied, threatened, and coerced computer makers and consumers to use its technology instead of Netscape's.

Within a year, Microsoft's share of the browser market jumped from 20 to 50 percent while Netscape's dropped from 70 to 50 percent. In another twelve months, Microsoft had won the business of about 76 percent of new browser users. That war was over.

Toward the end of 1998, while the trial was in its early months, Netscape was bought by America Online (AOL) for $10 billion in a deal that set the stage for an even greater battle of titans. That merger made AOL the king of the Internet, eclipsing the reach even of Microsoft. And AOL immediately announced that wasn't the end of its ambitions. It was joining with Microsoft's archrival Sun Microsystems to form an Internet service company that could challenge, and perhaps supplant, both Microsoft and IBM.

Although the Microsoft antitrust suit had not yet been concluded, that merger sent a clear signal throughout the business world. Even greater battles were looming for control of an industry that holds ever-increasing influence on the basic functions of everyday life. Whether any governmental action would—or should—be able to control these battles in the rapidly changing world of technology raised questions that were not resolved when the Microsoft verdict was rendered.

On November 5, 1999, the U.S. government won a sweeping victory over Microsoft. In delivering "preliminary findings of fact," the presiding federal district judge, Thomas Penfield Jackson, personally a Republican and a conservative, declared Microsoft a monopoly that stifled industry. He accepted virtually all the government's assertions against Microsoft, and his findings were delivered in unexpectedly blunt and unequivocal language. Not only did he agree with the government's version of the case, he also agreed with its theory about monopolistic practices Microsoft had employed to crush competitors. Subsequently, Microsoft was dealt a further blow when the government won a court order to break it in two. Microsoft then took its appeal to the highest tribunal, the U.S. Supreme Court, which sent the case back to a lower appeals court. Nearly a year later, at the end of 2000, it could be two more years before that appeals process produces a final verdict on Microsoft's fate.

Whatever the ultimate outcome, the antitrust case was a disaster for Microsoft's business empire, for Bill Gates personally, and even for the way the public viewed other titans of Technotimes.

◦ ◦ ◦

The Microsoft trial provided a window on the inner workings of the computer business world—not the public relations myths puffing the computer industry's leaders and companies as modern explorers and public benefactors, but the reality of a cutthroat business filled with egotistic, arrogant people. Out of the torrent of documents emerged a highly unflattering portrait of technocrats determined to do all in their power to destroy their competitors in order to achieve their material goals.

For all the talk of the wonders of technological change and the benefits it would bring to society, the trial demonstrated that greed, rivalry, and revenge inhabited this new business world no less than they did the old. Many of its principal players exhibited the same kinds of cockfight mannerisms displayed by the Jay Goulds and Jim Fisks of the Gilded Age who sought, sometimes successfully, to corner the sugar or copper markets. They engaged in a never-ending kill-or-be-killed game—the ultimate in masters-of-the-universe gamesmanship—in which their personal fortunes mattered most.

However damaging this picture was to the computer world and some of its leaders, the greatest harm was done to Microsoft and Gates. From Microsoft's corporate side, the practices disclosed made the popular saw about Microsoft being the "praying mantis of the business world" seem more true than canard: "First, Microsoft had sex with you. Then it devoured you." For Gates, the picture was even more damaging.

Until the Microsoft trial, most Americans during the boom did not view Gates, or Microsoft, as threats. Unlike the muckrakers during the early decades of the twentieth century, when the issue of predatory corporate power produced Progressive Era reform legislation backed by Republicans like Theodore Roosevelt and Democrats like Woodrow Wilson, Americans at century's end were not in a mood to mount another revolution for social justice.

There was no little irony in that change of attitude, for by the end of the Nineties the kinds of monopolies controlled by the old Robber Barons had been eclipsed in size and power by the greatest wave of corporate conglomerates ever. The new techno billionaires wielded even more power than the old industrial barons of the past, and their corporate conglomerates affected the lives of far greater numbers of people around the world. Interlocking blocs of power formed new combines in the auto, chemical, electronics, entertainment, health, pharmaceutical, publishing, and telecommunications industries. At the same time, newly deregulated U.S. financial markets produced trillion-dollar banking empires whose assets exceeded all but a few of the resources of the world's largest nations.

Far from fearing these massive new economic formations, Americans during the boom either took them for granted or concluded it was useless to protest against them; they were simply too big, too powerful, too beyond accountability. Besides, unlike Americans of the Progressive and New Deal eras, increasing numbers of people had come to distrust, or fear, big government as much as or more than big business. They also took for granted the reforms over the last century that provided guaranteed government benefits for medical care, Social Security retirement income, and unemployment. To many Americans, such societal reforms made it seem the great battles over social injustice and class inequities had already been fought—and won. Rather than being threatened by the specter of a Gates or a Microsoft, Americans admired them, enjoyed the benefits of Technotimes, and longed to share in its wealth as well.

Neither did most of the public equate the dangers of a John D. Rockefeller controlling the price and supply of oil, with its obvious harmful effect on society and people's ability to travel and to heat their homes, with that of Bill Gates and Microsoft dominating the cyberspace era in which everyone supposedly is free to share. And for all the similarities between Gates and Rockefeller in their mutual hunger and drive to amass power and wealth, the public images of them could not have been more dissimilar.

The public saw Rockefeller as old, remote, forbidding; he was the perfect kind of character to fuel conspiracy theories about a behind-the-scenes power broker wielding great influence. People saw Gates as young, impetuous—a wunderkind type whose rumpled hair, soft little-boy face, rimless glasses, and casual open-collar dress made him seem oblivious to anything around him. He appeared more a *Mad* magazine caricature than a threatening, manipulating Rasputin. He was the classic nerd absorbed in his books and experiments—different but not dangerous.

That public image was severely tarnished by the antitrust trial, and particularly by the new portrait of Gates that formed during some twenty hours of his videotaped deposition played in open court for all the world to see. The deposition showed a far-from-decisive Gates, slumped in a chair, not looking directly at his interrogator, compulsively sipping from a can of Diet Coke, and giving answers that were unresponsive, argumentative, or flat denials of what the government maintained could be demonstrated as true from his own verbatim e-mail messages.

That Gates was also capable of acting boldly and decisively and with foresight—and was often contemptuous toward underlings, rivals, or new acquaintances—had been documented abundantly in numerous books and articles about him. That he could also be

profane, insensitive, and ruthless was hardly surprising, or even noteworthy. More intriguing as a revelation of his true character was
the near-uniform account of insiders who described his behavior as
that of a spoiled, sullen boy, given to explosive rages and repeated
tantrums. His own father's testimony, as told to the *New Yorker's*
Ken Auletta, provided some insight into the figure who exercised
such influence on the United States, and indeed the world. "We
knew he was a smart kid," William Henry Gates, a courtly lawyer
and pillar of Seattle society, said of his son, Bill, the fourth in direct
line to bear the same name. "That was pretty evident. More than
smart, he was so curious about everything. He did not possess the
innate social skills that a lot of other kids come up with. He was
shy, and didn't have a lot of self-confidence."

Another intriguing aspect of Gates's personality is worth noting. Born into a family of wealth and standing—his mother was a
leader in Seattle's charitable endeavors—and given all material and
educational advantages from prep school to Harvard, Bill Gates,
whether enfant terrible or shy awkward boy, seems to have adopted
as a lifestyle and as a model a most surprising figure, especially considering his own affluent, privileged background. When Gates
built his sprawling 40,000-foot mansion compound on the shores
of Lake Washington, across from Seattle, with its twenty-four bath-
rooms, six kitchens, Art Deco movie theater, two dining rooms,
helipad, and underground parking garage for a hundred cars, all at
a cost perhaps as high as $100 million, he had inscribed around the
dome of the library these words:

> *He had come a long way to this blue lawn and his dream must*
> *have seemed so close he could hardly fail to grasp it.*

The line is from F. Scott Fitzgerald's *The Great Gatsby,* the jazz-
age novel that tells the story of Jay Gatsby, a poor boy without the
breaks who reinvented himself into a fabulous figure of wealth,
mystery, and ultimately tragedy. Gatsby has become an icon of that
era; his dream of attaining great wealth typified public attitudes in
the Roaring Twenties at the peak of *that* great boom.

What dreams fired Bill Gates are unknowable and, curiosity

aside, not that important, either. What *is* important about Gates and his times is the company he built, the influence it exerted, and what it represented for the future. But, on reflection, perhaps those dreams *are* worth exploring, for they offer clues into prevailing American attitudes during the last, and greatest, boom of the century.

If freedom was the motivating factor in creating the United States, then desire for improving one's standard of living ranks just behind it. Gatsby's golden dream of accumulating great riches, no less than Gates's, affects and reflects national life and character. In the America of the Nineties and beyond, for Gates, for Microsoft, and for countless millions more, that old dream assumed even greater dimensions.

·   ·   ·

When I began this book, I was told that of all the people I *must* see in order to understand the new technoworld, Microsoft's Nathan Myhrvold was that person. Often, that kind of buildup leads to a letdown. In a typical account of Myhrvold, I had read how he was described as a "Renaissance man, with interests ranging from cooking to paleontology." In another, he was said to be "an accomplished chef...an amateur photographer, a fly-fisherman, a race-car driver, and a bungee jumper." Happily, he turned out to be all these—and more. Far from being swollen with his own importance, Myhrvold is delightfully modest and self-deprecating, personal traits even more impressive considering the influence he exerted on Technotimes.

As the chief technology officer of Microsoft, the head of its critical research division, and, as a *Newsweek* profile once put it, "Bill Gates's favorite geek," or, in other accounts, Gates's alter ego, Myhrvold played a major role not only at Microsoft but in the technology industry. He became famous as someone who predicts the future, not just by propounding abstract theories, but by developing practical ideas that directly shape it.

Whether acting as Gates's guru or as Microsoft's resident genius, Myhrvold was responsible for many of the company's principal projects. When he came to Microsoft in 1986, after it bought his own small software firm for $1.5 million, Myhrvold was in his mid-twenties. A decade later his personal fortune was over $300 million, and rapidly rising. By then, his role had extended beyond developing Microsoft's technology strategy. Working directly with Gates, he became a central figure in forming Microsoft's overall strategy, both for business and technology.

In Myhrvold's view, one of the main reasons for Microsoft's success was the fact that Gates himself began as a technologist. Thus, Gates understood what was necessary to create further technological advances. Gates understood something else that was unusual among his high-tech competitors. That was the necessity to make a major commitment to basic research.

By the end of the Nineties, Microsoft was spending about $3 billion a year directly on research, and that commitment has been critical to the company's technological success. "Most Silicon Valley companies don't have research labs like we have here," Myhrvold says. "They crank stuff directly into products; research is something the next start-up can handle, or a university can handle. We thought it was very important to create technology ourselves and invest in it.

"Initially, people told us we would have two great failings. First, we'd never hire great researchers because they would want to work in a research lab like those of IBM that had been around for years or university ones like those at Stanford, Harvard, and MIT. Why would they go to some little company [like Microsoft] where there's no history? The second thing they told us was that even if we did hire great researchers, we wouldn't be able to create new products. It was widely accepted then that investing in research didn't pay off directly for the company.

"Well, we've proven to be one of the best research labs in computer science in the world. We now have many of the best people in the world in computer science ... and every Microsoft product has technology in it from our research here. So much so that last year I decided we'd triple the size of our research group, which is even more unprecedented than starting one. Research budgets are being slashed all around the world. The government is cutting back, which I think in the very long run is a terrible idea."

Nathan Myhrvold is a bouncy, ebullient man, filled with energy and humor. If the Microsoft culture stresses the casual look, he could be its poster person. Sitting in his cluttered office, wearing sneakers, unpressed slacks, and open sports shirt, with his full red beard and mustache streaked a bit with gray, his curly uncombed hair, rimless glasses, beaming expression, and somewhat portly figure, he appears more a cherubic Santa than a philosopher king of Technotimes. His enthusiasm is contagious. At the time we talked, he was finishing plans for a camping trip to Mongolia and bubbling over the prospect of adventuresome travel mixed with scientific exploration and his hopes of unearthing dinosaur bones.

Ask him about the future technology is creating, and he quickly says: "One of the changes that I like is we're breaking the tyrannies of geography, and changing distance. Distance in geography has governed human life for enormously long periods of time. There are still plenty of ways in which it does, yet the world keeps getting smaller. For example, you traveled all the way out here to talk to us [just] as Boswell and Johnson traveled to Scotland to see what the Scots were like. The time will come when that sort of travel will seem incredibly passé. It's already almost such. We couldn't have the same richness of discussion electronically, but we may well exchange e-mails after this. So that's one of the ways I like to think of as breaking the tyrannies of geography. It no longer holds us back."

He gets up, paces a bit, resumes his analogy. "When you take constraints away," he says, "a funny thing happens. You break the bounds, and people become very nervous about it. It's true with men who have been released from prison. Without the regularity of prison life, they have a hard time adjusting. They've become accustomed to a structured situation. You break the tyranny of geography, and all of a sudden you're terrified of all the consequences."

Myhrvold warms to his subject. He becomes even more animated, leaning forward and gesturing to make his points. "Here's an example of a theme: This country was founded on very fundamental bets on human nature. The bet that it was a good idea to let people say what they wanted. Freedom of speech. At that time, it was a radical notion. It became the law here in the United States. No place else had that. Sedition was a dangerous thing; you had to stamp it the hell out. The same bets on human nature are in the Bill of Rights: the notion that it's better to let some guilty people go free to avoid prosecuting the innocent. The same bets are in a whole variety of our rights. Only if you think people are mostly good, and will mostly work things out, would you ever consider placing those bets, because each one of them has a downside. They can be abused.

"Okay. Fast-forward 212 years or whatever. It's been an enormously successful experiment. We have the most open society in the world, and we are far richer from it in every sense of the word, monetarily and culturally. What the Internet, with all its new technological advantages, is offering us is a society that is even more open yet—more open because the controls that once existed no longer do. The government doesn't have the right to knock down my door. But it could. On occasion, it does, sometimes wrongfully. But if I encrypt something with the right encryption algorithm, all the king's horses and all the king's men can't break it down."

The "downside," as he put it, is that historically fears and concerns over rapid changes that seem to threaten moral and cultural values have produced periods of reaction, triggering legal repression against innocent citizens. That is also part of the American story, and Myhrvold worries that the United States is entering one of those reactionary periods now. He cites two seemingly disconnected societal threads: The explosion of pornography on the Internet has sparked attempts to suppress it through criminal sanctions. At the same time, public opinion polls repeatedly show a large majority of Americans would vote *against* passage of rights spelled out in the Bill of Rights. "It's a fascinating thing because the cyberworld offers us the same bets on human nature as those bets more than two hundred years ago when this society was being created," he says, "but they're cloaked just enough differently so that people who would never think of voting down the Bill of Rights will stand up, in a well-meaning but misguided way, and suggest that we yank them all back. So I think: Do we go ahead or do we react and go back? Do we reaffirm our belief in those fundamental bets, and take the great plunge, or do we shrink and fall back? *That* is one of the great dilemmas of our age. That's going to color a large part of the public debates in the coming decades."

He doesn't want to be misunderstood, he says. "Sure, you have to worry about criminals using the Internet just like you had to worry about Jesse James robbing trains and Bonnie and Clyde using the automobile to make their getaway. But the net result— no pun intended—is we generally have used the freedoms of our society to the enormous betterment of it. Those freedoms have allowed us to come up with ideas, including economic ideas, that make this society as vibrant as it is. I believe it's okay to take the bet and let it be richer still. That being said, there are going to be lots of fears, lots of reactions, lots of people saying, 'No, no, we can't do that because we have to protect our children.' But that doesn't mean you repeal the Bill of Rights, either."

The technological revolution, despite its great advantages, becomes frightening to many because it *is* so different. Thus, according to Myhrvold, "Things are up for grabs both because they seem different and to some extent *are* different. How we as a society react to that is going to shape much of the next century."

As for himself, Nathan Myhrvold is enormously optimistic, though all the while holding concerns about problems the increasingly connected and ever-more-complicated technological world will bring.

"The technological revolution is yet to occur," he says. "The really important things that will reshape the world haven't happened yet with the new technology. We are nascent. We're as early in this stage as industrialization was a hundred years ago. We have to think not in terms of what's happening in the next five months, but what will it be like in fifty years from now. As much as we might have thought we were industrialized in 1900, or as much as we think we're really computerized today, the *real* question is not 'Oh, my God, what has technology done to us?' but what is technology going to do in the future."

Two polar forces are converging. From the positive side, electronic commerce will continue to spread rapidly. New markets will open. Businesses will be brought into ever-closer contact. The system will become more efficient; it will create more economic opportunities.

From the negative side, the number of speculative markets will increase. Consequently, "basic commodities, things we all care about, are suddenly going to have speculative panics—scares, ups and downs, crashes. That's going to be very dislocating for people."

If Myhrvold is right, that means the American people and their political and business leaders are going to have to develop a different way of thinking about their future. They are going to have to learn to adapt more rapidly to change. They especially are going to have to understand more clearly the forces that are creating those changes and how to direct and control them.

As he says, things that involve technology move very quickly. Things that involve changing society by changing people's behavior and attitudes don't happen so quickly. It's a social phenomenon, not a technological phenomenon, that we have to understand. We still don't; we still haven't come to grips with that.

He gets up from his chair, moves toward a nearby glass case containing a mounted switch, points, and says: "This power switch is from the first computer. That's the switch that turned it on. Right there!" He steps aside to permit the reading of the inscription on the case:

From the first programmable electronic computer developed by John von Neumann at the Institute for Advanced Study at Princeton University 1939.

Nathan Myhrvold, still standing, turns back to his visitor and pronounces what passes for a benediction on the challenges of the future. "Our high-tech industry is so taken up with the rapid pace of change that it can't look farther ahead than six weeks," he says. "Two years is infinity. What's worse, the technology industry has no history. People in it don't consider the history of technological change interesting or valuable. I do consider it valuable."

Like Santayana, he's a believer that those who do not learn the lessons from the mistakes of the past are doomed to repeat them. I must add, that I don't mean to suggest that Nathan Myhrvold, for all his talents, is an infallible wise man. A number of his remarks seem either naive or wrong. His statement that America was founded on a positive belief about human nature—that people should have the freedom to say what they want—ignores the pessimistic view of human nature held by the Founders. The Bill of Rights had to be forced on them. Similarly, his optimism that the Internet offers us a society that will be more open than ever and therefore freer from controls is, at best, dubious. The growing evidence of the boom years is that our lives will be—indeed, already are—increasingly dominated by a few giant corporations. Even more doubtful, given the downward slide of the Internet economy, including the standing of Microsoft, is his belief that electronic commerce will continue to spread rapidly and that the number of speculative markets will increase. Still, with his enthusiasm, his sense of power to transform the future, he's both memorable and instructive not only in trying to remind others that the actions of the past affect the future, but in offering a fresh breath of optimism about the larger potential for technology to improve life.

· · ·

As Myhrvold would say, fast-forward:

By the end of 2000, a stream of high-level Microsoft executives had either retired or were joining other companies, confirming the internal fears expressed to me earlier about Microsoft losing talented top people. Within a span of only two years, four of the nine executives on Microsoft's executive committee were on long-term leaves of absence or "had sharply reduced their responsibilities." Among them was Nathan Myhrvold. But the greatest change involved the fortunes of Bill Gates, Microsoft, and the high-tech industry Gates had so dominated and personified.

At the end of January 2000, Gates stepped down as Microsoft's chairman, handing over the reins of leadership to Steve Ballmer. The reason given was Gates's desire to devote more time and energy to further development of computer software; but clearly other factors were at work. Like John D. Rockefeller before him, and perhaps driven by the same motivation to repair his image, Gates embarked on a philanthropical spree, dispensing a series of widely publicized major charitable gifts through his personal family

foundation. He became, again like Rockefeller, the greatest benefactor of the times. Gates was still the world's wealthiest man, but
in that two-year span the stock market had slashed his personal fortune from over $100 billion to $85 billion, then to $63 billion and
even less in 2001.

The change in Gates's and Microsoft's worth reflected the extreme volatility, including periods of market meltdown and panic,
that had affected once endlessly soaring technology stocks. In the
wake of the antitrust verdict against it, Microsoft lost more than
$200 billion in market value. By the end of 2000, Microsoft's stock
plunged from a peak of over $120 per share to below $40. It was
then trading 50 percent below its high at the end of 1999. One analyst, surveying the uncertain economic climate facing Microsoft in
2001, gave an assessment that would have been unthinkable earlier.
"It seems like a long, long time ago," he told the *Wall Street Journal,* "when Windows 95 had the whole world sitting on the edge
of its seat." Even the dominant and supposedly stable Intel, the
bellwether of Silicon Valley success, experienced plunging values.
Spooked investors dumped stocks and fled. Numerous dot-com
start-ups crashed. Many of the young hotshot beneficiaries of IPOs
also found themselves crashing—or, as they termed it among
themselves, "burning." They weren't immune from the historic
ups and downs of the market cycles after all. The boom and the
technological revolution that drove it continued, but some of the
golden glow had faded. It just didn't seem quite so glorious, or so
certain, anymore.

Chapter

4

Seeding the Future

"And the cutting edge of that cutting edge is genetics . . ."

olly the Sheep entered the world on February 23, 1997, seven miles from Edinburgh in Roslin, Scotland. She was plump and perfect, with a fine full coat, gently curling ears extending over her lovable lamb face, and two soft little eyes that seemed to be saying, *Here I am,* as she stared straight into the flashing lights of the cameras for her birthday pictures.

No sooner had the world got a good look at her and heard her creators murmur their pride in their Scottish brogue, than Dolly found herself embroiled in a nasty controversy. Dolly was a fraud, said one esteemed microbiologist in New York, reflecting the views of other scientists who doubted the evidence before their eyes and the claims of her creators, Dr. Ian Wilmut of the Roslin Institute and Dr. Keith Campbell of PPL Therapeutics, also in Roslin. She wasn't a clone, after all—certainly not the first mammal grown to adulthood to be cloned from an udder cell of a six-year-old ewe. Biologically impossible, the contrarians argued.

A little over a year later, in July, scientists in Hawaii made an even more dramatic announcement. They had created dozens of cloned mice, the one mammal that some biologists believed could not be cloned, because of the extraordinarily rapid development of the mouse embryo after fertilization supposedly made fertilization impossible. Not only had they cloned those dozens of mice, claimed Dr. Ryuzo Yanagimachi of the University of Hawaii and his postdoctoral student, Dr. Teruhiko Wakayama, they had even succeeded in cloning some of the clones they created. "Wow," Dr. Barry Zirkin, head of the division of reproductive biology at Johns Hopkins University in Baltimore, told the *New York Times,* in reacting to this news. "This is going to be Dolly multiplied by 22." He could have said 222 and made the same point: No one knew where this acceleration would end, but that it would keep on seemed certain. And from Princeton, a renowned mouse geneticist and reproductive biologist found the speed at which the mice cloning reportedly took place breathtaking; but the implications, he said, were even more astonishing. "It's absolutely incredible," Dr. Lee Silver remarked. He harbored no doubts about what this meant: "Absolutely, we're going to have cloning of humans."

Immediately, a venture capitalist in Hawaii announced plans to form a consortium of companies and academic scientists to make the cloning of adult animals a commercial possibility within a few years. No, he assured reporters, there were no plans to clone humans, adding: "We have no interest in cloning humans."

Other scientists quickly offered their opinions. The cloning of those mice, if it really happened, would spur new research in embryology and genetics, leading to commercial applications. It could mean producing such science fiction creatures as goats that make useful drugs in their milk or pigs that grow replacement organs for humans. More venture capitalists were heard from. One biotech venture capital firm announced plans to license the mice-cloning technology. Published stories began describing cloning engineering techniques such as "transgenic animals" (genetically mixed creatures) and "xenotransplantation" (putting organs from other species into human beings through nuclear transplants). That raised the prospect of taking a cell sample from a human being and growing new cells to correct genetic problems or treat diseases.

Less than three months passed before two teams of scientists working independently, one at the University of Wisconsin and the other at the Johns Hopkins University School of Medicine, said they had achieved one of the most coveted goals in biology: they had isolated a primitive type of cell that can grow in every kind of human tissue, including muscle, bone, and brain. These human embryonic stem cells multiply tirelessly in laboratory dishes, thus offering a self-replenishing supply to grow replacement tissues for people with various diseases. This development was hailed as "a landmark event with vast biomedical potential."

Just one week later, biologists in a small biotechnology company in Worcester, Massachusetts, Advanced Cell Technology, said for the first time they had made human cells revert to the primordial, embryonic state from which all cells develop by fusing them with cow eggs and creating a hybrid cell, part human and part cow. From this method, they reported, replacement body tissues of any kind could be grown from a patient's cells, "sidestepping the increasing scarcity of organs available for transplant and the problem of immune rejection." Nicholas Wade, writing in the *New York Times* on Nov. 12, 1998, described the potential of these stem cells in terms easily understood: "Should the technique work," he wrote, "it could well change the face of medicine by providing a universal spare parts kit for the human body. No longer would patients have to wait in line for scarce transplantable organs; their failing hearts would be injected with heart muscle cells grown from human embryonic cells. The cells would be genetically manipulated so as to avoid provoking the patient's immune defenses, and immortalized with a gene that enables cells to divide indefinitely instead of growing old after an allotted number of divisions."

Immortalized . . . divide indefinitely: suddenly, the implications were extending beyond sci-fi into the realm of remarkable reality.

The next month, Japanese scientists reported they had cloned eight calves from cells gathered from a slaughterhouse, creating eight identical copies of a single cow. After Dolly the Sheep in Scotland and the mice in Hawaii, cloning of the cow brought to three the number of mammal species to be genetically duplicated.

Commercially, the cow was the most important animal to be cloned thus far. In the Japanese case, eight cows were born from ten cloning attempts; in Dolly's case, she was the only sheep born after more than two dozen such attempts.

A few days later, South Korean researchers claimed they had cloned an embryo from the cell of a thirty-year-old woman. Their announcement was greeted with widespread skepticism in the scientific community because of lack of verifiable proof. But unlike the doubts raised at the time of Dolly's birth, no longer were leading scientists skeptical about the possibilities of cloning. "The question isn't whether they did it or not, but whether there is any scientific reason to believe any top human fertility expert can't try and even succeed in cloning a human," Randall Prather, an animal embryo researcher at the University of Missouri told the *Wall Street Journal.*

As Dolly's second birthday neared, a Chicago physicist ignited great controversy by announcing he was assembling a team to produce the first human clone. Largely overlooked about the same time, but "the scariest news of all," in the words of commentator Charles Krauthammer, himself a medical doctor and practicing psychiatrist, were reports from two obscure labs at the University of Texas and the University of Bath. In the last four years one group had created headless mice, the other headless tadpoles. Writing in *Time* under the wonderfully evocative headline, "Of Headless Mice . . . and Men. *The ultimate cloning horror: human organ farms,*" Krauthammer said:

> For sheer Frankenstein wattage, the purposeful creation of these animal monsters has no equal. Take the mice. Researchers found the gene that tells the embryo to produce the head. They deleted it. They did this in a thousand mice embryos, four of which were born. I use the term loosely. Having no way to breathe, the mice died instantly.
>
> Why then create them? The Texas researchers want to learn how genes determine embryo development. But you don't have to be a genius to see the true utility of manufacturing headless creatures: for their organs—fully formed, perfectly useful, ripe for plundering.
>
> Why should you be panicked? Because humans are next.

On May 27, 1999, three days after Dolly celebrated her twenty-eighth month of life, scientists made one more announcement about the fabled sheep. Dolly, it appeared, was aging more rapidly than other sheep her age. It seems she genetically shared the age of the six-year-old ewe from whom she was cloned. That accident of birth means she will die far sooner than expected, and earlier than other sheep born the same time. She won't become the first living organism to achieve immortality through genetic manipulation. Dolly's demise, when it comes, will end her story, but not the greater story of genetic engineering of humans and plants alike. That story will unfold even more dramatically in the new millennium. It is certain to bring great changes, and create great controversies, all of which were barely imagined when Dolly first strolled daintily onto the world stage.

. . .

Dr. Robert Cook-Deegan, one of the world's foremost experts on cancer research as well as on genetics, says, "The two most profoundly influential technologies of our times are computing and genetics. The 1990s have been the decade when these two technologies converged in a powerful way. People are only just beginning to realize how incredibly important computing has become in everything that we do."

Dr. Cook-Deegan speaks with authority. As director of the National Cancer Policy Board at the Institute of Medicine in Bethesda, Maryland, just outside the capital, as well as director of the Commission on Life Sciences at the National Academy of Sciences, he has had an influential role in the exploding techno-world of health care.* In the area in which he focuses most intensely, molecular biology and genomics (the study of genes and their function), the vital interaction between computer technology

*His *Gene Wars: Science, Politics, and the Human Genome* (New York: Norton, 1994) provides an indispensable account of the Human Genome Project and its long-term implications. He has also served as executive director of the biomedical ethics advisory committee of the U.S. Congress.

and medical research is creating a greatly expanding body of new information and knowledge.

The convergence of genetics with powerful computing has formed a natural marriage. "You tend to put your resources into the areas that are moving fastest," Dr. Cook-Deegan says, "and that's what has happened since the term 'molecular biology' was first coined in 1938. In its original meaning, that term referred to using techniques in physics and chemistry to study questions of biology. For many, many reasons it has become the dominant framework for doing medical research. And the cutting edge of that cutting edge is genetics."

Out of this emerged a project that stirred immense excitement in the highest scientific, technological, and industrial circles worldwide. Already that project has developed, as Dr. Cook-Deegan says, "incredibly important information linked to incredibly powerful computer technology. It truly is a completely different way of approaching biology. It's really a big deal and it promises a big spillover throughout *at least* the twenty-first century."

It's called the Human Genome Project.

. . .

On May 11, 1998, J. Craig Venter, one of the world's most watched—and most controversial—scientists, announced an audacious plan to map and publish the entire human genetic code within a mere three years.

Venter's statement boldly challenged not only the United States government with all its resources but the entire scientific establishment, as well as some of the world's most powerful businesses and foreign governments. It ignited one of the great races in history, with scientific and financial stakes that made even the rise of Bill Gates seem small. Venter was betting he would be the first to unravel the entire human genetic code—the biological road map to life. Not only would he do it first, Venter said, he would do it much faster and at a tenth the cost of the Human Genome Project, an extraordinarily ambitious federal government program intended to take fifteen years to map the human genetic code at a minimum expenditure of $3 billion. Furthermore, Venter pointedly said his small new private company, Celera Genomics, would accomplish its goal without taking a single public dollar.

At that moment, the U.S. government's vaunted Human Genome Project had already been at work for nearly a decade, and its publicly funded research was proceeding in 350 U.S. government and university labs that were working in conjunction with many others overseas. The Genome Project had been launched amid great fanfare as being so vast and complex a project that only the federal government had the financial resources, the scientific talent, and the bureaucratic infrastructure to complete it successfully.

The government's project was moving along toward its announced target completion date of 2005 when Venter hurled his figurative bombshell, touching off an explosive reaction around the world. "It was as if," Lisa Belkin commented in the *New York Times* magazine, "private industry had announced it would land a man on the moon before NASA could get there. As if an upstart company intended to build the first atom bomb."

Within months, the federal government revised its project and vowed to move up its completion date by two years; it also promised to finish deciphering half of the genome by 2001, the same year Venter said he'd complete his work. Later, some of the largest, most powerful pharmaceutical and biotech companies formed a consortium with the government for the purpose of beating Venter in the great genome race.

By the end of 1999, two major rival entrepreneurial companies were aggressively trying to beat Venter's Celera to the finish line. Each competitor was backed by some of the biggest powerhouses in American industry. In Cambridge, Massachusetts, Millennium Pharmaceuticals was operating through partnerships with such corporate names as Bayer, Monsanto, Pfizer, Eli Lilly, American Home Products, Astra, and Roche, with a capitalization of more than $2 billion. Near Celera, in Rockville, Maryland, outside Washington, D.C., Human Genome Sciences had more than $1.5 billion invested in its effort and had formed a partnership with SmithKline Beecham. Celera's market capitalization was less than half that sum, but, it, too had formed partnerships with such major firms as Novartis, Pharmacia & Upjohn, and Amgen.

These extraordinary reactions to Celera's challenge, and the subsequent marshaling of vast resources among its major competitors, were hardly surprising. Everyone understood the immense stakes involved and the implications for the future.

Success in charting the full human genome would revolutionize the practice of medicine and biology; reap untold trillions for drug companies and biotechnology firms able to profit from the new discoveries; create entire new industries; and provide knowledge permitting mankind to conquer deadly diseases, create new organs, clone new forms of life, even reverse the supposedly immutable process that causes cells to age and their organisms to perish.

By decoding the "book of life" that holds the key to understanding how every physical characteristic in the human body works, scientists could definitively answer how and when our species evolved, how to "correct" through genetic engineering defects in the body that cause disease, and how to alter the way humans and other species develop.

Decoding the genome would also instantly raise a host of disturbing ethical, personal, and political questions. Not least of them: Who would own or patent these discoveries, and what would having control of them mean for the future?

None of this would have stirred such intense interest and controversy had Craig Venter not been the agent of it. For all his many detractors, Venter over the years had repeatedly proved his critics wrong. In a major *Time* profile of him, in which he was presented as the centerpiece of the magazine's special issue on the future of medicine with emphasis on "how genetic engineering will change us in the next century," he was described as someone with "a genius for making the tools of molecular biology do big things. He has decoded more genes, and faster, than anyone else in the world. He pioneered the most widely used method of tagging bits of genes. And he was the first to sequence the genome of an entire living organism. Nearly half the genomes that have been decoded to date were decoded in his lab."

The objective record confirms these claims. However controversial, J. Craig Venter is a person of professional substance and distinction.

His personal history is intriguing. By his own account Venter is one of those classic Churchillian late bloomers: a poor, rebellious student, with a history of disciplinary problems and of refusing to take tests. Born into the rigidly moralistic Mormon world of Salt Lake City and raised in the permissive San Francisco of the sexual revolution of the Sixties, one parent an accountant, the other a painter, he left home after barely graduating from high school and headed for Southern California. There he became, as he acknowledges, a beach bum—surfing and sailing his young life away.

Vietnam became for him, as for so many of his generation, the defining crucible of his life. He was drafted and, despite his poor academic record, scored the highest on an intelligence test taken by 35,000 recruits. He ended up as a senior naval hospital corpsman in DaNang during the bloody Tet offensive of 1968, working around the clock for days in the emergency room tending the wounded. It was during this time that a navy physician befriended him and later persuaded him to go to college upon leaving the service. He did, with the new desire to become a doctor and work with patients in the Third World.

This was in the early 1970s. When he was in graduate school at the University of California, San Diego, he remembers being taught that it would be very difficult to make new discoveries in biology. Everything worth discovering already had been found. That Age of Discovery was over. Now, Venter told me, he could take every scientist and every Ph.D. student on earth and let them randomly pick something from an immense database his private company was developing through its extraordinarily powerful computers "and they could spend their entire career making major discoveries out of it."

That kind of statement is typical of Venter—sweeping, brash, boldly boastful—and helps explain why he became so controversial, even hated and feared.

After completing his training, Venter's star rose quickly. In 1976, he moved to Buffalo, where he taught on the faculty of the State University of New York. There, he met and married a graduate student named Claire Fraser, who also became a distinguished molecular biologist. In 1984, the young couple accepted offers to set up separate research labs at one of the world's premiere bio-medical research centers, the National Institutes of Health (NIH) outside of Washington. As the federal government's focal point for biomedical research in the United States, the NIH operated under a nearly $16 billion annual budget. There, Venter immediately began using his research budget to pursue genomics. Despite his considerable success with discoveries and breakthroughs, he also stirred controversy among colleagues. His critics resented not only his manner but, they said, his methods.*

In June of 1992, at the age of forty-six, Venter and his wife ac-cepted the offer of a venture capitalist to found their own institute. Venter then forged another chapter of overnight success, Nineties style: in just twelve months he went from being a government sci-entist with two thousand dollars in a savings account to becoming a new millionaire. Four years later, after having produced pioneer-ing work recognized worldwide and amassing nearly $40 million, Venter took his greatest—and riskiest—leap.

Another venture capital firm, Perkin-Elmer Corp. of Norwalk, Connecticut, was developing a supercomputer to permit much faster analysis of genes. By the spring of 1998, Perkin-Elmer agreed to put up hundreds of millions of dollars and to supply scores of their new machines for Venter to use in a new company. They named it Celera, short for *celerity,* or, appropriately for the new go-go times, "swiftness."

Venter himself is casual, low-key, thoughtful. For someone uniformly described as arrogant and pushy, he exhibits a wry sense of humor. His eyebrows arch when he smiles, causing the lines to crease on his forehead and accentuate his bald dome, which is rimmed by a fringe of closely cropped dark hair streaked with gray. Venter is tall, athletically built, with forearms that seem to have

---

*Nobel Prize–winner Dr. James Watson, who with colleague Dr. Francis Crick made the great discovery of DNA in 1953, headed the Human Genome Project at NIH when Venter began producing gene sequences at unprecedented rates there. During a congressional hearing, Watson publicly dismissed Venter's sequences as work "any monkey" could do, thereby ignit-ing a historic scientific feud.

been strengthened from sailing his eighty-two-foot yacht, *Sorcerer,* on which he won the Atlantic Challenge Cup. Far from exuding know-it-all self-importance, he's given to quick, deprecating jabs at himself and others. "Even though we consider ourselves enlightened," he said, at one point, "physicians are the most arrogant group of the population—except for politicians." When our long conversation began, he set the tone by first gently mocking the hype being made of the wonders his genetic discoveries could bring.

"Let's see now. It's the year 3000 and I'm still alive. I'd say those are pretty fantastic discoveries we've made." He smiled. Eyebrows arched. "I'm constantly asked to predict the future," he continued, eyebrows arching again, brow furrowing. "It's actually getting harder to do that."

Then he got serious. Think of a scene out of the early stages of electricity when lightbulbs were just being introduced, he said, in response to a question about how the current stage of genetic discoveries compares with past innovations. The computer wasn't even envisioned when they first began laying electric wires. People thought anyone who would buy electricity was nuts. Who's going to buy that? He settled back, folded his hands behind his neck, lifted his right leg and draped it over the arm of his chair. He was entirely serious now. "I don't think the capacity of the human brain can comprehend what we're getting ready to do," he said, "and we couldn't do it without massive computers. We're building the second-largest supercomputer in history here to try and analyze this biological information.★ Even a year ago, people in the computer world said you don't need one that big; now we don't have one big enough. Computing power is a thousand times greater now. We can't even imagine computers big enough to analyze and model the development of the human body. Ten trillion different cells. Eighty thousand different genes. Each of those cells expresses different combinations of the eighty thousand. Those are very big numbers. The power of this information is going to be immense. I

★The largest supercomputers, at the U.S. Department of Energy, are used to simulate nuclear blasts.

don't think it's because *we're* doing it that we think it's going to be the thing that changes society the most. We're doing it because we think *it's* going to do that—or help it do that."

The speed at which genetic discoveries were being made was the most dramatic aspect of the change. Since Watson and Crick described what they called the "double helical structure"—or more popularly, the double helix—of DNA, the search for what was being called the "Book of Man," or the universal language of life, was intensifying.

That 1953 discovery of the double helix brought startling news: coiled inside every human cell lies six feet of DNA—technically, the long-chain organic molecule deoxyribonucleic acid—which contains all the encoded instructions needed to create life. Prior to that breakthrough, how to unravel the source of life remained an unsolved mystery. Following the discovery of universal "hereditary elements" by Gregor Mendel, a Moravian monk, in the year after the American Civil War ended, scientists were essentially unsuccessful in understanding how traits were inherited or what the substance of genes might be. Not until 1944 when researchers at New York's Rockefeller Institute demonstrated that DNA in bacteria contained inherited genetic information did the process of discovery begin to accelerate greatly. But, despite enormous research efforts after Watson and Crick, the ultimate solution to the great mystery had not been found.

"Thirty years ago we didn't have a single human gene," Venter said. "The techniques that we use for DNA sequencing weren't even developed until the end of the 1970s. At the start of this decade, in 1990, we didn't even know the structure of 2 percent of the human genes. At that point we had two dozen known genes from the human brain out of thirty thousand. Two thousand out of eighty thousand genes in the human body. All of medicine, all of biology, is prescribed based on that limited view of known biology. Basically, this revolution of information only started in 1995. So what we're doing is the equivalent of coming out of the Dark Ages. It's truly the start of the Renaissance period. Yet it's based on almost no knowledge. So imagine the future when we actually *have* the knowledge."

At the moment, Venter explained, he and his Celera were working on decoding four genomes. First, insect—the fruit fly. Then, human. Then, mouse. Then, rice. "Those are going to be the foundation of all the future of agriculture and the future of medicine. The *Drosophila* genome, the first insect, overlaps with the human. Three to five million kids die every year from mosquito bites, from malaria. On the other hand, insects cause billions of dollars of damage to crops. So insects overlap those two, medicine and agriculture. We're very closely related to fruit flies. They have a nervous system, a brain, and their bodies develop exactly the same as we do. Your baby developed with some of the same genes. So we're at the earliest stage of understanding the information that's generated here and is going to drive what happens in medicine and agriculture for the next several centuries. I don't think there's been anything like that before."

Venter quickly acknowledged that the explosive rate in the acceleration of scientific and technological discoveries has become almost frightening in its implications. "The pace is scary even to us," he said, "because it's truly beyond human comprehension. It's certainly beyond my mental capacity. In 1995 we discovered the genetic code of the first organism, and now we're trying to do complex organisms like human that are ten trillion times more complicated. So I don't think it's [solving the biological mysteries] going to happen in the next five years. Even the most imaginative computer scientist can't comprehend the tools we need."

In the meantime, Venter continued overseeing his company, recruiting more Nobel Prize–winners and, equally important for success, more engineers and computer scientists to operate the most powerful computer technology available. When completed, Celera's operation would be the second-largest computer facility in the country.

Touring it was like stepping into the highest-security world of the Pentagon or the Central Intelligence Agency—indeed, I was told I'd be one of the last people ever to walk in there who didn't possess a special way of entry for this most sophisticated security system: a scan badge, a password, a biometric scan of the hand, and, if that wasn't sufficient, "a large, burly guard to accompany you." The reason for these extraordinary security measures? Because, I was told, "the proprietary information [those machines contain] will be so valuable." As the computer chief conducting the tour said, "It will all be very controlled."

Moving from room to room in that sterile, silent place, I found myself thinking it unbelievable that those small gray metal boxes possessed the power they did and also held the life-and-death secrets assigned them. Yet, despite the seemingly innocuous sight and sound of those inert faintly humming machines, I was told that here resided the data that literally could change the world. "If you were to digitalize the Library of Congress, you'd have twenty terabytes of information," the computer engineer said, as we walked along. "What you've got sitting here in front of you is about half of that. And we'll be several times that when we're fully up and running."

As we moved along, the engineer kept up a running commentary on what we were observing: Everything state of the art. Everything protected. Smoke and heat sensors that can instantly detect a potential problem. Power from two grids on each side of a room. Eighteen thousand pounds of batteries backing up the power in case of an outage. Dual sets of batteries to add further protection. Generators on-line ready to start recharging batteries if anything fails. Fuel for fifteen days. Backup fuel for another thirty days. Backups on the backups on the backups. Fifty tons of ventilation equipment to insure 100 percent constant circulation of air at a desired controlled temperature. Air ducts, all industrial strength. Everything designed to "run and run and run and run." Much more advanced equipment on the way; it would replace everything already being used there. "The densest, most sophisticated network in one building in the world," all linked, with more cable, more fiber optics, and enough electric power to serve half of the entire state of Maryland.

Plans called for Celera to have three hundred new sequencing machines, each costing about $300,000, to be housed in a laboratory and office building complex covering some 200,000 square feet. According to a sixty-page study about Venter and Celera published at the end of 1998 by the Harvard Business School entitled

"Gene Research, the Mapping of Life, and the Global Economy," projections called for Celera to become "the largest accessible commercial database in the world, by a factor of five."

Not coincidentally, Craig Venter and his high-tech race for the future perfectly exemplify the ever-more-closely connected relationship between scientific, technological experimentation and the practical new business techniques that were changing the way Americans worked and lived. The essence of the information age— and Technotimes—lay in the ability of a company like Celera to generate valuable new genetic information by using the most powerful new computing machines and then making that information available commercially through the Internet.

When I asked Venter to describe the relationship between the scientific side and the business side of Celera, he said, revealingly, "We set it up so they're equivalent. Instead of writing and selling books, we're developing a different knowledge base. Our book will be the supercomputer. You can access it through the Internet. So it's the same business model: Data. Knowledge. Information. Gates sells software. We're using software to sell knowledge."

"And without the increasing power of the computer, you couldn't do much of it?" I asked.

"Any of it," he replied.

Craig Venter won his race, what the Harvard Business Report had called "the greatest mapping adventure of the century," but he did so in a manner that permitted the principal rivals to share in the triumph.

Venter's Celera, which had seen a spectacular run on its stock after the company went public in late 1998, announced in late October 1999 that it had sequenced (determined the order of the chemical letters) the first billion units of human DNA. At almost the same moment, the publicly financed Human Genome Project, which had drastically accelerated the pace of its research to meet Celera's competition, announced it had virtually completed decoding the first human chromosome. Within weeks, a series of private talks between the rival camps produced an agreement to pool their efforts and further accelerate the rate of decoding the human genome.

The agreement to collaborate served both public and private interests: the public consortium's by not seeming to have been humiliated by an upstart private firm and thus imperiling its future government funding for basic research, Celera's by gaining goodwill in the scientific and business communities, sharing in the data produced by the consortium, and avoiding a potentially destructive fight over who had access to final DNA data.

Some six months later, on June 26, 2000, the erstwhile rivals basked in the praise of President Clinton at a White House ceremony hailing their historic completion of a working draft on the sequence of the human genome. Scientists worldwide immediately would have a road map to 90 percent of the genes of every chromosome with final data still to be completed a year later. All of the Human Genome Project data would be made available—free—on the Internet.

It was truly a monumental moment, ushering in what would certainly be a wealth of new discoveries leading to the production of new drugs and advances in genetics that promised to affect life in revolutionary fashion.

This moment was reached five years before the Human Genome Project's original target date and only two years after Craig Venter first laid down his audacious challenge. While all sides could rightly claim credit for an achievement that promised immense future benefits for society, Venter particularly could take pride in being vindicated.* But in another sense, the outcome of this scientific race illuminated a much more important point about how scientific and medical progress can be made for all of society. For this effort clearly demonstrated the payoff for the public in investing great sums for basic scientific and technological research.

As with computer technology development, the great burst in

---

*On Feb. 12, 2001, Venter reported he and his colleagues had identified about 30,000 genes, far fewer than the previously estimated 100,000. At the same time, the publicly funded consortium that had competed with Celera confirmed those findings. These discoveries pose a major mystery about life: How is it that humans, with their far greater complexity, seem to have only 50 percent more genes than roundworms?

funding for basic health research grew out of World War II and the succeeding Cold War period. From the late 1950s until relatively recently, the federal government committed increasingly huge sums to every agency with responsibility for overseeing health care. Similarly, the policies that evolved over time were predicated upon the idea that federal spending for research and development would form the core of the nation's health effort. From this, private companies could come in and, as in mining, stake claims, then extract and exploit the new information by turning it into products and services for the public. This process would now occur with the genetic discoveries of the future. Without the basic federal financial backing, the raw material would not be there for the private sector to mine. Result: Everyone, private and public, benefits. Private, through desired products—or through new information, as in Celera's example—that produce profits. Public, through products that enhance, and save, lives. Result, too: For half a century, the U.S. government generated an immense amount of scientific and technological information relating to health. This effort spawned a huge health-care economy—literally, one-seventh of the entire U.S. economy—and led directly to the Human Genome Project.

During that half century the United States outspent the rest of the world by a great margin in health research dollars. This massive commitment made the United States preeminent in medical innovation and produced signal health improvements for its people. From the beginning, by far the greatest amount of that money went to the National Institutes of Health, where Craig Venter first worked as a researcher on the very Human Genome Project that led to the successful effort to map and publish the entire human genetic code.

In this sense, everyone came out a winner, though Venter privately could rightly claim to be one of the greatest winners of Technotimes.

* * *

Technological/scientific change is progressing at such a rate that even the most distinguished scientists speak of the acceleration with awe. "I find the rate of progress in all the sciences—everything—utterly staggering," says David Baltimore, a Nobel Prize–winner who left MIT in 1997 to become president of the California Institute of Technology in Pasadena, an institution that has been in the forefront of scientific and technological discoveries for more than a century, and now plays an even greater role in creating fundamental new advances.★ "All the questions that are being asked, whether they have to do with the structure of the earth or the structure of the universe, the advances of biology, understanding the brain and bringing it into the social milieu—which is going to happen but hasn't yet—all of this is very exciting. It's also going to put enormous pressures on society. It does change the relationships among people."

By the time he became Caltech's president, Baltimore was described by the head of the university's board of trustees, the famous Gordon Moore, articulator of Moore's law of computing and cofounder of Intel, as being "perhaps the most influential living biologist, and surely one of the most accomplished." That was no exaggeration. Baltimore won the Nobel Prize for medicine in 1975, at the age of thirty-seven, in part for demonstrating, with another biologist, Howard Temin, a mechanism for copying RNA molecules into DNA—a process hailed two decades later as "a central tool of molecular biology, as basic as telescopes are to astronomy."

David Baltimore's selection as Caltech's president, where he became the only functioning scientist to head a major U.S. university, came in no small part because of his prominence in biology. As the first Caltech president from that field, coming after a long line of notable physicists, Baltimore's appointment signaled how critical bi-

---

★Caltech was ranked as the No. 1 U.S. university in 1999 in the annual survey of *U.S. News & World Report*. Caltech spent an extraordinary $192,000 on each of its only 900 undergraduates compared to No. 2 Harvard's $81,000. Every member of Caltech's incoming freshman class ranked in the top 10 percent of his or her high school class, a feat unmatched by any other school. Its student/faculty ratio of 3 to 1 was also unmatched. (Harvard's was 8 to 1.) Caltech for the first time ranked above its archrival, MIT.

ological research will be in coming decades as further advances are made in genetics and molecular behavior. At the beginning of the twentieth century, discoveries in physics dominated the scientific world. The twenty-first century will be the century of biology.

"It's no surprise that it's happened," Baltimore says of the ascendancy of biology over physics. "I grew up in molecular biology in the Sixties and Seventies. The one thing we didn't know was how we were going to solve the complexity of human cells, mammalian cells, higher organisms in general."

At Caltech, one of the most extraordinary explorations in its implications for the future involves new research into the human brain. "The biggest problem we face in biology, the one we don't know how to solve, is the problem of the brain," David Baltimore says. "What is the brain actually doing? How is it doing it? What's the code that it uses? How does it learn? How does it provide for memory? How does it provide for recall? We're learning a lot about these things, particularly in this campus. We have as good a group in neurobiology as anywhere in the country, and they are learning some amazing things."

In the present, what's at stake in the new frontier of biology is something more elemental than understanding how genes work, how they can be manipulated to fight disease or change behavior, or even how our brains function and may be changed genetically. What science and technology are already doing affects the most basic source that fuels human activity and enables us to live—food. New knowledge has enabled scientists to genetically alter the plants that produce the crops that put food on our collective tables every day, everywhere. "We can see our way to understanding all the genes," one of Baltimore's colleagues says, "to be able to manipulate them, to move them around to make new crops. The brain stuff hasn't come yet."

· · ·

Not long after Bob Shapiro became Monsanto's chief executive officer, he found himself in the midst of one of the most emotional—and bitter—corporate battles in history. The battle was not between economic giants over a merger or acquisition, not a federal antitrust suit, nor a criminal conspiracy to defraud customers. It was over nothing less than how science and technology—and the new knowledge of genetics—has affected the future of agriculture in the most basic way possible: by changing how people everywhere raise crops, the nature of those crops, and what people eat as a result of the production of those crops.

Shapiro himself chose this battle. He had bet the corporate future of Monsanto on the belief that biotechnology—specifically, genetically modifying crops to make them more abundant and resistant to disease—was the way of the future. Believing that the twenty-first century would be a world in which biotech plays an increasingly vital role, he decided to put virtually all Monsanto's energies and assets into that area.

How he came to this fateful corporate decision provides a telling counterpoint to the scientific wonders of Technotimes and the hard realities of the human reaction to changes that came to be seen, and perhaps with good reason, as profoundly threatening.

Several years before Shapiro became CEO, his predecessor had begun an intensive effort to understand what "is going on at the cutting edge of biology," Shapiro recalled in the summer of 1999 during an hours-long conversation we had at his corporate headquarters in Chicago. "The conclusion that my predecessor came to, the hypothesis that he chose to work on, was that something really profound was taking place in our knowledge of life systems. This very likely was going to create a set of tools that would enable us to address some very ancient and intractable issues in a new and more effective way."

After Shapiro took command of Monsanto, he pressed ahead with the study of biotech's possibilities. His predecessor was correct, he concluded: something so profound was occurring that Shapiro came to believe "the fundamental nature of three of the world's biggest industries—agriculture, nutrition/food, and human health—would change. When those kinds of things happen, the institutional structure that operates on old technology tends to crumble. Everything's up for grabs."

Not least in this crumbling of old methods of operating was the recognition of the increasingly critical role the computer played in creating new institutional and business arrangements. This was not a new discovery for Shapiro; he long had observed how closely connected computer-generated information technology was to life science industries like Monsanto. "Since the mid-Eighties it's been pretty clear that you can find genes whose functions you'll understand and place them in other organisms," he said. "In our case, the organisms of interest were plants. By 1991 or '92, when I was looking at this very, very carefully, I concluded that the technical issues were not going to be the barrier. From the mid-Eighties on, it became increasingly clear that the things that people thought of as barriers—How would you find the genes? How would you know what they did? Could you really transform crops with them?—those barriers were being overcome. Scientific or technological limitations would not make this impossible. In other words, the technology is not going to be self-limiting in terms of its ability to change crops."

Once those central premises were established, the questions became more operational. Was there a business system in place that would allow someone to bring these new crops to market? How would you get paid for doing it? The answers became complicated. "In principle," Shapiro explained, "what you're doing is selling a little piece of DNA. How do I do that? Do I put it in an envelope? Do I say, 'Here's a lot of DNA. Do something with it!' Depending on what that DNA does, it creates value to different people in the system. How do I collect from each of those players? It creates value for a farmer in the case of genes that enable him to produce more, more easily, and at lower cost. The assumption was that you put the genes in seeds, you sell the genes to the farmer at a premium price, and that captures some part of the value being created. Maybe that wasn't *that* complicated, although it did lead us to go out and invest about $8 billion in putting together a seed company capability—something you wouldn't have predicted four or five years ago. But suppose it creates value for somebody farther down the chain? Suppose it produces value for consumers in the form of foods that taste better or are healthier? How are you going to get paid for that? Who's going to pay? Those are the kinds of issues we've been dealing with."

Behind those and other questions lay a greater one: Did the potential benefits in terms of long-term profits and future markets outweigh the risks of failure? Monsanto, weighing the positives and negatives in as coldly calculating a bottom-line business assessment as possible, decided it did. *This* was their future; they could not afford to miss it.

"By '95 or '96 we said this is going to work," Shapiro recalled. "It's going to be big. It's going to change the way agriculture works in the world. And it's going to require a lot of investments on our part. So we better design our company to be able to do that. In '97, we ended up spinning off our chemical business. That's gone. We made our corporate future's bet on the future of biotech."

For several years, Monsanto aggressively went on what the *Wall Street Journal* accurately described as an "$8 billion spending spree on crop biotechnology." In a relatively brief time, Monsanto spent those billions buying seed companies. They did so to assure that the new genes its scientists were developing would move quickly to farmers, enabling them to produce better, more profitable, crops. To help gain the greater amounts of capital needed for its ambitious biotech plans, Monsanto sought to combine its resources with other, even larger, business entities. That effort collapsed when a projected $38 billion dollar merger with the American Home Products Corporation fell apart at the end of 1998, leaving Monsanto billions of dollars in debt from its new biotech acquisitions. Under pressure to raise new billions to pare its debts, Monsanto began selling off assets, including all of its lucrative food-ingredient businesses, among them the highly successful Nutra-Sweet, which Shapiro had headed before becoming CEO.

Shapiro and Monsanto were not deterred from their primary goal. They were still determined to move forward toward what they were convinced was not only their future, but the world's: the world of biotechnology.

As they did, their business and scientific planners carried with them a clear vision of that new agricultural world they would conquer. It went something like this:

Within a decade, the vast majority of crops, even trees, will be genetically modified—and modified in a variety of ways. Take corn, for example. Instead of having one kind of corn, the customer will be able to choose among varieties capable of offering a hundred different features. One type of corn will be genetically modified to make heart-healthy oils. Another will be modified to make proteins usable in animal feeds. Yet another will produce quantities of starches used in making high-fructose corn syrup. Each type of genetically modified corn will have certain "agronomic traits" that affect how a farmer grows it. The corn will contain genetic material that protects against insects. It will be changed genetically to be more resistant to weeds. From this will come greater benefits to farmers—and a bewildering new set of choices every farmer will have to make in raising crops. The entire industry will be affected by new ways of operating. The very nature of farming practices will also change through the use of such advanced technological techniques as earth-circling satellite mapping of soil conditions that will provide farmers with hitherto unattainable, reliable information about the composition, temperature, and moisture of surface and subsurface farming land. A new term to describe this technique already is being employed: *microenvirents,* a scientific study of the environment in which a particular plant is growing and how that environment differs from the way the plant in the next field is growing. From this new knowledge will come further modification to produce seeds that are most productive in a particular type of soil. Some scientists predict a fivefold increase in the productivity of farmland through this technique.

In short, we stand on the verge of a veritable revolution in the way humans provide their basic staples of life; a revolution that holds enormous implications for the future of a planet in which population is rapidly increasing, the land available for productive cultivation is being destroyed or depleted through topsoil erosion and environmental degradation, changes in the climate caused by manufactured pollutants threaten a global warming with potentially disastrous consequences. All these crises are occurring in a world that experiences an ever-widening disparity between its have and have-not populations.

On one side stands the postindustrial, technologically proficient developed world, much of which—and most singularly the

United States—enjoys the most prosperous period ever known. On the other stands the nontechnological developing world, holding the vast majority of the world's population, which experiences increasing poverty, disease, starvation. The gap between these two starkly different rich and poor worlds has never been greater, and becomes more so with each passing year. Statistics measuring death and disease in the poorer parts of the planet are startling for the story they tell about the dismal prospects of life there. Two alone must suffice: Ninety-eight percent of all children on earth who die before their fifth birthday live in the developing world. Ninety-five percent of all people who are HIV-positive also live there.

Historically these vast inequities have bred conflict and its inevitable companion, more mass suffering. They raise anew age-old Malthusian questions about the capacity of the human species to feed itself and to care for the health needs of the many ill and dying. Pessimists peering into the future offer dire projections of mass starvation, new diseases, the plague years revisited. Optimists foresee the prospect of improving the quality of life everywhere through new methods of farming that will provide more abundant, healthier food for increasing numbers of people and through striking advances in health care in both diagnosis and treatment.

Shapiro and Monsanto, naturally, held the positive vision of the future and were optimistic about their role in both shaping and profiting from it. That view collided cruelly with other deeply held beliefs that swiftly placed Monsanto's corporate future, and Shapiro's personal one, at great risk. Even as Monsanto plunged ahead with its biotech bet on the future, it found itself facing impassioned opposition worldwide to what became, in headline shorthand, the perils of "GM," or genetically modified crops. This battle extended far beyond bare-knuckle business competition over new markets and new profits. It unleashed ancient fears; lashed primitive passions; raised ethical, societal, and religious questions of immense complexity. As new technological methods were introduced, they triggered new protests. Biotech opponents spanned the ideological spectrum from radical eco activists on the left to traditionalists adamantly resisting any change on the right. Heads of state, environmental groups—notably the Greens in Europe—and

ordinary citizens joined in a battle that assumed the ardor of a cru-
sade, even at times of a terrorist campaign, aimed at stopping by
any means Monsanto's attempts to introduce genetically modified
crops into the marketplace.

At stake were not only multibillion-dollar conglomerates and
the profits they hoped to reap through biotechnology, but po-
litical/social/economic relationships worldwide. Leaders found
themselves embroiled in heated debates over how biotechnology
would affect their societies. Driving those debates were fears that
genetically modified plants and animals threatened the basic food
supply—and even life itself. The darkest views, passionately held,
raised the specter of a Frankenstein-like global nightmare: Fear-
some mutants. Emergence of deadly diseases resistant to traditional
forms of treatment. Environmental degradation threatening all life
on earth. Genetically modified food that poisons entire populations. Pesticides that produce new strains of crops by protecting
them while at the same carrying within them elements that destroy
other forms of life, causing entire species to become extinct. Test-
tube Dr. Jekyl and Mr. Hyde experimentation that transforms
harmless people into monsters by what they consume.

There was no neutral ground; the intense emotions made it
difficult to discuss the issues dispassionately and to appraise realisti-
cally both the benefits and the potential problems of the new technology. Critics denounced interference with nature on a large scale
as being unethical and immoral. They saw the biotech companies
as greedy combines profiteering from the human miseries that their
unsafe products created. Biotech defenders pointed to their research methods and their scientific safeguards, and touted these
products as providing greater health and nutrition benefits everywhere. They saw their critics as ignorant zealots sowing unfounded
public anxiety through fear-mongering tactics.

Against that backdrop, Shapiro and Monsanto were betting
their futures—and fortunes. Even as they believed themselves to be
developing promising new techniques, the biotech industry and
food manufacturers were rocked by a series of highly publicized
disasters that further inflamed opponents and raised concern about
the food supply. Among them: the outbreak of mad-cow disease in

Great Britain, an affliction that induced a fatal brain-wasting disease in humans caused, some scientists suggested, by people eating beef that had been fed dead livestock as a source of protein; evidence that a natural pesticide produced by genetically modified corn threatened Monarch butterfly caterpillars and some beneficial insects; the breakdown of crucial quality control measures at Coca-Cola bottling plants in Antwerp, Belgium, and Dunkirk, France, that made hundreds of people sick, leading governments to order Coke removed from shelves, and European newspapers to publish stories speculating that Coke cases shipped to consumers were contaminated with rat poison; the discovery of dioxin-polluted chicken in Belgium; the ban on sales of Monsanto's Roundup Ready soybean seeds—gene-altered to resist Roundup weedkiller—came as Japan decreed it would require labels on genetically modified foods; and more recently, Kraft's recall of taco shells containing genetically modified corn that had not been approved for human consumption.

Even though not all these incidents could be blamed on genetic alterations, for Shapiro and Monsanto the new controversies could not have come at a worse time. Their market position was being battered, the economic future for biotech seemed far less attractive to uneasy investors, and they saw their company turned into a symbol of the most greedy producers of genetically modified "Frankenfoods." Protesters had taken to waving placards branding them as "Monsatan."

Nonetheless, when I spoke with him, Shapiro still retained an outward optimism. He took the view that benefits from biotech would succeed in changing public attitudes, especially as even more dramatic positive developments began emerging in the marketplace out of the enormous research projects then underway at most pharmaceutical companies and many universities. "Implicit in all that," he said, "is the basic sentence that says: 'There are great developments happening in biology and they're going to have many implications for agriculture, nutrition, and health.' And that's where we want to be. That's where we want to position ourselves."

Wishes are one thing, fulfillment of them another.

By the closing months of 1999, a new form of Gresham's law about bad news driving out good was operating forcefully around the world: The greater the pace of biotechnological and genetic change, the greater the intensity of the opposition to it.

No longer was the opposition centered mainly in Europe. It spread around the globe. In Japan, the Kirin Brewery announced it would only use corn not genetically engineered; the next day its chief competitor, Sapporo Breweries, joined the ban. It, too, would use only traditionally raised corn. In Mexico, Grupo Maseca, that country's leading producer of corn flour, said it would stop importing genetically modified corn. In the previous year, Grupo spent $500 million importing U.S. corn. Similar actions came in other countries. And, not least, from inside the United States itself, Gerber and H. J. Heinz, rival producers of baby food, announced they would not use genetically altered corn or soy ingredients in their products. But the greatest opposition continued to be centered in Europe, as the European Union had stopped buying all American corn and where Europe refused to drop a ban on imported American beef raised with growth hormones. In reaction, the United States imposed a 100 percent tariff on some European food products, including Roquefort and foie gras—a move that ignited scorn and fury in France and spawned violent attacks on American enterprises there.

These reactions laid bare the deep anxieties and fears stirred by genetically modified foods. For American farmers, the developments were devastating. They alone in the then-booming American economy were already experiencing hard times, and they had paid premium prices to invest heavily in genetically engineered crops. Suddenly, as their fall harvest neared, they found markets they depended on being closed to them. The economics were unsparingly bad. U.S. farmers export a third of their crops, and in 1999 had planted some 60 million acres—an area equal in size to the entire United Kingdom—with genetically modified corn and soybean seeds. Unless changes in attitudes led to reopening markets, their future prospects were even more threatened. So were the hopes for the future of biotechnology and the prospects of U.S.

corporations at the forefront of the industry, like Monsanto, DuPont, and Novartis.

On one level, the furious reaction to the new technology was understandable. It was fear of the unknown, fear of being inexorably trapped by uncontrollable health- or life-threatening circumstances. Nor were those fears mere figments of imagination. Two generations had passed since Rachel Carson's *Silent Spring* had exposed the dangers of such pesticides as DDT and the horrors of thalidomide, a supposedly safe and beneficial drug, had been documented. Public awareness of these and other human-forged disasters had left a legacy of distrust. People are no longer quick to believe assurances of experts who said no harm could result even as people saw and heard stories that seemed to prove those claims wrong. They were also skeptical about biotech industry explanations that the most stringent safety controls were being applied to the production of genetically modified foods and that further scientific tests were being made to ensure their safety.

Shapiro had no illusions about the peril these attitudes presented Monsanto. "If the question is, Is it possible for this technology to be badly used and create dangers?, I say, Yes, it is," he said. "There's no doubt about that. So the question becomes, How do you prevent that from happening while moving ahead with the positives involved? But when you dispose of the fact questions, the opposition is still there. It is not about data. It is not about food safety. It is not about noble environmental risks, in my view. So, what's it about? One obvious parallel that occurs to everybody is that this, like nuclear power, is a sorcerer's apprentice issue. You're dealing with the fundamental forces of the universe. In this case, forces at the center of the universe that in most traditions are perceived as sacred. You're monkeying around with that. We know that you, like Dr. Frankenstein, have wonderful technical skills, but do you have any wisdom? I'll be blunt about this to the point of being offensive, but to Europeans the notion that an American corporation has wisdom is absolutely unthinkable. And by the way, I have a lot of sympathy for that point of view. If someone were to ask me where would I look for wisdom in the world, I probably wouldn't start with the Fortune 500. So that's understandable. So part of it is about the role of corporations and can they foist radical change, totally outside what's perceived to be a democratic consensus-building process and regulatory hurdles, just because they want to do it? The world's going to change because some American companies say it must?"

On Shapiro's last point, there can be no doubt. Behind this increasingly emotional conflict was a greater one that bears directly on the role and position of the United States in the new postindustrial, post–Cold War era. Stripped of all its niceties, it's about deep-seated resentment—envy, jealousy—of American power, American wealth, American values, American culture, American dominance, now and into the future. On this issue, as Shapiro said, "There really is a sense that the United States is throwing its weight around. We're telling the Europeans how to eat. And, boy, they don't have a lot of respect for our knowledge of what to eat."

"You've built the bomb and what else have you done for us, they're telling us?" I interjected.

"Yes," he replied, "but also, you've built McDonald's. What do you guys know about eating?"

At the time we spoke Shapiro continued to believe these issues could be resolved by adequate testing, by accumulation of definitive data, and by public education. He was convinced, or so he said, that sufficient guarantees could be given the public to assure them of the safety of the new genetically modified foods. He also knew that for many people, however, safety alone was not the main question.

Other sets of questions arose, more ambiguous and complex. Shapiro again: "We're talking about modification of plants, but the same set of technologies can be used in many other ways, like Dolly the Sheep. And what about our fantasies of cloning people, and what about gene therapy for people, what about obtaining new organs, and how is access to *that* going to be rationed? Who gets to decide? Mind-blowing questions!"

Speaking as a technologist, with a business stake to protect but also as someone with a lifelong interest in the forces that shape history developed during his days as a Harvard undergraduate majoring in history, he addressed the test facing society this way:

"The challenge for the next couple of generations is going to be to design a set of technologies that enables people to lead decent lives without destroying the ecosystem that we depend on. That means new technologies. And the ones that are at hand give you the best chance to do that. I don't know if it can be done, by the way. I'm not saying that my grandchildren are going to live in a better world. They may live in a crummy world. But if it's going to be done in a way that gives the outsiders a chance to lead a decent life without real destruction, it's going to have to be done based on technologies that exist today or are on the verge of existing today, because we don't have enough time to create a whole new technology. So if there's a message of hope, to me it first has to start with technology. And it has to say there's a way to give people a prayer for a better life without raping the planet in a way that just can't be continued."

In saying that, Bob Shapiro demonstrated why he had earned a reputation as being one of the most enlightened chief executives of a Fortune 500 corporation. However admirable his vision, though, it did not guarantee the success of his dream for Monsanto's future. Eventually, Monsanto's failing fortunes forced it into a merger with the giant drug company Pharmacia & Upjohn, in which Monsanto was clearly relegated to the inferior role. So was Shapiro: Pharmacia's chairman would run the new company with Shapiro watching from the sidelines. Immediately, the newly merged company announced plans to begin selling off a large portion of the agricultural business Shapiro had championed and to shift its primary emphasis back to the manufacture of the immensely profitable sure thing—drugs.

· · ·

When I asked Craig Venter what he thought were the risks facing society from the astonishing advances being made in genetics and biotechnology, he replied: "Probably the biggest risks in history."

He summed up his fears in one short word: *Misuse*. Misuse that poses a threat to civil liberties, misuse that reinforces prejudices and hatreds, misuse that leads to stereotyping and harassment of citizens because of their genetic makeup. "My biggest concern is bad science," he said, "people wanting to justify their lifelong prejudices with mis-scientific data or limited scientific data."

At the time we spoke, the war in the Balkans was producing daily televised scenes of the horrors created by old hatreds. "It's very disturbing for us to be doing this work [in mapping the genome] when this horrible new term of ethnic cleansing has been introduced to our language," Venter said. "It's such a grotesque term. The same things that happened fifty years ago are happening again, and they're only reported on the third or fourth pages of the newspapers now. It's somehow socially ethical. It almost sounds like a good thing. The irony about the groups trying to kill each other in Yugoslavia is that genetically they're more identical than you or I are. So it's not a genetic problem. It's another human deficiency problem."

He gives other examples of how the new genetic information can be misused. "In terms of society, we're very worried about pedophiles being released from prison, and now with the Megan's laws, they can be harassed. So a not unlikely scenario is, Why don't we genotype all the pedophiles in prison? Then we'll have the genotype of these individuals, and then we can start screening the even broader population. We can identify them from birth or before they actually do something evil. But what that assumes is even *if* there was an increased genetic tendency for something like that, that leaves out the notion that there's free will. I think what a lot of people are looking for in society is to have genetics absolve them from individual responsibility. *I can't help it, I'm obese. My genes made me that way . . . I can't help it, I'm a pedophile. I was born that way . . . I was born to smoke . . . I was born to be an aggressive killer.* I don't think it takes too much imagination to see that scenario developing. If you look under the lamppost you can find all those associations, but it doesn't mean all the people of that gene would be pedophiles or murderers. People constantly try to identify the criminal element. So I'm very concerned about the misuse and abuse of information. As a society, we're going to have to come to grips with this information."

Venter was expressing a point of view common to the varied leaders in science and technology with whom I spoke in separate conversations across the country over a two-year period. No matter what their particular field of expertise—doctor, ethicist, biologist, physicist, technologist—they all expressed concern over how society would use the new discoveries. They also voiced frustration at the failure of society and especially its political leaders to address the critically important and difficult new issues being created by the technological revolution.

Genetics alone holds potential landmines for society. Cloning, for example. No longer is it thought that a Dolly the Sheep is a fraud, or that the ability to clone is scientifically impossible. Cloning is being done rapidly everywhere and becoming easier and easier to accomplish. From there, the steps lead inevitably to manipulating genes and perhaps creating entire new species. As one noted biologist told me, "It's going to be phenomenally easy to go in there and manipulate genes. Society has to face up to that. You can't hide your head in the ground and say we're just not going to do that. It's going to be real easy to do. Somebody's going to do it. It's going to be done because it's doable. That's part of what human beings are going to have to face."

· BOOK ·
TWO

Teletimes

Chapter

5

Trial of the Century— Part One

"There is an absolutely, utterly macabre nature to all this.
Look at people rushing to the side of the road . . .
cheering him on, yelling, 'Go, Juice, Go!'"

I n the electronic era, no words cause the nerve endings of the nation to vibrate more than: *We interrupt this program to bring you . . .*

So it is that shortly after seven o'clock, Pacific time, on the evening of Friday, June 17, 1994, prime-time television viewers of ABC's *20/20* newsmagazine with Barbara Walters and Hugh Downs hear an announcer interrupt that program to say: "This is a special report from ABC News."

The familiar face of ABC's debonair Canadian-born anchor fills the screen. "I'm Peter Jennings at ABC News headquarters," he says crisply, broadcasting from New York. "Let's immediately go to a picture in Los Angeles. We're interrupting *20/20* so that the rest of you around the country can see this quite extraordinary scene unfold on one of the freeways in the Los Angeles area. We believe at the moment it is 91-West."

While he speaks, the TV cameras show live pictures of a large procession of cars following a single white vehicle on the freeway below.

"Down there on the ground," Jennings continues, "is a white Ford Bronco. It's just in the left of your picture, in the center, going from right to left, and it is not being chased, but it is being accompanied by a real phalanx of police cars on the ground, police helicopters in the air, and obviously the ABC News helicopter in the air, as well."

The cameras zoom in for a closer look. At an almost stately pace of about thirty-five miles per hour the procession moves along the Artesia Freeway, traveling west toward the Pacific Ocean. Dozens of police cars, their overhead red emergency lights slowly flashing, maintain what seems a respectful distance in evenly spaced ranks several hundred feet behind the Bronco. Twelve helicopters, the *chug-chug-chug* sound of their blades contrasting with the even tones of Jennings's voice, hover overhead. They keep pace with the Bronco. On both sides of the highway, large numbers of people can be seen lining the road. As the Bronco passes, many wave greetings and shout words of encouragement. Others hold aloft hand-printed signs and wave them jubilantly. Drivers whose cars are halted on the freeways by police honk their horns as if in celebration when the Bronco nears them. "This has been going on for some time now," Jennings explains, in his low-key manner. "Police believe they have located O. J. Simpson, wanted on two counts of murder, in this white Bronco going somewhere."

The cameras zoom in even closer. The same measured pace, the same precise distance, separates the Bronco from the Los Angeles County Police Department and California Highway Patrol cruisers trailing behind. All of the vehicles are bathed in the strong bright rays of a Southern California sunset.

There are conflicting reports about how Simpson was found, Jennings explains. One says he was tracked through cellular phone calls made from the Bronco. Another says that when seen he's in the backseat, holding a gun to his head, demanding to be taken to

his mother. "But this is not a chase," Jennings says. "This is basically an accompaniment, as people are pulled over to the side of
the road in various places. You can see what will happen at this one
particular location."

Again the cameras zoom in.

"All other traffic will be pulled over, waved over to the side of
the road," Jennings continues. "The exits onto the freeway have
been blocked by police, and at every stage along the way, you can
see traffic going in the other direction slowing down, clearly understanding that *something* is happening, and probably, given all of
the news coverage in Los Angeles today, understanding *what* is
happening" [italics added].

In an intimate, conversational tone, Jennings explains to his
viewers, "just in case you have been *completely* out of touch," what
this extraordinary scene means, on a day that from a news standpoint in all other respects has been ordinary. The Dow closed at
3,777, down thirty-five points, amid nervousness that the rising
price of oil could trigger further global inflationary pressures leading to higher interest rates. Eight members of a the Branch David-
ian cult were sentenced to up to forty years in jail for their roles in
a deadly shootout with federal authorities in Texas the year before.
These maximum sentences left the jury forewoman in tears; emerging from the courtroom, she told reporters the sentences were "entirely too severe." Senior U.S. generals forced the Army Secretary
to withdraw a memo to the Defense Secretary recommending that
thousands of female troops be permitted to join men in combat
units. Dropping of the planned request came, Pentagon officials
said, after a majority of the army's field commanders strongly opposed the idea of women in combat roles, basing their opposition
on considerations of "strength, morale, and privacy." A strike of
motormen and conductors halted service on the nation's busiest
commuter rail system, the Long Island Railroad, forcing 100,000
commuters to find other ways to work. As news, none of these developments come close to competing with the one event of this Friday that will be fixed in the nation's memory for years to come.

Earlier that day O. J. Simpson, the Hall of Fame football
player, known as the "Juice" by virtually all sports fans and known

to countless others for his roles in Hollywood films and his national TV commercials and sportscasts, was charged on two counts of murder for the gruesome killings of Nicole Brown Simpson, his former wife, and Ronald Goldman, a waiter who was her friend. Their bodies were found five days before, shortly before midnight, lying in pools of blood on the Spanish-style walkway leading to the $700,000 townhouse where Nicole Brown Simpson lived, just two miles from Simpson's multimillion-dollar estate in the quiet, tree-lined section of Brentwood, home of numerous Hollywood stars. Their bodies were savagely hacked and pierced by many knife wounds. Her throat was slashed through to the spinal cord, leaving a five-and-a-half-inch by two-and-a-half-inch gash running from the left side of her neck to her right ear. There were numerous additional wounds, including four in the left side of her neck and three punctures in the back of her head. Her black cocktail dress was ripped, her bloodied hands were in a defensive position as if attempting to ward off an attack. The body of the young man lay ten feet from hers, partly obscured by bushes. His neck had been slashed several times on both sides, and he had been stabbed three times in the chest, once in the abdomen, and once in the thigh. His hands, too, had been cut many times. In all, his body bore twenty-two separate knife wounds. Near his feet a brown leather glove, sticky and soaked with blood, was found. A blue knit Navy watch-cap and an envelope containing a pair of women's glasses also lay near his body.

Asleep inside at the time of the murders were the Simpsons' two small children, Sydney, eight, and Justin, six. The front door stood open. Lights were on throughout the house. A trail of blood, showing human footprints intermingled with those of a dog, led away from the doorway, down the steps toward the sidewalk.

A veteran L.A. homicide policeman, one of the first officers to arrive, calls it "the bloodiest crime scene I have ever seen."

Since discovery of the bodies, the Simpson story has dominated news coverage as few events before it have. For days, the nation's airwaves and newspaper front pages have been filled with rumors and leaks from police and prosecutors. The public is informed that a trail of blood led from Simpson's white Bronco, parked before his mansion, to his front door; inside the house, blood is said to have been found on his clothing.

Simpson, who flew to Chicago to keep a commercial commitment shortly before midnight that Sunday night, less than two hours after the murders were committed, returns to Los Angeles Monday morning to a media circus. As more damaging information about the murders is released surreptitiously by police to the press the next day, Tuesday, Simpson is officially named a suspect, though the public is not so informed. Still, he has not been arrested. He remains in seclusion in his Brentwood mansion—Brentwood, by now identified repeatedly in the media as home to such stars as Angela Lansbury, Dennis Quaid, Meg Ryan, Roseanne Arnold, Michelle Pfeiffer, Meryl Streep, Tom Hanks, among many others, as well as the mayor of Los Angeles, the district attorney of Los Angeles, and several prominent Los Angeles judges. It is a preserve of privilege and power; those who live there maintain a private homeowners' association with a special plainclothes security detail to watch over them and their grand homes. Brentwood also displays the symbiosis between fame, wealth, power, and politics, Hollywood style, that so typifies America in the Nineties. So does the media circus that forms instantly around the latest celebrity to be involved in a scandalous episode. In O.J.'s case, camera crews and correspondents dispatched from around the world mass outside his Brentwood mansion where they join others in an around-the-clock stakeout and wait for another morsel of news, or rampant rumor, to pass on breathlessly to the public.

As the media horde swells and more incriminating leaks are reported worldwide, Simpson attends the funeral services for Nicole with their two children and her family in a Roman Catholic church. He appears to be in deep mourning; friends tell the media how distraught he is during the service, and of the tears he sheds. Back to his Brentwood mansion he goes, followed by a caravan of police and media vehicles. TV viewers are told he's the man being shielded by a long coat as he enters his house. It's a ruse orchestrated by an off-duty L.A. police officer whom Simpson hired. After the funeral services, Simpson has slipped away from the church unseen and traveled to the "very large" home of a friend in

Encino, in the San Fernando Valley. Impersonating Simpson, and being shielded by the coat, is Simpson's friend, A. C. Cowlings. The media—and the police—are fooled. Still, Simpson is not charged with any crimes.

All that ends early Friday morning. Police inform Simpson's lawyer that warrants have been issued for his arrest.

Acting on Simpson's behalf, his lawyer negotiates a time for Simpson to surrender to police—eleven o'clock that morning—at police headquarters. The hour comes and goes. No Simpson. After waiting another hour, police dispatch a squad car with two officers to the residence in Encino where Simpson's lawyer tells police the football star is in seclusion, under medication, "and in a very fragile state." They're also told that doctors and a psychiatrist are with him in the house, treating him for exhaustion and depression. He's heavily sedated. Officials are assured the formal surrender and arrest will take place there. When the officers arrive, they discover Simpson has fled in a white Bronco with his close friend and former football teammate, Al "A. C." Cowlings. Simpson's lawyer, who was with the athlete only fifteen minutes before, is upstairs conferring with the doctors when police arrive.

Furious and embarrassed, police officials hold an angry televised press conference. Orenthal James Simpson is declared a fugitive from justice. Airport and border patrol officials throughout Southern California are alerted to be on the lookout for him. Anyone helping Simpson avoid the law will be committing a felony, police warn. An all-points manhunt is launched for O. J. Simpson, wherever he is, wherever headed.

It isn't long before police, but not the public, know where he is. At 6:25 Pacific time a passing motorist calls the police. "I think I just saw O. J. Simpson on the 5 Freeway," the caller reports. "He is heading north, and I got the license plate of the white Bronco, and he, like, stared us down like he was death."

Patrol cars are instantly dispatched to the freeway. They're aided in their search by signals they're picking up from cell phone calls being made from inside the Bronco. Soon, they have it in sight.

Minutes later, from inside the Bronco, Simpson's friend, Al Cowlings, who is driving, places his own frantic 911 call: "You've got to tell the police to just back off," he says. "He's still alive, but he's got a gun to his head."

A police bulletin is flashed: *Armed and dangerous.*

At 6:44 Bob Tur, a local helicopter pilot who earns his million-dollar annual living selling film footage and live reports of L.A. calamities, especially on the freeways, to local TV stations, spots the Bronco and reports from his chopper: "What we have here is a suicidal man, O. J. Simpson, with a gun to his head."

By now, the Great Bronco Chase is being broadcast live around the world. Everywhere, television stations interrupt their scheduled programs to go live to Los Angeles and the white Bronco on the freeway. In New York City alone, nine television stations preempt regular programming to cover The Chase. Talk-radio stations also interrupt all planned programming. A rising crescendo of voices fills the airwaves—voices of commentators, broadcasters, friends, acquaintances; voices of the people, interviewed at random, in cities large and small, in ballparks and inner-city playgrounds. Some of the voices sound disembodied; others are merely those of unidentified people. One, a woman, adds her plaintive comments to the swelling national chorus: "You watch this and you go, 'What's going on? What's going on with our whole society?'"

Wherever there's a television set, people stop to stare. In bars, offices, homes, the crowds gather around the ubiquitous tube. In health clubs, exercisers stare at TV sets covering The Chase while pedaling stationary bikes. In communities across America, streets are empty. Even West Hollywood's Sunset Strip is deserted; normally, as the weekend begins, it's thronged with nightclubbers and bumper-to-bumper traffic. In some areas, patrons keep Friday-night dinner reservations but bring miniature TV sets with them to watch the live Simpson drama from their restaurant tables. Implicit, but not spoken in the media-produced circus, is a tantalizing ingredient that compels ever more people to watch. They may witness a television-age first: the violent death of a celebrity, live and in color, on their screens.

As The Chase proceeds, still at the same measured pace, crowds spring up along the freeways, alerted in advance to the location of the Bronco by their TVs, their radios, their cell phones.

Now, the crowds are massive. People race ahead of the caravan, run up embankments, produce more and more handmade signs. They cheer, wave, scream. Reporters search and stumble to find the correct words to characterize their behavior. It's a "spectator sport," it's a "high-speed parade," it's a "Super Bowl scene," it's a "Roman circus." Most are hopelessly inadequate; pictures tell the story.

Peter Jennings, continuing his steady flow of commentary, captures the madness best. At one point, he interrupts another correspondent to say, "Look at all these people rushing, waving. There is an absolutely, utterly macabre nature to all this. Look at people rushing to the side of the road . . . cheering him on, yelling, 'Go, Juice, Go!' "

As Jennings's voice is heard, the cameras remain fixed on the Bronco and the armada of police vehicles still cruising slowly along the California freeway toward the huge orange ball of the sun, rapidly sinking over the Pacific.

The spectacle is greater than any TV soap opera that attracts tens of millions of viewers daily; greater than any supermarket magazine offering the latest scandalous—true or not—revelations about lives of the rich and famous; greater than any gossip column appearing in the daily papers. And it's all live, in color, in prime time, in everyone's home.

The cameras remain trained on the procession, still heading no one knows where.

On and on, minute after minute, hour after hour, the surreal procession proceeds. As it does, more people rush to the highways to witness it firsthand. They cheer Simpson on. *Juice! Juice! Juice!* Another chant sounds: *The Juice is loose. The Juice is loose.* And minute by minute the dimensions of the national TV audience keep increasing. Close to 100 million Americans are now watching. That's millions more than voted in the last presidential election in 1992, millions more, even, than watched the last Super Bowl, which each year attracts the greatest national TV audience and commands by far the highest price from advertisers paying to sponsor a national TV program.

All scheduled TV programs have been preempted for coverage of The Chase, including a national professional basketball championship game between the New York Knicks and the Houston Rockets. Howard Kurtz, the media critic for the *Washington Post* and CNN media TV commentator, describes the unfolding scene as "one of the top five television events of all time really—right up there with the shooting of Lee Harvey Oswald, the first moon walk, and the start of the Persian Gulf War." Other commentators compare it to John F. Kennedy's funeral, the explosion of the *Challenger* space shuttle that killed all aboard, the nationally televised rescue from a surface well of an infant dubbed Baby Jessica by the media. Still others liken it to the long-running TV serial of years past, *The Fugitive,* recently made into a movie starring Harrison Ford, or the popular chase movie of the mid-Sixties, *Bonnie and Clyde,* in which the doomed fleeing criminals meet their fate in a burst of police gunfire while sitting inside their car. One commentator calls it "a kind of male *Thelma and Louise* in reverse."

Now the caravan passes the Los Angeles International Airport, known by all who travel there by the initials LAX. Now the Bronco turns off the freeway onto Sunset Boulevard, a name synonymous with Hollywood in film lore and popular TV serials. In increasingly incredulous and excited tones, the ABC commentators, who know Los Angeles best, inform Jennings and the greater national audience that it appears as if Simpson is heading toward Brentwood, perhaps to the home of Nicole and their children where the bodies were found. Minutes later, in even more astonished tones, they begin to guess at the final destination. An excited Al Michaels, a celebrated TV sportscaster and friend of Simpon's, interrupts Jennings.

> Al Michaels: Peter, he's at his home. There's the satellite truck.
> Peter Jennings: My goodness, gracious me.
> Judy Muller, a local L.A. correspondent: He's come home, after all of that. After all of that.

So he has. The great runner has run home, or, as one critic puts it, made "O.J.'s last end-run home." Sports metaphors naturally leap to commentators' lips. Most popular are variations on "O.J. the gridiron escape artist" contrasting with "O.J. the fugitive from justice." Still, the drama isn't over.

Outside the gates, an enormous crowd gathers, with more people arriving by the minute.

For days, during the TV stakeout of the mansion, crowds have massed out front, standing alongside street vendors, who have been doing brisk business selling tuna melts and burritos. The atmosphere is carnivalesque. Two flight attendants from Minneapolis, drawn to the scene that Friday night as soon as their flight landed at LAX, take turns standing on a stepladder peering over the ivy-covered wall into the grounds of Simpson's mansion. "We love this stuff," one of them tells a reporter from the *Sacramento Bee*. She notes the signs of wealth and the kitsch that adorn the grounds: the Bentley in the driveway, the rock-studded swimming pool, the waterfall, the wishing well, the dollhouse, the ceramic figures of Snow White and the Seven Dwarfs near the eucalyptus trees, rose-bushes, and gladiolas. Her fellow flight attendant says, "We were on the Graveline Tour last year. They pick you up in a hearse and take you to the place John Belushi killed himself and the homes of other dead stars, including Marilyn Monroe." All of them took place in Brentwood, too. Of O.J., she says: "I hope he didn't do it. But every finger points to O.J. I used to have a crush on him. I don't anymore."

As night falls over the darkened streets of Brentwood, the crowds take on an ominous cast. Police throughout West Los Angeles are placed on tactical alert. Heavily armed SWAT teams are dispatched to the mansion. Paramedics are summoned, ready for an emergency. L.A. police air support officials warn news helicopters they will ask the Federal Aviation Administration to ban all flights there unless they increase their altitude by a thousand feet. More police vehicles, more officers, move in to try and maintain crowd control.

The situation grows even more tense, and wilder, as the white Bronco turns inside the gates. In the backseat, O. J. Simpson is spotted. He holds a blue steel revolver against his chin.

Outside the gates, the cheering and chanting increase. "We love Juice." "Don't lose the Juice." "Save the Juice." "Go O.J.!" More placards are waved. All signal support for Simpson. The crowds rush forward; police move to block them. Some police cars are rocked by the crowds. "Free O.J.! Free O.J.! Free O.J.!" comes the chant.

Witnessing the moblike atmosphere, one ABC correspondent tells Jennings it reminds him of a scene from a 1975 Al Pacino movie, *Dog Day Afternoon,* in which crowds massed before a bank to loudly cheer bank robbers holding hostages inside while outside police nervously prepare for a riot. Others, with more literary memories, think it resembles a scene out of Nathanael West's Hollywood tale, *The Day of the Locust,* a novel that brilliantly portrays the emptiness of Tinseltown and pathetic hunger of celebrity-starved mobs seeking meaning for their lonely lives through worship of synthetically glamorous film stars.

Now comes an incredible final act. As the nation prepares for a sensational, quite likely tragic, denouement, the TV ratings rise. Transfixed, people watch what television critic Walter Goodman later describes in the *New York Times* as a "strangely powerful and prolonged anticlimax" to the long, slowly developing scenes that hold "the imagination prisoner in a public dream."

As the Bronco stands inside the compound, its red hazard lights slowly blinking in the balmy June night, the driver's door opens. Cowlings emerges, walks inside Simpson's Tudor-style mansion, and begins angrily talking with police and members of Simpson's family. Alone in the back sits Simpson, still holding the blue revolver.

The minutes slowly pass. The cameras remain fixed on the darkened scene. The prospect that this spectacle will end either in a suicide or in a slaying by police, captured live and in prime time, increasingly seems more probable. Shadowy scenes of heavily armed police moving about the grounds reinforce the sense of impending doom. One moment adds a Keystone Kops touch: A member of the police SWAT team has disguised himself as a bush and crouches, weapon ready, aiming toward the Bronco.

Finally, after fifty nerve-racking minutes, bulletins are flashed worldwide. Simpson has surrendered. He's taken into custody at 8:56 P.M. California time. In the East, where the network anchors are still broadcasting live, the clock approaches midnight. After five hours, and a fifty-mile freeway chase, O. J. Simpson is heading toward jail and a trial on charges of murder.

For America, and for much of the world, all this is only the beginning.

· · ·

In keeping with an age of excess and short memories, the O. J. Simpson murder trial is called the Trial of the Century. It's a foolish label, but inevitable given the hunger of a mass media to hype the latest sensational newsbreak in order to attract higher ratings. Nor does the "Trial of the Century" superlative really fit, ignoring, as it does, others that transfixed the nation throughout the century.

In 1906, the trial of Harry K. Thaw for the murder of the great architect Stanford White laid bare the private demimonde arena of immense wealth and hypocrisy that typified the Gilded Age. In 1914, as America was about to step fully onto the world stage with the outbreak of World War I, the trial of Leo Frank for the slaying of little Mary Phagan in Atlanta—"the American Dreyfus Case"— triggered an outbreak of prejudice against Jews that led to the formation of the Anti-Defamation League of B'nai B'rith and the rebirth of the murderous Ku Klux Klan. In 1921, the murder trial of the immigrants Sacco and Vanzetti, "the poor fish peddler and the poor shoe cobbler," passionately divided Americans along class lines and sparked heated debates about the fairness of American justice. In 1925, the Scopes evolution trial exposed societal conflicts between science and religion, liberalism and conservatism, and pitted the agnostic lawyer Clarence Darrow against the aging, dying fundamentalist orator William Jennings Bryan. In 1948, the Hiss case personified fears about communist subversion that marked the new Cold War era and elevated Richard Nixon into a figure whose actions would deeply affect national political life for the next three decades. In 1951, the espionage trial of Julius and Ethel Rosenberg fueled conspiracy theories about traitors within and provided a backdrop for an era of character assassination known as McCarthyism.

Memorable as these are, they pale beside the first national media extravaganza to be called the Trial of the Century. That was the 1935 trial of Bruno Richard Hauptmann for the kidnapping and murder of the Lindbergh baby.

However the Simpson trial compares to these earlier events, it represents more than merely another spectacle that momentarily commands attention in the Nineties. Nothing in that decade, or in many decades past, shines a more merciless light on some of the most troubling elements of American society. Nothing better exposes the least attractive sides of mass American culture, and nothing illustrates more vividly America's racial and class divisions.

The O. J. Simpson trial is about race, sex, celebrity, wealth, power, privilege, and prejudice—both white *and* black prejudice. It is about police, judges, juries, crime, violence, and the administration of justice. It's about all of these things, all at once, and about something more that especially typifies the times—television.

In the O. J. Simpson trial, television found its perfect subject and aggressively exploited it in ways that influenced much of what followed in the Nineties and beyond. Television already had changed America, and with the O.J. trial television itself was changed—for the worse. The O.J. trial lowered existing TV standards. It dramatized sensational developments. It generated a vast audience and created a mass appetite for more of the same. It fostered a pervasive scandal culture. It highlighted conflict and confrontation over substance and seriousness. It diverted the public from consideration of other issues. It affected the practice of journalism, the legal profession, the criminal justice system, and the political system and diminished them all. It influenced attitudes about leaders and public figures, celebrities and cultural heroes. It cheapened public discourse. It deepened an already pervasive sense of public cynicism. And it set the stage for even greater focus on spectacle and scandals, further intensifying an already prevailing public belief that seeing was no longer believing.

The characters themselves seem drawn from central casting of a TV melodrama; each becomes a caricature of American types.

Yet when assembled for their final public portrait, each in his and her way provides evidence of the paradoxes that comprised America at the millennium.

Foremost among them is O.J. himself.

· · ·

Everyone thinks they know him: O.J., the good guy, always smiling, always charming, always nonthreatening, always ready to sign one more autograph for adoring fans; O.J., the all-American sports hero, the amiable athlete turned affable TV commentator; O.J., the entertaining pitchman who is constantly seen in American living rooms, courtesy of TV commercials, leaping over barricades and racing through airports as he promotes his sponsor's products; O.J., the loving father and family man; O.J., the black man, who plays so long and successful a role as an appealing, self-effacing American icon, that he's no longer seen as black. In fact, he doesn't see himself as black—he's often quoted as saying, "I'm not black, I'm O.J." In manner, language, lifestyle, and personal association he passes, seemingly effortlessly, into the upper echelons of the white world. Exposed to the good life, courted by the powerful, he belongs to the best country clubs, moves in the best social circles, golfs with the corporate elite, travels on private jets to fabulous resorts, rubs shoulders with influential deal makers.

So nonblack, so nonthreatening racially is O.J. that he becomes the first black athlete to be employed by corporate sponsors to endorse products not marketed solely to blacks. His marriage to a beautiful young blonde, a familiar golden girl of the Nineties type, draws none of the sneers and hatred and jealousies that mark other interracial celebrity marriages. In the American heart and mind, O.J. is colorless. His celebrity status is such that even some who resent him do not express their feelings publicly. That is particularly true, and particularly complicated, when it comes to blacks.

After O.J.'s arrest for the murders, some of these hidden emotions surface. "He forgot that he was black," one black woman says when she phones a Dallas talk-radio show devoted solely to the O.J. case. "He didn't show love to us that he should have showed. But deep in our hearts, all of us loved him. He left us years ago."

Another black woman calls the same show to say: "Even though he was with a white woman, he was something for our race to be proud of. I feel hurt. I feel hurt."

O.J., the reality, is infinitely more complicated. His life story illuminates the continuing struggles of African Americans to escape the obstacles that keep them separated from the predominant white mainstream. O. J. Simpson, the gentle sports hero fans think they know, is, in fact, the product of a tough, violent upbringing in the slums of San Francisco.

His is a raw and painful childhood, but typical of blacks like him who live in the dismal public projects of the inner city. In the tough Potrero Hill district where he grows up, 70 percent of the blacks are on welfare, and, as in O.J.'s case, a majority of the children are raised without a father in the home. It's common for them to join a gang, as O.J. does at the age of thirteen when he becomes a member of the Gladiators. A year later, at fourteen, he experiences his first arrest—for robbing a liquor store. In junior high he joins what he once described as his "first *fighting* gang," the Persian Warriors. With them, he participates in pitched battles, usually on weekends, with rival gangs. In their world, violence is commonplace; brawling and stoning cars are part of a normal weekend. Witnessing sudden death is also a common experience. Years later, O.J. remembers being in gang fights "where a couple of guys got croaked."

By the age of fifteen, he has earned a reputation as being especially good with his fists. "I only beat up dudes who deserved it," he once explained, "at least once a week, usually on Friday or Saturday night. If there wasn't no fight, it wasn't no weekend." It was then, barely into his teens, that he wins renown within the gang culture for the manner in which he beats an older, much feared, leader of a rival gang, the Roman Gents, the toughest gang in the city.

After he ends his pro-football career, O.J. recalls that fistfight in a 1976 *Playboy* interview that reveals much about him the public would later be forced to confront, but does not want to. His fight is with Winky, then a battle-hardened twenty-year-old. "One night I was at a dance in the Booker T. Washington Community

Center," O.J. remembers, "when, all of a sudden, this *loud* little sucker—an older O.J.—comes up to me and says, 'What did you say about my sister?' I'd heard of Winky—just about everyone had—but I didn't know that was who this cat was, so I just said, 'Hey, man, I don't know your sister. I don't even know *you*.' It wasn't cool to fight in the community center, so the guy started walking away, but he was still talkin' crap to me and I yelled back, 'Fuck you, too, man!'

"Well, a few minutes later, I see a whole bunch of Roman Gents trying to get this cat to be cool, but nope, he's comin' over to me and he shouts, 'Motherfucker, I'm gonna kick your ass!!' And then—bingo!—the music stops and I hear everybody whisperin', 'Winky's gettin' ready to fight.' *Winky!* Damn, I didn't want to fight *him*. So as he walks up to me, I say, 'Hey, man, I really didn't say *anything* about your sister.' But before I can say anything else, Winky's on me, and swingin'. Well, I beat his ass—I just cleaned up on the cat—and as I'm givin' it to him, I see this girl Paula, who I just loved, so *I* start getting loud. And as I'm punchin', I'm also shoutin': '*Muthafuckah! You gonna fuck with me??*' "

That's the O.J. his peers know. The public never does. Over the years, as he becomes enshrined among American sports heroes, O.J. adopts the style—and the speech—of the successful white world.

O.J. works hard at transforming his public persona from brawling street tough to smooth, confident member of the successful elite. Long before the brutal murders with which he's charged, Lee Strasberg, the acting coach who helped Marlon Brando, Marilyn Monroe, and other stars, and was then assisting O.J. in his Hollywood roles, says of O.J.: "He already is an actor, an excellent one." Strasberg adds, "A natural one." And once, while shooting a TV commercial with underprivileged black youths in Oakland, in the territory where he grew up, O.J.'s mask slips. Inadvertently, he begins employing the street language of his gang childhood. Furious, he announces he wants to redo the commercial. The second take goes perfectly. "That's what happens when I spend too much time with my boys," O.J. says afterward, in explaining his slip. "I forget how to talk white."

For O.J. and others like him, sports offers the surest path toward that road to success. It's not unusual for a black child from the slums with natural athletic ability to achieve an American Dream lifestyle and public fame. It is highly unusual, however, to become a superstar athlete in spite of suffering from a severe childhood physical handicap.

As a two-year-old, O. J. Simpson was afflicted with rickets. The disease, resulting from a lack of calcium in the bones, a commentary on his impoverished inner-city diet, withered his legs and left him bow-legged and pigeon-toed. To correct his disability, leg braces were required; his mother, however, couldn't afford them. So for the next three years, O.J. shuffled around his house with an improvised contraption that enabled him to walk, while strengthening his legs. For several hours every day he put on shoes connected to each other by an iron band and struggled to walk. Ultimately, his handicap was corrected. The tenacity and drive displayed by this bow-legged kid with rickets who couldn't walk unaided is as impressive as any of the subsequent stories in the Nineties that celebrate the determination to succeed of the new golden entrepreneurs of Technotimes.

From high school on, O.J.'s athletic prowess propels him up, and eventually out, of the world of his birth. His career ambitions are minimal. By his own account, he was "a lousy student" and "didn't exactly kill myself studying." He's so eager to leave the classroom that he thinks of enlisting in the Marines and fighting in Vietnam. By the time he graduates from junior college, he has smashed all existing football rushing records for that level of play and finds himself aggressively courted by recruiters of big-time collegiate athletics. They shower him with offers of full scholarships—and, typically, much more, most of it hidden.

Though ostensibly an amateur endeavor, guided by principles of good sportsmanship, American collegiate athletics, especially football, long since has passed into the realm of high-stakes, high-revenue professional sports. It's the biggest of businesses, awash in cash, commercialism, and corruption. It's all about money. A young potential superstar like O.J. finds himself at the center of a virtual bidding war. "A whole bunch of 'em were offering all kinds of

under-the-table shit," he recalls of the college recruiters who besieged him. "In addition to a regular scholarship, most of the
schools were talking about $400 or $500 a month and stuff like a
car. One school was gonna arrange for my mother to clean up an
office for $1,000 a month; another was gonna get my mother a
house."

Whatever the offer, O.J. more than proves his worth; he richly
rewards those who invest in him. Both in college at USC and then
in the National Football League with the Buffalo Bills, his athletic
ability attracts legions of paying fans. They fill the stadiums. With
them come the networks and the sponsors. They vie for the right
to telecast his games, not only locally, but nationally. These generate still more revenue.

Through it all O.J. soars. As his earnings multiply, he acquires
more of the taste for and the trappings of the affluent life. To an
adoring public, he becomes a self-effacing, beloved superstar. It's an
intoxicating role. O.J., like so many sports heroes who achieve
celebrity at an early age, takes the fawning worship of starstruck
fans and effusive praise of sports announcers as his due. So, too, he
takes as his natural right the physical gratification that comes with
quick fame and wealth—the easy and endless sexual conquests, the
eager girls on their knees, the constant adulation of faithful camp
followers. No wonder primal urges are unchecked. Stars believe
they can always get away with outrageous behavior, and often do.
In the Nineties, sports stars become involved in even more notorious cases, often resulting only in slaps on the wrist for the offenders, if that.★ Rules are made for lesser mortals.

★Violent behavior of sports stars, in both amateur and professional ranks,
reached such a point that commentators began describing the National Football League as the National Felons League. This, after two players (the Carolina Panthers' Rae Carruth and Baltimore Ravens' Ray Lewis) were charged
with murder, and many other episodes of violent acts stained boxing (Mike
Tyson, the former heavyweight champ, bit off the ear of the current champ in
the ring and was also convicted of rape); hockey (Marty McSorley, charged
with assault after savagely high-sticking an opponent on the ice); baseball
(pitcher John Rocker being let off with a $500 fine after uttering a string of

But the idea that such a supposedly familiar public figure could be capable of the monstrous murders with which he's accused is simply inconceivable. Indeed, the public can be excused for being so misinformed about O.J. and the life he's led. The O.J. the public has grown to love is wholesome, charismatic, uncomplicated.

As with so many other celebrities, especially sports celebrities, O.J. has led a charmed life, protected by a cordon of publicists, agents, producers, sponsors, sportswriters, and commentators, and protected no less by the police who treat him as an untouchable. Even in prison, O.J. demonstrates the power of special privilege accorded the celebrated star. In the Los Angeles County Men's Central Jail, where he is incarcerated along with sixty-four hundred other inmates, O.J. lives alone in a row of seven cells. Most of the other inmates are housed two to six in a single cell. They wait in line to use pay phones. O.J. doesn't. They must bathe in communal showers. O.J. showers by himself. Other prisoners are limited to visits of only twenty minutes a day. O.J. sees his lawyers and forty specially designated visitors for up to ten hours a day.

When asked about this special treatment, a sheriff's deputy justifies it by saying: "O. J. Simpson was living in a Brentwood estate worth $5 million, now he's incarcerated in a 9-by-7-foot cell. . . . It's all relative."

Not until the aftermath of the murders do other aspects of his life become widely known.

Within days, former friends and associates tell reporters of his jealousies, his rages, his record of hitting on women, casually, repeatedly, his blatant sexism and possessiveness. Once, in a fine Santa Ana restaurant where O.J. was hosting a group of his friends

---

bigoted, hate-filled insults in a national interview that would have resulted in immediate firing in the private sector); basketball (Latrell Sprewell, gaining a multimillion-dollar contract, a shoe sponsor TV deal, and facing no charges after choking and threatening to kill his coach). None of these incidents, nor other acts of loutish behavior such as throat-slashing gestures by players after touchdowns, produced outcries among fans or affected the broadcasters and advertisers who continued to feature the stars on lucrative telecasts that supplemented athletic salaries which rivaled those of CEOs.

and his wife for dinner, he grabbed Nicole's crotch and loudly proclaimed, "This belongs to me." On another occasion, as testimony later reveals, he boasted to an acquaintance how easy it would be to kill someone by slashing his neck with a knife—and demonstrated, with gestures, how he would do it. And once, according to information a former Hollywood associate of his tells prosecutors, when the subject of Nicole's boyfriends was raised O.J. angrily vowed to "cut their fucking heads off" if he ever finds them driving his cars.

The public knows nothing about this side of O.J. It certainly has no knowledge of the devastating record of desperate telephone calls Nicole Brown Simpson made to police emergency numbers over the years, both during her marriage and after her divorce, as she sought protection from a violent, battering O.J. Nicole's police emergency calls document not only O.J.'s explosive violent nature, but also the failure of both police and judicial authorities to take effective action to stop his abusive behavior. Not that such failure is unique in Nicole's case; police routinely fail the battered wife, as numerous court records show. That dreary kind of record was compounded by the circumstances of the O.J. case. O.J.'s a superstar; superstars receive different treatment.

All this changes after The Chase and the arrest. Police and prosecutors, or both, immediately slip tape recordings of many of Nicole's 911 police emergency calls to local TV stations and the networks. Along with the recordings are equally damaging leaked written reports of police investigations of those incidents.

One police report, on New Year's Day 1989, describes how Nicole, wearing only bra and sweatpants, runs from bushes where she's hiding after having called police from inside their Brentwood mansion. Badly beaten, her lip cut, one eye blackened, Nicole keeps telling officers, "He's going to kill me; he's going to kill me." Does he have any guns? police ask. "He's got lots of guns," she replies. Then she bitterly complains to the police: "You never do anything about him. You talk to him and then leave. I want him arrested."

At that point, according to the police account, O.J. appears in a bathrobe. "I don't want that woman in my bed anymore," he screams at police. "I got two other women, and I don't want that woman in my bed anymore." When warned he is going be arrested, O.J. yells: "The police have been out here eight times before and you're going to arrest me for this? This is a family matter. Why do you want to make a big deal out of it? We can handle it."

Ultimately, Nicole doesn't press charges, but a city attorney files misdemeanor charges of spousal abuse against O.J. He pleads no contest, is fined $970, ordered to perform 120 hours of community service, and attend counseling sessions twice a week for three months. He's also given two years' probation.

The incident attracts little news attention. It has no demonstrable effect on O.J.'s public popularity. "It was perplexing," a female former employee of NBC Sports remarks to *Sports Illustrated* immediately after the murders. "People at NBC Sports used to always remark about the beating, shaking their heads and saying, 'Here's a man who used to beat his wife, and none of America cares or remembers.' People refused to believe because they thought he was such a nice guy."

It isn't just the American people who don't believe or care about such behavior. Neither do O.J.'s corporate bosses. Three months after the 1989 incident, NBC signed O.J. to an annual $400,000 broadcast contract, and he got another contract for more than half a million dollars a year from Hertz rental car, the sponsor of his TV commercials.

Even more damaging are tapes leaked to CNN and then broadcast over that network in prime time days after O.J.'s arrest. These include a 911 call Nicole made to police from inside her home, after her 1992 divorce, on October 25, 1993. The transcript of that phone conversation frighteningly foreshadows her fate on the front steps of her townhouse less than a year later:

911 Operator: *911 emergency.*

Nicole: *Could you get someone over here now, to 325 Gretna Green. He's back. Please.*

911 Operator: *Okay. What does he look like?*

Nicole: *He's O. J. Simpson. I think you know his record. Could you just send somebody over here?*

911 Operator: *Okay, what is he doing there?*

Nicole: *He just drove up again. Can you just send somebody over?*

911 Operator: *He just drove up. Okay, wait a minute. What kind of car is he in?*

Nicole: *He's in a white Bronco. But first of all, he broke the back door down to get in.*

911 Operator: *Okay. Wait a minute. What's your name?*

Nicole: *Nicole Simpson.*

911 Operator: *Okay. Is he the sportscaster or whatever?*

Nicole: *Yeah.*

911 Operator: *Okay. What is—*

Nicole: *Thank you.*

911 Operator: *Wait a minute. We're sending the police. What is he doing? Is he threatening you?*

Nicole: *He's fuckin' going nuts.*

Her furious profane response doesn't stop the police operator from asking still more questions and still not responding swiftly.

In view of what happens later, Nicole's words as she tries to explain her fear of O.J. are especially chilling: "The kids are upstairs sleeping and I don't want anything to happen."

She explains to the police operator how O.J. came to her townhouse earlier, broke down her back door, went upstairs and pounded on her door until she fears it, too, will be broken. "Then he screamed and hollered," she says, "and I tried to get him out of the bedroom because the kids are sleeping in there."

The 911 Operator replies laconically, "Okay."

Nicole, her tone increasingly urgent, continues trying to describe the danger she feels. The operator interrupts, maddeningly, to say, "Okay. So basically you guys have just been arguing?"

At that point, the tape picks up the background sound of a male voice, roaring and shouting unintelligibly. The conversation continues:

911 Operator: *Is he inside right now?*

Nicole, desperately: *Yes, yes.*

911 Operator: *Okay, just a moment.*

O. J. Simpson: *[Unintelligible.]*

Through the sound of O.J.'s angry shouting, the tape clearly picks up a fragment of Nicole's plaintive voice: "—the kids. O.J.—O.J., the kids are sleeping."

The police operator interjects with yet another question: "He's still yelling at you? Just stay on the line, Okay?"

The conversation ends with a heartbreaking appeal from Nicole. "O.J. O.J. O.J. Could you please leave. Please leave."

After another frightening encounter with O.J., Nicole expressed the terror she felt in a tape recording. It's entered into the public record, along with a diary she kept in the years prior to her death. The diary details her fears of O.J.; it also describes the beatings and humiliations she suffers, including a vivid account of how O.J. once "beat me for hours." On the tape, she says that during O.J.'s rages he "gets a very animal look in him, his veins pop out and his eyes get black." Looking at him, she fears that "if it happened once more, it would be the last time." Eight months later, she's hacked to death.

· · ·

O.J., obviously, is the central character. The remainder of the cast, however, offers no less revealing glimpses into late-twentieth-century American society.

Nicole, for instance. In the voluminous accounts published about her after the murders, she appears as a classic California beach-bunny type: the beautiful young blonde who meets the famous sports hero while waitressing at a tony Beverly Hills watering hole, the Daisy Club. She's barely eighteen, just graduated from high school. They begin living together not long after but don't marry until eight years later. Her background is seldom mentioned, but it speaks to post–World War II American demographic and social changes.

Louis Brown, Nicole's father, is a native of Kansas. In the 1950s, at the peak of the Cold War, he finds himself stationed in Germany while working for the armed forces newspaper *Stars and Stripes*. There, he meets and marries Juditha Baur, of Rollwood, West Germany. A few years later, the Browns, now with two small daughters, Nicole and her older sister, Denise, leave Germany and follow the familiar path west for a new beginning in California. The only remnant of Nicole's German background is her mother's insistence that the little girl recite the Lord's Prayer each night in German before going to bed.

If Nicole had a career ambition, other than enjoying the good life, it is not known. Later, in the instant psychobabble commentary devoted to her, a patronizing tone comes through some of the analyses. She's described, condescendingly, as yet another vacuous Southern California type—the empty-headed hot babe whose fortune is her looks and whose sole ambition is to use her sex appeal to become a celebrity trophy wife with all the glittering material accoutrements such an arrangement brings. Just another on-the-make L.A. party girl. A bimbo. A gold digger. Feminist critics both pity and deprecate her; to them, she becomes the latest example of the way young women are sexually exploited by domineering men in the still-far-from-liberated America of the Nineties.

Blaming Nicole catches on quickly. Men, especially football fans, ask: Why didn't she leave him? Nobody forced her to stay. Supermarket tabloids begin running covers playing on women's worst instincts, the catty tearing down of the girl who has it all— and never deserved it, of course—and who just maybe has it coming, especially if she knows she's beautiful and flaunts it.

For all the talk of post-feminism, for Nicole and others like her who don't go to college, who don't pursue a professional career, the reality is that being Mrs. Somebody often appears a better choice than being Mrs. Anybody or Ms. Myself, especially if one aspires to the benefits of the affluent life. In her case, the choice becomes more complicated when compared to the path women in the Nineties were expected, and pushed, to follow: Go to college, have a career, have a family, "have it all." Wonder women! Supermoms! For whatever reasons, Nicole becomes one of those exceptions to the Nineties' expectations for Ideal American Women. She fills her life with other signs of accomplishments—a tanned and toned body, designer clothes, a celebrity marriage, and abundant funds to maintain the good life. Besides, the mass cultural message relentlessly beamed at the twentysomething Nicoles through television dramas, sitcoms, soaps, films, and seductive TV commercials and glossy magazine ads reinforces the desirability of following the material-girl path: Strive to be the glamorous Ideal Woman. Be desired and envied by all, but had by few. And the easiest way to achieve that status is to have enough money to make yourself irresistibly attractive—to enjoy regular workout time with professional physical trainers, visit fancy figure-enhancing salons, purchase killer designer wardrobes, all essentials for the successful material life.

Marrying O.J. provides more than the bountiful life; by any measure, it's a golden life. In her 1992 divorce petition Nicole describes the sort of life to which she has become accustomed, and which, she believes, two beautiful, popular, and wealthy people in west L.A. ought to have:

"We moved into a $5 million residence in the exclusive area of Brentwood. We had a full staff to help us. The house was extensively remodeled a few years ago, and no expense was spared.... We also spent our summers at a $1.9 million Laguna Beach house, which is situated on the sand. This house was never rented but only kept for our own enjoyment during the summer and at other times during the year. [O.J.] and I maintained a bicoastal lifestyle. We have an apartment in New York, which I used several times each year, sometimes for as long as one month at a time.... Whenever we traveled on commercial airlines, we flew first class. However, it was not unusual to travel by private jet, such as on trips to Las Vegas."

She describes their annual trips to glamorous resorts in Hawaii, Vail, Aspen, and Mexico; recalls chartering yachts in Florida at $10,000 a day for intimate cruises; tells how she receives $6,000 a month in "spending money"; mentions owning "Ferrari automobiles and other vehicles"; provides accounts of lavish entertaining at their Brentwood home, with celebrity guests frolicking in the pool or playing tennis on their court, and adds, unnecessarily, "The lifestyle that he and I shared was truly substantial."

To maintain that lifestyle, Nicole receives $9,000 a month in alimony. O.J.'s ordered to pay an additional $15,000 each month for child support. Her divorce settlement of $433,000 tax-free enables her to buy the Brentwood townhouse. Among other possessions, she gets to keep one of the Ferraris.

Her post-divorce life seems a caricature of a favored, self-indulgent, and empty L.A. existence. She jogs and shops, dines out and goes to bars, frequents boutiques of the stars and meets friends for lunch at the Daily Grill, at Toscana, at Mezzaluna. At Le Beach Club, an upper-crust salon, she tans herself. At The Gym, a Brentwood fitness center, she works out. At Renaissance, a Santa Monica nightclub, she dances till two o'clock in the morning each Thursday. "She'd work up a sweat dancing, do a shot of tequila, and then head right back onto the dance floor," the bartender there later tells reporters. Her well-toned figure is seen to best advantage in the black Lycra pants and leotard top she customarily wears. She attracts admiring men, many of them from the emerging twentysomething group straight out of an Aaron Spelling TV show.

One of them is Ron Goldman. He's twenty-five years old, handsome—"gorgeous," as numerous women later describe him—superbly fit, just the sort of charming, attractive young L.A. man who appears to be floating vaguely through life without fixed purpose but who nonetheless enjoys the pleasures of a comfortable, hedonistic existence. He too faithfully works out at The Gym, often side by side with attractive wealthy young women like Nicole. In his brief life, he's been a tennis pro, a male model, a paramedic, a restaurateur, a waiter. Once he appears on the nationally syndicated TV show *Studs*. It's one of several brainless Nineties programs that celebrate handsome hunks like himself as desirable dates and potential sexual mates, marriage not expected, for panting young women who compete to spend time with them. At least that's the insulting portrait of the supposedly typical young American woman the show depicts.

At the time of Ron's murder he's working at the fashionable Mezzaluna restaurant, a place Nicole also frequents, and where she dines with her family in the final hours of her life. Not that it matters much, but Goldman's actual relationship with Nicole never becomes clear. Obviously, it's more than casual. Neighbors report seeing him playing with Nicole and O.J.'s children; he's also often spotted driving her white Ferrari around their Brentwood neighborhood. It's Ron that Nicole calls at Mezzaluna to ask that he bring her a pair of sunglasses Nicole's mother inadvertently left there after dinner that Sunday night.

At 9:45 Ron leaves the restaurant carrying the glasses and travels the few blocks to Nicole's townhouse. Thirty minutes later a screenwriter who lives in the neighborhood hears a dog's "plaintive wail." Twenty minutes after that mournful sound, another screenwriter takes his dog for a brief walk through the neighborhood. He leaves his home after watching the end of the *Dick Van Dyke Show*; he intends to return in time to see the beginning of the *Mary Tyler Moore Show*. On the street he encounters Nicole's Akita. The paws are bloody, he notices, the tags missing. It is the Akita that leads to the discovery of the two bodies sprawled on the tiled walkway inside the dimly lit gateway to Nicole's townhouse.

· · · ·

Into this brutal, tawdry melodrama enters a major figure whose life provides another link between American past and present—and evidence of how far the United States has come in overcoming some of the prejudices that in the not-too-distant past mocked the promises of its equality-for-all founding creed.

Lance Ito, the judge named to preside over the Simpson trial, is the child of Japanese-American public-school teachers who met in a Wyoming internment camp. In the hysteria following the Japanese attack on Pearl Harbor on December 7, 1941, in one of the great injustices of American history, they and other loyal American citizens of Japanese ancestry are taken from their California homes by U.S. military personnel without judicial hearing or evidence of any wrongdoing and forced to live behind barbed wire under the watch of armed guards. If that experience scars them, it does not hinder their son Lance's American opportunities. After graduating *cum laude* with a political science degree from UCLA in 1972 and a law degree from Berkeley three years later, Ito's career path proceeds steadily upward. For ten years, he serves as deputy district attorney in Los Angeles where he earns a reputation as an effective, serious prosecutor of L.A. criminal elements, especially the inner-city gangs. Then, he's elevated to judgeships—first, to the L.A. Municipal Court, finally to the Superior Court.

When he's named to preside over the O.J. trial, Lance Ito, at age forty-three, already is regarded as a rising star of the judiciary; two years before, the Los Angeles County Bar Association named him its trial judge of the year. His appointment wins acclaim from leading members of the bar and bench; they're certain he'll preside with dignity, while maintaining proper judicial decorum and discipline that will ensure fairness. No circus atmosphere in Judge Ito's court. Even before the trial begins, predictions are being made that Lance Ito will emerge with greater public esteem and professional promise.

His life is instructive for another reason that also implicitly testifies to positive changes in American society. His wife, Margaret "Peggy" York, is white. Not once is this racially mixed marriage raised as a subject for public gossip or scrutiny. Despite the prejudices that affected his own parents in the 1940s, by the Nineties such unions are taken for granted. So is something else in Ito's story.

Peggy York is a career professional—and a professional in a calling that until very recently has traditionally been an all-male preserve. As one of the increasing number of professional working women in the Nineties, she illuminates the most dramatic change in the American workforce, and in the family—the rise of career women to something approaching parity with men. By the end of the decade more than half of the graduates of medical, law, and business schools are women, an astonishing advance from the comparable figures of female graduates just a generation before.* True equality is not achieved, of course; women still experience discrimination in pay, position, and promotion, but, as in Peggy York's case, significant breakthroughs are occurring.

At the time of O.J.'s trial, Peggy York is the highest-ranking woman in the Los Angeles Police Department, the head of its

---

*By the end of the Nineties, MIT reported that women represented more than 50 percent of its undergraduate science majors and that women were graduating there at a higher rate than men. By then, women comprised a substantial majority of *all* college students nationally, representing 60 percent of students, and the number of women enrolled in colleges and universities was increasing at a faster rate than that of men.

bunco-forgery unit and someone whose career serves as a model for other women. She and Ito met, with melodramatic appropriateness, at four o'clock in the morning at an L.A. murder scene to which she, as a young detective, and he, as a young prosecutor, were dispatched. They first saw each other while staring over a dead body.

In one more curious example of truth trumping fiction in the O.J. case, Peggy York coincidentally turns out to have once supervised the homicide detective who becomes the key prosecution witness against O.J.—and, eventually, by far the most controversial. That's Mark Fuhrman, the white L.A. homicide detective who becomes famous and then notorious as the cop who finds the incriminating blood-soaked leather glove behind O.J.'s mansion hours after the murders.

When Fuhrman first enters the scene, he appears to be an exemplary officer in the best Sergeant Friday tradition. He's a selfless, dedicated detective motivated by a desire to seek truth and justice. On the witness stand, testifying about his role in the case, he's well-spoken, well-mannered, well-dressed: the cool, crisp professional at work.

To the national television audience viewing the trial live from L.A., Fuhrman initially leaves a highly favorable impression. He appears believable, sympathetic even, as he describes to the jury how his heart "started pounding" when he realizes the glove he found at Simpson's mansion matches the one lying near the bodies of Nicole and Ron at the townhouse two miles away. As the trial progresses, though, this public impression changes radically. Simpson's defense team unearths damaging facts about Fuhrman, especially information contained in a disability benefits lawsuit he filed against his department in 1983. In that proceeding, the detective admits harboring hostile thoughts about blacks and other minorities—facts the Simpson legal team immediately makes widely known. They also make it known, through selected leaks to influential media, that they will brand Fuhrman as a racist cop who fabricates evidence to convict blacks of crimes they did not commit. He planted that glove, they maintain, to frame O.J.

By far the most explosive disclosure about Fuhrman comes when O.J.'s lawyers reveal the contents of twelve hours of private tape-recorded interviews with him by an aspiring young screen-writer seeking material about problems women officers face in a male-dominated police force like L.A.'s. On those recordings, be-ginning in the year 1985 and continuing into the trial year of 1994 itself, the lawyers claim Fuhrman repeatedly makes racial slurs of the most invidious kind.

When questioned under oath, Fuhrman denies ever having used the most offensive racial slur, "nigger," at any time in the past decade when referring to a black person. But when the tapes are played in court, Fuhrman is heard repeatedly using that epithet, and in the most despicably racist manner possible. He's also heard joking about police beatings of blacks. He hints that he has planted evidence at crime scenes. "We could have murdered people and gotten away with it," he boasts at one point to the screenwriter. He exposes his racial and sexual and ethnic prejudices even against members of his own police force, saying of them on one tape: "We got females and dumb niggers and all your Mexicans who can't even write the name of the car they drive." At another point, he says, "What you gonna do with some nigger with a knife? Go, En garde?" And, referring to his hometown in the state of Washing-ton, he says: "People there don't want niggers in their town. . . . We have no niggers where I grew up."

As if this were not enough to discredit him, a white woman he met ten years earlier claims that Fuhrman told her then: "If I had my way, all the niggers would be gathered together and burned." The jury also hears testimony that some years earlier Fuhrman bragged how he liked to "take niggers into an alley" and beat them with a baton until they "twitched."

Not since the days of racist sheriffs and policemen in the Deep South during the bloody civil rights revolution of the Fifties and Sixties had Americans been exposed to such naked evidence of in-stitutionalized police racism—and this time, a generation later, racism thriving in a major urban department.

The inflammatory Fuhrman disclosures have three critical ef-fects. First, they deepen the already strong belief in racial conspir-acies held by black Americans. America's racist legacy of injustices against blacks makes it easier for them to believe that a racist white cop, which Fuhrman surely is, manufactured evidence of the bloody glove in order to incriminate Simpson. Second, they unify black opinion about O.J.'s guilt or innocence. Initially, blacks were divided in their views about O.J., with women and those with higher education thinking him guilty. That changes after the Fuhrman disclosures. Third, they permit the Simpson defense lawyers to shamelessly exploit those beliefs in racial conspiracies by aggressively employing "the race card" at every opportunity, cynically playing upon the emotions of the predominantly black jury and also on the attitudes of millions of blacks beyond the courtroom.

Simpson's lead defense attorney, a flamboyant black named Johnnie Cochran, himself a former L.A. prosecutor, says Fuhrman's racism typifies what he charges is a vast conspiracy led by a racist rogue cop out to frame Simpson. In his emotional final summation to the jury, Cochran compares Fuhrman to Adolf Hitler and accuses him of being a "lying, perjuring, genocidal racist" with a long-time vendetta against O.J. But Fuhrman did not act alone, Cochran theatrically thunders at the jury. He was merely a central actor in a conspiracy in which other detectives, crime-scene experts, and the West L.A. police commander were all involved, and all hiding behind a "code of silence." "Stop this cover-up!" he shouts dramatically at the jury. "Stop this cover-up!"

Lost in such appeals to racial animosities is the idea that O.J. could be the victim of nefarious racist police tactics *and* guilty of the murders with which he's charged.

· · ·

In the aftermath of O.J.'s arrest, the revelations about Nicole's terror-stricken emergency calls and O.J.'s prior record for wife abuse initially focus great attention on the issue of domestic violence, particularly violence against women. For years, advocates and activists have been trying to place that issue squarely on the public agenda. Now O.J. gives them the perfect opportunity to get across their message about the prevalence of wife battering and spousal abuse. They appear on television to air the problem. They write op-ed pieces. Congress even holds hearings.

Black women are prominent among these advocates. Many speak out against O.J.'s violent behavior toward Nicole, as they have been doing repeatedly about violence committed against women by men, whether black men or white men. At this point in the O.J. story, they don't view the case along purely racial lines. As in the highly charged Supreme Court confirmation hearings for Clarence Thomas, women see abuse or harassment, while men see an "electronic lynching." Indeed, just a day after Nicole's body is discovered, the domestic violence issue appears to be creating further sharp divisions between African-American men and women.

On that Monday a female judge in Indiana refuses to free Mike Tyson, the former black heavyweight boxing champion then serving a six-year jail sentence for rape. Immediately, Indianapolis talk shows are besieged by angry callers. Women, many of whom identify themselves as black, praise her decision but add that Tyson should serve more, not less, prison time. Some believe he should have served the full sixty years to which he could have been sentenced after being convicted on two counts of criminal deviant sexual conduct and rape.

African-American male callers, by contrast, are outraged at the judge for not immediately freeing Tyson. The champ is "the greatest," they say. He never should have been imprisoned in the first place. He's the victim of a "scheming woman." The female judge's ruling proves how "the system" castrates and lynches black men. This becomes a theme that repeats itself ever more forcefully during the O.J. trial as the issue of domestic violence against women is overtaken by an even more combustible one—race.

The race card, as exemplified by the character of Mark Fuhrman, plays powerfully on deeply held African-American fears and resentments. It intensifies the already strong belief that blacks cannot expect fair treatment from the nation's criminal justice system. The O.J. case, in this reasoning, becomes an opportunity for racial payback to counter past wrongs extending back to the very beginnings of the American experiment—the wrongs of slavery, of murder, of rape, of castration, of lynching, of segregation, of discrimination, of injustice. Fanning racial flames even higher is the performance of the black press, where O.J. is portrayed as yet another "victim" of white racism perpetrated by the combined conspiratorial efforts of a white legal and police establishment and a "white media."

While O.J. becomes another example of a black hero being destroyed by the white conspiracy, Johnnie Cochran, O.J.'s lead attorney, is hailed as a new black hero. He's even called a new civil rights leader for his defense of O.J. In sharp contrast, one of the key prosecutors, Chris Darden, also black, is condemned in the same black press as being a contemptible "house Negro" for prosecuting O.J.

In the end, race, not domestic violence, not corruption at the core of professional athletics that inspires above-the-law attitudes among its pampered stars, not even murder, becomes the emotional touchstone of the O.J. case.

· · · ·

No thinking American in the Nineties could be surprised to learn that the United States still suffers from pervasive racial prejudice. Despite African-American advances in employment and income opportunities; despite the ending of legal segregation in housing, in schools, in the military; despite integration of previously all-white police departments enabling blacks to become chiefs of police in such former racial trouble spots as Birmingham, Alabama, and Charleston, South Carolina; despite anti-discrimination laws and affirmative action programs aimed at combating discrimination nationwide, racial suspicion and racial hostility still afflict the United States. The shock produced by the O.J. case comes not from the discovery that racial resentment and anger exist. The shock comes from the depth and virulence of them.

In the O.J. trial, everything is seen through the distorting prism of race. Blacks and whites examine the same evidence and draw starkly differing conclusions from it. One side sees a murderer; the other sees a victim. One sees clear and compelling evidence of guilt; the other sees a sinister conspiracy that seeks to convict the innocent.

Barely a month after O.J.'s arrest, 63 percent of whites answering a Time/CNN poll say they believe he will get a fair trial. Only 31 percent of blacks feel the same way. While 66 percent of whites believe he received a fair preliminary hearing, only 31 percent of blacks agree. Seventy-seven percent of whites believe the case against O.J. is either "very strong" or "fairly strong." Only 45 percent of blacks agree.

As the trial begins, the racial lines harden. Out of public view, the sequestered jury becomes riven with increasing racial tensions. Black and white jurors use separate gyms, watch movies in separate rooms. Black jurors complain that whites on their panel are given preferential treatment by sheriff's deputies. They suspect the deputies are secretly searching their housing quarters while they are on jury duty seeking evidence of bias against the prosecution or to discover they have been violating the judge's orders not to read news accounts about the case. One black juror, after being removed from the panel, goes on TV to accuse a white juror of kicking her. She says the same white stomped on the foot of another black juror in the jury box. Another black juror files a formal protest to Judge Ito about the racism perceived by black jury members.

Six months into the trial, a national survey by Lou Harris and Associates finds that 61 percent of whites believe Simpson is guilty. Sixty-eight percent of blacks think him innocent. Only 8 percent of all blacks surveyed believe that O.J. murdered Nicole and Ron. (Twenty-four percent of blacks polled say they aren't sure about O.J.'s guilt or innocence.)

By trial's end, a state of near total racial polarization exists across America.

· · ·

In the O.J. case, the public fascination with violent entertainment and courtroom drama mixes with the conjunction between capitalism and celebrity. The case becomes a prime example of how profit seekers can manufacture and exploit a mass audience and how television provides the perfect vehicle to promote that rush to profit. As Walter Lippmann observed in the wake of the TV quiz scandals that rocked the television world at the end of the Fifties, "While television is supposed to be 'free,' it has in fact become the creature, the servant, and indeed the prostitute, of merchandising." To that point, Lippmann thought television's major influence had been twofold: first, "to poison the innocent by the exhibition of violence, degeneracy, and crime, and second, to debase the public taste." He should have seen the Nineties.

In the O.J. case, everyone cashes in—the media, the lawyers, the judge, the jury, the publishers, the entertainment industry, the marketers of items bearing on the case. Even friends and foes of O.J. profit, eagerly, aggressively, shamelessly. Not least among them is O.J. himself, as well as supposedly bereaved relatives of the victims. The O.J. trial becomes not just "a rush to judgment," as his lawyer dramatically and repeatedly warns; it becomes a rush to capitalize, and capitalize in a way America has not witnessed before.

Not that the phenomenon of exploiting tragedy and sensation is new. American, indeed human, history is replete with notorious examples, from Roman emperors offering their subjects bread and circuses in the form of staged brutal spectacles to divert them from other matters to Hitler applying that ancient lesson well by staging massive torchlight rallies before wildly cheering crowds, diverting them from horrors being committed by the Nazi regime. In every age, crowds gather instantly at scenes of suffering, and the greater the calamity, the greater the celebrated figure, the greater the crowds. Always accompanying them are the merchants of misery cashing in on the latest scandalous spectacle.

Americans no longer are able to witness public executions,★ as they did in masses in the Old West, or flock to lynchings, as they did in the Deep South, but their response to scenes of tragedies resembles other times of mass hysteria such as when screaming mobs broke police barriers and rushed to draw closer to the casket at the funeral of movie star Rudolph Valentino at an undertaker's establishment on Broadway...or surrounded the courthouse and created daily bedlam in Flemington, New Jersey, during the Lindbergh trial...or trampled each other in the crush at Marilyn Monroe's house after she was found dead...or materialized instantly out of the darkness before the stately Dakota apartment building in Manhattan after John Lennon, the Beatle, was murdered...or massed outside O.J.'s mansion even as The Chase proceeded.

★But perhaps they soon will. Already, some TV executives are fighting for the "right" to telecast executions such as that of Timothy McVeigh.

In the O.J. case, the difference from the past lies not in the human instinct that lures crowds and hucksters to scenes of disaster. In the Nineties, the difference lies in the ability of everyone, everywhere, to participate vicariously in those scenes as they are occurring.

For this capacity, thank technology. Mobile TV minicams and earth-orbiting satellites provide the technical ability to go live, virtually instantly, from any scene of disaster or scandal. For the decision to bring more and more of these scenes into everyone's living room, credit a number of factors that converged in the Nineties. Intense competitive pressures among proliferating cable channels scrambling to wrest market share from the traditional networks created increasing demand to broadcast the latest, most sensational newsbreaks as they happen—and the more scandalous and lurid the better. As cable channels focused increasingly on the sensational and the scandalous, the old networks adapted by furnishing more of the same in an attempt to hold their declining audience.

The disgraceful attack talk-radio programs, with their growing audience and increasing influence, also affected the electronic and celebrity culture of the Nineties. With their daily airing of ideological conspiracies and preoccupation with scandals—proof never necessary and rarely even a consideration—the talk-radio shows demonstrated the impact, and the money, to be made by appealing to the worst in people. Television, especially cable, followed their lead; tabloid TV joined attack radio in filling more of the nation's airwaves. "Trash TV" was on the rise.

None of this readily explains the appeal of these offerings, however, or the paradox they present about American society in the Nineties. Americans, after all, were better educated, more sophisticated, more tolerant, more aware of subtlety and nuance and the imperfectability of public and private lives than ever before. They were, in the main, practical and realistic, generally hard-eyed, and not easily swayed by cheap appeals to emotion. So why were so many so captivated by such tawdry daily fare?

Part of the answer rests in the nature of the times. The best of times they may or may not have been, but they were certainly times blessed by an absence of crises—crises domestic or foreign, economic or social, environmental or medical. Freed from the kinds of concerns that compel public attention, Americans were also free to indulge in the titillation of gossip and scandal. They were free to be entertained by the spectacle of celebrities and public figures brought low. And with relentless, nonstop intensity, the electronic media dished up scandal in helpings that enabled every citizen to share in every gory, sordid detail.

Nor was O.J. the first of the great scandalous spectacles Americans witnessed in the Nineties. By the time of O.J., Americans were conditioned to witnessing a succession of long-running scandalous episodes. No sooner did one end than another took its place. Each attracted an immense audience; each received frenzied media coverage; each was treated as if it said something significant about American society and thus deserved intense attention; each fueled an appetite for more of the same; each became a springboard for a successor, happily supplied by producers of Teletimes who sought and supplied the latest scandal for public consumption, all in a breakneck race to boost ratings.

So many were there, and so rapidly did they replace each other, that it seemed as if the single most defining characteristic of America in the Nineties was an all-consuming preoccupation with scandal—scandal that over time merged into one continuous serial production.

The names of the players and the particulars about the scandals changed, but the object was the same—scandal, always more scandal: Dr. Kevorkian and the first assisted suicide; Rodney King, beaten viciously by Los Angeles policemen; Jeffrey Dahmer, "the homosexual cannibal," and horrific acts of mass murder; Anita Hill, Clarence Thomas, sexual harassment, and pubic hairs; Mike Tyson and rape; Tailhook and sexual assault in the Air Force; the Packwood diaries and sex on Capitol Hill; Michael Jackson and that young boy; Nancy Kerrigan, Tonya Harding, and their violent skating rivalry; Lorena Bobbitt and her husband's severed penis.

All these occurred before O.J. Other episodes flitted across the TV screens during and after the long period when O.J. dominated the national stage: Susan Smith and her two drowned children; the Menendez brothers; JonBenet Ramsey, the pathetic six-year-old pushed by parents to compete in beauty pageants by tarting up and acting like a budding Lolita, found murdered in the basement of her home; Louise Woodward, the young British nanny, and the death of the child in her care; the murder of fashion designer Gianni Versace in a Palm Beach oceanfront mansion, perfectly providing a spectacle that combined synthetic glamour and glitz with salacious tales of gay sex; the clearly deeply troubled teacher, Mary Kay Letourneau, and her sad, sick affair—naturally labeled by the tabs "forbidden love"—with her thirteen-year-old student; Dick Morris, a president's Machiavellian pollster, sucking the toes of that prostitute on a Washington hotel balcony; Marv Albert, the loudmouth sportscaster, biting a woman in another hotel room near the capital, providing perfect fodder for the televised celebrity and scandal culture, and being returned to the air as a sports commentator, apparently no less popular, or perhaps more so, than when his scandalous behavior created yet another mass spectacle.

Drawing the most intense media focus and public attention were the trials that resulted from many of those episodes. They were the easiest to cover and offered a convenient running plot line of scandal and suspense.

The cumulative effect of these events was to divert attention from the really great episodes of the Nineties, and especially from two that came into play with tremendous force then. One, as we've seen, was the revolution in science, technology, and medicine rapidly changing life on the planet. The other was the growing concentration of great blocs of power through the greatest wave of mergers ever, creating new entities reshaping the basic economic and social structures of the nation.

As time passed, few Americans could recall specific details of the various episodes to which they were exposed or their outcomes. Nonetheless, stamped in the collective public memory was a hazy montage of sensationalized scandals. While people professed to be repelled by media excesses and obsessive attention to scandal, they also took guilty pleasure in watching, and wallowing in, the spectacle.

Two weeks after O.J.'s arrest, CNN dispatched its cameras, and a correspondent, to a popular Atlanta fast-food restaurant, aptly named The Varsity, for a daylong sounding of public attitudes about the O.J. case. The slice of vox populi aired was highly revealing. Virtually every person interviewed expressed the same kinds of underlying ambivalence. They hated what they were seeing, or so they said, but they were watching all of it.

Citizen A, a man, tells the correspondent, Leon Harris, he thinks the media has ruined O.J.'s chances of getting a fair trial.

Q: So, you think O.J. Simpson is being victimized?

A: Yes I do, yes I do.

Q: Would you like to see more or less or the same amount of coverage as you've been seeing so far then?

A: I think it's interesting. I like the coverage, but I don't think it's good for the man's chances of getting a fair trial.

Q: So you don't like the coverage, but you have been watching it, why?

A: Just something to do really, basically. You know, I get off work from a hard day's work, I just like to come home and relax, you know, so I watch the news, see what's happening.

Citizen B, a woman, dining with her family, says she's troubled at feeling "like I'd become a voyeur into Mr. Simpson's dilemma. It's not a nice feeling to feel like a voyeur."

Q: You mean you're not comfortable with what you're seeing?

A: No, I'm not comfortable with it at all.

Q: And yet, you still watch.

A: Not as much as maybe some other folks. It's there, you just don't have the ability to turn it off, you're mesmerized by it.

Q: You think it's fair?

A: Probably not.

Citizen C, a man, complains about what he calls TV's "total saturation" on O.J.

Q: Is it fair to say that you've pretty much reached the saturation level, you don't care if you don't see anymore, one way or the other?

A: Pretty close, pretty close. I mean it's interesting to watch, but all four, five, whatever, networks is a little too much.

Q: Tell me, what you've seen thus far, in the past couple of weeks, has it forced you to make up your own mind about Mr. Simpson's innocence or his guilt?

A: It makes it look like he's guilty, but you really don't know, there are a lot of situations that they haven't explained yet. I really don't want to make up my mind right now.

Q: You want to see more or less?

A: Keep updated, let me put it that way, maybe not total saturation, but to be updated.

Television provided much more than "updates." It offered a new form of public entertainment—a live, free theater of spectacle and sensation. O.J. had it all, the serious and the sordid. It was irresistible. In the process, old news barriers and taboos were broken; practices previously deemed unacceptable by mainstream news organizations became acceptable amid rapidly evolving standards of the electronic age. Even some talk-radio hosts expressed concern about the negative impact they were having on the public. Days after the murders, during a Los Angeles convention of national talk-show hosts, one of them acknowledged to a *CBS Evening News* interviewer that with the O.J. case talk radio had "gone totally over the line." Then the talk host quickly added: "With each case we say that—and the line gets pushed further." That didn't stop the lines separating accuracy from rumor, fairness from unfairness, good taste from bad, from being driven farther and farther apart.

Distinguished news executives from both print and television bemoaned the lowering of journalistic values, the cheapening of reportage, the omnipresent "gotcha" aspects, the circus atmosphere that typified the coverage. "I don't like the idea that a murder trial has been turned into an entertainment special," Don Hewitt, the executive producer of *60 Minutes,* wrote in a *New York Times* op-ed article. "There are certain moments in American life that have a certain dignity." Not an O.J. moment, though, especially an O.J. moment that increases ratings and one that shows the growing public appetite for more of the same—more of the spectacular, more of the sensational, more of the scandalous. Which, of course, is what the public got.

. . .

From day one, the commercialization surrounding the O.J. affair knew no bounds. Months before the trial itself began, the scene outside the courthouse where the preliminary acts were taking place became a veritable daily *Vanity Fair* assemblage of humanity. Vendors, hawkers, hustlers, crusaders, quacks, paraded back and forth loudly and raucously, angling always to attract the myriad lenses of the TV cameras massed around them.

Here, a vendor's loud cry sounds: "*Whaddaya say, Whaddaya say, Whaddaya say. T-shirts, ten dollars. Don't squeeze the Juice. Whaddaya say, Whaddaya say . . .*" There, lines of women, placards waving, chant: "*Stop domestic violence / Break the code of silence. Stop domestic violence / Break . . .*" All of them, from Jesus freaks mingling with other protesters to those seeking in some unfathomable way to show support for the long-dead Geronimo, maneuvered around long lines of the public that formed before sunup each day in hope of gaining scarce public seating inside the courtroom. While this spectacle occupied part of the daily O.J. public stage, other scenes in the spectacle were occurring as the battle to benefit commercially intensified.

Within hours of O.J.'s arrest, Los Angeles talk shows were speculating about who would play him in the movie version. Within days, Fox became the first national broadcasting company to announce rapid production of a rip-and-read TV movie based on the daily headlines in the case. Within a week, the first instant O.J. book was published.

St. Martin's Press, already proudly promoting itself as the publisher that had cornered the Nineties' instant book market with publication of such works as *Lethal Lolita: The Amy Fisher Story; Bad Blood,* about the Menendez brothers killings; and *Dreams of Gold: The Nancy Kerrigan Story,* rushed *Fallen Hero* into print even before O.J.'s preliminary court appearances began.

"This book on O. J. Simpson has been written in a week," its then editor, Charles Spicer, explained to the media, "and the reason is the competition has become so stiff."

Other publishers rushed to produce O.J. quickies; other TV productions were being planned; other movie scripts being written. Fierce and frantic competition ensued. Commercial schedules were disrupted, planned events postponed. Eventually, one publicity director told her field reps, there would be a backlash. Eventually, people will say, "I'm tired of watching O.J."

They were not. O.J.'s trial went from being the most watched legal proceeding in U.S. history to the single event that received more coverage over a longer period of time than anything telecast before it. In its first year as a news story, O.J. attracted more TV coverage than the brutal war in Bosnia, more coverage than the election campaign for president of the United States, more coverage than a terrorist bombing of a federal building in Oklahoma City that took 168 lives. And O.J. received not just more coverage than any of those individual episodes; O.J. received more coverage than *all of them combined.*

So distracting from other important public events was the coverage that by the late summer of 1994, a bare two months after the murders, California's acting secretary of state petitioned Judge Ito to declare a two-week recess of the O.J. judicial proceedings before the November general election was held. He argued that "the intense media and public focus" on the O.J. trial, scheduled to begin September 19, would result in reduced election coverage, particularly over TV, and produce "a less informed electorate and lower voter participation." Ito denied the petition.

As the media melodrama continued, O.J. obsession spawned new cable TV programs and talk shows. To them flocked many of the players in the O.J. case. Out of the continuous coverage of the trial and the constant chattering commentary came a new term for Teletimes: "All O.J., All the Time." It was a formulation that would later be applied to another, greater, scandalous televised spectacular that attracted even more obsessive national coverage in the Nineties and led to the impeachment of the president of the United States.

The longer the O.J. story dominated the news, the greater became the frenzied efforts to capitalize on it. Within six months of the beginning of the trial, the value of goods and services being sold as commercial spinoffs of the case approached an astounding $200 million dollars—more than the gross domestic product of Grenada, as the *Wall Street Journal* carefully calculated.

Tickets were sold for a museum exhibit that featured a scale model of the courtroom, with an Akita dog on the witness stand. There were O.J. books in profusion, O.J. trinkets, O.J. golf clubs, O.J. golf balls, O.J. serials, O.J. statues. Just four weeks after his arrest, O.J. himself filed a request with the U.S. Patent and Trademark Office to trademark his initials in order to guarantee the exclusivity of these and a long list of other merchandise he planned to offer for sale to the public. Later, he entered into an agreement with the Florida Department of Citrus for use of the "O.J. brand"; that agreement allowed O.J. the right to trademark his name on "metal goods," "paper goods and printed matter," "clothing," and "toys and sporting goods" but not on "any goods or services" sold in Florida "in connection with oranges, orange juice, or orange juice products." Naturally, O.J. quickly wrote his own best-seller, *I Want to Tell You,* for which he received a million-dollar advance, and also commissioned for sale a bronze statue of himself in a limited one-edition run of 25,000 copies. Standing twenty-and-a-half inches high, the statues were offered at $3,395 apiece. He also starred in his own sixty-eight minute video, *Minimum Maintenance Fitness for Men,* pitched "toward men whose busy schedules don't leave them time for a workout." Still later, he earned a reported $3 million by telling his side of the murders in a video called *O.J. Simpson: The Interview.* It sold for $29.95.

The commercialization was shameless. O.J.'s great boyhood friend and fellow pro-football player, Al Cowlings, who drove the famous white Bronco during The Chase, let it be known during the trial that the year-old vehicle was up for sale.* Although the Bronco's Blue Book value was then listed at $15,000, Cowlings's

---

*Media myths notwithstanding, *that* Bronco was not O.J.'s. He had given it to his pal Cowlings, as he made gifts of other cars, including other white Broncos, to other friends. O.J.'s own white Bronco was recovered outside his mansion by police the night of the murders and was found to have bloodstains in it that subsequent DNA testing said matched samples of Nicole's and Ron's blood—and his own.

lawyer cagily used the media to inflate his client's sale price. Cowlings already had been offered $200,000 for the Bronco, the attorney slyly told reporters. Lest it be thought Cowlings was flagrantly cashing in on his friend's notoriety, the lawyer maintained that "Al really isn't able to use it because it's the most recognizable car in America"—as if it could be distinguished from countless other white Broncos then traveling the nation's roads.

In time, Cowlings's closeness to O.J. made him "financially secure," his publicist confided to a reporter. He lived in an exclusive Pacific Palisades enclave and, in addition to the sale of the Bronco, gained wealth after establishing a pay-per-call line that carried his descriptions of happy times with O.J. and Nicole.

Sales of Broncos spurted after The Chase. The Bronco, in reality a light truck measuring 184 inches long by 79 inches wide and weighing more than five thousand pounds, had already helped create America's sports utility vehicle (SUV) craze that symbolized the casual, mobile, and affluent boomer lifestyle of the Nineties.

With its powerful two-hundred horsepower V–8 engine and its sporty look, the Bronco was designed to appeal to the most enduring of American myths—the taming of the Wild West by intrepid American cowboys and (in the Nineties version) cowgirls (or, to fit the times, suburban housewives and single young professional women, who bought more SUVs than men).

Though SUVs were given western names other than Bronco—Wrangler, Blazer, Cherokee—to convey a spirit of pioneering dash and adventure that would appeal to a typical American's unfulfilled desire to experience rugged, free, wide-ranging western life, in reality the advertising marketers who pitched their sales over TV were appealing to a different kind of constituency that defined a large segment of society. Instead of a taming of the West, the SUV's real utility more accurately could be said to lie in a further conquering of the suburbs. As the marketers well knew from their demographic profiles of purchasers, the greatest market for America in the Nineties was a suburban one, and a market populated by precisely the kinds of Americans who best fit the demographics of typical SUV owners: white, college-educated baby-boomer professionals who married later in life, had two children, owned their own house in the suburbs and, thanks to the boom, could easily afford the $31,000 (and up) cost.★

Whatever the sales strategy, and regardless of whether they lived up to the advertising hype, the Bronco and other SUVs conveyed a message that singularly suited the times. As they barreled along the nation's highways and suburban biways, they projected a big, swaggering, get-out-of-my-way look. By the end of the Nineties, SUV sales were rising rapidly. In 1993, there were 7 million SUVs on the roads. By the end of the Nineties, more than 20 million of them were on the highways, and sales for the sixty or so brand-name models were still rising.

. . .

The craze to cash in affected more and more players. Nicole's supposed best friend, Faye Resnick, wrote a quick, sleazy tell-all memoir, *Nicole Brown Simpson: The Private Diary of a Life Interrupted,* that was published exactly two months after Nicole's death and became an immediate best-seller. In it, she wrote of Nicole's alleged drug use and casual love affairs, including a lesbian encounter she herself claimed to have had with Nicole. It was written, in collaboration with a columnist for the gossip and scandal publication the *National Enquirer,* after Resnick received a reported six-figure advance. Resnick, a three-time divorcée and admitted cocaine addict who had been treated at three rehabilitation programs over the previous eight years, piously told the press that

★A fascinating insight into the demographic slice of SUV owners and the marketing strategy created to appeal to them came in a lengthy article by Frank Ahrens, "Why Big Is Back: Interpreting America's Love Affair with the SUV," in the Nov. 7, 1999, issue of the *Washington Post Magazine.* According to Bob Casey, auto historian and curator of the Henry Ford Museum & Greenfield Village in Dearborn, Mich., quoted in the article, in addition to western names, marketers gave the vehicles wilderness and journey names— Yukon, Mountaineer, and Tahoe; Expedition, Explorer, and Excursion—evoking "an adventurous lifestyle that most SUV buyers are happy to experience solely via TV ads." That SUVs were gas guzzlers with the worst gas mileage ratings did not detract from their appeal in the live-for-the-moment big boom Nineties.

money wasn't her primary object in writing the book. She did it because she had promised Nicole to describe Nicole's abusive relationship with O.J. if anything happened to her. The tabs and the trash-TV shows jumped on these cheap, unproved allegations and trumpeted them as further titillating evidence of the scandalous lives led by the rich and famous. Later, Resnick posed nude for *Playboy*.

O.J.'s last girlfriend, Paula Barbieri, a twenty-seven-year-old model for, among other outlets, Victoria's Secret, the sexy lingerie label, was another of his friends who received the gift of a white Bronco. She produced her own book, which added virtually nothing to the story, after receiving a $3 million advance. As the O.J. trial began, pictures of Paula posing in *Playboy* appeared. She also played the part of a battered mistress in a film, *The Dangerous*. Later, her manager complained to the press that the publicity she received had been harmful to her financially. She was finding it more difficult to get modeling appearances. "It's been rough on her financially this past year," he said, adding sanctimoniously. "She could have sold out like everybody else." Instead, she turned down money from the tabloid press because "she has too much integrity."

Bit players worked no less aggressively to cash in. Brian "Kato" Kaelin, the tousle-haired aspiring actor in his mid-thirties who was a house guest of O.J.'s at the time of the murders and also a friend of Nicole's, provided comic relief at the trial with his vague, inarticulate Nineties-style way of speaking, but had no definitive answers about the murders. Kato immediately became the center of intense media attention—and began fielding offers. He hired an agent, made a quick guest appearance as cohost of a cable TV entertainment talk show, *Talk Soup,* weighed proffered film offers, appeared in one, *Life or Death,* and as an emerging new celebrity was said to enjoy the attentions of Hollywood starlets. "He is probably one of the princes and playboys in Hollywood," his agent told *People* magazine. "The girls love him." Another acquaintance of Kaelin's, a Hollywood producer, offered unsolicited advice on how Kato could best cash in: "If I were him, I would be writing a book right now. Why not? You can't ask for more publicity than this. God bless him, I hope he makes a million bucks." A TWA flight attendant, Tracy Hampton, dismissed from the jury in May 1995 by Ito after telling the judge "I can't take it any more," turned up months later in the pages of *Playboy*. With her bunny money, she quit her job, began acting classes, and auditioned for TV and movie roles.

Being a member of one of the bereaved families proved no barrier when it came to cashing in, either. Nicole's father earned a total of $262,000, *Time* magazine reported, after selling his murdered daughter's private possessions. For her diary, he was paid $100,000 by the *National Enquirer*. For selling and then narrating the home video of Nicole's and O.J.'s wedding, he received $162,500 from a syndicated tabloid TV show, *A Current Affair*. One of Nicole's sisters, Dominique, admitted in a deposition taken by O.J.'s lawyers that she had sold a series of pictures of Nicole, including one showing her topless during a vacation in Mexico, to a scandal journal for $32,000.

The tabloids had a field day. O.J. was their biggest story ever. By January of 1995, three months into the trial, the *Enquirer* had made O.J. its cover story for twenty-one of its last twenty-seven issues. The massive coverage paid off; its circulation increased by 500,000 copies a week. So aggressively did it and the other tabs pursue the story, showering cash for "exclusive" stories and pictures, that the State of California enacted three new laws penalizing witnesses or jurors who sold information about a pending trial. This came after it became known that some O.J. witnesses had sold their accounts to the *Enquirer*, which made no pretense of being troubled by the ethical implications of paying sources for stories or intruding into the judicial process. Its editor-in-chief readily acknowledged in a *Time* interview that a substantial portion of its $16 million annual editorial budget was paid to informants. One *Enquirer* reporter boasted of having knocked on two hundred doors before locating a former maid of Nicole's, and then paying her $18,000 for an account of Nicole's relationship with O.J. A salesman who sold O.J. a fifteen-inch stiletto for eighty-one dollars five weeks before the murders, confirmed that he and bosses had sold their story to the *Enquirer* for $12,500.

That was peanuts when compared to the literally millions upon millions members of the defense and prosecution teams collectively earned from publishing advances to write their memoirs after the trial.

They became media stars and instant celebrities, turning up with regularity on new cable TV talk shows spawned by the O.J. scandal. Their lives, too, became fodder for the tabs, the gossip columns, and the trash-TV productions.

Marcia Clark, the forty-year-old chief prosecutor for the people, found details of her private life extending back years—her divorce, her status as a single mother with two young sons, her social life—exhumed in the popular press. The way she dressed, the way she moved and walked, were examined and discussed as widely, if not more so, as the quality of her performance as a lawyer in a major murder case. Here was sexism still at work in the national media. Little note was taken, by contrast, of the two-thousand-dollar Armani suits worn by Robert Shapiro, the lawyer who had fashioned a highly profitable career defending celebrities—Linda Lovelace, the *Deep Throat* porn actress; Tina Sinatra, Frank's daughter; Johnny Carson, TV host; Robert Evans, big-time Hollywood producer; athletes Darryl Strawberry and Jose Canseco—before joining O.J.'s team.

Johnnie Cochran, the new civil rights hero, who had employed the race card so nakedly that he had guards from the black separatist movement, the Nation of Islam, escort him to and from court—all captured daily by the TV cameras and by the eyes of the jury—didn't escape the scandal net, either. Details of his marriages were unearthed and breathlessly published. So were the scandalous allegations made against him midway through the trial. At a highly publicized L.A. press conference, a woman identifying herself as Patricia Ann Cochran announced she had been Cochran's mistress since 1966, when he was her divorce lawyer, and was the mother of Cochran's son, then a UCLA law student. She had recently changed her name to match his, she explained before the cameras, and was filing suit to force Cochran to resume support payments of four thousand dollars a month, plus other expenses, that she claimed he had cut off in the last year.

However embarrassing the public airing of such private linen might have been to Marcia Clark and Johnnie Cochran, it obviously didn't depreciate their commercial worth as hot new media celebrities. It might even have enhanced their value. Both Clark and Cochran went on to appear as host or co-host on their own national cable TV talk shows and gain lucrative lecture bookings after collectively being paid a reported $7 million in book advances. Other members of the defense and prosecution teams also began appearing regularly on TV talk shows and making the lecture circuit for fees ranging upward from $25,000 an appearance after book publishers paid more millions for their stories.

Long before the trial even began, virtually all hope had vanished that it would provide for a watching world an example of the American criminal justice system at its best, a serious civic proceeding that exemplified the most cherished judicial attributes: dignity, decorum, and fairness. It quickly degenerated into a spectacle that demonstrated some of the worst characteristics of Teletimes. No one escaped unscathed. The lawyers fought among themselves, played as much to the TV cameras as to the jury, argued their cases in impromptu press encounters at every opportunity, leaked damaging information to the press, and also exhibited a taste for cashing in on the instant celebrity television had conferred upon them.★ One attorney on the defense team was involved in a New York trial that conflicted with O.J.'s. He asked for and won a postponement from that engagement; then he asked for another postponement from his New York obligations, in effect arguing it would be unfair to him financially if he were denied the chance to participate in the O.J. show, never mind his East Coast client. This time, his request was denied.

★After Simpson's lawyer, Robert Shapiro, attacked O.J. prosecutors for using the media to turn public opinion against O.J., a *CBS Evening News* network broadcast disclosed that Shapiro himself had urged that very strategy in a legal journal article a year before entitled, "Using the Media to Your Advantage." In it, Shapiro said attorneys shouldn't lie to the press, but should "come up with phrases you believe in and are comfortable saying." He also advised lawyers to "Repeat them continuously and they will be repeated by the media. After a while, the repetition almost becomes a fact." Throughout the O.J. case, both sides used the media, and the media willingly let themselves be used.

The jurors squabbled among themselves. At times, they, too, acted petulantly. Once, they even acted mutinously. On that occasion, thirteen of the eighteen remaining members of the panel showed up in court dressed in black. They threatened to refuse to enter the jury box until Judge Ito heard their protests about his decision to dismiss three of their guards in the wake of charges the guards gave some jurors favorable treatment. Though it wasn't known until after the trial ended, many of the jurors already had sold their stories to tabloid TV shows, granting exclusive interviews immediately after the verdict. And the judge, despite early expressions of confidence that he would live up to his reputation for ensuring scrupulous courtroom discipline and decorum—Ito would exercise his usual "cool manner and firm hand," one puff piece had predicted before the trial—showed himself to be petty and temperamental, given to angry outbursts against media excesses and threatening often to ban cameras from his courtroom.

Yet Ito himself astounded lawyers and law professors across the country when he permitted himself to be interviewed extensively by a local Los Angeles TV correspondent in the midst of the O.J. proceedings taking place in his court. Portions of the interview, conducted at Ito's home, were then broadcast each night beginning at eleven o'clock Sunday, November 13, for an entire week over L.A.'s KCBS Channel 2. More astonishing yet, these nightly airings on the local "Action News" broadcast took place during one of four so-called sweeps periods each year. The sweeps are the critical times when TV audiences are measured to help set advertising rates—the bigger the audience, the more money stations can charge their advertisers. Nor was the timing of Ito's interview with the commercially crucial sweeps period accidental. The channel promoted those nightly segments of his interview in full-page newspaper ads and in on-air promos, all intended to entice more viewers.

Reaction in the legal community, and in some press circles, was swift and strongly critical. In San Diego, a defense lawyer expressed typical consternation. "It's out of control," Elisabeth Semel told the San Francisco Chronicle's legal affairs writer. "The side show is obscuring the heart of the case." In Los Angeles, another prominent defense attorney, Harland Braun, voiced astonishment at how "a garden variety murder involving a celebrity" now is "going off into all kinds of side shows. Ito has become a side show. It's unbelievable."

And in New York, a respected legal scholar on judicial ethics at the New York University School of Law, Stephen Gillers, reflected sadly that "There's something about the big publicity monster. It co-opts everybody."

It certainly did in the O.J. case, nor was that trial the first that raised serious questions about the "publicity monster" that accompanied sensationally televised court hearings. By the time of the O.J. case, forty-seven states permitted TV cameras in courtrooms, and with the advent of the around-the-clock cable telecasts of CNN and such popular programs as "Court TV," the televised trial had become a staple of the electronic media. "Court TV" alone had been televising them for several years before the O.J. trial.

As for the rhetoric about how televising the trial live from Los Angeles would provide a great national civics lesson, and all the pretrial arguments from freedom of the press advocates who urged that cameras be allowed in the courtroom not only as a constitutional right but as a check on abuses, in the end the O. J. Simpson trial produced more public cynicism and disgust.

When it works as it should, the American criminal justice system is a noble, indispensable defender of freedom and individual rights. That it often does not work as well amid the scandal culture of Teletimes is only one of many lessons emerging from the Nineties.*

⁎ ⁎ ⁎

O.J.'s saga was over. After riveting the nation for a year and a half, after a trial lasting for nine months, after a jury was sequestered for

---

*As an ardent believer in and lifetime practitioner in the American free press, I reach this conclusion with sadness and sorrow. But as Justice Holmes rightly proclaimed in his famous ruling setting limitations on the absolute right of expression, no one has a right to yell fire in a crowded theater. Unless the bar, the bench, and the media agree on strict standards of conduct with strict sanctions to be imposed on offenders, the public interest is clearly better served by severely limiting or outright banning live TV courtroom coverage in trials, especially trials that promise to be so emotionally charged and intensely covered.

265 days facing a virtual ton of overwhelmingly incriminating evidence to assess, it took those O. J. Simpson jurors only three hours of deliberation before rendering their verdict on October 3, 1995—not guilty on all charges.

The same cameras and technology that brought O.J. live into people's homes and offices for all those months now captured America's reaction to the outcome.

Just as in the beginning, the cameras brought Americans together, and then sharply divided them.

In Los Angeles, a deathly silence settled over the courtroom when the clerk began reading the verdict beginning with the words, "In the matter of the people of the State of California versus Orenthal James Simpson, we the jury find the defendant..."

Screams of joy and cries of outrage rang out in the courtroom at the pronouncement "not guilty." Those same conflicting emotions were immediately displayed in televised scenes across the country.

Pandemonium swept black neighborhoods in Los Angeles. Worshipers in a black church there began jubilant celebrations. Others took to the streets amid wild cheering.

In Washington, D.C., in black neighborhoods along North Capitol Street in sight of the Capitol, young black men gleefully leaned out of passing cars and high-fived each other, some shouting, "The Juice is loose."

Outside public buildings where throngs gathered with people carrying portable TVs and radios, in public school classrooms where students listened over public address systems, in packed office conference rooms where workers watched the TV screen, news of the verdict showed blacks joyfully cheering and whites shocked into silence.

Of the many public-reaction scenes broadcast live that day and later repeated on evening network telecasts, two in particular showed the immensity of the racial divide the verdict exposed.

At Howard University, black law students, watching the verdict from the vantage of the school that more than any other has provided historic African-American leadership in the civil rights movement, spontaneously burst into prolonged cheers when they heard the words "not guilty."★ At the same moment, other cameras panning the faces of mostly white law students at Columbia University recorded stunned expressions and gasps of disbelief.

·　·　·

National reaction broke along the same racial fault lines. To an extraordinary degree, whites thought O.J. literally got away with murder because of a racially biased jury. Blacks believed the verdict just because they thought sufficient evidence existed of a white police frame-up that more than raised reasonable doubts about his guilt, or because a not-guilty verdict symbolized payback by blacks against whites for past acts of injustice, or because of a combination of these and other factors. Out of the torrent of commentary the verdict unleashed, one remarkable example emerged, though it does not seem to have attracted much notice, certainly not the Pulitzer Prize for commentary it deserved.

Writing on deadline immediately after the verdict that day, Michael Wilbon, a *Washington Post* sports columnist who is black, memorably expressed the greater dimensions and significance of the case. Under the title "A Celebrity Goes Free," Wilbon described the uniformly jubilant reaction of blacks across America and commented:

★Howard, founded two years after the Civil War in the nation's capital and funded by the new U.S. Freedman's Bureau and later by Congress, became the most influential university for black citizens, attracting a stellar African-American faculty in varied fields: Ralph Bunche, in political science; John Hope Franklin and Rayford Logan, in American history; Charles R. Drew, in medicine; E. Franklin Frazier, in sociology; Sterling Brown, in English. But its greatest, most lasting impact was in civil rights, where from the 1930s on, its law school became the heart of America's civil rights movement. Through the efforts of a brilliant faculty, including Thurgood Marshall, Charles Houston, Spottswood W. Robinson III, James Nabrit, William Hastie, and Leon Andrew Ransom, many of the cases which led to major U.S. Supreme Court civil rights decisions and congressional legislation were argued and threshed out in its law school.

All over urban America you could find these scenes yesterday. It was as if acquitting O. J. Simpson made up for Rodney King and Emmit Till. For all the black fathers and uncles and grandfathers who'd been jailed unjustly, for every brother who has been framed or railroaded, beaten into a confession or placed at the scene of a crime when he was a million miles away. You know what? It doesn't make up for it. I'm a lot less concerned with O. J. Simpson's guilt or innocence than I am with this unqualified embrace of a man simply because he is a celebrity.

He addressed the greater implications of America's obsession with and glorification of celebrity, placing special emphasis on the effect on black Americans. "All of America has become mesmerized by celebrity in the past 20 years," he wrote.

But nobody buys into celebrity, nobody's suckered inescapably into it like black people, my people, the people who can least afford it. You know what happens every single day in urban courtrooms in this country? Black juries, or predominantly black juries, convict people of crimes with no more drama than necessary. Ordinary, everyday people. But not the chosen ones. You know who the chosen ones are in black America? People who dunk, tackle or sing. Can't touch them. A black delivery man on trial facing the same evidence Simpson faced is a black delivery man headed to prison for life. . . . I worry that the people who feel overjoyed at Simpson's acquittal don't get it. Simpson is free because he played football, because he turned that into a movie career and he's rich. Period. This doesn't symbolize anything or portend great changes in the judicial system to somehow ensure a better shake in the future for African-American citizens. . . . I worry that we, black people, are so desperate for heroes we'll take the worst candidates on the face of the earth because they ran sweet or had a nice crossover dribble. In the last year we fawned over a drug user (Marion Barry), a convicted rapist (Mike Tyson), and a wife-beater (Simpson), as if those three somehow reflect the best of what we offer to society at-large or our own communities.

Wilbon wanted his readers to know he wasn't "naive about one of the primary emotions involved here: vengeance," adding:

A lot of black people could care less about Simpson and see him truly for what he is. They simply see this as payback, even if the score is still about 1 million to one. They feel the chickens might have come home to roost yesterday for all of our relatives and ancestors who've been beaten and raped and lynched and murdered by whites without any consequence whatsoever. . . . The bigger issue here, of course, is race. It's always race. What we've seen on television and heard on radio before and after the verdict only confirms that blacks and whites have a completely different reality when it comes to some things. You see evidence, I see a plant. I see a racist cop, you see a defense attorney's diversionary tactics. The lines aren't always that clear, but they were in this instance.

With disturbing eloquence, he posed the larger challenge arising from the Simpson case: "Until we as a nation begin to pay attention, those two separate realities will continue to exist. And in one of those worlds, a blind and undying love for anyone famous will continue to drain us of energy that ought to be channeled in another direction."

*　*　*

Even then, it wasn't over. Sixteen months later, on February 5, 1997, a largely all-white jury in Southern California's city of Santa Monica unanimously found Simpson responsible for the murders of Nicole Brown Simpson and Ron Goldman in a civil suit brought by the Goldman family. O.J. was held responsible on all eight counts for "wrongful death" and ordered to pay $33.5 million to the Goldman family.

Once again, the rendering of the verdict produced an extraordinary national television moment, though this time the actual reading of the jury's finding was not televised live. By sheer coincidence, but as if fated, the timing of the second O.J. verdict came precisely as the newly reelected president of the United States was about to give his most important speech of the year, the constitutionally prescribed State of the Union address, telecast live in prime time throughout the nation and the world. This is the one moment a president can tell his fellow Americans and their lawmakers what initiatives he proposes to undertake that will affect their destiny, and the moment when he can expect to command the undivided attention of even a cynical society largely uninterested in the workings of Washington and the government. For the people, it affords the only chance each year to witness the assemblage of all branches of their government gathered before the president in the congressional chamber on Capitol Hill.

This time, as the president entered the chamber, the nation's television screens were split into two segments portraying scenes occurring simultaneously a continent apart. One showed corre-spondents breathlessly chattering about the O.J. civil verdict about to be handed down in California as the clock passed six o'clock Pa-cific time, the other showed the president and legislators inside the national Capitol shortly after nine o'clock eastern standard time that night. All day, TV executives and producers had wrestled with how to handle these twin events, one on the set stage of the Capi-tol, the other the culmination of a breaking news story that again dominated media coverage and national attention.

To their credit, most network executives decided not to co-opt the president's address for O.J.; it was covered live, with words superimposed across the bottom of the screen giving bare news of the verdict. One network, though, gave the verdict equal standing with the president by employing a split screen and covering the two events simultaneously. No sooner was it over, though, than the coverage overwhelmingly shifted away from the president's speech—away from analysis of his appeal for the Senate to confirm a critical chemical weapons treaty then blocked by conservatives, away from his proposals to deal with health care and entitlement reform, away from other vital subjects addressed. For the rest of the night, once again, it was "All O.J., All the Time."

· · ·

When the O. J. Simpson case finally took its place in American cultural and social history, a great deal of moralizing ensued about its meaning. O.J. wasn't *really* that significant an event, some of the critics maintained, certainly not one that deserved the amount of coverage devoted to it. It didn't even reveal much new about the state of race relations, the arguments went, to say nothing of providing insight into other important aspects of American society.

There was truth in some of these judgments, but far more foolishness than wisdom. Viewed strictly as a news story, details of the brutal murders alone were enough to arouse morbid fascination in the least-interested news spectator. It was a great story, one that helped define the times and deserving of intensive coverage. Not, to be sure, much of the kind of coverage it received—coverage lacking in perspective and proportion; shameless, sensationalistic coverage that ignored the deeper significance of a case that displayed the powerful interaction between big-time sports, celebrity, spectacle, scandal, entertainment, power, money, commerce, and—yes—fundamental attitudes about race. Perhaps, most of all, O.J.'s was a story that illustrated the influence of television on American life.

By the mid-Nineties, when the O.J. obsession consumed America, television had been a dominant factor in national life for half a century. By then, it had become the primary influence in homogenizing American culture, but an influence that asserted its greatest power by appealing to the lowest common denominator. As the historian Daniel J. Boorstin observed, in a brilliant 1971 essay assessing television's influence after its first quarter century in a special issue of *Life,* nothing since the invention of the printing press had so radically altered life. (The same point, of course, as we have seen, was made later about the impact of the Internet.) The invention of printing, as Boorstin put it, democratized learning. The invention of television democratized experience. But it took five hundred years for the revolution of printing to create fundamental change in human society. Television had conquered America in less than a generation.

At its best, television enables people to witness the great events of their times, to be transported electronically to distant corners of the globe and even into space, to draw together in moments of national crisis and tragedy, and to enjoy splendid (if increasingly rare) offerings ranging from drama and concerts to news specials and documentaries. At its worst, television fragments society as nothing before, creating, in Boorstin's view, "mysterious island-audiences, newly separated from each other," fostering a blurring and vagueness to life that makes people feel accustomed to experiencing staccato, constantly changing scenes disconnected from reality, a world in which "everything becomes theater" and in which "any actor—or spectator—holds center stage."

If that was true in the early 1970s, as demonstrably it was, it was even more so a generation later. Far from being an aberration or an isolated event, the O. J. Simpson episode stands as one of the more significant events in contemporary American history. Its greatest legacy stems from the way it influenced the actions of other national institutions, from journalism to the law, and how it also affected public attitudes. Most of all, it illustrates the increasing impact of television in national life, especially since it fostered a public appetite for more televised spectacles featuring more celebrities entrapped in more scandals or disasters and an ever-expanding market to capitalize on them. For Americans in the Nineties, comprising as they did an audience even more fragmented in a nation even more divided, the worst was yet to come.

Cult of Celebrity

"In the future everyone will be famous for fifteen minutes . . ."

A ndy Warhol didn't make Felix the Cat famous. Felix became famous all by himself.

Felix's brush with fame came long before Warhol became famous for transforming a can of Campell's soup into something called American pop art, and then for predicting that we'll all have our fifteen minutes of fame courtesy of the TV cameras. For Felix, his moment arrived in the late 1920s in a midtown Manhattan studio. That's when RCA engineers focused their arc light on a crude papier-mâché statuette of Felix. His reflection was immediately picked up by a battery of photoelectric cells and sent whizzing all the way to Kansas. There, an excited group gathered in another studio reported seeing Felix's blurry likeness emerge over a screen. The television age was born.

From the perspective of future decades, Felix the Cat is barely remembered, if at all, but it's fitting that the first fleeting celebrity of Teletimes was a cartoon character. As the sardonic and unofficial philosopher king of the electronic age, Andy Warhol would have relished the irony of that combination of cartoon character and celebrity as he later enjoyed the manner in which celebrities swiftly arrived and swiftly departed from the American scene.

Time has affirmed the accuracy of Warhol's famous prediction about fame. By the Nineties, the hunger for instant celebrity was a mass phenomenon, one that afflicted everyone—from presidents and prime ministers to politicians and CEOs, artists and writers, actors and athletes, TV commentators and investigative reporters, pundits and campaign manipulators, and anyone else from kings to commoners. It also affected the most notorious, from Mafia chieftains to disgraced public figures, all of whom sought—and received—instant access to the cameras and, through them, the country and the world. The O. J. Simpson spectacle did not create this mass media phenomenon and the resulting cult of celebrity, but it did accelerate the process by which the famous and the infamous interacted with each before the cameras to provide the public more mass spectacles and more mass entertainment. While O.J.'s legacy affected the entire American culture, it had particular impact on the law and the media, as the experience of only two of the many players shows.

The Lawyer

Laurie Levenson's beeper gave off its chirping signal while she was on the golf course in Los Angeles that morning in June. ABC was calling. O.J. was going to turn himself in at the downtown L.A. police headquarters, they said. Could she come immediately and offer live commentary to explain the legal process and the legal implications of the surrender and the criminal charges about to be lodged?

Levenson, then in her late thirties, was no novice at giving legal commentary on television, but she hardly thought of it as a principal avocation. She certainly didn't think that a TV appearance would change her life, consume her every waking moment, elevate her into the realm of instant electronic celebrity, and compel her, eventually, to make an anguishing professional choice about turning her televised minutes of fame into a new career.

As associate dean of the Loyola University Law School in central Los Angeles, Levenson was a lifelong resident of L.A. who at that moment was doing exactly what she had dreamed of doing. She was fulfilling her youthful belief that the law "was a noble instrument that could raise up the downtrodden—the public interest idea, the Bobby Kennedy model—as one would say."

Citing "the Kennedy model" came naturally to Levenson, for Kennedy's life and death had influenced her strongly. Her father, a doctor, was one of the physicians who treated Robert Kennedy in the Ambassador Hotel kitchen after an assassin fired those bullets into Kennedy's skull that night in 1968, and her mother was active in community work in the black sections of Watts in central L.A. Laurie grew up a Sixties' liberal, motivated by a desire to make a difference in society. After toying with the idea of becoming a doctor, she went on to Stanford Law School and then began to practice law, first as a public prosecutor, later, in private practice. She loved every aspect of the law but found her greatest reward in teaching and training young lawyers.

When her beeper sounded that morning she was a respected figure in the Los Angeles bar and an admired teacher/administrator at the Loyola Law School. She also had recently been exposed to the world of legal commentary on television.

Two years before, in April of 1992, she was commissioned by the *Los Angeles Times* to write an op-ed article when an all-white jury in suburban Simi Valley, adjoining Los Angeles, rendered its verdict in the Rodney King case. King was the black man whose brutal beating by four white Los Angeles policemen was captured by a private citizen on videotape. The sight and sound of those blows from the police batons—*thwack, thwack, thwack, thwack, thwack*—striking King fifty-six times in eighty seconds as he lay helpless on the ground, played over and over in slow motion and in stop action on America's TV screens, burned that episode into the national consciousness. Emotions were inflamed in every black community, nowhere more so than in Los Angeles.

In a moment Levenson would never forget, she was sitting in her university office that April 29 when the King verdict was broadcast live over TV. The white jurors had acquitted the four white policemen. No sooner did that news flash over the airwaves than sirens began sounding across Los Angeles. In a spontaneous outpouring of rage, blacks took to the streets, burning and looting and shooting and firebombing in what became the single greatest spasm of urban rage in twentieth-century U.S. history, an urban eruption that left more than fifty people dead, more than fifteen hundred buildings destroyed, more than a billion dollars in property damage, and more than two thousand people hospitalized. Pillars of smoke billowed into the heavens. In a scene people would carry with them for years to come, huge clouds of black smoke nearly obliterated the sight of a setting sun burning red through the clouds over Los Angeles. Levenson, horrified, found herself and her law school at ground zero of the riots. Just a block away, every building was burned to the ground.

Her article about the acquittal of the white policemen in the King case attracted wide attention. A year later, with the coming of April, the nation was once more fixated on Los Angeles. A second Rodney King jury was deliberating the fate of the same four white policemen whose acquittal had ignited the great racial riots of 1992. This time, instead of an all-white, local, suburban jury, the second King jury was a racially mixed federal panel that had been convened in central L.A. On the eve of the verdict, the nation held its collective breath, the city and state mobilized their forces, and U.S. troops were placed on alert. When the verdict was announced, this time there was no blanket acquittal; two of the four policemen were found guilty. Los Angeles was spared another riot.

It was against that background that Levenson took the next step in her path from legal professor to legal electronic commentator. National Public Radio, recalling the article she had written after the Simi Valley acquittal, had asked if she would cover the retrial for them. Her initial response was negative. She had a life; she had a job. She really didn't want to do it. NPR was undeterred. At least come down the first day, she was urged. After all, you know the players; you know the judge; you know the prosecutor; you know the defense lawyers; you have been a federal prosecutor; you have been a trial attorney; you know the applicable law involved. Just come down that first day, NPR implored. Describe the legal lay of the land for our national listeners.

She did, though reluctantly, feeling as if she had been somewhat "suckered in." Immediately, she found herself hooked. "I was
the only noninvolved party down there who could explain to the
media what was happening," she recalled. "Besides, I love trial
work, always did, and this was a really important case to society. I
could do it; I could answer the questions about it."

Levenson was a natural. She not only knew the law, she had a
winning personality and could speak clearly, knowledgeably, and
extemporaneously about it. As that trial attracted greater national
attention, the TV networks began courting her. "Grab Levenson,"
a CBS producer urged his superiors. So she began doing network
TV commentary.

Here was a new challenge, as intimidating as it was exhilarating. "We're a bunch of novices," she says of herself and other legal
commentators. "You're fairly intimidated when you first start out.
They sit you down in front of a camera and tell you what to do. It
takes a real pause to say, 'No, that's not what I'm going to say, that's
not what I'm going to do.' You're really outside of your regular element." At the same time, she discovered she liked it. It was "seductive." "You're almost making up the rules as you go along,"
she says. "No one trained me to be a legal commentator. Very early
on I decided my only key rule is to tell the truth. Call it the way it
is. I also found it can eat up your life. It's all-consuming. You think
these trials are the whole world, and once you say something you
can't take it back. That's different for lawyers and professors who
are used to writing out their notes, reviewing them and having
someone else review them. Doing live TV is one of the scariest
things in life. Having said that, I found that actually it was a very
rewarding experience. The journalists I met were good people. By
and large, they didn't deserve any bashing. Still, once the King case
was over, I thought I'd be done with this. I thought it was my
once-in-a-lifetime experience."

Other highly publicized trials arose, however, creating other
calls for her services. Inexorably, Levenson found herself being
tempted to become more intimately involved as a player in the
celebrity game. She offered commentary on murder cases, some

relatively minor, others, like the trial of the Menendez brothers, whom she defended, of national interest. Then came the beeper on the golf course calling her to a case that, as she says, was in its own category, being played in a league beyond them all. "Suddenly," she says, "O.J. no-showed, and all of life changed."

That day of the Bronco chase she found herself live on air with Peter Jennings before the entire nation. Instantly, she was caught up in the frenzy that gripped Los Angeles and the country. Several networks competed for her time and talent. Eventually, she signed an exclusive contract with CBS; for months on end she was seen across America as one of the new O.J. legal network commentators.

"It was really intense," she recalls to me. "It was just like a roller coaster. It's hard to describe how all-consuming it was. Let me put it this way: Forty of us were basically living in a trailer at Camp O.J. to get out the news. It was so intense we had a dead rat in our watercooler for a week, and nobody noticed! For me, it was hard to watch the case. It was like watching a car accident in slow motion. I had been a prosecutor. I tell you, they should have won that case. They blew that case. On the other hand, the jury comes out with that verdict, and here I am, somebody who's devoted to the system of law. To be honest with you, I knew, based on the evidence, that the guy was guilty. And one of the toughest times personally was when Dan Rather turned to me when the verdict came out, 'So, Professor Levenson, in the ten seconds remaining, what does this verdict say about justice?' And I remember saying, 'Dan, that's an issue that will be debated for years to come.'"

Looking back on her experience, she is struck by the conflicts involved, especially for herself as a lawyer and as a person. "There were highs. Wow, this is exciting. The media, all these people watching—everybody! I've got the inside scoop. And all these people want to know the inside scoop. Then there were the very lows. The reality that this isn't a game, that there's so much at stake, and it's being treated like a game. The public's treating it like a game. The media's treating it like a game. For someone like me who went into the criminal area because I really believe you can get justice out of it, it was demoralizing—downright demoralizing. Then we saw the second half of it, the O.J. civil case. That was surreal. Here's O.J. walking through the courtroom and everyone's playing the game. Everybody knows he's a murderer. And everybody's kissing up to him."

The lessons to be learned were profound. On what she calls her "good days" Levenson focuses on how much "we learned about women and how they're treated in our society as trophy wives and battered women. We learned, what we already knew but didn't want to remember, that there's an enormous racial divide and it's most pronounced in the criminal justice system. We learned that we have to confront our doubts about the jury system. We learned that we have to talk about celebrity justice and whether there is equal justice when someone is famous and has more money. If you focus on all those, this was a heck of a learning experience. I know I learned a lot. The public would also learn a lot if it could actually remember something for more than a nanosecond. But the public doesn't remember. Why? They move on to the next TV program. If the public stopped and paused and said, 'Wow, trial of the century! It may take more than five minutes to learn something from that.' But they don't pause, because the next one's around the corner."

More troubling were the implications of what impact the expanding celebrity culture was having on society. As Levenson says, "I have seen a blurring of entertainment and news. That's quite clear. Everyone blames the media for it, but actually I think it's as much the public's fault as the media's. I think the public gets so used to it that they love the high-speed car chases. Why? They like to be part of a happening. Why did tens of thousands of people line up to cheer O.J.? I think most of those people actually thought he was the murderer! And yet we have this spectacle that people want to be part of. So they join. It doesn't matter what's the cause; whether it's good or bad, they join the cause, and the media offers them the allure. Everyone wants their ten seconds of fame. These publicized trials have become participatory sport. In fact, it was pretty amazing during the O.J. case that the hardest decision for the networks was whether to leave their soaps to carry the trial live. It was unclear whether people loved the soaps more than they loved the O.J. trial. It's amazing how many people in the middle of the day take their time out to watch these soap operas. We've become a soap opera society in a way I had never realized. I thought people went to work during the day. As for the blurring of entertainment and news, I think it happens because it's all about ratings. The Simpson case was pack reporting. Everybody's chasing the same story. So in order to get ahead, they pick the angle, even if it's irresponsible or they don't have as many sources as they ordinarily would. That was one way to break out of the pack. And they brought in entertainment people to help them do it. That's their job—to make it sexier."

Laurie Levenson is not the kind of person who criticizes other professions and other individuals but excuses herself and her own profession from critical scrutiny and judgment. Over time, she came to be especially troubled with the new role lawyers were playing. She and the entire nation saw how other lawyers (and journalists) discovered that their appearances on network news shows and on many of the new "All O.J." cable talk shows benefited them both professionally and commercially. They were in a buyer's market. The new shows needed a stable of "expert" analysts; the more familiar the face, the greater the demand for appearances. The greater their celebrity value, the more clients could be attracted to their firms. They began making extra money, and, as we have seen, reaping additional profits from lectures, talk shows, and book contracts.

Levenson saw firsthand how the public exposure was affecting the lives and careers of other high-profile lawyers. She watched as they hired personal hair stylists and makeup artists. She listened as they asked others to grade their televised performances—*"How did I do? How did I do?"*—and discussed their next deal. To Levenson, all of this, as she puts it, "gave the legal profession a black eye. It makes people even more cynical about the law. It feeds into the philosophy that people are just in it for themselves. They're in it for the image. They have no real dedication to the process. The trials are just launching pads for entertainment careers. People are picked for television because they have a quick wit. It's all about appearance and a quick wit. And there's plenty of evidence for that."

She also came to believe that if the public became more cynical about lawyers and the law, "lawyers are more cynical about themselves."

Levenson doesn't say that with any sense of moral superiority. She learned, as well as anyone involved, how enticing the process was. Calculating the payment she received for her time on television, she found "it added up to a lot of money." Her prospects of earning greater sums were excellent. She had offers to become a permanent member of other programs. A major talent agency wanted to represent her. She had a fundamental career decision to make. As she says, "I remember clearly after the O.J. case having to consciously make the choice, 'What are you, Laurie? Are you an entertainer or are you a law professor?' It would have been much more lucrative to choose another path, but I had a lot of doubts. I also thought about that piece I wrote right after the first King verdict when I said I thought there were dangers, all sorts of dangers, for lawyers becoming commentators. One of them is not realizing you're picked to do your commentary because you have a certain point of view. Lawyers are used to being advocates in the courtroom. Now they're doing it on TV. They unabashedly participate. They've completely abandoned the idea that they're supposed to be objective."

She had to make a choice, and Levenson is honest enough to say it wasn't easy. She talked it over with her husband, also a lawyer in private practice who does much pro bono work for the American Civil Liberties Union. It's your decision, he told her. In the end, after much soul-searching, she decided not to cash in. It wasn't the way she wanted to live her life, and besides, she says deprecatingly, "Others may have been much better at it than I." Laurie Levenson went back to being, as she puts it, "just a nerdy law professor" and mother of two small children.

That wasn't the end of her learning experience, however. Back in her law school, she discovered young future lawyers were also affected by what they'd been seeing on television. For the first time in fourteen years of teaching and counseling students, she found them asking a question she'd never heard before when they wrestled with what career path in law to pursue. "They'll come to me and actually ask," she says, "'How do I become a legal commentator?'"

They were interested in practicing celebrity law.

The Journalist

Before O.J., The Chase, and the trials manufactured a new mass audience surpassing the soaps; before cable TV and the high-voltage, low-content accusatory cable talk shows became omnipresent; before the Internet granted instantaneous access to people everywhere with the click of a mouse; and before the blossoming of the celebrity/scandal culture created greater pressures on broadcasters to telecast more and more of the same, Dana Millikin had already seen the nature of her business change forever. She even remembers the time and place when TV news, as she and others in the business knew it, was radically transformed—and not for the better, to put it most kindly.

It was in the early Reagan years of the 1980s, and she was a young news executive at KHOU–TV, a local Houston station. Millikin was a pioneer of sorts, for when she began her career in the mid-Seventies, TV journalism, like print journalism, was still solidly a man's world. In quick succession she learned her trade, working her way up from minority (female) trainee intern to news camera-woman to on-air reporter to program producer to assignment editor to head of the station's investigative unit. In the Eighties, after moving up to management, her TV world suddenly changed.

She was standing in her newsroom that day when a wire service story out of Washington clattered into the office over a teletype machine. Millikin looked at the wire copy, ripped it from the machine, and began moving rapidly through the newsroom, waving the story and shouting to her colleagues: "Mark this down. This is the day it will change. This is the day it will all change."

The U.S. Federal Communications Commission (FCC), the wire service reported, had just deregulated television, freeing broadcasters from regulations affecting everything from public-service programming to advertising.

"Once they did that, TV news changed," Millikin says. "Until then, what counted was the local station's reputation in the community, and that helped them with their relicensing applications with the FCC. They poured big bucks into local coverage. It was a prestige thing. It was almost like giving to charity. Then all of a sudden it was no longer necessary to have the news. The new broadcasters, the cable businessmen, started coming in. Now we had to make money to do what we were doing. So suddenly everybody's scrambling. They have to make money, which means they have to have ratings, and that completely changes everything. If you have to have ratings, you have to get more people into the tent, what they call 'getting more suckers under the tent.' When ratings become more and more important, you start lowering standards and going down. It really started sliding downhill."

That FCC decision not only changed broadcasting, it also affected the electronic culture and American society for years to come.

The FCC deregulation decision grew out of a deeply rooted conviction about the proper role of government in American life that crested with special force in the Reagan years. Under Reagan, deregulation of society, from its markets to its regulatory agencies, was a central goal in the drive to shift the nation away from dependence upon the federal government. Now America would be "privatized." Profits, not public interest, became the operative principle. In furtherance of these goals, Reagan appointed a strong proponent of deregulation to head the FCC. "Television is just another appliance," said Mark Fowler, a former lawyer for broadcasters, after being named FCC chairman. "It's a toaster with pictures." As chairman, Fowler took the position that "It was time to move away from thinking about broadcasters as trustees [for the public]. It was time to treat them the way almost everyone else in society does—that is, as business."

Under Fowler, everything governing broadcasting changed. Scrapped was the system of public licensing of the airwaves that had been in effect since 1934. Abandoned were FCC regulations requiring broadcasters to devote a minimum portion of their airtime to news and public-service programs, a requirement that affected religious, children's, and public affairs programming. While this substantially reduced the amount of time devoted to such programming, in some cases by more than half, it greatly expanded the potential for profitability for the stations and owners. Stations were permitted to increase the amount of advertising they could run each hour, resulting in a sharp rise in the number of spot commercials aired and a drastic reduction in the amount of air time each took. Almost immediately the fifteen-second spot became a new TV standard, one that further shortened the nation's attention span and led to snappier, simplistic political and commercial slogans. Fowler's FCC also abolished a log-keeping requirement that enabled groups advocating better public service to evaluate and challenge programming offered by TV license holders. And they abolished the FCC's long-standing fairness doctrine, which required broadcasters to air controversial issues by providing a balanced presentation of public issues.

These changes alone represented a revolution in the way broadcasting was permitted to operate; other Reagan-era changes produced even greater differences. At Fowler's urging, Congress raised the limit on the number of TV stations a company could own from five to twelve. This ignited a flurry of attempts to take over local stations—and the networks. Capital Cities Communications acquired ABC; General Electric, NBC; the Loews Corporation, CBS. One of many consequences was a direct impact on the network news departments. To secure additional capital needed to fight a take-over attempt, CBS, for instance, was forced to increase its debt substantially. Deep cuts in its news department resulted.

Driven by a bottom-line mentality, the networks and their new corporate owners eagerly responded to the new climate of deregulation. They gave the public what they thought it wanted—and, not incidentally, got what *they* wanted—greater audience ratings and higher profits. Increasingly, the news divisions were dragged into the ratings battles and pressured to make more money.

The forces unleashed in the Eighties, a time characterized by a mergers-and-acquisition fervor, accelerated during the boom years of the Nineties. The big consolidation craze to combine greater and greater blocs of economic power found tempting targets in big media. Disney took over ABC; Viacom gobbled up CBS; Microsoft began broadcasting operations with NBC; and, ultimately, in the biggest of all corporate mergers in history to that time, the

Internet provider America Online took over Time-Warner with its string of TV and radio outlets and interlocking major communications/entertainment companies.

So rapid were these changes, and so great their scale, that it was easy to overlook their consequences, especially amid the lulling glow of the boom. Seen in perspective, however, they raise vexing questions about how to regulate the workings of a free society while simultaneously ensuring equality of opportunity. These basic conflicts raised the oldest questions about the public vs. the private interest; about a belief that government governs best that governs least and one that holds the people are best served by a government empowered to act as a neutral umpire to protect their interests; about a laissez-faire, anything-goes, society and one that plays by rules that reduce predatory blocs of power that stymie competition.

For much of the twentieth century, these questions kept recurring as American leaders attempted to fashion policies to deal with the rapidly expanding use of the airwaves and the equally rapidly expanding opportunity for the abuse of those airwaves. And from the beginning these policies revolved around a critical question: Were the airwaves a public resource to be protected in the interests of all the people or a resource that could, and should, be exploited by private interests for private gain?

The first attempt to address those questions grew out of the disaster of the sinking of the unsinkable ocean liner *Titanic* in 1912 when radio operators were unable to bring relief to the doomed vessel. The Radio Act of 1912 empowered the U.S. Secretary of Commerce to issue licenses for radio stations to U.S. citizens. Within another decade, spurred by further technological advances that ignited a rapid increase in the number of unregulated radio stations, it became clear even in the permissive anti-government climate of the Roaring Twenties that intolerable chaos and incidents of abuse were occurring.

Out of that background the then–secretary of commerce convened a series of national conferences between representatives of the radio industry and the government. The idea was for them to agree upon a voluntary system of self-regulation. But the commerce secretary took that occasion to express in the strongest terms a principle that he believed should be adopted as transcendent—when it came to the public airwaves, the public interest must be paramount. With no small irony, the government official who first pronounced the primacy of public over private interest was Herbert Hoover, viewed by succeeding generations of Americans as a staunch conservative. In reality, he was a great public servant who by the early Twenties had forged a reputation as a foremost Republican progressive and humanitarian in the tradition of Theodore Roosevelt. Certainly the words Hoover delivered to radio owners in that first 1925 conference could not have more eloquently raised the public standard.

"The ether is a public medium," Hoover told the broadcasters, "and its use must be for a public benefit. The use of a radio channel is justified *only* if there is public benefit. The dominant element for consideration in the radio field is, and always will be, the great body of the listening public, millions in number, countrywide in distribution."

Two years later, one of the authors of the seminal Radio Act of 1927, a Maine Republican named Wallace H. White Jr., further advanced the primacy of the public interest. "Licenses," he said, "should be issued only to those stations whose operations would render a benefit to the public, are necessary to the public interest, or would contribute to the development of the Act." He also said: "If enacted into law, the broadcasting privilege will not be a right of selfishness. It will rest upon an assurance of public interest to be served."

In that act, broadcasters were deemed "public trustees" who were "privileged" to be granted use of a scarce public resource. While the government permitted them to manage and operate their stations under a federal license, they had to understand that "the station itself must be operated as if owned by the public."

There it was, in the peak of the boom years of the Twenties before the crash, long before the New Deal and the assumption of a greater federal role in national life, the articulation of a fundamental belief that the public airwaves were just that—*public,* owned by everyone. Licenses granted for their use were solely for a public purpose; such grants were "a privilege." They were not to be employed for selfish interests. A year later, when Hoover occupied the White House, government regulators declared the public interest must be applied to the *content* of the programming to be broadcast: "The emphasis must be first and foremost on the interest, the convenience, and the necessity of the listening public, and not on the interest, convenience, or necessity of the individual broadcaster or the advertiser."

Establishing the dominance of the public interest over "a right of selfishness" formed the background for succeeding decades of debate about the government's role in regulating the airwaves. Broadcasters in the United States operated under two established principles. They had to meet specified public-interest standards in offering programs to the public. They had to demonstrate that their programming met the public-interest requirements before the government, through its regulatory powers, would renew their licenses to continue broadcasting. The burden was on the broadcasters to show they were serving the public interest.

Measured against that background, the changes produced by Reagan-era deregulators were startling. Now the public-interest standard gave way to the actions of the unfettered marketplace. As Reagan's FCC chairman Fowler proclaimed early in his term, "I believe that we are at the end of regulating broadcasting under the trusteeship model. Whether you call it 'paternalism' or 'nannyism'— it is 'Big Brother,' and it must cease. I believe in a marketplace approach to broadcast regulation.... Under the coming marketplace approach, the Commission should as far as possible, defer to a broadcaster's judgment about how best to compete for viewers and listeners, because this serves the public interest."

Here was a totally different philosophical approach, one that set in motion many of the electronic changes that coursed through the culture. First, Fowler deregulated radio. Then he deregulated television. Struck down, among many other public-service requirements, were guidelines banning the broadcast of false, misleading, and deceptive commercials. The burden had shifted. Now the *government* had to prove that the public interest required retention of such guidelines.

The broadcasters had won. Laissez-faire practices had triumphed over those of the neutral umpire state. The consequences were felt immediately at television stations. Call it a "dumbing down" of the medium, a cheapening of its content, a rush to exploit scandal and spectacle, an unseemly selfishness that made pursuit of profits its highest value, or any other pejorative that comes to mind, the essence is the same: The standards, and the practices, had changed irrevocably. The changes applied to broadcasting content, its bottom-line motivations, and its collective judgment about what constitutes news. Technological innovations assisted the rush to attract greater audiences—and greater profits—by broadcasting more spectacles and scandals attracting greater audience ratings.

Long before O.J. and the low-speed freeway chase provided the biggest spectacle yet, local TV had learned to use its new technological adaptations to lure even bigger audiences. TV cameras, for instance, could now be attached to helicopters instead of being handheld by cameramen struggling to focus on a scene below. Soon, local stations in L.A. were dispatching helicopters to patrol the highways daily, looking for new disaster pictures live and in color. Not one of them maintained news bureaus and correspondents in the state capital, Sacramento, to cover the governor and the lawmakers in America's mega-state, the place whose economy and lifestyle so strongly influenced the rest of the nation and much of the world.★ Instead, they trolled the L.A. freeways for the latest sensational development—a suicide, as it unfolded; a spectacular crash; an earthquake; a mudslide; . . . another disaster.

★In the 1998 California governor's race *all* statewide local news broadcasts devoted only an average of *four minutes* each night to coverage of the race. A total of 8,664 hours of local news was broadcast during the three months before election day. Only *0.3 percent* (27 hours) was devoted to the campaign. The candidates filled the gaping news gap with paid TV political commercials; inevitably many contained negative attacks and many were misleading. The public received virtually no help in sorting fact from fiction from a news media whose mission supposedly is to provide accurate information and perspective on key public events.

The desire to have the latest sensation live—and, better, exclusive—for the evening news spurred greater competitive efforts. Technological advances enhanced the spectacles; the audience vastly expanded. In the new, instantaneously linked electronic world, producers talked to producers and correspondents halfway around the globe, all reacting to the same live televised disaster scenes. The evolution from Disaster TV to Shock TV to Trash TV to Scandal TV was inevitable.

In the entertainment TV business, producers analyzed Nielsen ratings to find who was watching what program in what time period. From those raw figures, they determined the size of a target audience and then aggressively went after it by crafting a program to appeal to it. In the TV news business, the function of determining what is news and what should be covered is supposed to be the domain of the editor or the producer. Now that role had been enhanced, if not usurped, by the findings of the same kinds of focus groups used by entertainment producers to learn what kind of shows people want. Now, across the country, local stations used focus groups to decide what kinds of news their viewers wanted to watch. Disasters that trigger a rush of adrenaline, a primal tingling sense of danger and excitement, alerting the eye and the brain to focus on what follows, always scored high. That led to "the disaster psychology" that permeated newsrooms. That psychology could be summed up in the simple catch phrase "If it bleeds, it leads."

"It's a syndrome," said producer Dana Millikin. "You have to watch. Why? Because your body says you'd better pay attention to this because it's bad. It's gonna get you. So you get in close to it, get the disaster, and get the ratings. Ratings! That's why the focus groups are so important. It's not so much to tell me what's the news. The focus groups are so important because the stations are so desperate to get the demographics that their advertisers want. Take, say, A&E [the cable Arts and Entertainment channel]. They were getting us old people and they needed to grow because their stockholders wanted them to grow. So friends of mine who have done A&E profiles were told: 'We've got to tits-and-ass it up, man. We've got to get some younger viewers in here.' Now T&A is second to action. That's still the real hot rod—real action."

From there, it becomes a scramble to find a new event that tops the last in dramatic impact.

"Okay, there's the riot," Millikin explained to me, "It doesn't get worse than that. Then there's the flood. Oh my God, it doesn't get worse than that. Then there's the earthquake. Oh—my God, it doesn't get any worse than that. Then there's the fire, oh—my—god... Topping, topping, topping, topping, topping, always topping. You've always got to top the last one. That's why when you had O.J. and the low-speed chase, everybody here was saying, 'What will top the low-speed chase?' Then you have the Princess Di crash. Oh, my God, Princess Di! Then Princess Di dies! And everybody's crying, and everybody's bringing flowers, and everybody's asking where's my nearest British Embassy to go down and give flowers and sign books...."

After Di, there's always something else—the impeachment of a president; a war in a place called Kosovo; the death and funeral of "America's royal prince," JFK Jr., featuring a cast of his entire celebrity family; and then, well, then, there's always something new, a Cuban boy seeking freedom plucked from the sea, perhaps, or, always, a breathlessly delivered update on JonBenet, even if no new news has actually occurred.★

·   ·   ·

In both print and electronic journalism, the same trends could be seen—a blurring of the line between news and entertainment, a greater focus on scandal and celebrity, a "gotcha" philosophy of investigative reporting run amok, a lessening of standards long in

---

★The JonBenet case provides a perfect example of excessive exploitive TV coverage. A study in *Brill's Content* shows that the TV network programs *20/20, 48 Hours, Hard Copy, American Journal, Dateline NBC, Entertainment Tonight, Extra,* the weekend edition of *Extra,* and *Inside Edition* aired 438 JonBenet segments during a a 22.5 month period from Jan. 1, 1997, to Nov. 19, 1999, while another 195 segments were aired that same period on *Geraldo Rivera* and 44 on *Larry King Live.* That doesn't include the plethora of JonBenet stories (literally hundreds) published in the tabloids and traditional newsmagazines in that same period.

place in mainstream publications about accuracy, gossip, rumor, plagiarism, privacy, fact-checking, using multiple sources, breaking confidences.

The examples were numerous; in the Nineties, they occurred with distressing frequency. Two prominent columnists for the respected *Boston Globe* were fired after admitting having fabricated characters and quotes. The *Cincinnati Enquirer* was forced to apologize to Procter & Gamble for unethical tactics employed in an investigative report. The *Los Angeles Times* initiated a corporate-mandated policy to blow up the wall separating the news side and the business side of the paper, with a result that the paper secretly entered into a profit-sharing deal with a new L.A. commercial sports center that was to be featured in a major article in the newspaper's magazine. The financial arrangement between the sports center and the paper called for the two sides to split the advertising revenue in the magazine, thus raising a serious conflict of interest compounded by failure to disclose the deal to readers who rightly could be suspicious about the integrity of the article. A joint journalistic investigative effort between *Time* magazine and CNN came a cropper after its first highly publicized venture about the supposed American use of nerve gas in Vietnam could not be substantiated. Its producers were dismissed.

The result of such media mistakes and excesses was mass media that brought more discredit upon themselves and that reflected a more negative, certainly more cynical, view of society's leaders and institutions—and a media that focused more on trivial concerns, on scandals and celebrities.

How great a change in news values had occurred was documented in a national study of media trends by the Committee of Concerned Journalists. In the twenty-year period from 1977 to 1997, the study showed, coverage of "hard" news dropped significantly in newspapers, major magazines, and national TV news programs. At the same time, there was a corresponding increase in the number of "soft" news stories. The greatest change involved coverage of celebrity and lifestyle news. It had more than doubled over that twenty-year span. Analysis of newsmagazine cover subjects was even more revealing. In 1977, nearly a third of all *Time* and *Newsweek* covers portrayed political figures. Twenty years later that figure had dropped to only one in ten covers. The percentage of covers devoted to national issues dropped by a third over that period. In that same time, the percentage of covers featuring celebrities increased by a third.

Even a new national magazine supposedly devoted to chronicling political trends and issues, *George,* edited by John F. Kennedy Jr., put sexy celebrities on its covers thus tacitly emphasizing entertainment priorities over those of serious public affairs and political news. Supermodel Claudia Schiffer wore nothing but a Clinton/Gore sash on one *George* cover, Kate Moss posed nude in shadows on another, and on yet another, Christy Turlington appeared nude with a television set between her legs and a caption that read: "We Like to Watch."

Gossip columnists were on the rise, most notably in electronic journalistic outlets. They offered daily doses of unsubstantiated scandalous tidbits, with repeated allusions to the alleged immorality of public figures, especially political figures.

Many factors contributed to these developments; virtually all of them could be explained by changes in public attitudes toward public service, politics, leaders, and institutions in the wake of assassinations, Watergate, Vietnam, riots, random violence, and conspiracy theories formed during years of national turbulence. All of them affected the way Americans viewed their society in Teletimes.

The question was a chicken-and-egg one: Was the media responsible for the cynicism and disbelief that afflicted America even in the midst of the boom? Or was the media merely holding up a mirror that perfectly reflected the society before it?

More critically, what part did television play in this process? Should it be blamed for shortening the nation's attention span... for lessening a sense of national perspective... for fragmenting society instead of bringing it together... for glorifying excessive violence... for pandering to sex and prurient interests... for emphasizing now and today over the lessons of yesterday or the needs of tomorrow... for accentuating the negative over the positive... for presenting froth more enticingly than substance... for elevating sports, entertainment, and political figures into celebrities... for further blurring the line between fantasy and reality . . . for creating a desire for illusion and escape . . . for transforming Americans from participants into spectators . . . for intensifying acclaim for society's winners at the expense of its losers . . . for focusing on attainment of fame and fortune as the most desired life achievements . . . for accelerating the connection between false advertising techniques and political candidates . . . for worsening the corruption of money in politics through the high price of commercials . . . for cumulatively contributing to more public distrust, more cynicism, more conspiracy theories, more erosion of the notions of the public service and the public weal?

Finally: Were Americans becoming what they saw over their screens?

· · ·

At the very time that TV demonstrated its capacity to generate huge audiences for spectacles and scandals, affecting how people feel about their society, fewer and fewer Americans were watching TV, especially TV news. After rising steadily during the years of television's rapid penetration into American homes—98 percent of all households had at least one TV set in 1999, while 40 percent of households had three or more—network viewership was declining. The hours people spent before the tube stood about where they did three decades earlier when the U.S. population numbered some 75 million fewer people. Then, the average American spent seventeen hours a week watching television. By the end of the Nineties, the average weekly viewing time had actually dropped to sixteen hours and twenty minutes. This was hardly surprising, given the greater competition for public attention TV faced with the proliferation of new electronic gadgets that quickly filled American homes, providing more outlets for home entertainment through VCRs, CD ROMs, and, of course, the Internet.

Professional newspeople bemoaned what many feared was the decline of their business. The electronic entrepreneurs who played a great part in igniting and expanding the long boom viewed the concerns of these professional newspeople as irrelevant to the far greater story of our times: How people will communicate and share news and information in our interconnected future—a future, thanks to them, that already had arrived.

Look, they said, the days of the traditional TV newscast, to say nothing of the traditional newspaper rolling off massive, outmoded printing presses, belong to the past. The future—the Internet future—is theirs. Its power to reach and influence people has never been equaled, and its greatest period of influence is yet to come. Think what the power of electronic communications has demonstrated. By letting people everywhere see what is going on elsewhere, the new communications technology already has changed the course of history. You can argue, for example, that the American military kept the peace during the long Cold War period. But you cannot argue it was American military might that brought down the Berlin Wall, or dismantled the Soviet Union, or opened China to the Western world. Much of the credit for those remarkable events must be given to the new ability of people, even in the most closed and repressive of societies, to see and learn how others live, and want to emulate them.

Michael Bloomberg is one of the electronic entrepreneurs who makes that argument, and his personal experience gives it special weight. By the end of the Nineties, Bloomberg had emerged as one of the biggest winners in the boom. When he was let go as head of the equity trading desk at Salomon Brothers, Inc., at the age of thirty-nine, Bloomberg parlayed his partnership settlement into an investment that made him a billionaire. Through a rapidly expanding network of more than eighty Bloomberg offices worldwide, he built a fabulously successful electronic financial data news service that became indispensable to traders operating in different time zones around the globe. He saw the impact of the new communications technology on international events as being irreversible. "It's hard to see anyone going back to closed societies," he told me. "You can't put the genie back in the bottle. We can't keep others from seeing what's going on. I would argue that the Internet is, for the first time, a communications system that has one characteristic different from all previous ones. That is, the average person can spread ideas to lots of people, economically, quickly, without any government interference. That is a fundamental difference from anything that's come before. I'm skeptical now that politicians can ever stop you from knowing what other people are saying, what other people are doing."

He was dismissive of fears expressed by practitioners of the Old Media that the New Media, which he represents, will affect society negatively. "I don't know if I got better information when we had these gods of the media in the past," he says. "They had stuff they didn't tell me. What right did Bill Paley or Walter Cronkite or the *New York Times*, for that matter, [have to] tell me what's fit for me to know about? That's an elitist thing. I find it offensive when the media sits there and decides what's in my best interest. Who made them God?"

But Bloomberg acknowledged that one result of the new technology was fragmentation of the broad national audience major TV networks and national newspapers used to reach. Now, in this rapidly changing world in which new outlets of information compete fiercely for a smaller segment of society, the operative word among broadcasters is *niche*.

Programmers have learned that they have to fragment their audience by what they call "narrow-casting": that is, targeting a specific audience and charging advertisers to reach that segment, or niche, of society. As result, there no longer is "an American audience." There are many American audiences. Nothing connects them to each other.

"The audience for sports and movies, that's what people really care about," Bloomberg believes. "People talk about working at home, but they come home and they have the problems of the kids screaming and the spouse being a pain in the ass, and they just want to get a beer and watch television. They like a little relaxation. They like to dumb out and become couch potatoes. You set your mind on something that is pure entertainment. So the big demand for information is still going to be entertainment; the number of people who care about serious news isn't going to grow."

This phenomenon affects society in other ways. In a time when more people express disbelief in their leaders and institutions, the public views the information it receives from the media with increasing distrust. "People see the press now as part of the big corporate consumer product," said Steven Brill, a lawyer and founder of *Court TV,* the cable program that broadcast the O.J. trial and attracted a significant niche audience in the Nineties, and, later, founder of the journal of media criticism, *Brill's Content.* "They see celebrity Washington reporters as being the same as the people they're writing about. It's sad. It diminishes everything."

Here, again, the symbiotic relationship between celebrity and media affects journalistic performance and the public response to it. Brill is another one who points to the O.J. case as a prime example of the mass phenomenon; he, of course, televised it endlessly. "The person who got the biggest advance for the book in the O. J. Simpson case was the woman who lost the easiest case to win in history," he notes, referring to Marcia Clark, "and now she's on MSNBC substituting for Geraldo." As for the O.J. case, it contributed strongly to sowing even more public cynicism and, because of the mass audience it attracted, even more examples of cynical media behavior. "We see so many 'Trials of the Century,'" Brill said, "and the next one could be tomorrow because you have this [mass media] machine that needs it. One of the results is that so many people feel more estranged than ever. That's why the media's going to be more important. That's why the media being cynical about all these people has greater consequences. There's nothing to counter it."

The solution, the new entrepreneurs say, is self-evident. The Internet.

*   *   *

Before the Internet burst upon the world, respected print and electronic journalists were already concerned that their core professional standards of ethics, accuracy, and fairness were eroding. The Internet promised to make this even worse. Indeed, it already has! For one thing, the new electronically connected world instantly places non-journalists on equal footing with professional practitioners. Fine, many will say. About time, others, rightly disenchanted with journalistic performance, will add. Democracy in action, still others will remark. The reality is not that simple.

On the Internet, everyone can play the role of reporter, commentator, investigator, editorialist. Everyone can claim to present supposedly reliable information as "news" when in fact it may be inaccurate and uninformed—just gossip, innuendo, or conspiracy masking as news. Already, the Internet has become the province of cranks, sensation seekers, professional hit men, rumor-and-hate mongers, maliciously motivated ideologues. Their aim is not to inform, explain, analyze, or lend perspective and meaning to important events. Their aim is to promote their own agendas and causes. If accomplishing that requires the destruction of reputations through dissemination of false information, or leveling false charges to make their case, well, so be it. The Internet is a place where some news practitioners, such as the online gossip purveyor Matt Drudge, boast of attaining an accuracy rate of 80 percent—a figure that even the least reputable newspaper or TV broadcast would deem unacceptable. That kind of woeful journalistic performance would be grounds for instant dismissal.

Even this does not address more serious questions about the Internet. Despite their major effort to offer responsible news and information, the new online dot-com operations like those of the *New York Times,* the *Washington Post, MSNBC,* and others, have yet to demonstrate the public is responsive to them. Their collective audience remains small—19 percent of American households had access to the Internet in 1999, a figure that is expected to rise to 40 percent in 2001. These figures illustrate the dramatic rate at which Internet access is expanding; but that doesn't mean people are using the Internet *news* offerings in great numbers. Americans are turning more readily, and in growing numbers, to other offerings of the Internet—to gossip, entertainment, and, most of all, pornography. As Michael Bloomberg said, "On the Internet, being able to chat and to see pornography are the overwhelming things. In our lives, we want to chitchat with each other, and we want sex. Those are two of the basic instincts, and the Internet facilitates them."

Where that leaves the state of the democracy in the new millennium is an open question, one that will become significant as the Internet age expands and the traditional information era declines.

The events of the Nineties suggest that gossip, scandal, sports, sex, and the cult of celebrity will form even more of the wave of the future.

That's a pessimistic view. There's a more positive one. That is, in the midst of the boom, Americans naturally felt less inclined to pay attention to news and current events; the glow of good times inoculated them from that need. An absence of a sense of danger or crisis, whether resulting from war, terrorism, depression, or natural calamity, coupled with a belief that the boom would endlessly roar on, created a collective soothing lull and sense of security. It is an illusion. New economy or not, as dot-coms crash and layoffs rise, the ebb and flow of market forces have not been checked; crashes are not impossible; the new technoworld, whatever its boosters said, is not fashioning a recession-proof economy the likes of which has never been seen; the forces of history have not been repealed; human nature has not changed. When new crises arise, as inevitably they will, people will immediately turn to a Jennings instead of a Geraldo, a *New York Times* instead of a Matt Drudge. In the end, those who produce the most reliable, trustworthy news products will prevail.

That, at least, is the hope, and the record of the past supports that hope. Not always, however. Consider what happened to those soaring hopes that the advent of television would usher in a cultural golden age producing better-informed and wiser citizens.

·   ·   ·

Decades before the basic nature of TV programming changed, television, that cultural wasteland of conformity, had been losing its battle between providing quality programming and mass entertainment. In a famous speech in 1958, Edward R. Murrow, the most admired television newsman of his era, warned against the growing trivialization of television. He urged the networks to "get up off our fat surpluses and recognize that television . . . is being used to distract, delude, amuse and insulate us."

What he would have said after witnessing the shameful spectacles that television provided to further "distract, delude, amuse and insulate us" four decades later hardly needs to be imagined.

By the end of the Nineties, in the midst of the biggest boom ever, television seemed to have fulfilled one prophecy. Over a century and a half before, Thoreau warned that people were in danger of becoming "the tools of their tools." As Americans become more captivated by their wondrous new technological and electronic tools, it is an open question whether they are controlling their tools or their tools are controlling them.

· · · *Chapter* · · ·

7

Dream Factories

*"That's what show business is for, to prove that it's not what
you are that counts, it's what they think you are."*

I n our age of celebrity, when everyone dreams of being famous,
David Geffen's American experience can top all the dreamers.
He's a celebrity, though he doesn't like to think of himself as
such, courted by countless other celebrities whose faces are far
more familiar but who possess infinitely less power.

By the end of the Nineties, Geffen stood at the apex of the entertainment industry. When he joined forces in 1994 with two
other Hollywood powerhouses, the film director and producer
Steven Spielberg and the animator and former Walt Disney Co.
studio chief Jeffrey Katzenberg, to form the multimedia production studio DreamWorks SKG with the hardly immodest goal of
creating "the digital entertainment studio of the twenty-first century," PR agents hailed this "merger of moguls." Their hype machine became even more active after DreamWorks announced
plans to build an $8 billion state-of-the-art studio, the first such
new studio in Hollywood in sixty years—a dream that was not fulfilled amid the changing economic conditions at decade's end.

Geffen at that point was already firmly established in the public mind as a preeminent Hollywood power; in the gossip and entertainment items published or broadcast about him, he was invariably referred to as the "billionaire dealmaker," the "music industry force," the "Zelig of the entertainment world," or the "entertainment mogul." He was all these and more, but "mogul" best describes him literally and psychically. His life is a testament to the oldest American dream, the rags-to-riches rise of a child of humble immigrant stock to a position of great wealth and influence.

That very dream drove Geffen from the streets of Manhattan west to Hollywood. Indeed, Geffen could easily adapt his own story to film and finance and produce a familiar up-from-obscurity Hollywood tale of an American Everyman struggling and achieving success through his own wits and determination.

Geffen was born in Brooklyn, the son of immigrants. His Ukrainian mother, Batya Volovskaya, had fled the Russian pogroms before the Bolshevik revolution, emigrating to Palestine where she met Abraham Geffen. After their marriage, in the depths of the Great Depression, they moved to the United States. They were poor when they arrived—his mother found work making brassieres; his father struggled to find anything rewarding—and remained poor. As a boy, Geffen was taken by his parents on weekends to see double features on Forty-Second Street, just off Times Square. Quality entertainment was not the lure; the movie tickets were cheaper there. He began frequenting the theaters on his own. One day, on his way to the movies, he stopped by a secondhand bookstore on a Forty-Second Street subway landing and bought a book for fifty cents: *Hollywood Rajah,* a biography of the Hollywood magnate Louis B. Mayer, a cofounder of Metro-Goldwyn-Mayer.

Young Geffen had never heard of Mayer and knew nothing about people who ran Hollywood studios. He read the biography. "Wow," he thought, "that's the coolest job. Wouldn't it be great to have a job like that?" He mentioned his new dream to his parents. "Don't be ridiculous," they said. "What are you going to do in Hollywood?" Geffen dropped the subject, but not the dream. As he became increasingly dissatisfied with his life, he thought, "If I have to take the subway to work every day of my life, I'll kill my-self." He vowed not to remain in New York because, as he ex-plained, "I hated it. So I graduated from high school, moved to California, and had a series of horrible, horrible jobs."

Then, in keeping with a Hollywood happy-ending plot line, Geffen got one of those fabled "breaks." Film lore notwithstand-ing, stories about stars and starlets being discovered innocently eat-ing ice-cream cones in Norman Rockwell Main Street parlors are mostly creations of the press agent's artifice. In Geffen's case, the break was real. An acquaintance got him a job as an usher at the CBS Television Studio in Hollywood. "The very first day I went to work I was an usher on the *Judy Garland Show* with Ethel Mer-man and Barbra Streisand as guests," he recalls. "My first day on the job, and I thought this was the greatest thing I'd ever seen. I'd pay *them* to let me be an usher there. So that's how I started."

Geffen never looked back. He made entertainment business connections and used them to hustle and claw his way to the top. Like another future Hollywood megapower, Michael Ovitz, he worked in the mailroom of the William Morris talent agency and from that springboard, Geffen became a talent agent, manager, and investor, all in the exploding music business. Step by step, he moved up. By the Sixties, while still in his twenties, he was man-aging the careers of some of the biggest names in the music busi-ness—such stars as Joni Mitchell, Neil Young, and Crosby, Stills, and Nash. By 1970, having forged a reputation as a shrewd and tough bargainer—the consummate big deal maker of the entertain-ment business—he had founded Asylum Records, with a roster of artists who were the major names in music at that time, including that icon Bob Dylan. In what became a familiar entertainment-business pattern, more deals and mergers followed: Asylum was bought by Elektra, which was sold to Warner, which formed the hybrid WEA. In the process, Geffen's entertainment role ex-panded. He had his own film company, rose to vice chairman of Warner Brothers Pictures, and then to executive assistant to the chairman of Warner Communications. By 1980, when he founded Geffen Records, he was presiding over a great entertainment empire and exercising a major influence in film, in theater (where he was a key financier of *Cats*, the most successful musical in Broadway history) and in the recording careers of more stars, like John Lennon, Guns N' Roses, Elton John, Don Henley, Whitesnake, and Aerosmith. As their careers soared, and their personal wealth increased, so did his. In 1990, then in his late forties, he sold his business to MCA, winding up with nearly three-quarters of a billion dollars in cash. From there he rode the great stock market boom of the Nineties, watching as his fortune, and his entertainment-industry influence, multiplied.

Along his way to the top, David Geffen realized a prince-and-the-pauper dream that applied especially to Hollywood and Geffen's version of an entertainment-industry mogul.

While living in a small house on Angelo Drive in Beverly Hills, his route home took him past the huge closed gates and the high wall shielding the grounds of the most fabulous of the great Hollywood estates, the place where Jack L. Warner had lived lavishly and entertained royally in the years he was the dominant force in Warner Brothers. Geffen, passing those gates daily, was starstruck. He imagined what it must be like inside that closed and guarded compound, and dreamed of being there.

One day as he and a friend were driving by the estate, they noticed the gates were open. "Let's drive in and say we've made a mistake if they stop us," Geffen said impulsively. They did. Immediately they were surrounded by guards. Police were summoned. Ignominiously, they were thrown out. Geffen never got to see anything beyond the entranceway to the gates and the curving roadway which trailed off through acres of manicured grounds leading, somewhere up there, to the crest of the hill and the great mansion at its very top.

One morning, much later, Geffen read that Jack Warner's widow, Ann, who lived in the mansion, had died. Geffen immediately called her lawyer, whom he knew personally and dealt with professionally through his own Warner business role. He had always wanted to see the mansion, Geffen explained. He had always wanted to get a better insight into how the people who invented the movie industry lived during the peak years of their influence. Geffen wasn't interested in buying the property, he quickly explained; he merely would like to see what it was like. Of course, the lawyer instantly invited Geffen to join him for a tour of the grounds and the 13,600-square-foot mansion.

Geffen was not disappointed. "It was amazing," he says. After thanking the lawyer, and again saying he wasn't interested in buying the property, he left.

The next day Geffen received a call from Barbara Warner Howard, Jack Warner's daughter. The family *really* would prefer to sell the estate to someone in the entertainment business, she explained, someone who would keep the property together instead of selling it to a developer who would subdivide it, destroying its special character and place in movie history. If Geffen was interested, she said, speaking on behalf of the family, he could have it with all its original art and furnishings included.

Geffen's highly attuned business-deal antenna vibrated. "Hmmm," he thought. "Maybe there's a deal here." As he later recalled, "I got caught up in the deal of it all."

Soon he received another call, this from a movie-producer friend.

"I heard you saw the Warner House," the producer said.

"Yeah," Geffen replied.

"I know everything about that house," the friend went on.

"Really?" Geffen said.

"Yeah. Can I see it?"

Geffen made another appointment to inspect the property; this time he was accompanied by his producer friend. As they toured the mansion, with its Tara-like front columns and its grand rooms and vistas, its sweeping staircases and private movie-viewing room, the friend began pointing out its magnificent appointments.

"See this wallpaper," he said. "This is from the Imperial Palace in China. They paid $150,000 for it in 1932. It must really be worth a fortune now. See over there? That's a Chippendale."

On and on, he continued to point out the priceless furnishings in the great mogul's mansion high in the Hollywood Hills overlooking the entertainment capital of the world.

As they continued their inspection tour from room to room, floor to floor, Geffen was thinking, "Oh, great! I'm going to steal this thing. I'll sell everything in it and get the property for a bargain." He bought it for $47.5 million, and, as he says wryly, "It turned out to be one of the dumber things I ever did."

As soon as the sale was completed, Geffen brought another friend, an expert on antiques and furnishings, to inspect the property and assess what to do with its belongings. "I gave her the same tour," he recalls. "I said, 'This wallpaper was from the Imperial Palace in China.' She says, 'What are you talking about? It's French wallpaper. 1890. It's worth about $15,000.' So I said, 'What?!?!' Then I said, 'This is a Chippendale.' And she said, 'You're not going to yell at me, are you?' And I said, 'It's *not* a Chippendale?' She said, 'I think it's Warner Brothers.'"

Only in Hollywood.

Geffen sold everything in the house, watched as the real estate market collapsed, then spent eight years renovating and restoring the mansion. Finally he was satisfied. He had transformed it, he says, exposing his deep competitive side, into something "infinitely better than it was when Jack Warner had it."

Looking back, Geffen has no regrets at all his effort and expense. In the end, owning that mansion became a fulfillment of a dream. "It was a creative endeavor, and I could afford it," he says, "and I wanted to honor what the house meant to me, what the idea of those Jewish immigrants who came to California and invented the movie business meant. So I wanted to honor the house. It's a beautiful house, and I feel privileged to own it."

It is a privilege that he takes pride in sharing with famous names—presidents and power brokers, stars and financiers—whom he invites to private functions there.

He does so while harboring a secret. "What the hell am I doing in this big house?" he says, smiling. "I'm a single guy. I actually live at the beach [in Malibu] in a small, little house, and I come here for meetings, for business or dinners. I don't even spend much time here. But I do have my art here."

His modern art collection ranks as one of the finest: works by Jackson Pollock, Willem de Kooning, Jasper Johns, and others, all world-famous paintings, all arranged for viewing in room after room. "These are all de Kooning's, 1960, 1949, 1950," he says, while escorting a visitor on a tour of the great collection. "Look at this one. It's really gorgeous. And here's another Jackson Pollock. This was the room with the wallpaper. They used to have dinner parties for thirty and forty people. The room was so big that I built a library in it to make it smaller."

As for his preference for modern art, Geffen offers this explanation: "This reflects the art of the time I was alive. That's what interests me."

Egotism aside, no one witnessing the pleasure with which he discusses his art could doubt the sincerity of his passion. A student of Hollywood lore, however, could be excused for suspecting that another motive fires his desire to acquire and display great works: He sees them as emblems of his discriminating taste, proof he possesses the high culture that befits a modern Hollywood mogul. As Bogart says in the penultimate line of *The Maltese Falcon,* they are the stuff that dreams are made of.

· · ·

For nearly a century American culture, as produced and distributed by the entertainment industry in Hollywood, has been the greatest export of America to the world. In virtually every country, movies are *American* movies, music is *American* music, television is *American* television. However one defines American culture, Hollywood's influence in shaping it has been surpassing. Over the years, the picture that Hollywood presents to the world becomes *the* American portrait, and no matter that it's invariably based more on myths than on realities.

The American types spun out of the Hollywood dream factories have changed notably in look and in character: from the strong, silent, noble cowboys and spunky damsels in distress of the silent film era to the wisecracking, cocky, but endearing characters grinning and bearing it as they sing and dance their way through the Great Depression to the selfless, courageous men who go to war to preserve the "American Way of Life" and their loyal, loving sweethearts who sustain them on the home front of the

Forties, and on through the succession of heroes and villains, good guys and bad guys, gangsters and molls, innocents and depraved alike that have followed. They have all affected how Americans think of themselves, and how the rest of the world thinks of Americans. And through those decades, Hollywood has come to symbolize all the attributes others long for: glamour, fame, wealth, beauty, sex, fashion, and the luxurious lifestyles that naturally go with them.

From the initial small studios that relocated from the East to spring up in what had been a tiny suburb of Los Angeles, giant film factories proliferated. By the Twenties Hollywood had become, as the screenwriter and wit Wilson Mizner put it then, "not a place" but "a state of mind"—a mythical land where stars and magnates lived the fabulous lives others dreamed of living in mansions and beach bungalows spreading far beyond Beverly Hills and Malibu through the canyons crisscrossing Los Angeles. Hollywood then, as Richard Griffith and Arthur Mayer write in their sprightly history, *The Movies,* "was garish, extravagant, ludicrous, acquisitive, ambitious, ruthless, beautiful—which was just what its world public wanted it to be. Its very unreality was protective of the illusion. Dream worlds are not supposed to be lifelike."★ Still, to the great majority of those far more innocent Americans, the lives of Hollywood's beautiful people were most real and most desirable. They were lives to be envied and emulated. After all, were they not surrounded by the trappings of luxury complete with princely salaries, midnight pool parties, Dom Perignon, and seductive, willing stars and starlets? Of course they were.

Or so went the myth. However false, it was a fable Americans hungered to believe.

At the peak of its power as film capital of the world, through its profusion of epics and blockbusters and romances and adventures and musicals and mysteries and westerns, Hollywood put its stamp on the American mind as nothing before it had. It was the

---

★Their book was first published in 1957. In re-reading it for this work, I found it interesting that the theme of "dream worlds," later turned into "DreamWorks," was so familiar a metaphor for Hollywood.

great American storyteller, and the American story Hollywood told created a common picture of the country in which every American, even the most oppressed member of the black minority, shared. Silly, saccharine, naive, unrealistic, even untruthful though most of the movies were, with their obligatory happy endings and their one-dimensional characters who were depicted either as saints or sinners, they nonetheless painted a picture of an America that was reasonably faithful to the idealized one most people carried in their heads: Optimistic. Patriotic. Proud. Energetic. Courageous. Fair-minded. Democratic (with a small "d").

Though filled with double-entendres and openly sexual and violent in content and manner, Hollywood's films were produced under a strict code that supposedly protected the morals of the country—and which met the objections of local censors and religious groups nationally. Under the self-policing imposed by its own appointed "czar," Will Hays, a political hack who was postmaster general during the corrupt Harding administration, Hollywood took the high ground—and produced, among others, biblical epics whose storylines called for portrayals of violent combat and much bare flesh but in which good always triumphed over evil.

In the early days of the Great Depression, facing more criticism and threats of boycotts from church-backed "legions of decency," Hollywood adopted a Motion Picture Production Code governing all films produced by members of the Motion Picture Association of America. Under it, Americans were noble; women were happiest at home, always hunting husbands while at work; minorities, for the most part, were not unhappy or exploited: they could even be viewed as noble savages; Asians were cunning, slant-eyed, and treacherous; Hispanics were nonexistent, except as paunchy, murderous *banditos* and sullen Rositas displaying ample cleavage; African Americans were deferential, if not obsequious, happiest when singing and dancing. *My Mammy!* Men and women weren't filmed sleeping in the same bed. Abortion, incest, sodomy, homosexuality, lesbianism, nymphomania, or acts of bestiality or depravity weren't ever mentioned, except in the most peripheral way and with a knowing wink and a nod. Nudity couldn't be shown, and certainly not frontal or—horrors!—total nudity. Whores that filled the bordellos in every dismal dusty cowboy town weren't prostitutes; they were lovable "dance-hall girls." If a prostitute was presented as such, but never called that, she had, of course, a heart as good as gold. Even mention of sin was a sin; in 1934, censors forced the change of a Mae West film title from *It Ain't No Sin* to *I'm No Angel*. Profanity was banned. When Clark Gable was permitted to tell Scarlet, after intense debate, "Frankly my dear, I don't give a damn" in the closing moments of *Gone with the Wind*, realists hailed it as a breakthrough and moralists denounced it as further evidence of national decline. As late as 1959 Chicago censors attempted to suppress mention of women's panties and the words *rape, semen,* and *contraception* from the film *Anatomy of a Murder*. Otto Preminger, the director, refused to delete the offending references and won his case in court.

As Hollywood matured, however, its portrait of America broadened, deepened, and darkened. Films explored, however timidly, such hitherto taboo areas as anti-Semitism, *Gentleman's Agreement*; religious demagoguery, *Elmer Gantry*; racism, *Home of the Brave*; sexual exploitation, *Sadie Thompson*; political corruption, *All the King's Men*; alcoholism, *The Lost Weekend*; psychosis, *The Snake Pit*; military madness, *The Caine Mutiny;* alienation, *Rebel Without a Cause*; police violence, *Detective Story*.

In the process, Hollywood largely ignored one of the greatest American stories—itself. While novelists addressed the increasingly influential role Hollywood was playing in shaping American attitudes and values—Budd Schulberg's *What Makes Sammy Run,* F. Scott Fitzgerald's posthumously published and unfinished masterpiece, *The Last Tycoon,* Nathanael West's *Miss Lonely Hearts* and *The Day of the Locust*—Hollywood shied away from the subject it knew best. But not always. On occasion, Hollywood dealt, tentatively but memorably, with its own culture and its larger-than-life celebrities and stars. A notable example was the first filming of *A Star Is Born* in 1937, followed four years later by Orson Welles's great *Citizen Kane* and its account of a most powerful media baron whose pernicious influence extended into the entertainment capital itself.

The morality codes were still in existence in the immediate postwar era, but they didn't stop production of a classic look at Hollywood's on-the-make hustlers and neurotic egomaniacs who combine to produce tragedy. In the 1950 production of *Sunset Boulevard,* still perhaps the best movie about Hollywood, the young hack and gigolo, Joe Gillis (William Holden), says to the delusional, faded old star, Norma Desmond (Gloria Swanson): "You were a big star once." And she replies: "I'm still big. It's the pictures that got small."

And pictures were what Hollywood provided, celluloid, not video, pictures. For Hollywood was still the unrivaled *film* capital of the world. It was at that very point when Hollywood's supremacy was severely threatened.

The advent of television sent shock waves through Hollywood. What would happen to box-office receipts when television offered to bring around-the-clock entertainment into everyone's living room free of charge? Not since the technological innovation of talking pictures had Hollywood tycoons been so shaken. Once again, however, they demonstrated their ability to adapt. Ultimately, they won the battle with their new electronic competitor by a simple technique. They co-opted it. The major studios retooled and began producing the bulk of America's television fare on their own back lots. They also discovered that television offered them another source of quick and easy profits; they could sell their backlog of old films to the TV networks, thus reaping new profits on the almost cost-free reshowing of old movies—and then reshowing and reshowing and reshowing them, adding more revenue with each viewing over the tube.

By 1968, the turnaround was complete. Television was then supplying a third of the film industry's domestic revenues, and its TV business side was exploding faster than its traditional film side. To feed the rapidly growing TV audience, Hollywood produced made-for-TV films exclusively for the electronic market. Revenues soared. In another major change during this period, Hollywood's decades-old morality codes were abandoned. On November 1, 1968, the old code was scrapped. A new, voluntary grading system was implemented. It ranged from G (suggested for general audiences), up to R (restricted) and on to, *beware*, X (persons under sixteen not admitted).

The old Hollywood era was over. In its place was a new Hollywood, bigger and more powerful than ever.

Among many changes that transformed the old film capital, the role of television was transcendent.

David Geffen, as shrewd an observer of the way Hollywood affects America as anyone, describes television's far greater influence on movies this way: "When I was a kid, the movies were inspired by the literature of the previous twenty, thirty, forty years. Now people don't read and the movies are inspired by television as opposed to the reverse when television first came out. Then the movies inspired television. Television has dumbed everything down."

Geffen made those remarks before television had demonstrated how low it could go with the outbreak of the TV quiz shows *Who Wants to Be a Millionaire,* in which guests could win that amount on air, and *Greed,* a program that featured the snapping of crisp dollar bills under the noses of contestants. These became such immense instant hits that they spawned other imitators. Their success ignited a new quiz-show mania on television, all of which perfectly reflected the money-mad character of the Nineties.

Even worse TV fare—if that were possible—was to come with such prime-time "reality" shows as *Survivor* and *Big Brother* attracting tens of millions of viewers and celebrating the lowest of low programming. In keeping with the ethos of the Nineties, the winners became millionaires by demonstrating their ability to form alliances, then betray, cheat, and turn on each other in order to outlast their fellow contestants, who in the game were supposed to be members of the same tribe, or family.

Perhaps the dumbest show of them all was a much-hyped special, *Who Wants to Marry a Multi-Millionaire?* In that garish spectacle, fifty young women competed in beauty-pageant style to marry a mystery multimillionaire hidden from their sight behind a large screen stationed appropriately on a Las Vegas stage. This prime-time Fox network show, combining all the worst elements of the TV culture, attracted nearly 30 million viewers. The denouement provided even more fitting elements for an era in which "making it," and chutzpah, were most prized character traits. The husband, "Rick Rockwell," not his real name after all, turned out to be a self-proclaimed motivational speaker and bit sitcom TV actor with a history of being a self-promoter. Once, he had told jokes for thirty straight hours in order to get into the *Guinness Book of Records*. He also had been the subject of a restraining order as a result of charges that he physically threatened a girlfriend. Even his supposed great wealth became, upon closer scrutiny, questionable. The bride, Darva Conger, an emergency-room nurse, won the contest after performing a quivering little soliloquy into a handheld mike in a scene witnessed by millions of Americans. "If you feel that I am the perfect woman for you and you choose me to be your bride," she said soulfully, dressed like other finalists in a long pure-white wedding gown, veil, and train, "I will be your friend, your lover, and your partner for throughout whatever life has to offer us. We'll have joy, maybe a few tears, but more ups than downs. And you will never be bored."

Rick chose Darva, knelt, proposed, kissed and wed her on stage, and off they went on their sponsor-paid honeymoon cruise. It lasted barely a week. Turns out each had signed a prenuptial agreement stating that either could annul the marriage if they chose. Turns out, too, they didn't, as it was delicately put in subsequent interviews, "consummate" the marriage; they occupied separate cabins. Then came the inevitable: each appeared on round after round of infotainment shows to tell their side of a sad story of unfulfilled love. Darva couldn't understand what she had been thinking, she maintained. But she intended to keep the diamond ring and the sports utility vehicle that were part of the sponsor's package to the happy bride. After all, she had new expenses and had been forced to hire a publicist. Maybe she'd tell her full story in a book, she hinted. Later, she showed up at the Academy Award ceremonies in Hollywood, wearing a designer gown and accompanied by her new talent manager. She got her fifteen seconds of fame when she was shown on TV being turned away from the celebrity-filled *Vanity Fair* party after the awards were presented.

Naturally, she also posed nude in *Playboy*—tastefully, of course—
and talked of having a career as a talk-TV host.

Rick, in best boom style, talked about renewing his comedy
career, accepting bookings for lectures, maybe also working out a
book deal. Asked on one prime-time show, *Dateline,* what factors
he considered in choosing Darva to be his bride, he explained that
it was much like buying a computer: "Have you ever heard someone talk about buying a computer?" he asked. "You buy a computer, check them out, find all the features you want and then you
make the best decision you can, based on the amount of money
you have and the best information you have at the time. And you
don't look back because the next day there's going to be another
computer that is more powerful and better. And I kind of tried to
approach my decision on that basis."

Thus, the perfect marriage of Technotimes and its computers
and Teletimes and its celebrity fixation was brought live into every
living room courtesy of the entertainment culture and its tabloid
TV arm.

· · ·

By the Nineties, the entertainment capital stood at the center of a
vast corporate state, multinational in character, controlled by conglomerates that combined into fewer and fewer entities and exercised greater and greater power. Disney, Time-Warner, Sony,
Viacom, General Electric, NewsCorp, AOL, AT&T—more and
more big players entered the field. They presided over empires that
would have astonished the old Hollywood moguls.

Amid rapidly changing ownership patterns, old lines that once
delineated individual fields of film, television, and music became
blurred and entangled. Now, the entertainment industry embraced—
and controlled—all forms of media under the rubric of entertainment: motion picture and television production studios; recording
studios and music publishing operations; local radio and TV stations and network and cable news franchises; cartoon, music, sci-
fi, and entertainment networks; general circulation newsmagazines
and sports and entertainment journals; professional sports franchises,

baseball, basketball, hockey, worldwide wrestling; book publishing houses and Internet sites. The list goes on—and on.

None of the entertainment entities stood alone. A movie studio no longer meant only movies; now it encompassed all media. Television no longer meant only the networks; now it encompassed an exploding number of channels offering individually tailored entertainment. Differences that used to exist between the studios had long since evaporated. Hollywood old-timers could wax nostalgic about the day when MGM was known for a certain kind of film, and Columbia and Paramount for another; when Warner's made gangster movies and Universal, monster films. Now that belonged to the musty past. The ownership that financed film, TV, and other productions essentially competed to offer the same kind of fare. The difference was the extraordinary production expenses involved: $50 million for a film, with a $30 million advertising budget. As a result, fewer films were being produced and the stakes involved in getting a winner were greater. But the payoff was also greater; even medium-sized hits generated extraordinary sums of money.

The influence exerted by this empire over what people saw on TV or movie screens, heard on recordings or radios, read in various fields of publishing, and even thought, felt, and formed opinions about was, literally, incalculable, and, as the merger mania continues, it is destined to become more so.

With the old barriers removed, the content offered by the entertainment industry changed dramatically. In the new, anything-goes entertainment world, television and movie screens portrayed violence for violence's sake, brutality for brutality's sake, blood for blood's sake, sex in its rawest and most anatomical form, death most barbarous and disgusting.

Gone were the days when debates raged over uttering *damn* on the screen. They were as remote as the old peep shows of the penny arcades in the early 1890s when movies were struggling to be born. Now, every four-letter word known to man, and some previously not coined, became commonplace on the tube and in the movie theaters—so much so that the frequency of profanity per each six-minute segment became a subject for scholarly tracts in the late Nineties.★

No human function, however biological or gross, was left unexplored—or undepicted. No act, however violent or savage, was left unexamined, with every element of butchery, torture, gore, and mass violence shown. No body parts, however private, were left to the imagination, and all were seen performing, in color, as nature had designed them from the beginning. On film, nothing was out of bounds. On television, scenes of explicit sex and violence that would never have been broadcast just a decade before were broadcast daily into 100 million homes, representing 98 percent of all American households.

Even some of the most sophisticated and talented Hollywood people were appalled at what they were seeing. Gary Ross, the film writer and director, and his wife, Allison Thomas, each prominent in Hollywood circles, and by self-definition political liberals, remember reacting strongly to the latest movie offerings shown at a Toronto film festival they attended in the late Nineties. "Seventy-five percent of the movies were about violent, dark, extreme things—pedophilia, dismemberment," Thomas recalls. "Everyone thinks that's the cool hip thing. In fact, people in the Gen X crowd have developed this defense where they'll say, 'It doesn't bother me,' but it reminded me of talking to fourth-grade teachers after the guy in L.A. shot himself on the freeway over TV. You remember, it was a big scandal when they didn't blip it out of TV coverage. It happened at four in the afternoon when kids were

---

★Profane language was being used once every six minutes on network TV shows, every two minutes on premium cable shows, and every three minutes in major motion pictures, according to a study by the Center for Media and Public Affairs released in March 2000. The study examined 284 TV series episodes, 50 TV movies, and 189 MTV music videos that aired during the 1998–99 season, as well as the 50 top-grossing feature films released during 1998. Researchers identified 4,249 scenes with profane or crude language, including 966 scenes with "hard-core" profanity, such as the "f-word" and the "s-word," as the study delicately put it.

watching TV at home. So teachers were prepared to deal with the reaction of their students when they came to class the next morning. They expected the kids to be basket cases. After all, they had witnessed someone killing himself in a spectacular fashion. And the kids' first reaction in class the next day was, 'No, it was so cool. I mean, did you see the blood flow?' This was their number one response. It was very familiar to them. That was the hip way to react. After the teachers told them they found it so upsetting they had cried, the kids went, 'I cried, too, and I couldn't sleep last night.' After they drew pictures and told stories in the afternoon about what they had seen, all their other reactions came out."

Thomas became so concerned about the nature of what was being shown in the movies and on TV, that she reacted in a way she never would have imagined in years past. "When we had our kids, we unplugged all the TVs because there was no way we could watch with them in the room and have any confidence that what was on was worthwhile," she says. "It's very disturbing. I check with Gary periodically, 'Have I turned into this wildly moralistic person?' We're not religious, but I'm so aggressively protective of my kids in this society. We're so under attack. Families are so under attack. There's so many weird things that come through that TV, and I think it's disturbing."

Carole King, one of the great recording stars of the last generation, who with movie producer-director husband Phil Robinson is part of another prominent Hollywood couple with liberal political views, reacts equally strongly about the entertainment being offered. "Our television, our film, our daily diet of violence routinely desensitize our sense of reality. Reality becomes as something removed," she says. "Directors are trained to deliver very short bursts of sound and action. They're not storytellers, they're visual effects experts and button pushers. They know how to get a visceral response from you."

As Phil Robinson says of the new style: "People aren't interested in a well-told story anymore. Movies have become rides. Like the new generation of rides in an amusement park, you sit in a chair that doesn't go anywhere but it moves in sync with a movie that's projected. So movies have become rides, and rides have become movies. The audience has been conditioned to having a visceral experience, and it doesn't matter if they go to a movie theater or a theme park for it."

Heated debates broke out in the Nineties about the effect all this was having on American culture and the character of the country. Across the ideological spectrum, critics blamed Hollywood's TV and film fare for perverting American values; for contributing to the spiral of mindless violence, especially among the young; for desensitizing people to pain and suffering through presentation of scenes of unremitting grossness, shock, and horror; for exploiting sensation and spectacle out of all proportion to their true significance, as the news business had done with O.J. and other garish scandals; for cheapening, coarsening, and dumbing down the society.

The debate was as old as film, and probably as old as all forms of entertainment. The same kinds of charges had been leveled at Hollywood in decades past when the morality codes were implemented. That didn't mean the current charges were without merit. While many superb TV serials and movies came out of Hollywood in the Nineties—among them TV's *ER, Law & Order, NYPD Blue, The Sopranos, Seinfeld, Mad About You,* and such films as *Schindler's List, The Usual Suspects, Unforgiven, The English Patient, Fargo, As Good as It Gets,* and *American Beauty*—the overall level of entertainment was mediocre at best and, at worst, unworthy of serious commentary. Much of it was as bad, and as degrading, as Hollywood's most vociferous critics said it was.

Congress held hearings. Moralists were troubled. Thoughtful citizens expressed deep concern. Whether motivated sincerely or ideologically, the critics could agree only on the broadest generalization: something was wrong with Hollywood and its entertainment industry. Aside from repeated expressions of concern about the erosion of standards, the lowering of taste, the decline of excellence, no national consensus emerged on what action could—or should—be taken. The critics were left to sputter in outrage and frustration and take refuge behind such familiar, and inadequate, American refrains as: *Do something!* And: *There oughtta be a law!*

In truth, as we have seen, no critics of the entertainment industry were tougher than many of the most respected, and talented, leaders of that industry. They, too, expressed concern about the constant pressures to succumb to base instincts, to attract large audiences by sleazy, formulaic scripts aimed at appealing to the lowest common denominator. There used to be more collegiality among writers, directors, producers, and agents, some of the veterans would say. Sure, cut-throat competition always characterized Hollywood, but even during its most dog-eat-dog days, there used to be more consideration for artistic achievement. Now the competition had become meaner, the financial stakes greater, the personal and professional pressures more intense.

And make no mistake, they would go on to say, it's only going to get worse. That was the signal being sent as control of the industry shifted to fewer, and more powerful, competing entities. Carole King expressed a common theme among Hollywood people when she said, "I'm troubled by the fact that the world has become a corporocracy and corporations are not personally accountable for anything they do, good or bad. I'm troubled by the fact that everything, including the arts, is market driven. The arts traditionally should have a place *not* to be market driven—a place for people to express their hopes and dreams and innermost feelings." She, and others, make the point that the studios used to be owned by people in the movie business. Now they're owned by multinational corporations who have no passion for movies. The same is true for radio and music and television. As one prominent producer put it, when speaking of the effect of the mergers and consolidations sweeping Hollywood, "If you're NBC and you're owned by General Electric, or if you're ABC and you're owned by Disney, or if you're a cable company owned by Fox, you can't tell me the pressure to conform to their desire for greater profits doesn't affect what you produce."

It was hardly surprising, if no less shameful, then, that networks cut back their news coverage of the Republican and Democratic presidential conventions in the summer of 2000. Nor, shamefully, was it surprising that two networks, NBC and Fox, decided not even to broadcast ninety minutes of the first presidential debate televised that October. NBC opted to broadcast a baseball game over its network that night; Fox gave its viewers a prime-time enter-

tainment program. That decision of the two networks, coming in
the midst of the closest presidential election in forty years, meant
that the number of Americans who watched the debate sank to
the lowest ever, some 40 million fewer than watched the first
Kennedy-Nixon TV debate in 1960. The reason for this abroga-
tion of corporate good citizenship? As they say, a no-brainer: desire
for profits checked any sense of rendering a public service over the
public airwaves.

. . .

However much the making and the content of films had changed,
they still remained basically the same business. That was not true of
television. "I don't think there's anything as explosive as television
in its expansion since the advent of cable," says Ted Harbert, for
twenty years a top entertainment TV executive for ABC before
taking over TV production for the new DreamWorks combined
operation toward the end of the Nineties. "Now there are hundreds of television channels. The good news is that in many ways
television is better. The bad news is that in many ways television's
worse. The best of television is better than it ever was. But now we
have things that are so much worse than we could have imagined
in the Sixties and Seventies, things that I think do real damage to
the society."

Harbert recalls, with a certain sense of awe, the halcyon days
when ABC's entertainment division became the number one network through the success of such shows as *Happy Days* and *Lav-
erne & Shirley.* "I remember getting the overnights," he says.
"There were just three markets of overnights then, New York,
Chicago, and L.A. Now there are twenty-five or thirty overnight
markets that you can get reports on each morning. The networks
had 98 or 99 percent of the audience then. You couldn't help but
make money because there was such a huge audience even if you
had a failed show.

"The guys that created that entertainment business, the Gold-
ensons, the Paleys, the Sarnoffs, they were businessmen, but they
really loved the shows. They were really willing to put on what
they thought were 'good shows.' Any artistic endeavor has to

come from a vision, from someone's passion, from someone saying, 'This is good, let's do it!' That's true for music, books, everything."

Attention to the bottom line certainly was critical in that era, but so was a willingness to take risks and even reject the findings of focus groups so vital in determining what airs—and what doesn't.

People in Hollywood love to tell of those times when the artists triumphed over the telemarketers and produced shows that went on to become fabulous successes. That, supposedly, was the case with such TV programs as *Hill Street Blues, All in the Family, Taxi,* none of whose pilots fared well in tests with focus-group audiences.

That's so unusual now that the occasional example of a show being produced, to say nothing of becoming a hit, after flopping in audience test groups becomes an object of instant legend, passed on in conversations as a morale-building example that talent still sometimes triumphs over crass materialism. Harbert cites the enormous hit *Seinfeld* as being the rare counterexample to the prevailing trend. In the first pilot test results for *Seinfeld,* he says, nobody liked it. Nobody liked the character of Seinfeld. Nobody liked Kramer. "Nobody liked any of it," he recalls. "It was never going to be a hit. Terrible. Terrible. Terrible."

His point being that in the old TV network era the businessmen who ultimately made the final programming decisions were willing, on occasion, to take those risks. By the Nineties, when the cable channels began proliferating, raising the prospect of multiple avenues for new profits, they became irresistible targets of takeovers. As Harbert says, "When they got to be such good businesses, they got taken over."

The new TV ownership breed took its place alongside the new Hollywood brand, and often became part of the same overall controlling entity. By and large, the old-line businessmen who were willing to take risks to produce shows because they liked them were gone. The nature of programming changed with their departure—a new development, but part of an old story.

In Hollywood, as in all forms of public entertainment, there always has been, and always will be, a fundamental conflict between the purveyors of Art and Commerce. Hollywood's leaders recognize, better than their critics, a larger truth about the nature of, and interaction between, the entertainment industry and American society. As David Geffen says pointedly, "The entertainment business reflects the culture. That's all it ever does. It reflects the culture."

In the Nineties, there was no question which side was winning *this* cultural war. It's Commerce over Art, hands down.

· · ·

However Hollywood has changed, some things remain constant. One mordant remark, cited many times in separate conversations with producers, directors, and writers as an example of the entertainment capital's underlying character, calls to mind the looking-out-for-number-one traits that motivated ruthless, backstabbing Sammy Glicks of legendary Hollywood past: *They don't root for you here until you get cancer.*★

A second saying, also heard often, conveys a truth as enduring in Hollywood's present as it was in its past: *Give me numbers, that's what success is.* Numbers, that is, TV ratings and box-office receipts.

In the new entertainment culture, the bottom line—what generates the greatest profits—matters even more, and not only in Hollywood. During the boom, the bottom line was enshrined as America's dominant ethic. "All of the culture is controlled by this obsessive need to deliver to Wall Street a profit statement showing that this quarter was larger than the last," says Norman Lear, one of

---

★Personal note: Although I never worked in Hollywood, a bit of this classic hyper Hollywood style became a favorite family story. In 1949, my father sold the film rights to his *New York Sun* newspaper articles, *Crime on the Waterfront,* for which he won a Pulitzer Prize, to Hollywood. When the first of what became numerous scripts was circulated among potential producers, the agent who handled the deal called up excitedly to gush: "Malcolm, it's the greatest! It's gonna be a smash!" Days later, after the material had circulated among more producers there, the same agent called back. This time, he shouted into the phone: "Malcolm, it stinks from coast to coast!" Five years later, after much travail, the movie *On the Waterfront* was finally released. Hollywood notwithstanding, and thanks to Budd Schulberg, Elia Kazan, Marlon Brando, Leonard Bernstein, Boris Kaufman, and many others, happily it didn't stink from coast to coast.

the most respected figures in the entertainment business and creator of the classic TV series *All in the Family,* starring that seminally American character, Archie Bunker. "In television, that's a kind of insanity that I've always equated with sweeps week. Sweeps week is the week when every independent station across the country or every network station, first for its local market then for the national network, tries to do something extra. If it's a story on rape, we'll do it. If beating women, or abusing women, catches the eye, we'll do that. The independent stations will do five nights of investigations of this or that or the other thing—always about sex, violence, or abuse—to catch the ratings. When sweeps week is over, they'll do marketing based on that week and bring in numbers for their spots based on that week. Then they'll go back to old programming, which doesn't have the sensationalism of sweeps week. And I think, 'What fucking idiot sold this to us?' You can't talk to any Madison Avenue guy, or any sponsor, or *anybody,* and have them disagree with you. They'll all look you in the eye and say, 'Lunacy. It is lunacy.' But they're in this trap. Why? I don't know, except that the name of the game is a profit statement larger than the last, all at the expense of every other value. It's great for the economy. It's great for the world of conglomeratizing and everything else. But is it insane? Nobody will tell you it *isn't* insane. It's fucking insane."

It's the same with the rapidly proliferating cable TV shows. "They were all started for the sake of that bottom line," Lear says. "They deliver as inexpensively as they can to make the most money to satisfy that same bottom line. So they rationalize what they're doing by saying, 'This is what people want, so this is what they get. We've got to give them what they want.' That's bullshit! We've got to take some responsibility for leadership. I will stop with the next fellow to look at an accident. And I'll laugh with the next fellow at someone falling on a banana peel. I'm prone to all the same bullshit. But there's also enough evidence that people can be weaned away from that crap to do better things. But not in the short term, not in the short term."

When it comes to targeting the audience, there's no mystery about what the advertising sponsors want. They're after youth. Therein lies a paradox. America is increasingly an aging society, with far more citizens at the upper age level than those at the lower, but advertisers place greatest emphasis on targeting youth through products and programs that appeal most to them. "The advertisers care about one thing," Harbert says. "It's better if you're affluent and have money to spend on advertised products, but it's more important to the advertiser to get 'em while they're young. Procter & Gamble and Johnson & Johnson can prove to you that women will make their decisions for life about what packaged goods they're gonna buy, and packaged goods practically run this entire business. They want you buying Tide and Campbell's soup. If they're advertising football, they get men to buy beer and tires. They want to get the young men. I'm only forty-three so I'm still in the classic eighteen-to-forty-nine market age group. They don't really care about me. They know my habits are pretty well set. If they can get me, fine. The car makers, yes, but CBS's plea to Madison Avenue that, hey, look at our older audience and all the disposable income we have, just doesn't matter. They don't care. They just don't believe you're gonna change brands. And advertising is about getting people to make a choice."

Something else sells—celebrity—and the entertainment industry dishes up celebrity in ever-greater quantities. "Jesus," Norman Lear sighs, "think what we've made of celebrity. *People* magazine comes along and suddenly there are eleven clones of it. *Entertainment Tonight* comes along, and there are two dozen clones of it."

As David Geffen says, "Celebrity has become virtually the most important thing in everybody's life. People are obsessed with famous people. People want to be celebrities more than anything in the world. So at the end of the twentieth century we're in a culture of celebrity." Recalling the Warhol adage about fame, Geffen says "Warhol turns out to be the great genius of the last half of the twentieth century. What he said about celebrities turns out to be infinitely more important even than his paintings, which are important because he understood the iconic value of certain things— of Campbell soup cans, and Coca-Cola bottles, and disaster. He did paintings of electric chairs and car crashes. He understood people's fascination with death and celebrity. So that's where we are at the end of the twentieth century, and it's only going to get worse."

America's hunger for celebrity and Hollywood's capacity to capitalize on it through film and movies and TV and sports and gossip are reminders of the fears expressed years before by the writer Norman Cousins. Surveying the scene during the early rise of television, he worried that the pablum it offered the public would lead to the trivialization of America. In the celebrity-crazed Nineties, an entire culture was being trivialized.

· · ·

In our increasingly fragmented culture, as we have seen, the common bonds that people once shared are either being sundered or cut entirely. Instead of a united society, we live more in a separated society. As we've also seen, more and more people are sharing things only with very small groups—the niche factor again, the targeting of specific demographic groups for advertising, entertainment, and news programming content. One consequence is that Americans are exposed to fewer ideas and experiences.

This cultural phenomenon presents the entertainment industry with a special challenge that extends far beyond the quarterly bottom-line profit reports and weekly box-office receipt tabulations. As Neal Baer was saying to me with quiet passion and obvious concern, "The arts can either connect or disconnect people. Great works of the past, from Sophocles on, guide us, show us our commonality, help us relate to and understand one another. Culturally, they helped people at all ages in all times. They were ways of making sense of things, and of the times. Stories are what kept cultures alive, what kept people connected."

Neal Baer's observations about culture and the arts are neither abstract nor academic. They grow naturally out of his unusual experience. He's a professional storyteller, as well as a Harvard-trained medical doctor who relocated to Hollywood (where he continues to practice by tending patients at Children's Hospital in Los Angeles) under the wing of another Harvard M.D. turned author and film producer, Michael Crichton. In Hollywood, Baer became an award-winning writer for the top-rated prime-time TV entertainment program *ER* a program that springs naturally from Baer's medical experience. The stories he crafts for *ER* have proved an exception to those programs that target smaller slices of the society for their success. *ER* has succeeded in appealing to a broad range of viewers because, as Baer says, it's about good stories, "tough stories about decisions in this high-tech world, but also about the world of life and death. It has sexy actors. It has ethical choices. It's a workplace kind of show, so you appeal in all kinds of areas. In some sense, that's the way our brains work. We use them as templates to guide our lives. That's why the great works of the past guide us and show us our commonality."

For all its success, and Baer's understandable pride in helping create it, he sees *ER* succumbing to the larger trend of the niche-targeted audience programs. "Our show," he says, "which I think is the last you'll see of this sort of phenomenon in terms of viewership, has between thirty and forty million viewers. The numbers are going down slowly. Ultimately, they will go down more."

Then he says, as he continues to muse about the outlook for more broadly based entertainment fare in an increasingly disconnected society, "I wonder if in providing so many choices we've somehow disconnected ourselves. Now with the Internet we have even more ability to disconnect. Soon we'll have fiber optics in place so you can dial up and not even go to a movie theater, which is a very communal activity. There's nothing like going to a movie and sitting with an audience, something that we're sharing. A shared experience."

America's cultural disconnect, of course, exists in many areas other than Hollywood. Some of the starkest scenes of that other side of our separated society lie within an hour's drive from the dream worlds of the entertainment capital.

· · ·

From the street, the fence surrounding the school yard in South Central Los Angeles appears unexceptional, until I look closer and see—nothing. Slatted and painted red, the fence effectively obscures the view of the yard beyond. Even the bright sunlight of a Los Angeles afternoon fails to illuminate what lies inside. Upon approaching the small cement school building beyond the fence's perimeter, I notice something else unusual. Though it's lunchtime, no students are in the yard. They avoid it. The reason quickly becomes evident: the school itself is pockmarked with bullet holes, the result of habitual drive-by shootings by rival gangs in the neighborhood. The slatted fence and the paint are pathetic attempts to shield the children who attend school there. At this public school, students have learned to accept the unacceptable. They have no choice; the world in which they live is a world of gangbangers, where possession of weapons is routine and sudden violence part of the rhythm of daily life. Before they reach their teens, many have seen friends and relatives shot or killed. Their lives are punctuated by funerals. They have learned not to expect much out of life, except what they can take in the brief time they think they have before them.

Inside the school, a group of students—all teenagers, all black—are talking about the violence and the unstable nature of their world in such a matter-of-fact manner that the very casualness of their remarks becomes chilling. They, and countless others like them in America's impoverished inner cities, are the truly disconnected from mainstream American culture.

Off to the side, one of them is telling his story. His name is John; he's slim, wearing a bright red shirt and chinos. He's sharp, quick, and speaks with a figurative shrug when he describes sporadic gang warfare in the neighborhood. Asked if any of his friends have been hurt, he says nonchalantly, "Several killed." Is fear part of his life? "No, I'm gonna die some day. We all gonna die. There's certain things I can do that's gonna rush it. Certain things I can do that's gonna slow it down. Once you go through so much, you get used to it. Know what I mean? Like the first time I ever saw somebody get popped, I was like, 'Dang, he got popped.' I felt for him, but now it happens every day."

Asked if carrying weapons is part of his life, he says, "You have to. It's your life or their life." As for himself, he says, "I'm the kind of person that I kind of keep to myself. People tend to tell me I have an attitude problem. They tend to say I joke too much and I don't take life serious. Probably I don't, because I look at it like you're only here once, so you got to make the best and have fun.

And there's no guarantee I'm going to be here tomorrow. I don't even know if I'm gonna make it. So you just have today."

The conversation continues:

Q: How old are you, John?

A: I'm seventeen.

Q: Where did you grow up?

A: Right here.

Q: Tell me about your background.

A: My dad is, I don't know, I don't know where he ever hang out. I seen him like, I think, my graduation was the last time I seen him. My ma, we pretty cool. I did live with her. My brother—I'm not close to none of my family you could say. I'm just like on my own.

Q: What does your mom do?

A: She doesn't do nothin' no more.

Q: And your father's gone. You don't see him, you don't know what he does?

A: I don't even like him.

Q: Why is that?

A: Because it's hard to—for me being young, I was cool at first with it. But now I don't like him because he did some things. You know, some things that I really, I can't forgive. If there weren't no such thing as hell, I'd have probably killed him. That's the only thing holding me back. Like if I knew I was gonna die and go to heaven even if I killed him, I would.

Q: What did he do that made you so angry?

A: He took like disciplining too far. Just too far.

Q: He beat you?

A: Too far.

Q: What would he do?

A: A normal whuppin', you whup a kid with your belt and send him on in the room. It's over with. But sometimes he don't know what he's doin'. Sometimes he might be under the influence of somethin' and he take it a step—you know what I'm sayin'. There's a scar on my back right now for the rest of my life. That's just—every time I wake up, when I be in the mirror, I'm gonna see that. I'm gonna remember he did it. So, you know, that's nothin' I could forgive him for. So I don't like him.

Q: Tell me about the gang-banging. How important is that to you, the gang?

A: Important.

Q: Very important?

A: Yeah. I don't know, it's fun. I enjoy it.

Q: You enjoy it?

A: Yeah, I get a kick out of it.

Q: What is it about the gangs; why do they fight each other?

A: Just the way things been for generations. Just the gang. It's been goin' on so long, nobody really know why.

Q: That's what I'm trying to understand. It makes you wonder. How do you stop it?

A: You don't. I don't know. Like the dinosaurs. Maybe they die out. An asteroid hit us. Armageddon. I don't know.

Q: I'm writing a book about America today. How do you feel about America?

A: I think it sucks to a "T." For one, I think the government's spending too much money helping other people when we ain't got nothin' ourselves. You read about the education; we's the lowest ones, the lowest ones. There's people that told me they went to other countries. I have a friend out by the airport, he told me they got computers in elementary schools, and we ain't got 'em in some of our high schools. God, you gotta stop that. It's like if we had them, we would do it. But we don't have it, so we don't care.

Q. What does it mean to be an American?

A: Nothin' to me.

The longer we talk, the more John opens up. He wants it to be known that he's not stupid, and clearly he's not. "People probably don't think I pay attention to the news. I read newspapers. But you wouldn't knows if you was to look at me. 'His pants is hangin'.

He's a gang-banger.' But I read. I hear all the bull." Indeed, he is surprisingly well informed about national events and expresses a shrewd and deeply cynical view of news events in Washington, including offering contemptuous opinions about the personages making the news there from the president to the members of Congress.

He talks about Michael Jordan, the one person he admires most, recalls how Jordan's father was killed, and how Jordan went on from there. "I think he got values and whatever," he says. "I got values, but not really, you know." Asked to explain, he says, "Values, but none that mean nothin' to nobody. Probably mean somethin' to me, but the next person don't care."

When asked what he wants to do with his life, John suddenly becomes defensive and more voluble. His words pour out in a torrent.

"I got skills," he says. "I can do things. I can do a whole lot of stuff. My brother worked contractor. I worked with him. I know how to do that. I can draw, do music, write my own music. I can do stuff. I know how to work the music equipment at the store. I can do stuff. You put me in a room with a pencil and a paper and a radio, I'm cool. I don't even need a TV. I be in my own little world. Sometimes I don't like something and I snap. That's my problem. I get mad real quick."

From all this, does he know what comes next? Is there something he wants to do, some job or career he'd like to have, say, in the next ten years? Again, he says how he likes music and to draw. "If I had a choice, I'd like to work like in some—," he begins, then suddenly stops. He glares and says sharply, "It's all about if they gonna hire me. They see my ponytail, you know, they say we ain't gonna give our business to this guy. They need to think a little bit more before they make their judgment. There's a lot of rich people, there's a whole lot of people that are very ignorant to me. They see me and they don't know what they talkin' about. My first reaction is to get mad."

Still, John says he intends to move away from his "hood." "It all depends on what you tryin' to do in life," he says. "I want money, somethin' like that. I like money, and I'm not tryin' to go to the pen. I'm tryin' to stay on the street." He pauses a moment, then says, "I'm goin' from here. I'll probably go to Arizona or Chicago, that's where I'll go. I just know I'm gonna do somethin'. I can't be out here too long because I'm not gonna waste the rest of my life. Nothin's forever, nothin's forever. It's like a circle, you on that way till the end."

An afterthought, a final question: "When you go to school do you study the history of this country? Do you get a sense of what's happening in this country?"

John, who says America means "nothin' to me," then gives an answer that shows how American he really is. Despite all his disadvantages and all the many ways in which he's disconnected from society, whether technologically, educationally, economically, or racially, John expresses an underlying faith about opportunities and the future. In words that could have come just as well from the hottest young Silicon Valley entrepreneur, he says, "We shouldn't be spend so much time dwelling about the past. Worry about the future. Because when we study the past, history repeats itself. If you do somethin', you can change it."

· · ·

David Geffen was in a philosophic mood. At the end of a long conversation ranging over his life and what connection it had to Hollywood past and Hollywood present, he raised a question about the future. He wondered if all people, as they start to reach their fifties, sixties, seventies, and eighties, long for the way it was. "Maybe that's the human condition," he reflected, as he turned back to think about the way Hollywood was. "You know, when I was a kid everyone said, 'Isn't it too bad we weren't around in the Thirties and Forties when Hollywood was really exciting?' And someone said to me last night, 'Maybe they'll say that about the Nineties.' I guess maybe they will. The future will be so pathetic that the Nineties will look attractive because of it.

"Look at the output of any movie studio, or all of Hollywood, in 1938, '39, '40, '41. You can see an enormous amount of films from that period that today you think are quite good. But I don't think ten years from now we'll look back at the output and say, 'Look what great films came from this period of time.'"

And ten years from now, what does he think the state of the culture will be then?

"I think we'll be saddened by how much dumber culture is than it is now," he replied. "Sadly and inevitably that is going to be true. Books will be less well written. Television will become even dumber than it is now, and it just seems to get dumber and dumber. And movies will be less good. Every once in a while there will be a terrific movie, and we'll be astounded by it. We'll celebrate the person who makes it because it will be so much rarer than it once was."

Scandal Times

Chapter

8

Bill's Story

I. "A VERY TOUGH SITUATION"

*"I want to say one thing to the American people. I want you
to listen to me. I'm going to say this again. I did not have
sexual relations with that woman. . . . These allegations are false
and I need to go back to work for the American people."*

Harold Ickes was flying away from the storm that suddenly
broke over Washington after the final edition of the
Washington Post came off the presses early that Wednesday morning on January 21, 1998. Ickes is on his way to San Francisco where he's scheduled to speak later that day in Palo Alto to
the Knight Fellows at Stanford University, and he is more than
glad to be removed from yet another crisis and another controversy
involving the president of the United States.

Ickes already has been through crisis after crisis with this particular president and has reason enough to feel embittered about
him. Over a twenty-five year span no one has a better claim to
being a close political confidante and adviser to both the president
and the powerful first lady; no one has been more loyal; no one has
undertaken more difficult political tasks than Ickes, whose official
title as deputy White House chief of staff in the president's first

term doesn't begin to express the range of his influence and power. However, instead of being promoted to chief of staff, a position he had every reason to expect and every reason to believe he had earned, Ickes was cynically cast aside in a move aimed at pacifying the president's political opponents in the second term. Yet angered though he is at that move—"pissed off," as he says in private—Ickes is remarkably, but typically, tough-minded and philosophic about such shabby treatment. Despite all the president's shortcomings, Ickes still admires him.

It isn't an act, either. Harold Ickes understands the brutal nature of politics as well as anyone; it's a knowledge acquired by long political experience and a knowledge quite literally implanted in his genes. Before helping this president fulfill his political ambitions, Ickes has advised and worked for at least ten other prominent Democrats, among them Ted Kennedy, Walter F. Mondale, Edmund Muskie, Morris Udall, Jesse Jackson, all of whom wanted to be president. He has engaged in some of the hardest-fought political battles of his generation, from civil rights, where he lost a kidney after being beaten by segregationists in Louisiana, to antiwar protests during the tumultuous Vietnam era, when he led students on behalf of the candidacy of Eugene McCarthy in 1968.

If these cumulative experiences aren't enough to toughen his skin, his own family background provides other forms of armor. He is the son of the legendary Harold Ickes, the Theodore Roosevelt progressive who became the ardent New Dealer serving Franklin Roosevelt as secretary of the interior and suffering slights and insults at the hand of that political master. He dutifully recorded them all in his secret diaries which, when published after his death, provided a treasure of insights into the realities of the life and times of that supposedly more idealistic Washington. Harold Ickes, the son, learned his lessons well from that legacy. After knocking about the West as a cowboy herding cattle and roping calves in rodeos and on ranches, after roaming the country restlessly exploring other avenues of American life, after engaging in a series of political causes and movements aimed at shaking the establishment, after completing years later his undergraduate and law studies at Columbia and Stanford, he emerged as a political practitioner with a profound distrust of the ways of Washington.

By the time Ickes finally worked for this president in Washington he had forged a reputation for being tough, able, intense, combative, and profane—and breathtakingly blunt—a reputation that evokes memories of the curmudgeonly character of his father.

All this is past history that January morning in 1998 as Ickes, now a private citizen and lawyer, flies west across the continent to California.

He's well aware, of course, that a new crisis confronts the president. It's all there on the front page of the *Post:* allegations that Bill Clinton has been having a longtime sexual affair with a young White House intern named Monica Lewinsky.

Ickes instantly recognizes the name, recalling her as the flirtatious intern who was sitting near his office, and the Oval Office, during a fateful shutdown of the government led by Republican members of Congress in the winter of 1995. While reading that story after his plane took off from Dulles Airport, two thoughts flashed through his mind. First, only two people know what, if anything, happened: Bill Clinton and Monica Lewinsky. Second, anything could have happened between them. After all, virtually everyone in the White House inner circle knows well that Clinton, as one of them puts it, "always had an eye for the ladies. He loves women; he's a very sexual man."

Upon landing in San Francisco, Ickes checks his office for messages. The president has called. He wants Ickes to get back to him as soon as possible. Immediately Ickes returns the call from the airport.

"This is a very tough situation," the president says. "What do you think about it?"

"I just got out here," Ickes replies. "I'm about to give a speech."

"Well, I really need to talk to you," the president says.

"I can turn around and come back if you want me to," Ickes tells him. "I can get on the next plane and come back."

"No, no, no," the president says. "Wait until you get back. Don't short-circuit your plans."

Ickes promises to return as soon as he finishes his scheduled appointments over the next few days, then continues on to Palo Alto. But he can't escape the sex scandal that instantly consumes the president, the capital, and the nation.

Three times, only hours apart, Ickes watches and listens as the president responds to the allegations in separate media interviews. The first is a televised interview in the White House with Jim Lehrer, moderator of the PBS prime-time program, *The News-Hour with Jim Lehrer,* originally agreed to by the White House before the Lewinsky story broke as an opportunity for the president to set the stage for his annual State of the Union message to Congress and the country scheduled for the next Tuesday night. The Lehrer interview is a disaster. Clinton has a haunted, cornered look; he acts—and sounds—evasive. When Lehrer asks the president whether he had an affair with Monica or asked her to lie, he answers: "That is not true. That is not true. I did not ask anyone to tell anything other than the truth. There is no improper relationship. And I intend to cooperate with this inquiry. But that is not true."

"No improper relationship," Lehrer says, "define what you mean by that."

"Well, I think you know what it means," the president replies. "It means that there is not a sexual relationship, an improper sexual relationship, or any other kind of improper relationship."

"You had no sexual relationship with this young woman?" Lehrer asks.

"There is not a sexual relationship; that is accurate," the president answers. "The—we are doing our best to cooperate here, but we don't know much yet. And that's all I can say now. What I'm trying to do is to contain my natural impulses and get back to work. I think it's important that we cooperate. I will cooperate. But I want to focus on the work at hand."

Later in the Lehrer interview, responding to a question as to whether he asked his close friend, the powerful behind-the-scenes political player Vernon Jordan, to get Monica to lie, Clinton says: "I absolutely did not do that. I can tell you, I did not do that. I was, I did not do that.... Now, I don't know what else to tell you. I've,

I, I don't even know, all I know is what I've read here. But I'm going to cooperate. I didn't ask anybody not to tell the truth. There is no improper relationship. The allegations I have read are not true. I do not know what the basis of them is, other than just what you know. We'll just have to wait and see. And I will be vigorous about it."

In a second interview, with Mara Liasson of National Public Radio, the president continues in that same vein: "I don't know any more about it, really, than you do. But I will cooperate. The charges are not true. And I haven't asked anybody to lie."

"Mr. President, where do you think this comes from?" Liasson asks. "Did you have any kind of relationship with her that could have been misconstrued?"

"Mara," the president says, "I'm going to do my best to cooperate with the investigation. I want to know what they want to know from me. I think it's more important for me to tell the American people that there wasn't [sic] improper relations, I didn't ask anyone to lie, and I intend to cooperate. And I think that's all I should say right now so I can get back to work of the country."

In his third interview that day, with the Capitol Hill newspaper *Roll Call,* Clinton responds to the question, "Was it in any way sexual?" by saying: "The relationship was not sexual. And I know what you mean, and the answer is no."

The next day, Thursday, during a midmorning White House photo opportunity in the East Room with the Palestinian Liberation Organization leader, Yasser Arafat, the president continues his unequivocal public denials. He stands before a fireplace mantel, faces the assembled press pool and television cameras, and tells them and the American people: "Let me say, first of all, I want to reiterate what I said yesterday, the allegations are false, and I would never ask anybody to do anything other than tell the truth. Let's get to the big issues there—about the relationship and whether I suggested anybody not tell the truth. That is false. Now there are a lot of other questions that are, I think, very legitimate. You have a right to ask them. You, that, you and the American people have a right to get answers. We are working very hard to comply, get all the requests for information up here. And we will give you as many answers as we can, at the appropriate time, consistent with our obligation to also cooperate with the investigations. And that's not a dodge; that's really what I've, what I've, I've talked with our people. I want to do that."

He makes them a promise: "I'd like for you to have more rather than less, sooner rather than later, so we will work through it as quickly as we can and get all those questions out there to you."

Three more days of denials continue before Ickes flies back to Washington late Sunday night and immediately goes to the White House where he is expected by the president.

It's nearly one-thirty in the morning that Monday when he enters the East Gate and immediately goes upstairs to the Solarium atop the White House and the presidential family quarters.

The president is just finishing a conversation with his Arkansas friend, the Hollywood producer Harry Thomason, and another close friend, the writer Taylor Branch, author of the Pulitzer Prize–winning history *Parting the Waters: America in the King Years.* Clinton's brother, Roger, also is present; he's walking about the room when Ickes arrives. After Ickes enters the Solarium, they all leave.

Ickes and the president are alone.

"What do you think?" the president begins.

Ickes says he has seen and heard him on those media interviews, especially the television conversation with Jim Lehrer.

"Look, Mr. President," he says, "I'm not your attorney. I don't want a lawyer/client relationship with you, and I don't want to know what the facts are because I'm going to be talking to the press and I don't want to know any more than anybody else knows."

"Well," the president cuts in, "I didn't do it."

"If you didn't do it, and your lawyers are comfortable, that's good enough for me," Ickes tells him.

When the president presses him for what Ickes thinks of his situation, Ickes, in typical blunt, profane fashion, answers.

"You are in a complete nosedive and you're three feet from ground zero. You're about to collapse. I've never seen anything as bad as this. The press doesn't believe you. If you watched yourself on Jim Lehrer, you wouldn't believe yourself. You look like a fuckin' dog that's been running all night and someone just kicked the shit out of you. I've never seen such a performance in my life. Nobody believed you. And there's no reason they should. You've got about three feet to go and then you're gonna hit the fuckin' sidewalk and splatter all over the street.

"Now, if you didn't do it," he adds, "and if your lawyers are okay with it, and you're comfortable with it, you'd better tell the American people, look them straight in the eye, and tell them that you didn't do it. There's no other remedy. Nobody can spin their way out of this one."

They continue talking upstairs in the Solarium, still alone for another hour and a half as the clock inches on toward three in the morning.

The next morning, in the presence of his wife and cabinet officers, the president delivers the strongest, most ringing defense of his innocence in the Monica Lewinsky affair. This time there's no trace of equivocation or dissembling. This time he speaks passionately, intensely, his body rigid, his eyes stern, his voice hard and strong. He begins vigorously pointing his finger directly into the lenses of the assembled cameras while uttering words that ring with righteous indignation, words that become engraved into America's consciousness.

"I want to say one thing to the American people. [Finger wagging forcefully]. I want you to listen to me. I'm going to say this *again*. [Stronger gesture]. I did *not* [Vigorous downward slash with finger] have sexual relations with that woman, Ms. Lewinsky. I never told anybody to lie, not a single time, *never*. [Said defiantly with finger pointing]. These allegations are *false* [Most emphatic gesture of all] and I need to go back to work for the American people."

· · ·

It has been said that twenty-four hours can be a lifetime in the life of a president, and for William Jefferson Clinton, forty-first president of the United States, the twenty-four hours that begin on January 21, 1998, set off a chain of events that change forever the nature of his presidency, affect his historical standing, and leave an imprint on the political life of the nation that will have consequences for years to come. The public lies that Clinton repeatedly and forcefully tells the American people are matched by similar private lies he tells to virtually all his closest presidential aides, including his vice president, to key leaders of Congress, and, not least, to his wife and daughter. Once uttered, he can never really take them back, never quite atone for them and for the agony they cause both his strongest supporters and the citizens he serves, who have entrusted him with the responsibilities of the most powerful office on earth. Having made his choice on what course to follow, either public denial or public acknowledgment and explanation, that first day he repeatedly tells essentially the same story in public and private.

The private assurances he gives to the leaders of his administration and his White House staff create a common front and ensure a common response that will be played out daily for the next seven months before a watching nation and world.

These assurances begin early and continue late.

At six-thirty that morning the president phones his friend, the most influential Washington attorney and political power broker, Vernon Jordan. Jordan, one of the heroes of the civil rights movement, has become the most prominent black figure in the capital, legendary for his highest-level political connections and his ability to broker deals, not least helping place people in need, or in trouble, in public or private sector jobs.

Jordan's in New York on business, asleep in his room at the St. Regis Hotel, when the phone rings. "There's a story out," he hears the president say. "And the story is about Monica Lewinsky and myself." The president quickly adds, "It's not true."

At that point Vernon Jordan is well aware of the Lewinsky story in the *Post*—and of Lewinsky herself and of her connection, if not her affair, with the president. He was awakened little more than five hours earlier, at one that morning, by a phone call from David Bloom of NBC–TV who asks Jordan for a comment on the Lewinsky story just then being distributed in Washington. Jordan, as he later recalls, "was stunned that some damn fool reporter would call me at one in the morning to ask me about anything." He has no comment, Jordan tells the network reporter, and says he's going back to sleep. Then he hangs up. But he's hardly unfamiliar with potential problems Lewinsky poses for the president. In fact, Jordan has been working for months behind the scenes to get Lewinsky a job in the private sector; he even spoke to the president about that job search before going to bed that night, and the day before he was at the White House for a session alone with the president in the Oval Office in which Lewinsky's job status was discussed.

He has also met face-to-face with Lewinsky privately, and from the way she talks about the president concludes "quite honestly that I was listening to a bobby-soxer who was mesmerized by Frank Sinatra, who was quite taken with this man because of who he was, because he was tall and he was handsome and because he was president." After listening to Lewinsky, Jordan asks her point-blank whether she and the president have engaged in a sexual relationship. When she says no, Jordan remembers, "I took her word on that," partly, as he later confesses, because "there's some questions you ask you don't want to know the answer to." He also takes the president's word early that morning that there has been no sexual relationship, again in part because he wants to believe he's been told the truth, and perhaps because he doesn't want to disbelieve a direct assurance from the president of the United States.

That pattern of private presidential denial and private acceptance of the presidential word continues throughout the day.

Before his daily breakfast meeting with his top advisers, the president looks directly at Erskine Bowles, his new chief of staff, and two other top assistants, John Podesta and Sylvia Mathews, as they are just about to enter the Roosevelt Room. The president addresses them. "I want you to know that I did not have sexual relationships with this woman, Monica Lewinsky," he says, in the line he has formulated and will keep repeating. "I did not ask anybody to lie. And when the facts come out you'll understand."

That night, after his round of media interviews and private sessions with other political aides and advisers, he meets in the Oval

Office with another assistant, Sidney Blumenthal, a talented former writer for the *New Yorker* and the *Washington Post*. As one of Clinton's most ardent supporters and someone with influential media contacts, Blumenthal can be relied upon to pass on privately to selected people in the press the presidential side of the Lewinsky controversy.

By the time they meet, Blumenthal already is primed for what will become his role in defending the president.

Earlier that day, Blumenthal met privately with the first lady, Hillary Rodham Clinton, with whom he has forged a close relationship. She is distressed, she tells him, that the president is being attacked for political motives because of "his ministry of a troubled person." The president ministers to troubled people all the time, she goes on, and has done so "dozens if not hundreds" of times "out of religious conviction and personal temperament."

She adds, "If you knew his mother, you would understand it," explaining: "The president came from a broken home and this [made it] very hard to prevent him from trying to minister to these troubled people."

Hours later, when Blumenthal meets alone with the president, he relates his conversation with Hillary Clinton. "I understand you feel this way," Blumenthal tells the president, "that you want to minister to troubled people, that you feel compassionate, but part of the problem with troubled people is that they're very troubled. However, you're president and these troubled people can just get you in incredible messes. I know you don't want to, but you just have to cut yourself off from them."

Clinton responds, "It's very difficult for me to do that given how I am. I want to help people."

Blumenthal persists in offering advice that he hopes Clinton understands is motivated by Blumenthal's obligation to be frank and candid when speaking to the president. "You can't get near anybody who is even remotely crazy," he says. "You're president..."

At that point, Clinton tells him, "Monica Lewinsky came at me and made a sexual demand on me."

The president says he rebuffed her, then confesses to Blumenthal that, "I've gone down that road before, I've caused pain for a lot of people, and I'm not going to do that again." The president says Lewinsky threatened him. She says she'll tell people they'd had an affair in part because among her White House peers she is known as "the stalker," and she hates it. If she has an affair, or says she has an affair, she won't be "the stalker" anymore.

Then the president of the United States tells his aide: "I feel like a character in a novel. I feel like somebody who is surrounded by an oppressive force that is creating a lie about me and I can't get the truth out. I feel like the character in the novel *Darkness at Noon.*"

Its melodramatic aspects and its falsehoods aside, this bit of presidential self-analysis contains a larger truth. Bill Clinton *does* resemble a figure out of fiction, but whether one created to highlight tragedy or portray farce is an open question. In his case, the answer is probably both, for his actions that day launch the nation upon a farcical episode that contains nonetheless genuine elements of high tragedy that will extract a heavy toll, leaving personal and political wreckage in its wake.

There's one notable exception to the president's lies that day, and it's revealing of him both personally and politically. It involves someone who represents the hidden, dark side of his character— Dick Morris, part political savant, part utterly cynical political manipulator.

Morris handled Clinton's first race for governor of Arkansas in 1978, and, in a stormy relationship as pollster and Machiavellian strategist, continued to work with Clinton on subsequent election campaigns in 1982, 1984, 1986, and 1990. When Clinton is in trouble, he instinctively turns to Morris, whose counsel both he and his wife, Hillary, value highly.

Even though Morris has no scruples about working both sides of the political fence, counseling conservative Republicans and Clinton haters and liberal Democrats and Clinton defenders alike, and profiting handsomely from each camp, the Clintons continue to seek his advice in moments of crisis. They turned to him after the shocking loss of the Congress in the 1994 political earthquake that brought Newt Gingrich of Georgia to power, a stunning electoral turn of events that seemed to signal the advent of a new conservative Republican era lacking only control of the White House to make it complete. That change in political fortunes prompted the wise people in Washington to predict the Republicans should easily be able to wrest the White House from a weakened Democratic president in the 1996 presidential election, a view that many members of Clinton's party privately shared.

Morris began working behind the scenes at formulating the strategy that resulted in the surprisingly easy Clinton reelection triumph. Morris was not able to bask in the glow of the victory he helped fashion, however. During the Democratic convention that summer in Chicago, his sordid and highly publicized relationship with a prostitute in Washington's elegant Jefferson Hotel became public in excruciatingly humiliating fashion. Aside from his penchant for indulging a fetish for toe sucking, Morris was shown to have demonstrated his powerful personal connection to the president only four blocks away in the White House by phoning Clinton while simultaneously engaging in sexual by-play in his hotel suite. He quickly gave up his Svengali political connection with Clinton and the first lady, neither of whom uttered a critical public word about him after his affair with the prostitute became known.

Dick Morris is still working in Washington on January 21, and has never heard of a Monica Lewinsky until he reads about her in the *Post* early that morning. He immediately calls Nancy Hernreich, one of the closest of the president's aides, a fellow Arkansan whom others view as something of a stern mother figure who can exert discipline over the undisciplined president, especially keeping him removed from easy temptation. As director of Oval Office operations, Hernreich tightly controls access to the president—so much so that a close Clinton associate privately expresses the belief after reading about Monica Lewinsky that, if true, it can only have happened when Nancy was not present; which, indeed, proves to be the case. Now Morris asks Hernreich to deliver a succinct message to the president: "If he wants to talk, call me."

Not long after, Morris is riding the Washington Metro to work when the White House pages him. He waits until his subway stop to get off and call back.

"He wants to talk to you," Morris is told. Moments later the president comes on the line.

"You poor son of a bitch," Morris tells him. "I've just read what's going on. I've been there. . . . I've gone through it and my heart just goes out to you incredibly."

"Oh, God," the president replies, "this is just awful."

Morris listens intently as the president says, "I didn't do what they said I did, but I did do something. I mean, with this girl, I didn't do what they said, but I did do—did do something."

He repeats this mantra twice.

"Look," Morris tells him, employing the inside argot of the political operative, "you may have to play this thing outside the foul lines."

He means, he quickly explains, transcend the legal process, go over the heads of the investigators, "and go to the public and ask them for forgiveness. Tell them what you did and ask them for forgiveness. There's a great capacity for forgiveness in this country and you should consider tapping into it."

During their conversation, Morris confides in the president and elicits a far-more-candid response than Clinton has given others. "It occurred to me," says the pollster, "that I may be the only sex addict you know and maybe I can help you."

"You know," the president replies, "ever since the election I've tried to shut myself down. I've tried to shut my body down, sexually, I mean . . . but sometimes I slipped up and with this girl I just slipped up."

"I know," Morris says. "You know addicts fall off the wagon. This is an addiction just like drugs or alcohol and you just have to recognize it and fight it."

Morris suggests a typical course of action that underscores the cynical side of politics. Rather than urging the president simply to tell the truth, he advises that they take an instant poll. The president immediately agrees. Once again, Morris puts aside his other business and goes to work privately for Bill Clinton.

At 11:15 that night, Morris receives the results of his poll and calls the president at the White House. The president excuses himself briefly as he switches to another, more secure, phone to talk. Morris tells him that his instant poll results dash the idea for a public confession in which the president asks for forgiveness. "They're just not ready for it," Morris says of voters just surveyed.

"Well, we'll just have to win, then," the president says.

"You bet your ass," Morris replies.

The battle is on. For the next seven months, while Clinton's supporters go forth daily to defend what in the end turns out to be indefensible, the political world and the American people are put through a degrading experience without parallel in the nation's history, one that further disconnects the people from their political process and encourages them even more to focus on the diverting side of the Clinton age, the still-expanding long boom.*

· · ·

Looking back, countless questions keep arising. One is, Why did so many intelligent, sophisticated people follow Clinton so blindly over the cliff, to their own discredit and to the damage of their own reputations?

The answers, while complex, are not that complicated. They all wanted to believe. They all thought the president wouldn't be that reckless in jeopardizing his office and his prospects for reelection. They all thought that, even if tempted, he wouldn't be able to engage in the kind of repeated, extensive sexual conduct that he did right off the Oval Office, literally surrounded, day and night, by a cordon of people whose primary goal was to safeguard him. They

---

*In the first days after Monica, three of Clinton's spin doctors make the following public statements: Ann Lewis, White House communications director: "I can say with absolute assurance the president of the United States did not have a sexual relationship because I have heard the president of the United States say so." Paul Begala: "I'm like Dale Evans with the Bible—God wrote it, I believe it, and that settles it." James Carville, chief spinner: "He's denied it to his staff, he's denied it to the news media, he's denied it to the American people and denied it to his cabinet, and denied it to his friends. Can't be any more emphatic than that." These kinds of statements are made repeatedly over national TV for the next seven months.

all also knew that the president and the first lady had been subjected to a long and relentless campaign mounted by zealous ideological enemies and that none of the many charges brought against them—from Whitewater land deal corruption to firing of travel office employees to misuse of FBI files—had been proved true.

There's another belief, held deeply by those who thought they knew the president best, that he would never be so wildly irresponsible as to take actions that would so devastatingly hurt the people he cares most about, his wife and his daughter.

"Everybody in that White House knew of his eye for women," one of his close associates says. "And I'll tell you I don't think anybody thought that oral sex was being committed. . . . You know why? Because no one thought he could do it without being observed. *Nobody thought he could do it without being observed!* That's why people suspended disbelief. They thought maybe a little hanky-panky. Maybe a smooch here or there. Maybe he was copping a feel. I think people thought he would've done it if he could do it, but nobody thought he could do it. So people were prepared to say, 'Okay, the guy said no. He is the president. He's looked me in the eye and said no.' That's number one. Number two is he couldn't do it back *there*. Then you get to the issue of the recklessness of his actions. Forget how they affect the nation. Okay? Forget the nation. Forget the staff. Think about two people he really *does* love, and he loves them deeply: Hillary and Chelsea. People say he wouldn't have done this because of Chelsea and Hillary. There's another thing people thought when they started putting all these pieces together. [One female staffer] who he's flirted with, puts his arm around her, she told me, 'There's no way he could've done it. You know why? He wouldn't jeopardize his reelection.' That's her rationale. So everybody had their own set of rationales."

They also were highly attuned to making sure that the president's contacts with women were free of any actions that could arouse suspicions of impropriety.

Leon Panetta, who, after a long career as a respected member of Congress from California, became Clinton's White House chief of staff during the president's first term, was hardly naive about how power and sex pose eternal problems in the political world, especially in the world of a president like Clinton. Panetta took extra precautions to see that all senior White House personnel were sensitive to situations between Clinton and women that might raise questions. "I was aware of the rumors and the allegations that involved the president—beginning as governor—with Gennifer Flowers and Paula Jones, and, you know, just the general rumors that had surrounded the president," he says. "As a result we took particular precautions to ensure that there was never the appearance of the president being with somebody so that it could be misinterpreted by the public or anybody else. On trips, for example, if an acquaintance wanted to drive with the president we would say 'no.' If there was a female acquaintance who wanted to greet the president, we would say 'no.'... We wanted to protect the president's office and protect his integrity."

The president, Panetta recalls, "was always very cooperative.... Even if it was an old friend from his days as governor, we would say, 'You know what the problem is, it creates the wrong appearance and it shouldn't happen.' And he would say 'fine.'... He never resisted." That was true no matter how celebrated the person who sought to meet the president. Panetta cites the actress, Barbra Streisand, as an example. If Streisand wanted to meet him at a certain place, he recalls, "and we thought it was not appropriate, we would tell him so, and he would agree."

Panetta's deputy, Evelyn Lieberman, kept a sharp eye on the behavior of young female White House interns, so sharp that she was seen by some of them as an overly stern school principal rigorously enforcing their deportment and the official dress code. Lieberman, an English teacher before entering political life, took pride in her overseer role, recalling how she would rebuke interns for behavior she thought beneath the dignity of the White House and the Office of the President. "Very often the kids would hang around [the Oval Office premises]," she says, "and I didn't like it.... They weren't supposed to be there and I made them go away." She was especially annoyed at the behavior of intern Monica Lewinsky. "I saw her where she shouldn't be," Lieberman remembers, "and said, 'What are you doing here?' She would say occasionally, 'Oh, I'm making a delivery,' and I would say, 'If you don't have work here, go back to your post.' If she told me that she needed to wait, I said, 'Wait some place else.'"

Lieberman also didn't like Monica's appearance. "Her skirts were probably too short," she says. "It doesn't take much more than that to set me off."

She was bothered enough by Monica that she mentioned her concern to her boss, Panetta, and warned Nancy Hernreich that Lewinsky is "a clutch," meaning, she explains, someone who hovers about the president or other White House principals. Finally, after having seen Monica "out of place" five or ten times in the presidential area, Lieberman says: "I decided to get rid of her. I talked to her supervisor . . . and I said, 'Get her out of here.'"

Monica is banished to the Pentagon, where she's befriended by an embittered former White House staffer named Linda Tripp.

· · ·

Once Monica's name surfaces publicly, the familiar pattern recurs as senior staff and administration leaders swiftly unite behind their president, still finding it inconceivable he has lied to them all. As Ickes says, "None of us knew the facts. I didn't press him. I don't think any of us knew the facts until August 17 [when the president acknowledged his relationship with Monica in a televised speech to the nation]. I don't think his lawyers knew the facts."

Like Ickes and Vernon Jordan, they don't press the president for more details or assurances.

"The guy who I've worked for looked me right in the eye and said he did not have sexual relationships with her," Erskine Bowles, Panetta's successor as chief of staff, recalls. "And if I didn't believe him, I couldn't stay. So I believed him." He, and the others, are united in another belief. They don't believe the repeated rumors about Clinton and women that they read about in the press. They view them as part of the continuing political war in which they are all engaged.

This attitude is true of those who know Clinton far better personally and for a far longer period of time. Harry Thomason, for instance. Just before Ickes joins the president in the Solarium in those early hours Monday, Thomason tells Clinton he should have been more emphatic in his TV interview with Jim Lehrer; he needs to make "a more forceful denial." But Thomason gives that advice without attempting to discover whether the president has anything to hide. He never asks Clinton if the allegations about Monica are true, "because that's what I wanted to assume and that's what I actually believed in my heart. . . ."

Of all those who begin strongly defending the president in public, convinced by his assurances that he has been falsely accused, none is more persuasive, more convincing, and, in history's hindsight, more poignant, than his wife, Hillary. In an extraordinary early-morning live television interview on NBC's top-rated *Today* show on Tuesday, January 27, the day the president delivers his prime-time State of the Union address, her words, and most important, her demeanor, probably save her husband's presidency.

Watching her, as millions do, it's hard to see how even the most unforgiving Clinton hater can fail to believe that she speaks out of an absolute faith in the truth of her husband's denials. She is neither strident nor defensive; she strikes exactly the right note, responding easily and fully to lengthy and probing questions from interviewer Matt Lauer.

Later, her appearance comes back to haunt her. It inflicts unimaginable humiliation as she learns her belief in Bill has been tragically misplaced, her words defending him mocked, her responses derided as yet another cynical "stand by my man" performance, her characterization of their enemies both extreme and wrongheaded. But on that morning she transcends the political and speaks in a compelling personal way that touches people's hearts.

She has known her husband for more than twenty-five years, she says, and in referring to their twenty-two years of marriage gives a description that every married couple can relate to: "I have learned a long time ago that the only people who count in any marriage are the two that are in it."

She speaks of his generosity and kindness, adding, "Everybody says to me, 'How can you be so calm?' Or, 'How can you just, you know, look like you're not upset?' And I guess I've just been through it so many times. I mean, Bill and I have been accused of everything, including murder, [by] some of the very same people who are behind these allegations, [so] from my perspective, this is part of the continuing political campaign against my husband."

For years, she goes on, she and Bill have been targets of people motivated by an intense political agenda. In attempting to describe their opponents, she gives a characterization of them that will subject her later to endless ridicule. They are part of what she calls "a vast right-wing conspiracy." Pressed to explain, she says, "This is what concerns me. This started out as an investigation of a failed land deal. I told everybody in 1992, 'We lost money.' People said, 'It's not true,' you know, 'They made money. They have money in a Swiss bank account.' Well, it *was* true. It's taken years, but it was true. We get a politically motivated prosecutor who is allied with the right-wing opponents of my husband, who has literally spent four years looking at every telephone . . . call we've made, every check we've ever written, scratching for dirt, intimidating witnesses, doing everything possible to try to make some accusation against my husband. . . . It's not just one person, it's an entire operation."

In hindsight, this exchange becomes especially painful:

Lauer: Let me take your husband out of this for a second. . . . If an American president had an adulterous liaison in the White House and lied to cover it up, should the American people ask for his resignation?

Hillary: Well, they should certainly be concerned about it.

Lauer: Should they ask for his resignation?

Hillary: Well, I think if all that were proven true, I think that would be a very serious offense. That is not going to be proven true.

She quickly adds: "I think we're going to find some other things. And I think when all this is put into context, and we really look at the people involved here, look at their backgrounds, look at their past behavior, some folks are going to have a lot to answer for."

· · ·

Subsequent events prove Hillary Rodham Clinton's belief in her husband's truthfulness profoundly wrong—and her belief that he

and she have been targeted by a heavily financed conservative conspiracy profoundly correct.

Whether the conspiracy was "vast" or not or whether it meets
the literal definition of "conspiracy" when perhaps "a loosely connected network of enemies" better applies are semantic distinctions
of little significance. The Clintons, in fact, are confronted by a determined group of enemies, heavily financed, operating secretly,
often sharing information, who attempt to discredit them both
through dissemination of information in the media, who surreptitiously collaborate with prosecutors, and who work ceaselessly for
years to destroy the president.

Understanding why the president inspires such hatred becomes
crucial to gaining insight on the political destructiveness that America experiences in the Nineties, for in most respects Bill Clinton is
an unlikely candidate for unremitting enmity.

Personally, he's affable, easygoing, almost always appearing reasonable. Nothing in his background suggests venality, or a hunger
for material gain. Money seems unimportant to him. Until he becomes president, he never earns more than $30,000 a year. He
doesn't even own his own home. Politically, he's not a radical. He's
not the leader of a political movement that challenges the status
quo. On the contrary, he's a moderate, consensus-seeking politician from a small border state whose record as governor shows him
customarily dealing with both Republican and Democratic legislators in an essentially nonpartisan manner. Indeed, many members
of his own party, and especially liberals, fear he lacks boldness and
will be too eager to compromise with Republicans after being
elected president. Even his election poses no threat to the political
order; he comes to office after winning only 43 percent of the
votes cast in a three-man race and commanding the smallest congressional majority of any president elected in the twentieth century. The problem Clinton poses is both more simple and more
complex than any of these factors. The problem is that from the
beginning people simply don't trust him—and his behavior, long
before Monica, confirms that view constantly.

By the time he becomes America's leader, he already is being
targeted as an illegitimate president, viewed by his enemies as per-

sonally corrupt and the head of the most corrupt administration in American history.

Understanding the inherent contradiction posed by the reality of the nonthreatening Governor Bill Clinton of Arkansas and the threatening President Bill Clinton seen by his enemies requires examination of the deep cultural and ideological divisions that exist in the country when he comes to power, and intensify throughout his presidency. In the best political analysis of him, Lars-Erik Nelson* writes of Clinton: "By nature and by circumstance, he was a centrist—a position that, for him, made political sense because, in his eyes, the Republican Party had moved so far to the right. There was a great political middle to be grabbed, and Clinton grabbed it. To the Republicans, however, Clinton was no centrist. He was a pot-smoking, draft-dodging, anti–Vietnam War liberal with a socialist wife who wanted to strip away your right to see a family doctor."

Nelson rightly points to another factor that strongly contributes to the scandal climate engulfing Clinton and the country: the impact of the unbridled ideologues of the electronic culture. "It was Clinton's great, and as yet unexplored, misfortune that he was the first Democratic president to take office since the astonishing rise of the demagogic radio talk-show hosts and their counterparts on cable television," he writes. "A caricatured view of Clinton as a dangerous, even subversive, liberal was broadcast for three hours a day, every day, from coast to coast by Rush Limbaugh and echoed by his small imitators across the country. They questioned his patriotism and his right to be commander in chief."

The relentless scrutiny and continued attacks come close to bringing down an American president, but Clinton's enemies would never have gotten anywhere had not the president recklessly fallen into the trap they set for him.

In the end, Bill Clinton is the agent of his own political fate. As Richard Nixon morosely acknowledged of *his* actions during

*Lars-Erik Nelson's sudden death at the end of 2000, at the peak of his career, with his best work still before him, deprived the nation of one of the most talented, and admired, journalists of the times.

another historic presidential scandal, he gave them the sword and his enemies proceeded to run him through with it.

Even as Hillary and the president's allies are vigorously rushing to his defense, in the top White House circle only one person knows the full extent of the dangers that lie ahead. That is the president of the United States himself.

· · ·

By the time Monica Lewinsky's face and name become emblazoned on television screens and newspaper front pages around the world, Bill Clinton already had become the most investigated president in American history.

Not that other presidencies were free of scandal—or scandalmongers. From the beginning, American presidents have found themselves accused of immorality and corruption. Each of the first three presidents, those alabaster figures referred to reverentially by generations of Americans as the greatest among the Founding Fathers, railed against the excesses and unfairness of the assaults levied against him. Typical of the abuse these men endured were the comments of a Philadelphia paper, the *Aurora,* shortly after George Washington delivered his famous farewell address. "If ever a nation was debauched by a man, the American nation has been debauched by Washington," the paper declared. "If ever a nation has suffered from improper influence of a man, the American nation has suffered from the influence of Washington. If ever a nation was deceived by a man, the American nation has been deceived by Washington."

Washington went to his grave despising the attacks of those seeking to tie him to scandals. His immediate successors, John Adams and Thomas Jefferson, shared his feelings of outrage at being maligned, especially at the hands of political enemies operating through a constitutionally protected partisan press. Adams, after completing his only term, wrote of the performance of the press: "If ever there is to be an amelioration of the condition of mankind, philosophers, theologians, legislators, politicians, and moralists will find that the regulation of the press is the most difficult, dangerous, and important problem they have to resolve. Mankind cannot now be governed without it, nor at present with it." And Jefferson, that patron saint of the free press, ritually praised over the years for having said before his presidency that given a choice between a government without newspapers or a press without a government he would unhesitatingly choose the press over the government, came to a totally different view during his White House years. "During the course of this administration, and in order to disturb it," he said in his second Inaugural Address, "the artillery of the press has been levied against us, charged with whatsoever its licentiousness could devise or dare. These abuses of an institution so important to freedom and science are deeply to be regretted, inasmuch as they tend to lessen its usefulness and sap its safety; they might, indeed, have been corrected by the wholesome punishments reserved and provided by the laws of the several States against falsehood and defamation; but public duties more urgent press on the time of public servants, and the offenders have therefore been left to find their punishment in the public indignation."

Nor did the press operate alone in exhibiting a taste, and a talent, for seeking out scandal in the White House. The people, no less, thrived on it. They seemed to relish the sight of their powerful leaders being brought low.

Dickens, during his travels across the young nation in the 1840s, was struck by what he called "the one great blemish in the popular mind of America." That was "Universal Distrust." The American people, he concluded, were so given to feelings of jealousy and distrust that they carried them into every transaction of public life. "It has rendered you so fickle," he wrote, in friendly warning to a people he had come to much admire, "and so given to change, that your inconstancy has passed into a proverb, for you no sooner set up an idol firmly than you are sure to pull it down and dash it into fragments."

He feared this trait was so destructive that it would affect the ability of future American leaders. "Any man," he predicted, "who attains a high place among you, from the President downward, may date his downfall from that moment."

Over the generations, the American presidency continued to be afflicted with allegations of corruption, immorality, and, at the least, mismanagement. Dickens notwithstanding, these reactions were probably inevitable, perhaps even useful if not carried to the extreme. They were expressions of the deepest underlying emotions at the heart of the American experiment in democracy: Distrust of authority; fear of tyranny; desire to check wherever possible abuses that come naturally with the grant of power over people's lives.

Despite all the controversies that surrounded them, of the forty presidencies before Clinton's, few were marked by genuine scandals—most notably, Grant's in the pervasive corruption of the Gilded Age and Harding's in the let-'er-rip payoffs of the Roaring Twenties—and of those scandals, none exposed a president actively engaged in criminal acts to benefit himself or to betray the nation. They were scandals characterized more by presidential omission than commission. Indeed, corrupt though those administrations were shown to have been, that knowledge did not become known until they were over. Harding, for example, was hailed at the time of his sudden death in office in 1923 as rivaling Lincoln for the presidential tone he set. Andrew Johnson was castigated, investigated, impeached, tried, and found not guilty by a single vote, but the charges of "high crimes and misdemeanors" brought against him in 1868 grew out of post–Civil War political clashes over Reconstruction policy and ideology, not allegations of personal corruption. Only Richard Nixon and Watergate can truly be said to meet the historic standard of a scandal that involves both personal presidential corruption and willful abuses of presidential power that threaten the nation's constitutional system of government and the rule of law. And it is against Nixon and Watergate in the Seventies that the Clinton scandals of the Nineties have to be measured to be understood.

The Watergate crisis that forced Nixon to resign his office under threat of impeachment was a complex and multifaceted historic drama.

It arose out of a background of an America experiencing the greatest turbulence and divisiveness since the Civil War—and, in many respects, the bitterness and hatreds that helped produce Watergate were as intense as those that divided the nation a century before. By the time Nixon became president, America was a nation in danger of falling apart. Racial riots and assassinations of political leaders at home combined with increasingly violent protests against the undeclared war in Vietnam. Confronted by that cauldron of growing public discord and disillusionment, Nixon fell back on his own form of siege mentality and paranoia. Acting from a conviction that he faced enemies determined to destroy him, he used the full powers of his office and of the government to combat those he believed threatened himself and his presidency. This led him to initiate actions, both secret and illegal, that ultimately brought about his own destruction. In the aftermath of his departure, Americans took pride in the belief that their system and the central elements in it—the prosecutors, the courts, the Congress, the press—performed properly and brought about a peaceful bipartisan resolution of a tormenting national crisis.

But America doesn't emerge unscathed. The turbulent Sixties and Watergate and its aftermath leave a legacy of deeper public distrust of public leaders. In part, it is spawned by conspiracy theories about the assassinations of John and Robert Kennedy and Martin Luther King Jr. and suspicions about the roles of the Central Intelligence Agency, the Pentagon, and J. Edgar Hoover's Federal Bureau of Investigation. Another factor, perhaps most important of all, is the growing appetite for scandal exhibited by the mass media.

In journalism, a lasting effect of Watergate and the magnificent example of dogged and meticulous reporting set by Bob Woodward and Carl Bernstein of the *Washington Post* has been the elevation of the role of the "investigative reporter."★ In the aftermath of Watergate, news operations across the country set up investigative

---

★As a second-generation journalist, I have always regretted the adoption of this term. It gives a special cachet to a form of journalism that needs no title other than reporter; reporting, by its essence, involves an investigation for truth, and the presentation of significant events factually, fearlessly, in context, in perspective, and without embellishment so their importance can be clearly understood by citizens. The designation of a special category creates an impression of the intrepid reporter as an "investigator," as a gumshoe, and entangles reporting with the deplorable kind of "gotcha journalism" that has become so prevalent after Watergate.

units. Naturally, given the fame, fortune, and Hollywood celebrity
status won by Woodward and Bernstein for their Watergate reporting, others began to compete to find the next scandal that
would unseat a president. While notable examples of journalistic
investigative reporting occur, too often coverage is motivated by a
desire to become the next Woodward and Bernstein, to discover
scandal where in fact none exists.

From Watergate—a genuine and extensive criminal conspiracy
that led not only to the fall of a president but to the criminal convictions of a score of White House aides and administration officials, including an attorney general of the United States and the
head of the FBI—flows a succession of supposed grave scandals in
administration after administration. All are designated by the sobri-
quet, -gate. In the generation that follows, the search for scandal, for
more -gates, accelerates: Billygate, Peanutgate, Koreagate, Irangate.

Most are minor indeed; few will be remembered beyond their
intense, but brief, time on the public stage. Few citizens can even
identify what scandals the respective -gates are supposed to have
represented; all that's left in the public mind is a vague memory
that some sort of scandal affected yet another presidency.*

*Iran-Contra is the presidential scandal that comes closest to Watergate in
raising grave constitutional questions about abuses of power. Its essence: That
in 1985 President Reagan approved three secret shipments of arms to Iran,
specifically transferring 526 U.S. missiles to that terrorist state even though
such transfers were illegal, in exchange for release of Americans held hostage
there. Then, after overcharging the Iranians by 38 percent, part of the profits from the arms-for-hostages deals were distributed for the personal gain of
the covert U.S. operators in charge of the arms sales and part were diverted
to provide cash to buy arms secretly for the Nicaraguan Contra forces being
backed covertly by the U.S. government in their fight against Communist
Sandinista rebels in Central America. These secret and expressly illegal operations led to congressional hearings that exposed what amounted to creation
of a secretly financed and unaccountable "government within a government"
using public funds to mount other "off-the-shelf" covert operations worldwide. All this raised the prospect of Reagan's impeachment, but Reagan's
personal popularity, the fact that he was nearing the end of his second term,
and the glow of good times in the Eighties combined to save him.

As each supposed scandal is designated the latest heir to the truly historic Watergate scandal, the constant repetition depreciates the meaning of scandal. By making all seem equally scandalous, by trivializing them, the media leaves the public with the belief that scandals are to be expected in all presidencies; that the American political system is corrupt; that all in it, from top to bottom, are tainted; that leaders are not to be trusted, not least among them presidents.

In Clinton's years the scandal pace intensified. Now, the public was treated to a succession of galloping -*gates*: Nannygate. Hairgate. Travelgate. Troopergate. Filegate. Finally, at the end of his first year in office, came the greatest -*gate* to date: Whitewatergate, involving Bill and Hillary's investment sixteen years earlier—even before he became governor of Arkansas—in a private Arkansas land development called Whitewater.

The story, clearly, was not new; nor was it a story that failed to attract public notice, or scrutiny from official investigators. They examined the complex steps that led from the $69,000 original investment of the Clintons through involvement with savings and loan banks and, ultimately, to the failure of the land development and bankruptcy.

Since 1984, the Whitewater deal had been investigated by the Federal Home Loan Bank Board and other federal regulators, all part of the savings and loan banking scandals of the 1980s that resulted in the failure or merger of some two hundred S&Ls, causing a loss to U.S. taxpayers of more than $200 billion.

The Whitewater story first surfaced in the national press early during Clinton's 1992 presidential campaign when a *New York Times* article by Jeff Gerth disclosed the involvement of both Clintons in the failed investment. The story did not spark others, and quickly faded from the public screen. But it did attract the attention of federal bank regulators, prompting an investigation by the Resolution Trust Corporation's Kansas City office.

By September of 1992, two months before Clinton's election, that investigation led to a criminal referral from the Kansas City RTC office to the U.S. attorney in Little Rock. The referral, about a check-writing scheme in the original savings and loan bank involved, mentioned the Clintons as witnesses. But it made no allegations about them and attracted no public attention. There it slumbered during Clinton's first presidential year until it resurfaced dramatically when it was joined fatefully with two other episodes that focused intense attention on Bill and Hillary, their Arkansas connections, and the Whitewater deal.

The first was the suicide that summer of Vincent W. Foster Jr., deputy White House counsel, lifelong friend of Bill Clinton, former Little Rock law partner of Hillary Clinton, and the official who served as the president's and the first lady's personal counselor.

His body was found in an isolated Civil War fort on federal parkland north of Washington overlooking the Potomac. He had been shot in the mouth; a vintage Colt pistol was still clutched in his right hand.

The mysterious circumstances of Foster's death—no suicide note was found, and nothing initially to indicate the reasons for his death—plus his extraordinarily close connections to both Bill and Hillary, immediately sparked lurid conspiracy theories in conservative publications and on talk-radio shows. These theories became a staple of Clinton's ideological enemies, nurtured and passed on publicly from that date until the end of the Clinton presidency.

In totally unsubstantiated rumors broadcast and printed about him, Foster was said to have been Hillary's secret lover and even to have been murdered because of his connection to her.★

Foster, forty-eight years old, a self-effacing man of honor, widely admired in Arkansas for his legal credentials and character—first in his law school class, first in the bar exams, leader of the state's most prominent firm—never worked in Washington and

---

★The lowest of numerous low moments came when the conservative radio talk-show host Rush Limbaugh, reaching a daily audience of millions, broadcast a wild rumor he had received in a fax from a conservative newsletter published in California that Foster had been murdered in a Washington apartment owned by Hillary Clinton and his body secretly moved to the park in Virginia where it was found. That outrageously irresponsible, and totally false, rumor caused the stock and bond markets to plunge sharply. Limbaugh defended his broadcast by saying, "That's what it said in the fax."

was unprepared for the experience, especially for the political controversies that keep engulfing the White House and which his official position required him to address. He found himself subjected to unaccustomed public criticism, especially in the sternly ideological editorial pages of the *Wall Street Journal* where he was depicted as a shadowy figure of influence, part of the Little Rock "legal mafia," pulling strings on behalf of his clients, the Clintons, perhaps covering up misdeeds and, it is implied, illegalities. Foster was the kind of private person who took his responsibilities with utmost seriousness and who prided himself on protecting those he served. Finally, he broke.

His death immediately triggered a series of investigations that over time ranged far beyond the initial police inquiries. It led to the appointment of the first special prosecutor, or independent counsel, charged with investigating allegations involving the Clintons—a step that powerfully affected the political life of the nation and Clinton's presidency for the remaining six years of the Nineties and even beyond into the new century.

Though Foster's death was ruled a suicide after extensive investigation showed he was deeply depressed, felt overwhelmed by the controversies affecting the administration during its early months, and was seeking medical help, that official finding of suicide by the U.S. Park Police never satisfied the Clinton enemies. They succeeded in forcing other investigations, in which two independent counsels and teams of other investigators continued to look into the circumstances of his death.

Years passed before the final official record was closed on the Foster case, with the same results as the first police inquiry: Death by suicide. Entered on the public record was a poignant handwritten note of Foster's. It was found in his briefcase, torn into twenty-eight pieces, overlooked during the first search of his White House office after his death. In it, Foster expressed his deep frustration and disappointment with his experience as a public servant in the capital. "I was not meant for this job or the spotlight of public life in Washington," he wrote.

In one scrawled sentence, Foster unwittingly provided an epitaph for an era of scandal: *"Here ruining people is considered sport."*

Vince Foster faded into history's shadows but the controversies his suicide sparked did not.

The "murder of Vince Foster" became an ideological rallying cry in stories, broadcasts, and speeches produced by right-wing Clinton haters. These intensified both in virulence and volume when two other events linked the Foster case to other supposed acts of criminality or immorality involving the president and the first lady. Both occurred in December, six months after Foster's death.

First came the disclosure early that month in the conservative *Washington Times* that certain files, including some dealing with the Clintons' Whitewater investigation, were removed by presidential aides from Foster's White House office the night of his death. Now Foster and Whitewater were inseparably joined in the latest, and greatest, search for scandal.

No sooner did reports surface that Whitewater files were removed from Foster's office than House Republicans on Capitol Hill escalated the assault on the "Clinton scandals." Sensing an issue that could embarrass and possibly seriously wound the Clinton administration—another Watergate, GOP political strategists began saying privately—they seized the opportunity to turn the tables on congressional Democrats who had long investigated Reagan and Bush administration figures for *their* alleged misconduct when the Democrats controlled Congress.

Early in December, Republicans requested congressional hearings on the failure of the S&L at the center of the Whitewater land development deal. Democrats instantly rebuffed them. Republicans upped the ante. They demanded access to investigative Whitewater files held by federal regulators that mentioned the Clintons. Then they publicly began calling for appointment of a special prosecutor to investigate Whitewater. This prompted Jeff Gerth to reexamine the long-dormant Whitewater issues in a *New York Times* article published December 15, thus setting the stage for other stories to come.

Days later, on December 19, the final strand in the web being woven to ensnare the president was in place. It had dramatic effect.

That night, CNN broadcast excerpts from allegations in an article to be released the next day by the *American Spectator,* a right-wing publication that had been attacking the Clintons incessantly from the day they entered the White House. Titled, "His Cheatin' Heart. Living with the Clintons: Bill's Arkansas Bodyguards Tell the Story the Press Missed," it was a hyperbolic exposé of Bill Clinton's supposed extramarital affairs while he was governor of Arkansas.

The author, David Brock, based the article upon charges made by four Arkansas state troopers, two of them anonymous. Later, Brock said he regretted his role in the article, and even wrote a public apology to the president; but the damage had been done, and it proved to be historic. Among the allegations made by the troopers is that Clinton, as president, sought to discourage them from speaking out about his alleged sexual affairs by offering them federal jobs. If true, that could make the president liable for criminal charges of seeking to suborn testimony from potential witnesses.

Troopergate was born. The Clinton scandal circle was closed. Now, the Foster death, Whitewater, sex, and allegations of presidential criminality were forever joined.

·   ·   ·

In an atmosphere of increasing distrust and suspicion, the president and the first lady bowed to the rising political pressure for appointment of a special prosecutor to investigate the complex, nearly incomprehensible, Whitewater situation. That action was implemented on January 12, 1994, when Attorney General Janet Reno named the first independent counsel to investigate allegations of "Clinton scandals," including Whitewater and the Foster death. This, in turn, initiated what became years of nonstop and fruitless investigations by the Office of Independent Counsel and House and Senate congressional committees, all at a cost in excess of $100 million; none resulted in prosecutable charges of criminality or misconduct against the president or the first lady. But now Whitewater assumed a life of its own.

The Whitewater scandal quickly turned into an example of mass media madness, all driven by the prospect that the long-ago land deal would lead to the fall of another president. Whitewater leaped out in daily front-page headlines, blared from television sets, boomed on radio talk-show commentary, and formed the stuff of even more venomous conspiracy theories about Vince Foster, the Rose Law Firm, the Clintons, sex, and attempts to silence critics. Most notorious were the production and dissemination of two videos, *Circle of Power* and *The Clinton Chronicles*. In *Circle*, a disgruntled former Arkansas state employee, whom Clinton fired after it was disclosed he had made nearly seven hundred phone calls from his official state phone to leaders and supporters of the Nicaraguan contras, told, without any proof, of "countless people who mysteriously died" after having opposed Clinton in Arkansas. The companion video, *Chronicles,* purported to link the then-Governor Clinton to secret drug-running flights and money laundering at a private airfield used by the CIA in Mena, Arkansas; it darkly suggested these unproved smuggling allegations were somehow connected to the "mysterious deaths." Both videos were played and promoted on televangelist Jerry Falwell's TV program; as a result, they became widely shown during services of other evangelical churches which, like Falwell's Moral Majority network, formed a major force among far-right political groups.

One other strand in the web was not immediately apparent, but it proved to be the most fateful of all.

In the *American Spectator* article, a woman identified only by her first name, "Paula," was reported to have claimed that Clinton sent state troopers to bring her to a Little Rock hotel room where the then-governor crudely propositioned her for sex.

Two months later, on February 11, 1994, "Paula" emerged at a Washington news session arranged by the Conservative Political Action Conference. She was Paula Corbin Jones, a former Arkansas state employee.

Paula stood before the TV cameras, a clearly ill-at-ease young woman with huge bows in her "big hair," flanked by a phalanx of conservative activists and attorneys. In a pronounced Arkansas twang, she nervously announced that she was filing a civil lawsuit against the president. She alleged that Clinton sexually harassed her, exposed himself, and asked her to "kiss it," and violated her civil rights in that Little Rock hotel room five years before while Clinton was governor and thus her boss.

Never had a president of the United States been the subject of such demeaning legal action. Never had a president's private life been discussed so intimately, and so humiliatingly, literally before the entire world.

Paula's legal brief even included a sworn affidavit in which she claimed to be able to describe "distinguishing characteristics" of the president's genitals.★ While the affidavit was never released to the public, knowledge of it was fueled by repeated leaks to the press. Naturally, these stirred salacious gossip across the nation and raised speculation about the incredible prospect that the sexual organs of a president of the United States might have to be examined physically in a legal proceeding.

By the time Paula came forward, the Whitewater scandal coverage was dominating all other news. In one week in mid-March, at the peak of the press frenzy, the nation's seven largest newspapers published 92 Whitewater stories. During that one month, the three TV networks aired 126 Whitewater stories. By comparison, from the first of that year to the end of March, the three networks aired 107 stories on the bloodshed in Bosnia, 56 on the Middle East, and 42 on the most important of the Clinton presidency's domestic issues, the attempt to reform the nation's health care system and provide universal coverage for all Americans.

★This assertion implied the president was deformed. Her actual anatomical description was unremarkable, describing "Mr. Clinton's penis," as "circumcised . . . rather short and thin . . . five and one half inches, or less in length, and having a circumference of the approximate size of a quarter or perhaps slightly larger. The shaft of the penis was bent or 'crooked' from Mr. Clinton's right to left if the observer is facing Mr. Clinton. In other words, the base of Mr. Clinton's penis, to an observer facing Mr. Clinton, would be further to the left of the observer than the head of the penis." The most extraordinary aspect of this disgraceful incident is what it tells about the depths to which Clinton enemies were willing to sink in their campaign against him.

What was not made clear to the public, either then or four years later when Monica Lewinsky entered the public stage, was the extent to which all the seemingly disconnected Clinton scandals and Clinton investigations were connected.

· · ·

Long before Bill Clinton ran for president, political enemies in Arkansas were working to discredit him and block his obvious ambition to hold higher office someday. These efforts intensified as the 1992 presidential year and Clinton's impending candidacy approached. Clinton's reputation for womanizing made him an easy target; his Arkansas enemies began working to find women with whom Clinton might have had sexual affairs, and then planted stories about them in the press. Private detectives were hired, and extensive contacts established with both national scandal tabloids and mainstream journalistic entities. These efforts produced enough success—most notably Gennifer Flowers, who emerged at a press conference arranged by one of the tabloids to tell of her twelve-year affair with Clinton—that Clinton political advisers had to conduct their own investigations of his possible extramarital relations in order, famously, to counter "bimbo eruptions." These continued to plague him during the presidential primary process and on to his election. By no means were his opponents all centered in Arkansas; virtually from the beginning of his national political rise, his home-state foes were powerfully assisted by a network of outside conservative groups and individuals.★

★It is not my intent to attempt a definitive account of this most complicated tale here, but interested readers will find a number of recent books and articles helpful. Among books, former prosecutor Jeffrey Toobin's *A Vast Conspiracy* (Random House, 2000) is the clearest and most comprehensive. Best on the Arkansas connections is Joe Conason and Gene Lyons in their *The Hunting of the President* (St. Martin's, 2000). The most detailed and coherent account of the Ken Starr prosecutorial operation is Susan Schmidt's and Michael Weisskopf's *Truth at Any Cost* (HarperCollins, 2000). Michael Isikoff adds telling information on the press, the scandal, and his own major reportorial role in *Uncovering Clinton* (Crown, 1999). David Maraniss's *First in*

Central among them was the extraordinary influence wielded by a reclusive Pittsburgh billionaire named Richard Mellon Scaife, an heir to the Pennsylvania Mellon bank and oil fortune. Over a span of four decades, Mellon showered money on conservative groups, think tanks, magazines, and newspapers to such an extent, and with such impact, that he rightly was credited by Newt Gingrich after the 1994 Republican takeover of Congress as being among those who "really created modern conservatism."

Scaife's powerful role in financing right-wing political efforts was first documented effectively in a page-one *Wall Street Journal* article by Phil Kuntz on October 12, 1995. Kuntz estimated that Scaife's grants and gifts to fund New Right and ultraconservative groups total $400,000 a week, or more than $200 million over twenty years. These included funding of such prominent groups as the Heritage Foundation, the Cato Institute, the Manhattan Institute, Accuracy in Media, and the *American Spectator* magazine, which played a central role in promulgating the Whitewater/Vince Foster conspiracy theories and—most critically—in bringing forward Paula Jones.

In the definitive analysis of Scaife's largesse as leading patron of conservative causes, in extensive special reports in the *Washington Post* in May 1999 that were the product of months of reporting and research by Robert G. Kaiser and Ira Chinoy, the total amount of his giving was raised to more than $600 million over a forty-year

---

His Class is essential for its account of the Arkansas years. The single best article on this entire aspect of the Clinton presidency is Anthony Lewis's superb lengthy essay, "Nearly a Coup," in the *New York Review of Books,* April 13, 2000. Lars-Erik Nelson's essay, "Clinton and His Enemies," on Jan. 20, 2000, in that same publication provides splendid political analysis. Also noteworthy in the April 27, 2000, issue of the *New York Review of Books* are the fascinating lengthy exchanges between two eminent legal figures writing from totally differing views of the Clinton scandals: "Richard Posner vs. Ronald Dworkin: An Exchange on Impeachment," prompted by Dworkin's March 9 critical review of Judge Posner's *An Affair of State,* which concluded legal grounds existed for Clinton being prosecuted and found "guilty of perjury and related offenses growing out of his affair with Monica Lewinsky and the ensuing investigation."

period. "By concentrating his giving on a specific ideological objective for nearly forty years," Kaiser and Chinoy wrote, "and making most of his gifts with no strings attached, Scaife's philanthropy has had a disproportionate impact on the rise of the right, perhaps the biggest story in American politics in the last quarter of the twentieth century. His money has established or sustained activist think tanks that have created and marketed conservative ideas from welfare reform to enhanced missile defense; public interest law firms that have won important court cases on affirmative action, property rights and how to conduct the national census; organizations and publications that have nurtured American conservatism on American campuses; academic institutions that have employed and promoted the work of conservative intellectuals; watchdog groups that have critiqued and harassed media organizations and many more. Together these groups constitute a conservative intellectual infrastructure that provided ideas and human talents that helped Ronald Reagan initiate a new Republican era in 1980, and helped Newt Gingrich initiate another one in 1994."

In history's measure, though, Scaife is likely to be remembered most for his role in largely bankrolling the opposition to Bill Clinton that resulted, years later, in the second impeachment trial of a president of the United States. "We're going to get Clinton," Scaife told a startled private luncheon group on Nantucket in the summer of 1994. This promise came at a time when the Clinton scandal coverage and official investigations were being eclipsed by the "All O.J., All the Time" media madness, and years before all the actors were in place for the historic impeachment denouement.

What Scaife didn't make known at that vacation luncheon was the wide range of groups he was funding to bring down Clinton and the way they all became connected in the end.★

★The luncheon was first reported in the 1999 Kaiser-Chinoy *Washington Post* special reports. They told how one of the luncheon guests, the New York book publisher and prominent member of the Louisville, Ky., media family, Joan Bingham, was startled by Scaife's remark. Scaife never denied making the comment after the first *Post* article appeared that spring.

A key example was "The Arkansas Project" that Scaife backed with a grant estimated at more than a million dollars. This secret project was launched with one objective: to dig up dirt on Clinton in his home state. It resulted in the Troopergate article in the Scaife-supported publication, the *American Spectator,* beneficiary of more than $8 million in Scaife money over a two-year period. That, in turn, led to Paula Jones and her suit and later directly to Monica Lewinsky. The *Pittsburgh Tribune,* which Scaife owns, and its affiliated Washington Journalism Center, played a leading role in pushing the Vince Foster murder conspiracy allegations through articles, broadcasts, and books.

Other Scaife grants that proved significant in the effort to "get Clinton" included those to Judicial Watch, a leading exponent of conservative conspiracy theories involving the Clintons and their administration—including the wild accusation that Clinton's secretary of commerce, Ron Brown, killed in a plane crash, was actually shot in the head and murdered to keep him from disclosing Clinton scandals. Judicial Watch used Scaife money to sue members of the administration in at least eighteen separate matters, forcing everyone named in dragnet subpoenas, from secretaries and interns to top officials, to pay extensive legal bills out of their own pockets. Another influential conservative group that received millions in Scaife money, the Landmark Legal Foundation, provided early legal assistance to Paula Jones and her attorneys and later to Linda Tripp. Members of the Scaife-backed Independent Women's Forum assumed leading roles as attackers of Clinton "scandals" on numerous cable-TV talk shows, appearing so regularly that they became known as the bottle-blond conservative brigade. Some of these women lawyers, among them Barbara Olson and Ann Coulter, produced popular anti-Clinton books.

The lines between the Scaife-funded groups and other Clinton opponents were twisted and convoluted, but ultimately they came together in the common effort to destroy the president. Thus, a group of young conservative lawyers working in private firms and calling themselves "the Elves" provided secret legal work in drafting a Supreme Court brief for Paula Jones. It argued that her sexual harassment suit not be postponed until after Clinton's presidential term. This led to a unanimous Supreme Court ruling that a sitting president had to deal with a civil suit while in office. The justices reached that finding by promulgating their theory that the president's official work would not be affected. They further concluded that the Paula Jones suit was "highly unlikely to occupy any substantial amount of [the president's] time."

This decision proved to be one of the most wrongheaded in Supreme Court history, and one that set a terrible precedent for future presidents.

The Elves also provided a perfect example of the way anti-Clinton groups converged and then conspired, with powerful political effect, in their multiple behind-the-scenes efforts to get the president.

While Scaife was bankrolling the Arkansas Project, one of Clinton's strongest longtime political enemies in the state, a lawyer named Cliff Jackson, was representing the state troopers and attempting to get them a book deal to tell their stories about Clinton's alleged infidelities. A year before, Jackson dealt with another of those wealthy ideological conservatives seeking to uncover Clinton scandal material, a Chicago investment banker named Peter W. Smith. Throughout 1992 Smith spent a reported $40,000 unsuccessfully trying to prove and then promote a story that Clinton had fathered a child by a black prostitute. Jackson's connection with Smith led him to ask the Chicago banker to recommend a writer to tell the troopers' story. Smith suggested David Brock, who wrote the *American Spectator* article, but not a book.

When "Paula" emerged in Brock's article, Jackson again turned to the Chicago financier for help. This time he asked Smith to recommend a lawyer for Paula. Smith, helpful as ever, suggested a young attorney with strong conservative credentials: Richard Porter, a new member of the Chicago firm Kirkland & Ellis, previously on the staff of Vice President Dan Quayle.

Porter became the first of the Elves. He recruited other young lawyers in Philadelphia, New York, and Washington who shared his ideological convictions. Together, they began their secret efforts on behalf of Paula Jones.

Another attorney at Kirkland & Ellis, far more prominent than Richard Porter, also consulted many times with the Paula Jones legal team and even considered writing an amicus brief on her behalf before the Supreme Court. This was Kenneth W. Starr, formerly a solicitor general of the United States. Later, Starr became the second, and historically by far the most important, of the Clinton special prosecutors.

Despite all the efforts to obtain incriminating material to force Clinton from office, none of the investigations, public or private, found the fatal silver bullet. Years passed before the widely expanding—and wildly divergent—cast of characters succeeded in finally setting in place the trap that ensnared the president, or, equally correct, the trap in which the president ensnared himself.

In the end, it came down to Monica Lewinsky and her supposed new Pentagon friend, Linda Tripp.

· · ·

Long before Linda Tripp meets Monica Lewinsky in April of 1996, when both are working in the Pentagon, Tripp is contemplating writing a book exposing what she believes are scandals committed in the Clinton White House. She worked there as a leftover from the Bush administration before being banished, like Monica after her, to the Pentagon. Tripp was encouraged in her book project by Lucianne (or Lucy) Goldberg, another memorable right-wing figure who worked for years in Washington before becoming a New York literary agent.* Tripp becomes her client.

At Goldberg's urging, Tripp wrote a synopsis for her proposed scandal book and, also at Goldberg's behest, hired a ghostwriter. It ought to bring Linda somewhere between $200,000 and $500,000 in a publisher's advance, Lucy advises her. That book never materializes.

---

*Linda and Lucy were first introduced by Linda's "dear friend and colleague" from the Bush White House, Tony Snow, another young conservative who became a syndicated newspaper columnist and cable-TV talk-show host on the Fox network, thus one of the young conservatives who forged media careers—Laura Ingraham of MSNBC and Ann Coulter are others—while being closely involved with anti-Clinton groups.

Linda puts aside the project, in part because she fears the loss of her government job after publication, in part because she believes her financial reward not worth that risk since she has to split royalties with her ghostwriter. Then she meets Monica.

Monica is heaven-sent. Linda finds the young former intern working in the office of the Pentagon spokesman; her desk is adorned with pictures of the president, and from her conversation, she is clearly infatuated with him. Linda immediately befriends the younger woman and quickly begins trying to find a way to use her to build a case against the president. She urges Monica to return to the White House to work, telling her she's the kind of woman the president would like. Having an affair with the president would be a "neat thing to tell her grandkids" someday, Linda encouragingly tells Monica. For months, as she continues building a friendship with the younger woman, Linda keeps hounding Monica about getting back to the White House where she can connect with the president. Finally, Monica blurts out, "Look, I've already had an affair with him, and it's over."

An even greater scandal presents itself to Tripp. Immediately, she resumes contact with Lucy Goldberg in New York, telling her about the secret she's just uncovered from Monica. Again, Lucy plays a fateful role: It's she who suggests Linda secretly tape-record her phone conversations with Monica. "You've got to really rat and you've got to tape," Lucy tells Linda on one of their taped phone conversations.

In September 1997 Tripp begins what turns out to be her historic secret taping. Two months later, after her attempts to plant media articles about the Clinton-Lewinsky affair fail to produce any stories, Tripp asks her friend and confidante, Lucianne, for help in passing on the story about Monica and Bill to Paula Jones's lawyers. From her close contacts in the conservative network, Goldberg is urged to call Peter Smith, the Chicago financier. He puts her in touch with Richard Porter, the original Elf.

From there, the story of Clinton's adulterous affair swiftly passes from hand to hand. Finally, through the active efforts of the Elves, Goldberg, and Tripp, it lands in the camp of Paula's lawyers of record. They quickly issue a subpoena for Monica, an action that stuns and frightens the young former intern.

Monica first becomes aware her secret relationship with the president is in danger of being exposed when Clinton phones her sometime around two-thirty in the morning on December 17, 1997. The president gives her disturbing news: He's just learned that Monica's name appears on a list of potential witnesses in Paula Jones's suit against him. Monica is "upset and shocked." Seeing her name on the witness list "broke his heart," the president tells her, but that doesn't necessarily mean she'll be subpoenaed. It's merely a possibility. In case she *does* receive a subpoena, however, she should call his secretary, Betty Currie, to let her know. When Monica asks what she should do about the subpoena if she receives it, the president suggests, "Well, maybe you can sign an affidavit."

Their conversation lasts about half an hour. It's now around three in the morning, but Monica can't go back to sleep. As the clock nears four, after another hour of anguish, she picks up her phone and calls Linda Tripp.

Monica knows Linda received a subpoena in the Jones case a few days before, not as one of the women linked sexually with Clinton but as someone whose name surfaced in an article about an earlier Oval Office incident in which a woman named Kathleen Willey supposedly was groped by the president—an incident that opens further avenues of exploration for a Jones team seeking to prove a pattern of presidential behavior. What Monica *doesn't* know is that for months Linda has been tape-recording their conversations and avidly telling Lucy Goldberg, among others, about them. She doesn't know that Linda is keeping notes and memos of their private conversations, and in a steno pad is compiling dates of Monica's sexual contacts with Clinton. Even more stunning, in terms of the massive betrayal underway, she doesn't know that Linda *already* has been talking to the Jones lawyers for at least two months, providing them with details of the affair and the tape recordings—and far from feeling trapped by a surprise subpoena from the Jones lawyers, Linda already has agreed to be a witness for them and tell what she knows in sworn testimony. Monica doesn't know that some ten months before, Linda and Lucy helped the *Newsweek* journalist Michael Isikoff with his article exploring the Willey incident. She doesn't know that after being encouraged by Lucy, Linda has been talking secretly for months to Isikoff—who uses the code name "Harvey" when he leaves messages for Linda—and that since November, Linda and Lucy have actively plotted to use Linda's tapes to expose Clinton. They've even offered to let Isikoff listen to some of them at the home of Lucy's son, Jonah, in Washington. But their hope of seeing Isikoff expose the Clinton/Lewinsky affair in *Newsweek* has not materialized. No Isikoff story appears. Monica Lewinsky remains anonymous.

Although she still believes Linda to be her trusted confidante, Monica for the first time is becoming concerned about her friend after Linda begins hinting she might just tell all if forced to testify. As Monica recalls, "I had been trying to convince her not to tell, that it's not anybody's business."

Now, in the dead of night, and in great turmoil, Monica places yet another call to Linda. She hopes Linda will understand that she's no longer alone in facing the possible ordeal of giving public testimony about the president. As she tells Linda, after relating the president's information about Monica's name also being on the witness list, she, too, faces the prospect of becoming a witness. She hopes her stance of not admitting anything will strengthen Linda's resolve. As she tells Linda, you won't be out there alone in saying, "I don't know anything about any kind of relationship between the President and Monica." Together, they can form a united front of denial.

Their conversation ends with Monica telling Linda they'll talk more about the Jones case during the day.

Two days later, on December 19, at about four o'clock in the afternoon while Monica is working at her Pentagon desk, her phone rings. The caller, a man, informs her he has a subpoena for her. She reacts furiously and fearfully: What's going on? Why is she being subpoenaed?

After pulling herself together, Monica tells her superiors she has an emergency, leaves her office, and goes to the Metro stop in the Pentagon area to meet the process server of the subpoena. Monica glances at it and bursts into tears.

It's addressed to her as "Jane Doe # 6"—one of the series of "Jane Does" the Jones team, aided by Arkansas informants and private investigators, subpoenaed to testify about any sexual relations they might have had with Clinton going back years before he was even governor. The Jones team, in other legal filings, also commands the president to supply "the name, address, and telephone number of each and every individual (other than Hillary Rodham Clinton) with whom you had sexual relations when you held any of the following positions: a. Attorney General of the State of Arkansas; b. Governor of the State of Arkansas; c. President of the United States." (He declines to answer these on the grounds they are intended solely to embarrass and humiliate the president of the United States.)

Her subpoena, Monica recalls, is "very scary . . . just sort of my worst nightmare." What most alarms her is the specificity of the items requested of her. Among a list of gifts she's ordered to turn over is a hat pin. That "screamed out at me because this was the first gift that the president had given me and it had some significance."

Now, as she's excruciatingly aware, somehow someone has been passing on explicit, detailed information about her and the president to the Jones team.

She calls Vernon Jordan from a nearby pay phone to tell him about receiving the subpoena, sobbing so much into the phone that Jordan can barely understand her. Pull yourself together, he tells her, and come to my office. Within an hour she meets the power lawyer for the second time there. This meeting intensifies her frantic search for a job away from Washington, an effort that began months before after she and the president ended their sexual relationship, and which, thanks to Jordan, is now about to end successfully. Jordan also arranges for her to hire an attorney, calling him personally from his office and making an appointment for Monica to meet him the next Monday. This session results in her preparing an affidavit, in which she states in careful lawyerly language: "I have never had a sexual relationship with the President, he did not propose that we have a sexual relationship, he did not offer me employment or other benefits in exchange for a sexual relationship, he did not deny me employment or other benefits for rejecting a sexual relationship. I do not know of any other person who had a sexual relationship with the President, was offered employment or other benefits in exchange for a sexual relationship, or was denied employment or other benefits for rejecting a sexual relationship."

The subpoena produces a feeling of extreme paranoia in Monica. At first, she thinks someone has broken into her office computer and read her e-mail. That's particularly alarming; as she knows, she keeps "a stupid spreadsheet on Microsoft Excel" stored in her office computer that lists all her sexual contacts with the president.

It's a secret list she has shared with Linda Tripp. But at that point she still doesn't suspect her friend of betraying her. On the contrary, "Linda always told me she would always protect me and she would never tell anybody and keep my secret." Even after Linda begins hinting she might have to tell the truth about Monica and the president, Monica naively attributes part of Linda's new attitude to possible resentment that Monica will get a high-paying New York job in the private sector through the influence of her powerful connections. She wonders if Linda, who has been talking to her lately about wanting to leave government service for a private sector job, also perhaps in New York, might be jealous on other grounds: That Vernon Jordan has helped Monica find a well-connected attorney while Linda feels she no longer trusts the lawyer with whom she then deals. In fact, as Monica knows, Linda even then is just about to retain another attorney; naturally, one recommended by her friends in the conservative anti-Clinton network and with the assistance of the Elves under Lucy's guiding hand, who makes calls to Los Angeles, Chicago, and New York on Linda's behalf.

Monica consults a Pentagon computer specialist about the security of the computer system. It would be very difficult for anyone to break into her computer, he advises her. Then she begins fearing her phone is tapped. She shares her growing sense of terror with Linda, still not suspecting the ultimate treachery. Now, when speaking on the phone with Linda and continuing to try and persuade her not to reveal her real relationship if called to testify, Monica begins speaking in code. Instead of saying, "I received the subpoena," she'll say, "I received the flowers." Instead of calling Linda by her real name, she adopts a code name of "Mary." Linda suggests she use it in their phone conversations. Instead of using her home phone to call Linda, she begins using pay phones all the while keeping up a stream of conversations with Linda.

The Christmas holidays become a time of increasing anxiety—and not just for the now-panicky Monica, but at the highest level of the White House as well.

· · ·

From the time he learns that the name of "Monica Lewinsky" appears on the Jones witness list, the president engages in a desperate attempt to keep her true relationship with him from being disclosed. He knows, within a few hours of its service, that she has received her subpoena; Vernon Jordan informs him of that when the two meet the night of December 19 in the White House, just hours after Monica and Jordan conferred. The president and Jordan also discuss where Monica's job search then stands.

As the new year of 1998 begins, other elements in the still subterranean melodrama quickly begin falling into place. A key moment occurs in Philadelphia on the night of Thursday, January 8. There two of the Elves, Richard Porter and Jerome Marcus, meet with their friend and former law school classmate, Paul Rosenzweig, who has just become a member of Ken Starr's legal team investigating the president. During dinner they tell Rosenzweig about Lewinsky, Tripp, and the tapes. The next day, Friday, Rosenzweig informs Starr's top deputy, Jackie Bennett Jr. Since Starr is away from his office until the following Monday, the twelfth, Bennett cannot inform him about the dramatic new development until then; Starr immediately gives approval for his office to get Linda as a witness.

In classic circular conspiratorial fashion, the final, fatal events in the Clinton scandal chain are swiftly set in motion: Rosenzweig reaches Marcus, one of the Elves, who contacts Lucy Goldberg, who calls Linda. "Phone prosecutor Bennett that very night," Lucy urges Linda during their Monday night conversation. "He's in his office right now awaiting your call." Tripp makes the call. She tells Bennett about Monica, the tapes, the president, and Vernon Jordan's help in Monica's job search.

As soon as they hang up, Bennett, accompanied by two other prosecutors in Starr's office, drives from Washington to Tripp's home in suburban Washington. At a quarter to midnight, they take a lengthy statement from Tripp that isn't completed until two o'clock that morning.

Linda tells all, repeating much of what she's already informed the Jones lawyers: How and where Monica and Clinton's sexual relationship began . . . How his secretary, Betty Currie, acted as the facilitator when Oval Office administrator Nancy Hernreich was away . . . How Monica arranged to send the president notes and messages in the White House via a private courier service. What Linda tells is crucial, but far from complete, information for gathering evidence. Linda does *not* tell them that she herself urged Monica to adopt that messenger service on the advice of Lucy Goldberg, whose brother's half interest in the service would allow them to document all of Monica's transactions in sending gifts and messages to Clinton in the White House. She also doesn't tell them the Jones legal team has already been informed about that courier service and its records have been subpoenaed by them.

Most critically, she *doesn't* tell them that she's been talking to the Jones lawyers for three months about Monica and Clinton.

A crucial revelation that Linda gives the prosecutors concerns a navy blue dress of Monica's that was stained with the president's semen during a sexual encounter in the Oval Office. Linda tells them that a photograph exists of Monica and the president together in the White House in which Monica wears the same dress.

Monica "won't have the dress dry-cleaned to this day," Linda says, making it seem as if the lovesick girl treasures the dress as a memento. Linda doesn't say *she's* the one who urges Monica to keep it safe and not to dry-clean it, because she might need it for proof of her relationship someday. "Store it in a zip-lock bag and put it in a safe deposit box, because it could be evidence someday," Linda tells her after Monica says she's going to have the dress cleaned and wear it for Thanksgiving dinner. The dress makes Monica look fat, Linda says; then, helpfully, offers Monica one of her own dresses to wear. Nor does she tell them how she implores Monica to preserve the dress, saying that's the same advice she'd give her own daughter if the daughter ever found herself in a similar situation. So anxious is she that Monica keep the dress with its incriminating semen stains that at one point she and Lucy Goldberg actually talk about stealing the dress to keep it as evidence.

Linda discloses how the president told Monica she could "deny, deny, deny" if called in the Paula Jones case, and paraphrases the latest secret tape containing the following dialogue:

Tripp: I'm going to tell the truth... you're going to lie.
Lewinsky: If I lie and he lies and you lie, no one has to know...

The prosecutors learn that Linda has given some twenty tape recordings of her conversations with Monica to her lawyer for safekeeping. They're also told how Vernon Jordan was assisting Monica in her search for a job away from Washington. Tripp doesn't tell them that it was *she* who urged Monica months ago in October to contact the president's great friend Jordan for help in finding a job. Linda informs them that Monica has told her about all the gifts she received from the president; Linda describes them for the prosecutors and also lists Monica's gifts to the president.

Finally, Linda tells them Monica wants to meet her in just a few hours to continue talking about what they might do if they both have to testify in the Jones case. Linda says she's been thinking about secretly tape-recording that face-to-face conversation with Monica. Their meeting is set for early afternoon, either in the piano bar area or the back dining room of the Ritz-Carlton Hotel near the Pentagon, across the Potomac from Washington.

Linda also raises with the prosecutors her "concern over her own immunity from prosecution by Maryland state authorities" because, to adopt the language and the form of the prosecutors' account of this meeting with Linda in which every name is uppercased, she "did not realize recording her own telephone conversations with LEWINSKY was [sic] a violation of state law at the time she did it, and that she was just trying to protect herself from the eventuality that LEWINSKY would lie and TRIPP would be attacked for telling the truth."

This is patently untrue. Linda's lawyer already has warned her that recording phone conversations in Maryland is illegal. When he learns from her that she has continued making the secret recordings, he goes "ballistic" and warns her she could go to jail. She agrees to stop taping, but doesn't entirely, and begins seeking another lawyer because she suspects her first one is allied with the White House. She certainly didn't begin the taping out of fear she "would be attacked for telling the truth" in a legal proceeding in which Monica would lie. Her recordings begin months before either she or Monica are involved in *any* legal proceedings, or even face the prospects of such proceedings through their eventual subpoenas by the Jones team.

As her conversations with Lucy Goldberg make clear, some of which she also records, Lucy and Linda are eagerly seeking material to "get Clinton." After she tells the Starr prosecutors her fears about being prosecuted for making the recordings, they say while they can't "provide immunity from the State of Maryland should it choose to pursue charges relating to the taping," she nevertheless "would be granted federal immunity by the OIC [Office of Independent Counsel] for the act of producing the tapes to the OIC."

Linda doesn't tell them that at least nine of the twenty tapes turned over to the prosecutors have been duplicated. This isn't discovered until after an FBI laboratory examination, a finding that leads the agents to write in an advisory memo: "If Ms. Tripp duplicated any of the tapes herself, then she has lied under oath before the grand jury and in a deposition." Nor does Linda say she's given Lucy Goldberg tapes that Lucy takes to New York where she makes duplicate recordings of them—the purpose not determined.

Hours later that January 13, having been granted federal immunity for secretly recording someone else's conversation, Linda Tripp meets a female FBI agent in a bathroom at the Ritz-Carlton Hotel. There, she's fitted with a hidden body wire taped to her inner thigh and signs and initials a handwritten form signifying she's participating voluntarily "in the consensual monitoring of Monica Lewinsky." At 2:29 P.M., the device is activated. Linda leaves for her rendezvous.

Exactly four hours later she returns to the hotel, meets the same FBI agent in the same bathroom, has the wire removed, then immediately hands the recording device to agents stationed outside the door. That's not the end of her day's work; she proceeds to give another lengthy interview with FBI agents assigned to the Starr prosecuting team in a hotel room.

Under their questioning, she adds additional information about her involvement. She tells them she was put in touch with the Starr office "indirectly with the assistance of her literary agent from New York, New York." They also want to know, in what becomes a re-curring refrain among the Starr prosecutors in the next few critical days, about her and Lucy Goldberg's involvement with *Newsweek* reporter Mike Isikoff. Has Isikoff been told of Linda's meeting with Monica that day? Answer: "To Tripp's knowledge, reporter Mike Isikoff is not aware she met Lewinsky today in cooperation with the FBI." About Lucy's knowledge of Linda's and Monica's meet-ing, the answer: "Tripp's literary agent and attorney are aware of her meeting and cooperation."

Amid the stilted bureaucratic language of the agents' written report, one poignant note appears: "LEWINSKY asked TRIPP whether the truth shouldn't be for good, not to hurt."

There's no account of Linda's response, if she gives one.

Nor is this the end of Linda's extensive contacts with the pros-ecutors. Over the next sixty hours, she has at least ten separate per-sonal interviews or phone exchanges with them. They're initiated either by the agents or, increasingly, by Linda as she keeps them regularly informed about meetings with, and phone calls from, Monica.

Linda's first contact after her body wire experience comes early the following morning.

At 6:15 on January 14 an agent phones her at home to say the recording she made of Monica ended before their conversation was finished. Again, the agent presses her about "whether she or her

friend were advising reporter Mike Isikoff of the investigative activity" of the Starr prosecuting office. Linda's response, couched in
the same official jargon: "Tripp advised she had been advised by
her friend that: the friend did not tell Isikoff about the 1/13/98
meeting between Tripp and Lewinsky; Isikoff was already doing a
story about the relationship between Lewinsky and President Bill
Clinton; Isikoff was getting his information from other sources."

If Linda tells the agents that she and Lucy already have tipped
off Isikoff about Monica, have even let him listen to provocative
fragments of Linda's secret tape recordings, it does not appear in
the official record. In fact, as Lucy well knows, Isikoff is then
preparing a *Newsweek* article exposing Monica's affair with the
president scheduled to be published that next Sunday night, the
eighteenth. Indeed, the day after Linda's 6:15 phone conversation
with Starr's investigators, she receives a "frantic phone message"
from Lucy alerting her that the "shit was going to hit the fan."
Linda needs to talk to Isikoff that day, Lucy says.

In the final hours before the Monica Lewinsky affair is trumpeted around the world, amid a frantic flurry of activity between
Linda and the prosecutors, Linda adds other salacious material to
the official investigative file—nearly all of which quickly finds its
way into news reports after being leaked to media organizations:
Monica refers to Clinton as the "Big Creep." She and Clinton had
"phone sex a lot" and Clinton "was more concerned about their
phone sex being revealed than anything else." Clinton "kept a calendar and would mark days he had been 'good,' meaning days he
did not have sex with anyone other than [redacted, presumably
Hillary]." She also gives the Starr prosecutors the name and address
of Monica's attorney in Washington, Francis D. Carter, whom Jordan recommended she retain.

One item isn't leaked: that Monica pleads with Linda not to
expose the affair. Pathetically, Monica doesn't want the president to
know she has betrayed his trust by talking about them to anyone
else.

Out of this mass of information, proved or not, Starr moves
rapidly to expand his legal authority for investigating the president
beyond Whitewater, Vince Foster, Filegate, and Travelgate. Now

he wants to include Lewinsky, Tripp, and related matters under his jurisdiction.

On Thursday, January 15, after meeting with Eric Holder, the number-2 person in the U.S. attorney general's office, Starr's top deputy, Jackie Bennett, receives the official go-ahead on Monica and Linda. The potential suspected crime is not sex; it is possible criminal obstruction of justice and perjury.

Neither Holder nor his boss, Attorney General Janet Reno, is told of the contacts between the Jones lawyers, the Clinton enemies, and a member of the Starr prosecutorial team. On the contrary, Bennett assures the Justice Department that, "We've had no contacts with the plaintiff's [Jones's] attorneys." As Anthony Lewis later writes, this is false: "Bennett knew that Paul Rosenzweig of the Starr office had learned about Tripp and Lewinsky from three lawyers who were working for Jones. Moreover, before becoming independent counsel, Starr himself had consulted half a dozen times with Jones's previous lawyers. . . . If Holder and Attorney General Janet Reno had known of these connections, he would surely have referred the Lewinsky matter to a different independent counsel, if any; but they were not told."

The timing of this official green light is crucial. Two days later, on Saturday, January 17, the president is to be deposed in Washington by Paula Jones's attorneys. While he's becoming increasingly concerned about the problems he might face there, especially after Monica's name appears as a witness and she's subpoenaed, the president has no idea of the extent of the information the Jones team has about his relationship with Monica, certainly no hint they have known about the secret tape recordings and even their contents for months, and no knowledge whatever that he's not only involved in a messy civil suit but is also now being investigated in a criminal proceeding by the full power of the U.S. government through the special prosecutor's office. Nor does he know that both these investigations are now inseparably joined; both sides are sharing the same humiliating, if not incriminating, material provided by the same sources that are determined to bring him down.

By the end of the day on Friday, January 16, all the elements in the trap that will ensnare the president have been set. Earlier that day, Tripp's new attorney delivers seventeen of her tapes to Starr's office. That night Tripp herself meets with the Jones lawyers—and the Elves!—in a room at the Four Seasons Hotel. There, the Jones team is told Starr's prosecutors are investigating Monica and trying to get her to cooperate with them. They are thus fully informed about all elements in the case and based on Linda's remarks to them are able to tailor their planned questions for the president the next morning.

If that's not enough, the Jones lawyers and two of Starr's key prosecutors, Jackie Bennett and Bruce Udolf, meet that midnight in another hotel—incredibly, but fittingly, in the same Howard Johnson's motel from which the Nixon conspirators set out twenty-six years before on their infamous bungled break-in of Democratic Party headquarters in the Watergate directly across the street.

· · ·

Hours before the presidential deposition is taken in downtown Washington by the Jones lawyers, the trap is sprung. Linda Tripp sets it in motion.

It begins when Linda calls Starr's office to inform them that on Friday the sixteenth Monica wants to meet her about 11:30 A.M. in the Pentagon Mall food court area of the Ritz-Carlton. Call her back, she's told. Change the meeting time to 12:30 P.M. That gives them time to arrange a classic "sting" operation, with Monica as the target.

Monica arrives on time and nervously waits for Linda, who is late.

When Monica finally sees Linda coming down the escalator, she begins walking toward her. Then she sees Linda motion behind her. As she does, two FBI agents, accompanied by three officials from the Starr investigating team, step forward and flash their badges. As Monica quickly understands, to her consuming terror, she has been set up. "They told me I was under some kind of investigation," she later testifies secretly before the grand jury, "something to do with the Paula Jones case. That they wanted to talk to me and give me a chance I think maybe to cooperate maybe, to help myself. I told them I wasn't speaking to them without my attorney. They told me that was fine, but I should know I won't be given as much information and won't be able to help myself as much with my attorney there. So I agreed to go. I was so scared." [Here, nearly a year later, she begins crying as she recalls that scene.]

Adding to Monica's terror—and confusion—is an incredible closing scene with Linda. Even as the agents surround Monica, Linda attempts to give the impression that she, too, is a victim of a setup. As the agents lead Monica away, "She said they had done the same thing to her and she tried to hug me and told me this was the best thing for me to do," Monica remembers. Linda, she thinks, in her way is still trying to help her, even if she obviously has co-operated in this incredible entrapment. Though Monica quickly learns from the prosecutors that Linda wore a wire at their lunch there three days before, she doesn't find out about the extensive tape recordings of their phone conversations until she reads about them in the newspapers.

Her heart racing, her mind assailed with a jumble of disconnected thoughts, Monica thinks the prosecutors forced Linda to record their earlier luncheon conversation. She knows she can no longer trust Linda, but still doesn't believe Linda to be her mortal enemy.

What follows is a disgraceful episode in the annals of American jurisprudence, an abuse of power that never adequately receives the serious public attention it deserves.

For nearly twelve hours, from about one in the afternoon to shortly before one early Saturday morning, Monica Lewinsky is under the control of Starr's deputies and FBI agents and faces intensive interrogation by them. After being taken to Room 1012 of the hotel, she's repeatedly questioned, as well as being informed intimidatingly and frighteningly about the consequences she will face if she doesn't agree to cooperate to the extent of wearing a wire and secretly trying to trap both Vernon Jordan and the president in recorded conversations. She's told she'll receive immunity from prosecution if she cooperates—a flat violation of a Justice Department rule holding that a government attorney may not discuss immunity "with a represented person" without "the consent of the attorney representing such person." If she doesn't cooperate and rejects immunity, she's told she faces twenty-seven years in prison for filing a false affidavit. In fact, her affidavit had *not* been filed with the Jones lawyers at that time.★ All of this questioning continues, for hour after hour, and all of it takes place without her attorney being present—or even being contacted.

From the beginning, Monica tells them—and keeps repeating—that she wants to call her attorney and her mother, to whom she's especially close. Repeatedly, she's discouraged from placing the calls. After she keeps insisting on calling her mother in New York City, prosecutor Bennett tells her, chillingly, "You're twenty-four. You're smart. You're old enough. You don't need to call your mommy." Then he tells her something that induces further spasms of fear, as she recounts, "I should know that they were planning to prosecute my mom for the things that I had said she had done."

Again, in recalling that conversation a year later secretly before the grand jury, Monica begins sobbing.

★This becomes a critical point and, ironically for the prosecutors, one that backfires on them. Through their briefings from Linda, they've been led to believe the president was attempting to remove a potentially damaging witness by arranging for a job. In return, Monica will sign an affidavit lying about their sexual relationship. At the same time, Linda keeps pressing Monica *not* to sign the affidavit until she has the job in hand: "Monica, promise me you won't sign the affidavit until you get the job," she tells her. "Tell Vernon you won't sign the affidavit until you get the job because if you sign the affidavit before you get the job, they're never going to give you the job." When Linda tapes Monica at the "body wire" lunch, Monica tells her she hasn't yet signed it, when she *has,* and also says she doesn't have a job yet, when she expects she'll be getting one that day. Monica's reason: "I didn't want her to think I had gone ahead and done anything without her and that I was leaving her in the dark. I wanted her to feel...sort of Linda and myself against everyone else because I felt like I needed to hold her hand through this in order to try to get her to do what I wanted...." That is, not reveal her affair with the president. By having signed the affidavit *before* getting a job, Monica unwittingly undercuts the legal case of obstruction of justice. There's no quid pro quo: In the end, Linda and the prosecutors are stung by their own sting. From Monica's standpoint, she also doesn't want her affidavit to help give the Jones lawyers "a little snag of yarn so they could pull the whole sweater apart."

As for contacting her attorney, she's told he will not be of help to her because he's a civil rather than a criminal lawyer. That is untrue. Frank Carter, the attorney, is an experienced criminal lawyer and for six years headed Washington's public defender service.

Disturbing as Monica's own sworn account of her ordeal is, the stark yet bloodless rendering of that same scene as contained in the official chronology and log of the investigators is even more so. It confirms, and in some ways strengthens, virtually all of her later testimony. It describes how the agents first approach her: "LEWINSKY was advised she was not under arrest and agents would not force her to accompany them to the hotel room. LEWINSKY told [redacted name of special agent] he could speak to her attorney. [Redacted agent's name] advised the offer to discuss her legal status was not being offered to her attorney, but to LEWINSKY alone. [The agent] explained to LEWINSKY she was being offered an opportunity to meet with OIC attorneys and agents and hear them explain why they felt she was in trouble without being required to make any statement. [The agent] further explained LEWINSKY would then have an opportunity to ask clarifying questions of the attorneys and be better informed as to whether she wanted legal counsel before making any statements, or whether she thought it better to cooperate with the OIC."

After entering the hotel room at 1:05 P.M., her interrogation immediately begins.

The log describes, without stating so in words, her growing state of hysteria.

1:10 P.M. Lewinsky given bottled water.
1:33 P.M. [Redacted name] began reading Lewinsky her rights as found on form FD-395, "Interrogation; Advice of Rights." [Agent] was unable to finish reading the FD-395.

No explanation is given, but as the log clearly implies by the following entries Lewinsky was breaking down; in fact, she was sobbing hysterically and bawling loudly all this time.

1:40 P.M. Lewinsky was offered a towel.
1:47 P.M. Lewinsky was offered water.

2:02 P.M. The air conditioning in room 1012 was turned on at Lewinsky's request.

2:13 P.M. Lewinsky was offered water.

2:15 P.M. [Redacted name] arrived in room 1012.

2:29 P.M. Lewinsky stated, "If I leave now, you will charge me now."

2:30 P.M. Lewinsky said, "I don't know much about the law."

2:33 P.M. Lewinsky asked and was allowed to go to the restroom.

2:36 P.M. Lewinsky requested and was given her second bottle of water. Lewinsky said she is not diabetic and no medication was necessary.

2:50 P.M. Lewinsky suggested she take a taxi to her attorney's office. Lewinsky advised she understands our risks.

3:10 P.M. Lewinsky asked if she could be escorted to New York to see her mother, Marcia Lewis.

Months later, after spending hour after exhausting hour trying to understand all the twists and turns in the complicated tale, the grand jurors are almost finished taking Monica's secret testimony. They raise one last line of questioning they're determined to explore: What precisely happened when Monica was first approached by agents and prosecutors that January day in the Ritz-Carlton food court near the escalator? In meeting privately among themselves earlier, they've agreed that this is a critical area they want Monica to describe.

Immediately, the lead prosecutor inside the locked and guarded grand jury room tries to forestall that line of questioning. "What areas do you want to get into?" Michael "Mike" Emmick asks the jury foreperson. "Because there's—you know—many hours of activity—"

The jurors are not deterred. Specifically, one of them quickly says, they want to know when Monica first learned that Linda had been taping her phone conversations.

Not until she read it in the press, Monica replies, then recounts how Linda tried to hug her and said "they had done the same thing to her."

As soon as Monica completes her answer, prosecutor Emmick again attempts to cut off that line of questioning.

"Any other specific questions about that day?" he says. "I just—this was a long day. There were a lot things that—"

Jurors, acting virtually as a Greek chorus, interrupt him. They *insist* on hearing what happened to Monica when she's interrogated by the Starr team in Room 1012.

"We want to know about that day," a juror interjects.

"That day," another chimes in.

"The first question," still another speaks up.

"Yes," a fourth adds.

"We really want to know about that day," a fifth states.

Emmick relents. "All right," he says.

Monica begins to reply, then asks if a female prosecutor, Karin Immergut, whom she has come to respect, can lead the questioning—and lead it in the absence of prosecutor Emmick, who had taken a leading role in her hotel room interrogation in January. Now, Monica addresses Emmick and says: "Can I ask you to step out?"

Emmick: "Sure. Okay. All right."

Prosecutor Immergut attempts to continue while her colleague, Emmick, remains nearby. "I guess, Monica," she says, "if Mike could just stay—do you mind if Mike is in here?"

Monica nods affirmatively. She *does* mind.

"Okay," Immergut continues. "Would you rather—"

Again, the stenographic transcript shows Monica nods affirmatively. She wants to proceed without him.

Emmick leaves. The jurors finally get to hear what they clearly most want to hear: How Emmick introduces himself in that room and tells her "that Janet Reno had sanctioned Ken Starr to investigate my actions in the Paula Jones case; that they knew I had signed a false affidavit; they had me on tape saying I had committed perjury...that I could go to jail for twenty-seven years; they were going to charge me with perjury and obstruction of justice and subornation of perjury and witness tampering and something else."

They want her to cooperate. When she asks "what cooperating meant," they explain it means agreeing to be debriefed, wear a wire, then place phone calls or make personal visits to Betty Currie, Vernon Jordan, "and possibly the President." They warn they "would be watching to see" if she attempts to signal any of those prospective "targets" that she's surreptitiously recording their conversations. "They kept telling me that . . . I couldn't tell anybody about this, they didn't want anyone to find out. . . ."

One of the reasons they don't want her to call her attorney, Frank Carter, is "because they were afraid that he might tell the person [Vernon Jordan] who took me to Mr. Carter."

Another juror presses her on this point. "Did they ever tell you that you could not call Mr. Carter?"

"No," she replies. "What they told me was that if I called Mr. Carter, I wouldn't necessarily still be offered an immunity agreement."

"And did you feel threatened by that?"

"Yes."

The jurors are especially interested that she's being questioned without her attorney present; they keep returning to that fact. "So if I understand it," a juror says, "you first met the agents . . . around 1 P.M. and it wasn't until about 11 P.M. that you had an opportunity to talk to a lawyer." That's correct, she says, adding she was also told that "Frank was a civil attorney and so that he really couldn't help me anyway. So I asked if at least I could call and ask him for a recommendation for a criminal attorney and they didn't think that was a good idea."

"Sounds as though they were actively discouraging you from talking to an attorney," another juror asks.

"Yes," she answers.

"Is that a fair characterization?"

Again, "Yes."

All the while, more and more agents and prosecutors enter the hotel room. By now, Monica is in a state of near-hysteria, crying, sobbing, trying to think, control herself, respond. "They kept saying there was this time constraint, there was a time constraint, I had to make a decision . . . then Jackie Bennett came in and there were a whole bunch of other people and the room was crowded and he was saying to me, you know, you have to make a decision. I had wanted to call my mom, they weren't going to let me call my attorney, so I just—I wanted to call my mom, and they—" It's at that point that Bennett cuts her off by saying "you don't need to call your mommy."

Another juror, listening to all this, asks again: "During this time in the hotel with them did you feel threatened?" When she says yes, the juror continues: "Did you feel that they had set a trap?" She replies, stammeringly: "I, I, I did and I had, I didn't understand, I didn't understand why they, why they had to trap me into coming there, why they had to trick me into coming there. I mean this had all been a set up and that's why, I mean, that was just so frightening. It was so incredibly frightening."*

Three hours and fifteen minutes after arriving in the hotel room, she's permitted to call her mother in New York. At first she fails to reach her, then calls again an hour later. Marcia Lewis, her mother, is stunned after hearing a hysterical Monica say she's been trapped and held by the FBI in a hotel room and needs her mother to come to Washington immediately. After speaking briefly with Monica, her mother asks to speak to one of the attorneys. She tells Emmick that she's taking the Metroliner train to Washington. Hurriedly, she leaves her New York apartment accompanied by her

---

*In *Truth at Any Cost,* written with the cooperation of Starr and his prosecutors and based on hundreds of hours of interviews with them, thirty hours with Starr alone, my colleagues Susan Schmidt and Michael Weisskopf conclude that the hotel interrogation with Monica "has been spun and distorted," claiming that Monica "was not mistreated...if anything she turned the tables on the prosecutors and had them rushing around trying to placate her" using "the same theatrics and emotionalism she used on President Clinton." They also say, accurately, that the prosecutors "did not need a warrant or a subpoena to seek her cooperation that night, and she did call her attorney that night." They acknowledge the prosecutors "did discourage her from calling an attorney," and justify it because they thought her lawyer "was involved in obstructing the Jones case." I disagree with these conclusions. I believe the more the events of that night are examined, the more outrageous the prosecutorial abuse of power becomes.

eighty-year-old mother, who suffers from Alzheimer's and knows nothing of what is happening, and her sister, Debra Finerman. Four hours later Marcia Lewis phones from the train to say it's experiencing travel delays. Finally, she arrives in Room 1012 at 10:16 P.M. "I was terrorized," Marcia Lewis recalls. "I found my daughter weeping, surrounded by FBI agents and prosecutors, being told she might go to jail." (Months later, during her own secret testimony, the mere recollection of that scene causes Marcia Lewis to break down, leave the grand jury room crying loudly and exclaiming, "I can't take it, I can't take any more. I can't stand it." She receives emergency medical treatment and is escorted out of the courthouse.)

After a brief but stormy conversation with Monica outside the hotel room—they argue loudly in the hallway, with Monica telling her mother "I'm not going to be the one who brings down this administration"—her mother is taken to another room on the same floor. For the next forty minutes she meets there with the entire prosecutorial team, an encounter duly recorded in the same sterile but powerfully revealing way on the official chronology:

"Lewis advised that this was an emotional experience for Lewinsky. Lewis advised Lewinsky was younger than her chronological age. Lewis asked if tapes were admissible in court. Lewis advised she wanted to protect Lewinsky. Lewis asked how she would know Lewinsky would not be charged if she cooperated. Lewis asked about Lewinsky's safety if Lewinsky cooperated. Lewis advised that Lewinsky talked about suicide six years ago. After Lewinsky's parents divorced, Lewinsky saw a therapist, but she is not currently seeing one. Lewis advised she alone could not take responsibility for convincing Lewinsky to cooperate with the OIC."

Her mother then calls her ex-husband, Bernard Lewinsky, in Los Angeles. Monica's father speaks with Emmick; the summary of that conversation shows the following topics discussed: "Cooperation, an interview, telephone calls, body wires, and testimony were mentioned."

Before hanging up, Bernard Lewinsky says he'll call back from a pay phone shortly. When he does, he informs Emmick that

Monica is now represented by an attorney, Bill Ginsberg, whom he has just retained for her. In a separate brief conversation with Monica, her father also tells her about Ginsberg, whom she knows as a longtime family lawyer and friend.

Even this news does not seem to satisfy the prosecutors. A minute later, Emmick asks Monica if she has an attorney. Yes, she says, naming Ginsberg. He then tells her "she did not have to accept an attorney she did not select."

Some twenty minutes later Ginsberg calls from Los Angeles and speaks to Emmick, "who advised Ginsberg he was uncomfortable with the relationship between Ginsberg and Monica Lewinsky." More talk ensues, with Ginsberg holding on the phone, and Emmick questioning Monica. When asked again if she has an attorney, Monica says it's Ginsberg.

At 12:17 in the morning, eleven hours and twelve minutes after being brought to the hotel room and interrogated without the presence of an attorney, Lewinsky finally gets to speak to a lawyer—by phone, three thousand miles away, for exactly four minutes. Ginsberg tells her not to say anything else until he meets with her personally after flying from Los Angeles to Washington. Monica never does speak to Frank Carter, her only attorney of record during the entire time she is interrogated in the hotel room.

After Emmick ends her brief phone conversation with Ginsberg, he tells Monica and her mother they're free to leave.

Still struggling to control her conflicting emotions—fear, anger, humiliation, uncertainty, dread—Monica finds herself thinking *I just don't want to cooperate, I'll just say I made it all up,* all compounded by other intense feelings: "I couldn't imagine doing this to the President. And I felt so wrong and guilty for having told Linda and that she had done all this." What she doesn't know is that even as she is experiencing her ordeal in that hotel room, Linda Tripp has left the same hotel to brief the Jones lawyers in another hotel room about what is taking place between Monica and the prosecutors.

In the end, Monica doesn't break. Before leaving with her mother at a quarter to one that Saturday morning, January 17, she tells them, "I'm letting you know that I'm leaning towards not cooperating." She has only a relatively few hours left before the most intense public spotlight begins to shine on every aspect of her life.

II. THE TRAP

Across the river, inside the White House, the President also has only the same few hours left before he faces the most pitiless and humiliating public scrutiny any chief executive has experienced. He, too, quickly discovers he's entered a trap—albeit a trap he himself helps fashion by his failure to act responsibly as an adult and as a president. However it's fashioned, the trap begins to swing shut after he arrives by official limousine later that Saturday morning, January 17, in downtown Washington to give his sworn deposition in the Jones case.

However difficult the president suspects that deposition might be, nothing can possibly prepare him for what he encounters. The set of questions that he's asked after he takes the oath is unprecedented in presidential annals. The Jones lawyers spell out for him what constitutes the definition of "sexual relations" they will be using in the deposition: Knowingly engaging in or causing "(1) contact with the genitalia, anus, groin, breast, inner thigh, or buttocks of any person with an intent to arouse or gratify the sexual desire of any person; (2) contact between any part of the person's body or an object and the genitals or anus of another person; or (3) contact between the genitals or anus of the person and any part of another person's body."

Contact, the lawyers helpfully define, "means intentional touching, either directly or through clothing."

When asked about whether his "contact" with Monica meets those criteria, the president denies, under oath, having any such contact. They have him, it seems, on perjury.

The very fact that he's testifying at all underscores what the Paula Jones case has become.

Nearly four years have elapsed since Paula first announced she was suing the president. By 1998, after all the legal maneuvering, all the headlines, all the leaks, and all the rumors, Paula's life has changed along with the cast of characters surrounding her. In 1997, she rejected a settlement her first set of lawyers, Gilbert Davis and Joseph Cammarata, had negotiated with the president's lawyers. After she did, they expressed their frustration by writing to her that, "Your focus has changed from proving you are a good person to proving Clinton is a bad person. That was never your objective in filing suit."

By the time the president gives his deposition, however, Paula's initial legal team has been replaced. Now Paula has been taken under the wing of yet another bleached-blond conservative activist, Susan Carpenter-McMillan, who becomes a familiar figure nationally assailing the president and defending Paula on the cable-TV talk network.

Carpenter-McMillan is a Southern Californian and a singularly Nineties kind of person, especially a Teletimes kind of person. She thrives in an age of scandals and celebrities. "I knew that if I came on the team, they were short-lived," she says of Paula's first attorneys of record. "I wanted them out. I wanted them gone the first week because I knew they were incompetent—not incompetent lawyers, but incompetent for *this* team."

She means, as she explains, she was bringing together a team that handles all the elements of a national case like this one: media, public relations, political attack strategy, enhancing poll returns. As she says, "When I work for a client I bring in A to Z for the whole thing. I mean, I need big guns." She herself plays the public role of biggest of guns, and the role she assumes fits the times, too. She's all over the talk shows, clearly glorying in her fifteen minutes of fame. Asked how many times she appeared on national TV talk shows on Paula's behalf, she says, "Every day, every day, sometimes two and three times a day"—and that doesn't include, she boasts, the number of times the appearances are rebroadcast, sometimes as many as ten times a day.

Paula, she says, becomes a cause; she, Susan, becomes the spokesperson for, and public defender of, "a young naive Arkansas girl" who thanks in part to her tutelage becomes "a very savvy bright California woman." She describes Paula, patronizingly, drawing out a fake accent, as her "leetle seester."

Carpenter-McMillan came to her position after riding the

media controversy trail, having been for thirteen years spokeswoman for the largest California right-to-life group, a position she
held until it was publicly reported she herself has had an abortion.
She attracted more notice, she recalls, after "ABC–7 in L.A. called
and said we're looking for a conservative commentator," which
she becomes, all the while being identified with "victim-oriented"
groups. In that guise, she meets Paula through a mutual friend,
and, as she tells it, out of friendship and compassion begins talking
to Paula about how she's not faring as well as she should in the
battle for public opinion: "So I looked at the case, evaluated it,
spent some time studying it, and said: Here's my condition. If I
come on board, I will not take a salary. My payoff will be any new
journalistic [and books and TV] deals [that] will be used down the
road to further our projects with children. . . . I evaluated it quietly
to myself and said, These lawyers are nice guys but they're not cut
out for a high profile case. I looked at Paula and how they'd been
dressing her and doing her hair, and I said, This has all got to go!"

She takes over, and, in her hardly immodest telling, looking
back from the vantage of nearly a year after the president's deposition, sketches not Paula's story but her own: the story of a female
Mr. Deeds who comes to Washington to slay the evil dragons. "I
can spend four hours telling you about my journey with Paula, one
of the most fascinating journeys I ever experienced. I mean, young
woman, kind of in-your-face California gal, realizing that what I
saw in Washington didn't play. Meeting the president of the
United States's lawyers, interacting with that whole group, that's a
whole other story, whole other journey."

As she says, it snowballs. She becomes known as "the Soundbite Queen," adding: "This is kind of an old sexual, nasty kind of
phrase, but apparently I give good TV."*

*In the car driving back to her San Marino home after several hours of mem-
orable, and repellent, conversation in the summer of 1998, she describes
what it's like to be, in her view, at the center of history, even shaping history,
and raising public questions unlike any others in U.S. experience. "Can you
imagine what it would have been like if anyone had asked about the size of
Abraham Lincoln's penis?" she says.

Out of that hot media background, Susan accompanies Paula to Washington for the president's deposition that Saturday morning. Though she's not permitted to be in the building itself when the president is questioned, and has no idea whether the session has gone poorly or well for Paula, typically she works to create a soundbite for the media that conveys success. "I thought, 'We're going to make it look like it went wonderful,'" she says, "so I said I need a restaurant nearby that's not too expensive but a cut above. I need a window. That's it."

She books a window table at the famous Old Ebbitt Grill, a block from the White House, leaks the setting to the media, then tells Paula and company as they emerge from the deposition that they're going to act as if it has been a signal success. Actually, as she says, "Paula could not tell me anything that had happened in the depo because I was not a part of the team and there was a gag order. So I never knew, even though people thought I did, which I loved; I loved them to think I knew when I didn't. So I said [to Paula and company]: 'I don't know if it went good in there or bad in there, and I know there's a gag order, but I'll tell you what, you're going to sit your butts down, you're going to put on the biggest smile on your face, and you're going to order champagne.'"

Which they do. Their smiling faces, their champagne glasses lifted high, their celebratory manner, are all captured in TV shots and newspaper photos that night and the next morning.

As it turns out, those pictures don't lie. Paula's side has as much reason to celebrate as the president has to despair.

· · ·

Bill Clinton returns to the White House shaken. The range of questions, the specificity of the details, the knowledge that horribly embarrassing revelations are likely to be made public soon, all send the president into a defensive, reactive mode. He cancels scheduled plans to occupy the presidential box at the Kennedy Center that night with Hillary, Vernon Jordan, Erskine Bowles and their wives, and begins a desperate attempt to regain control of a situation that appears to be perilously slipping out of control.

Starting around 5 P.M., he initiates a flurry of calls, three of them to Vernon Jordan within a short space of time: one to Jordan's home, one to his mobile phone, one to his office. He also calls his secretary, Betty Currie, at her home that night. They need to talk, he tells her urgently. Come to the White House the next day, Sunday, he commands.

Around 6 P.M. another shoe drops, but one of which the president at first is unaware: *Newsweek* decides not to publish the Isikoff story about Clinton and the intern until further checking is completed. This frustrates Lucy Goldberg's expressed belief that "the shit was about to hit the fan." However, this news, like so many other scandal items, is quickly leaked to the Internet gossip, Matt Drudge, who has been in constant contact with the network of people working to bring down the president.

At 11:27 that Saturday night, under the headline BLOCK-BUSTER REPORT, Drudge posts the following breathless item on his Web site, *The Drudge Report*: "At the last minute, at 6 P.M. on Saturday evening, *Newsweek* magazine killed a story that was destined to shake official Washington to its foundation: A White House intern carried on a sexual affair with the President of the United States!"

Drudge further reports: "reporter Michael Isikoff developed the story of his career, only to have it spiked by top *Newsweek* suits hours before publication. A young woman, 23, sexually involved with the love of her life, the President of the United States, since she was a 21-year-old intern at the White House. She was a frequent visitor to a small study off the Oval Office where she claims to have indulged the president's sexual preference. . . ."

The "White House intern" is not named.

From there the story spreads through the ether of the new interconnected electronic world and enters the news stream inhabited by ultra right-wing opponents of the president.

At 2:23 A.M. Sunday, the Drudge item is repeated verbatim on alt.current-vents.clinton.whitewater.newsgroup, a Clinton enemy Web site. By early Sunday afternoon it spreads to the alt.impeach.clinton site, then to alt.politics.clinton and to talk.politics.misc.

Sunday brings a day of even greater presidential anxiety.

When Betty Currie arrives at the White House, she finds the

president on the putting green with his dog, Buddy. Once in the Oval Office, he says he needs to talk to her about Monica Lewinsky in view of questions he was asked during yesterday's deposition. He needs to "refresh his memory" about Monica. The president rapidly proceeds to list a string of questions, one after the other, that he wants Currie's reaction to; they're phrased both as questions and statements at the same time: *I was never alone with Monica, right? You were always here when she was here, right? Monica came on to me, and I never touched her, right? You could see and hear everything, right? We were never alone, right?* To each of his question/statements, Currie answers, "Right." Clinton also tells her, "Monica wanted to have sex with me and that I cannot do."

After Currie returns home Sunday night, she's awakened around midnight by another call from the president. Never before has he called her that late. He asks if she's seen the latest news about Monica. When she says no, he says she can read about it on the Internet over her home computer in *The Drudge Report.* It's not good, he adds. Then he instructs her to get in touch with Monica.

As ordered, and despite the hour, she pages Monica twice. The first brings no response. The second, later, brings a most upsetting response: Monica says she can't talk to Betty. Then she hangs up.

Nothing better captures the panic that descends on the top level of the White House than the official White House phone log the following morning, Monday, the nineteenth, tracking the desperate frantic dialing of Betty Currie as she places call after call after call trying to reach Monica on behalf of the president. Currie, employing the code name "Kay" that she uses in contacting Monica, leaves the following pager messages that are retrieved later by the special prosecutor:

7:02 A.M.: "Please call Kay at home at 8:00 this morning."

8:08 A.M.: "Please call Kay."

8:29 A.M.: Currie calls Monica's home.

8:33 A.M.: "Please call Kay at home."

8:37 A.M.: "Please call Kay at home. It's a social call. Thank you."

8:41 A.M.: "Kay is at home. Please call."

8:44 A.M.: "Please call Kay re: family emergency."

8:50 A.M.: Currie calls Clinton.

8:51 A.M.: "Msg. From Kay. Please call. Have good news."

Currie is not the only person consumed by the looming Monica scandal. Vernon Jordan is another swept up in the by-now fast-moving developments. He first learns of *The Drudge Report* during a lunch on Sunday with Bruce Lindsey, the president's counsel and perhaps the closest to Clinton of the Arkansas group that comes to power with him. Lindsey shows Jordan a printout of the report. From that point, the Monica business appears to obsess Jordan.

The next morning, Monday, January 19, Jordan calls the White House. He gets cleared to see the president where, during lunch, Jordan gives him a copy of *The Drudge Report*. The two then adjourn to the Oval Office where they meet, face-to-face, alone. Jordan briefly discusses Monica's situation: Monica now has a job. She has signed the affidavit.

After he leaves, Jordan makes a number of phone calls: to Betty Currie at home; twice to Frank Carter, who informs him he has been replaced as Monica's attorney by a California lawyer named Ginsberg; to Bruce Lindsey, unsuccessfully, wanting to talk about Ginsberg. In one hour, he pages Monica three times; each time she fails to return his pages. Later, he denies those pages have anything to do with the Matt Drudge report. Although by then Drudge has updated the story and named "Monica Lewinsky" as the president's alleged paramour, Jordan later can't recall why he's trying to reach her. "I don't know why I was calling Monica Lewinsky," he tells the grand jurors during his secret testimony. "I'm certain I was not calling to ask her about *The Drudge Report*."

In all, Jordan's phone logs show thirty-one calls between himself, Monica, Frank Carter, the White House Counsel's Office, and the president that Monday.

Though reports about a White House sex affair are whizzing through cyberspace and being pushed especially hard on conservative talk radio, the mainstream media still has not carried the story. One brief exchange about *Newsweek* having dropped a story about an alleged White House sex scandal occurs Sunday on ABC's

This Week with Sam Donaldson and Cokie Roberts, but no details are mentioned.

The story remains in the scandal rumor category. The American people are still almost entirely unaware of it.

By Tuesday night, the twentieth, Drudge is reporting more leaked details: Federal investigators possess taped phone conversations that substantiate rumors of the presidential affair; Linda's tapes are finally making their way into the public conversation.

Late that night all this changes when, in a later edition, the *Washington Post* carries its page-one story about the president and the intern.

The bonfire has been lit. A media maelstrom sweeps the nation and the world, a maelstrom that eclipses even O.J., even Princess Di. Now it's "All Monica, All the Time."

Two notes, culled from countless pieces of scandal debris that immediately begin littering the political landscape:

When the president awakens Vernon Jordan at 6:30 that Wednesday morning in Jordan's room in the St. Regis Hotel in New York to tell him about the *Post* story, Jordan asks the president if he's told the first lady about it. He has not, he says, she's still asleep. He'll speak to her when she wakes up. Jordan counsels him, "Take it easy. We'll get through it."

The next day, while the media firestorm rages and the president makes his repeated public and private denials of knowledge about Monica and sex, Jordan receives a call from another of his well-placed friends—Peter Straus, formerly head of the U.S. Voice of America and once married to a member of the Sulzberger family, owners of the *New York Times*. Straus, as Jordan knows well, lives in New York and is soon to be married to Marcia Lewis, Monica's mother.

"Where's your girlfriend?" Jordan asks after taking Straus's call. Straus replies: "She's on suicide watch with her daughter."

* * *

It was probably inevitable.

Even before Monica, all the factors were in place for "the mother of all scandals" to erupt and preoccupy the nation, if not the world. The convergence of the combustible new forces of the electronic culture—the Internet; the rise of twenty-four-hour talk cable networks; the decline of network news viewership and newspaper readership; the fierce new competitive battles to attract wider audiences; the lowering of journalistic standards of what's "fit to print"; the further blurring of the lines between public and private lives; the new acceptance of, if not hunger for, public discussion of intimate details once thought unmentionable; the nonstop coverage of O.J.–type spectacles that rivet public attention on scandal and raise ratings and attract readers; the absence of critical news events to refocus public attention on other issues; the long and ceaseless effort by right-wing elements to "get Clinton"; the tempting target of perhaps the most reckless, and vulnerable, of presidents, William Jefferson Clinton; the increasing ideological divisions that poison the political system—all combine to elevate the Monica story into a new dimension.

Then there's the oldest of stories, the story that always sells the most—sex. It's all about sex—sex and power, sex and betrayal, sex and exposure, sex and scandal, sex and the fall of the mighty—and all told in the most embarrassing and titillating detail possible. Nothing's left to the imagination. Every twist and turn is relentlessly offered to and shared by the widest audience in history.

The story of the president and the intern isn't the first sex scandal to engulf the White House, of course. Even in the earliest days of the Republic, Americans were treated to an astonishing number of scandalous accusations involving their leading public servants. Jefferson, Hamilton, Burr, Jackson, all endured stories—proved or not—about their mistresses, their adulterous affairs, their illegitimate children. Jefferson was accused of "frolicking with his 'Congo harem' and adding to the labor force at Monticello by an annual increment of mulattos"; Hamilton and Jackson of adultery by rival papers. Cleveland's fathering of a child out of wedlock produced the taunting public chant: *Ma, Ma, Where's Your Pa/Gone to the White House, Ha, Ha Ha.* Harding and Nan Britton cavorting in the White House closet; FDR and Lucy Mercer; Ike and his British driver; Jack Kennedy and Marilyn, Fiddle and Faddle, gangsters' molls and prostitutes, all are not that unfamiliar to the public, especially a public conditioned to the scatological soap operas of the scandal/celebrity culture of the Nineties.

But none of them, whether viewed alone or in combination, comes close to approaching the omnipresent impact of "All Monica, All the Time."

Credit for that, at least in part, goes to technology. By the time Monica occupies center stage the electronic techniques to best exploit the latest scandal are also firmly in place. They've been tested and perfected. Again, the O.J. phenomenon provides a model.

In Monica, as with O.J., television screens are constantly filled with montages of faces whirling into place. Finally, they close on one scene: People pressed against a rope line, jostling each other, clamoring to reach out and shake hands with the passing president. And there, always in the center, right up front, stands a chubby young woman in a black beret. Ecstatically, she throws back her head and opens her arms to embrace the fleeting figure of the Leader of the Free World.

Freeze scene. Slow dissolve to black.

This scene, played over and over, around the clock, day after day, month after month, takes place against the backdrop of a drumroll of dramatic theme music. Superimposed letters flicker across the screen spelling out the words: *Crisis in the White House* and *Scandal in the Presidency.*

As it was with O.J., but even more so with Monica, scandal talk takes on a life of its own. It comes complete with its own graphics, theme music, and titles. It even comes with many of the same characters—Johnnie Cochran, Marcia Clark, Geraldo Rivera—who became so familiar night after night as they heatedly argued the guilt or innocence of a celebrated athletic superstar. Now the cast expands, entire new shows are added—*Hardball with Chris Matthews, Hannity & Colmes*—louder, more hyperbolic, more polemical, more accusatory, more outrageous as their new target becomes the president of the United States and a sordid sex scandal.

In terms of hours of constant coverage, of updates, rumors, allegations, new conspiracy theories, continual predictions of the outcome (most of them wrong), sheer domination of every form of the news media, the story of the pathetic intern and the wayward president turns into the most closely observed news event of all time.

It's like a huge furnace that needs to keep burning to function. It doesn't matter what it burns. It will burn firewood, or valuable furniture, or money. The fire treats all fuel the same. As it keeps burning, it needs more and more fuel to sustain itself, more scandalous tidbits and sensations to keep the blaze roaring. So one scandal becomes like any other, more fuel for the furnace.

Bill and Monica blend into Iran-Contra which merges into Watergate; they're all the same, all light up the national territory. Of course, they're not all the same; but who's to say otherwise? Even if someone *does* point out the differences, who pays attention amid the diverting glow of the great bonfire? Whatever people say about their dismay at witnessing the conflagration, they want to see more and more fuel fed into it. They become the furnace. The latest scandal becomes the greatest.

With Monicagate—the last of the galloping-*gates* of the Clinton years—the country has a scandal that tops them all in terms of personal behavior, or, more properly, *mis*behavior.

Monicagate, with its assorted characters and miserable, petty revelatory details, takes its place as the American *Vanity Fair.* The events it encompasses become elements in a modern version of *A Journal of the Plague Years.*

· · ·

It's a rich, but ignoble, cast of characters: defenders, investigators, fixers, haters, enablers, conspirators, blonds, talking heads, operatives, spinners, hustlers, liars. Of them all, the most important, and revealing, are The Intern. The Plotters. The Prosecutor. The President. Like them or not, they are much more representative of the country and their times than the crude caricatures drawn of them for the public, and far more telling for what they say about America during the boom.

The Intern

The first thing to know about Monica Lewinsky is that she's from Brentwood, the very same Brentwood that reigns as America's celebrity capital, home to O. J. Simpson as well as countless Hollywood icons. The second, not so significant except for its obvious irony, is that she comes to Washington and lives at the Watergate, thus neatly, and surely unintentionally, linking the two worlds of scandal and celebrity, Hollywood West with Hollywood on the Potomac, the entertainment capital with the political one.

How intertwined those capitals have become by the Nineties is shown by the way Monica herself makes the bicoastal move and gets her White House internship. Her benefactor is a wealthy Democratic donor and family friend by the name of Walter Kaye, a bubbling wisecracking character who comes over in his grand jury testimony as a Borscht Belt comedian/insurance salesman—a successful Willy Loman type.

Kaye made a fortune selling insurance and parlays his money into purchasing the highest political access. His portrait of how he became involved in politics perfectly captures the insidious connection between money and politics—if not actually corrupt, as it often is, at the very least corrupting of the ideals of a democracy that supposedly seeks to provide equality of opportunity for everyone, not the inequality of opportunity that comes only to the wealthy and the most powerful. Walter Kaye demonstrates how you can buy your way in, and then buy your way up. Money buys access. Access buys influence, and sometimes much more. It earns him a brief footnote in Scandal Times: Betty Currie and Monica adopt his name for their code, *Kay,* when Betty contacts Monica.★

His political influence odyssey begins, he recalls to grand jurors in typically exuberant manner, when he's talking to a woman official at the Democratic National Committee. "Listen," he tells her,

---

★The code names used would be childish if they weren't so serious—in the same way the humorous dialogue in the hit TV drama *The Sopranos* makes the violent acts portrayed seem innocent. Monica's mother and aunt, Marcia and Debra, call Vernon Jordan *Gwen;* Monica calls Hillary *The Babba* (grandmother), and the president *Her* when exchanging messages with Betty Currie. Monica also refers to the president as *Henrietta* during phone conversations with her therapist in Los Angeles; Linda asks Monica to call her *Mary;* Isikoff calls himself *Harvey* when contacting the Plotters, Linda and Lucy, and so on.

sounding like a parody of the junkman wheeler-dealer in *Born Yesterday,* "I'm an excitement nut. I like excitement. You offer me some exciting times, I will give you a contribution."

The official replies, "If you like excitement, Mrs. Clinton is speaking at the Mayflower Hotel. Why don't you come as our guest?"

He does. "And as soon as I get there, I know, boy, they're really working on me," he gurgles. "They put me in the first row for the lecture and it's the Women's Leadership Forum and very intelligent people, you know, bank presidents, college presidents, attorneys, judges. . . . And then when I walked into the luncheon, what do you think they do? They seat me at her table! I'm telling you, they should be in the insurance business. Don't laugh, because I got too much pain. And boy, I love it. I'm telling you the truth, you know. I heard her speak. I was very impressed. Never looked at a note, you know, and very interesting. And here I'm sitting right at the same table with her. So I say goodbye to her, you know, and we start talking, you know, and I suppose you realize it's easy for me to make conversation. That's my business, really. And right after the luncheon, David Wilhelm [then Democratic Party chairman] appeared on the scene. He took me upstairs, and I gave him a contribution. I became a managing trustee of the Democratic Party. That was my first check, I remember that."

And not the last one, either. Your political system at work, thank you.

He goes on to describe for the jurors how he's become friendly with Hillary, telling them he picks up the phone to call her whenever he needs to reach her. If Linda Tripp's sworn testimony is to be believed, Monica tells her that Kaye "pays for all of Hillary Clinton's parties." He exchanges gifts with the Clintons. He gives Chelsea Coca-Cola stock.

In a line that deserves a note in the social history of the boom, he explains, "The reason I like to give stock to young kids was they get the dividend every three months, you know."

He adds, articulating the creed, "I'm a great believer in common stocks."

He also becomes invaluable to the president for another reason after the scandal breaks: he discovers a loophole in an insurance policy that allows the president to receive attorney fee payments in connection with the Paula Jones suit.

Kaye's connection with Monica comes through his long friendship with Monica's aunt, Debra Finerman, and her husband.* Through them he meets Marcia Lewis, Monica's mother, and Monica. He helps Debra and Marcia get Monica her internship by "making some calls" to the White House and also assisting Monica with her application.

So Monica Lewinsky comes to Washington and becomes part of history, settling somewhere between a footnote and the main text of the times.

Though many of the portraits of her drawn during the scandal make her seem an aberration—the "stalker," the conniver, the insufferably imperious young queen—she's much more complicated, and much more representative of a significant slice of her generation than seems to be understood.

In some ways, she's a perfect Nineties Beverly Hills Girl–type: affluent, selfish, accustomed to getting what she wants, with a surpassing belief in her sense of entitlement. She's the product of a broken, if not dysfunctional, home, who grows up, as she tells Linda Tripp memorably on one of those tapes, learning from an early age how to lie. "This is how my family is," Monica tells Linda when trying to persuade Linda to lie if necessary about the affair with Clinton. "I would lie on the stand for my family. That is how I was raised as a family...." While Monica is sure Linda was "raised in a very sort of honorable family," she on the contrary, "was brought up with lies all the time....How you got along in life was by lying." She explains: "Once my parents were divorced,

*And perhaps something more than "friendship," if other sworn grand jury testimony is to be believed. The jurors are told that Kaye and Finerman had a sexual affair lasting twelve years. Debra, who frequently writes political gossip with her sister Marcia, also clearly is not averse to cashing in on Monica's notoriety. Because Monica is interested in Christian Science, Debra meets a book agent about doing a book on the subject. She's told that Monica's the only one who could make money from that project.

if I wanted money from my dad, I had to make up a story. When my parents were married, my mom was always lying to my dad for everything. Everything. My mom helped me sneak out of the house. I mean that's just how, that's just how I was raised."

Her relations with her parents are depressing.

Monica on her father: "He's a person that you say something and then it's a week or so later that he pulls you in for a conference." She resents his second wife, constantly referring to her as "the step-monster." When she was having trouble in school as a teenager, she says her father didn't come to her aid. According to Linda, she also feels he doesn't give her sufficient financial support, a long-standing grievance that causes Monica to blame him for his unwillingness to pay for her to go to a more expensive university when she attends Santa Monica Community College. "Every single time Monica talked to me about what her dad didn't do for she [sic] and her mom it was always about money. 'He didn't get me this for Hanukkah because the [redacted] says it's too much money.' 'He didn't do this.' 'He didn't send me a ticket because he's spending it on something else.' 'He didn't care about me as much as he cares about my step-monster.'"

Monica's mother, Marcia, has a permissive parenting style that doesn't let her interfere with her daughter. She doesn't challenge or discipline Monica; she doesn't want to know details about the affair, perhaps because testimony states she herself has had affairs with married men; she doesn't chastise her or lay down the law about ending it. She "humors" her. Monica does as she pleases.

> Prosecutor: Did it strike you as unusual that your daughter, a 23-year-old, was writing Valentine's Day messages to the President and referring to him as Handsome?
> Marcia: Yes.
> Prosecutor: And did you ask her about that?
> Marcia: At that time, no.

She comes over as ineffectual, somewhat helpless, but "a nice Jewish mom." If anything, Monica seems more like the parent than Marcia. Monica is aggressive, assertive; Marcia defensive, tentative.

Monica is highly materialistic and self-indulgent. She's constantly

shopping, constantly buying presents for people from casual acquaintances to best friends. Gifts are a large part of Monica's life. She
gives them to everyone. Her idea of a great time, as she enthuses in
an e-mail to a friend in Tokyo, is a daylong self-indulgent shopping
spree: "Ohhh how I long for the time when we can just spend a day
together," she writes, sounding like a parody of the affluent Nineties
lifestyle of the young, "starting w/coffee at Starbucks...shopping...
lunch at somewhere yummy...maybe a movie...more shopping...and then getting drunk on margaritas!!! Whooo-hoooo!"

Whether she buys all her gifts to bolster her "friendship" with
acquaintances or in an attempt to purchase their respect and assistance are unknown, but no matter, she's always passing out the
gifts—to former boyfriends, to a Secret Service agent, to a White
House butler, to fellow interns, to Betty Currie, to Linda, to the
president. Nice gifts, too: facials, Tiffany hand creams, antiques,
sterling silver cigar holders, designer ties (to the president alone ties
from Hugo Boss, Feraud, Ermenegildo Zegna, Calvin Klein), rare
books including, for the president, ones on Theodore Roosevelt
and a volume about Peter the Great published in 1802.

Cost is of no consequence. She just charges it all on a credit
card. Even in the midst of the affair, she still acts like a little girl,
using her mom's credit card and never asking "How much?" And
why should she? Her mom's pattern is to keep bailing her out
when she runs up unpaid phone bills in college or charges hundreds of dollars beyond her credit-card limit.

From Linda's tapes, when she and Monica are talking about a
Christmas present Monica buys for the president:

Linda: Was it expensive?
Monica: I don't know. I used my credit card, or my mom's
    credit card.
Linda: You should at least find out how much it is. [sigh]
Monica: I'm sure it wasn't much.

Even if it is "much," that doesn't seem to matter. Monica describes going shopping with her mom in New York. They want to
buy a fox fur coat for Marcia and a shearling coat for Monica. They
get Monica's coat for $1,995; originally it cost, $5,790. A bargain.

Deep down, Monica's a snob.

The tapes:

Monica: Well, these people, you should have seen their house and lifestyle.
Linda: Really?
Monica: Oh, disgusting. They used to put Cheez-Whiz on everything.

She's not only bold and brazen—brazen enough to lift her skirt and flash her bikini thong at a president she doesn't even know—but she possesses chutzpah in abundant quantity. She treats people of the highest standing as her equal. In a note to Vernon Jordan, she writes: "It makes me happy to know that our friend has such a wonderful confidante in you." As if, it hardly needs be noted, she and Vernon Jordan have the same level of friendship with the president. Nor does she lack for pushiness. Her memo to Jordan describing the kind of job she wants says volumes about this side of her character: "It is important to me that I be engaged and challenged in my work; that I not be someone's administrative/executive assistant." Her memo also includes three "tabs" with salary requirements and résumé.*

Monica Lewinsky, age twenty-four, barely beyond internship level, no real professional credentials, no conscious career ambitions, is nobody's secretary. This from someone who, though ob-

*Jordan tells the jurors that Monica, in effect, has caught a common Washington disease: a minor functionary becoming puffed up with self-importance solely because of exposure to lofty figures. As he says, recalling Monica's demeanor as she offers advice on how the president should settle the Paula Jones lawsuit: "She is not the only person that you see afflicted with—based on a tad of exposure to process and people they somehow believe that they're experts.... It's like when I was a civil rights lawyer in the South. You'd get a reporter from New York who came to the South and stayed one day and he thought he was an expert on race relations in the South. You understand what I'm saying? And so I think this exposure to the White House...made her feel like she was a substantive person and could have substantive conversations." Thus, Monica's "disease"—emblematic of celebrity worship on the political level.

viously intelligent and energetic, is woefully undereducated, can't spell or punctuate or "write a coherent sentence," who doesn't read, whose grammar is miserable, who pleads with Linda Tripp to help her use "a big word" to slip into a note or letter to the president, and is so wrapped up in her own little world that she doesn't seem to reflect on the larger currents around her.

But beyond the brashness, the chutzpah, the pushiness, the arrogance, beyond the toughness and the exalted sense of herself exists another Monica. This is the sad, lonely, pathetic child/woman who all her life has struggled to be accepted, to be admired, and is continually rejected and disappointed.

Clearly bright, perhaps with a talent for drama, with a pleasant voice that some earlier teachers thought could be trained for singing, from childhood she fights a losing battle with her weight. She feels disliked and disparaged by classmates. From her teens on, she embarks on what becomes a series of pathetic affairs. One of her first extramarital encounters shows how seriously disturbed and starved for affection she is. It also illustrates why later some of her White House colleagues think of her as "the stalker."

This affair involves Andy Bleiler, whom she meets at Beverly Hills High where he produced sets for the theater group in which Monica becomes involved. He's married when their affair begins, and it continues after Monica attends Santa Monica Community College. When Andy wants to end the affair—his wife is then pregnant—Monica threatens to tell the wife and also to turn up at Andy's classes in the high school, according to his grand jury testimony. In a pattern that repeats itself in Washington, she starts sending him gifts and cards. Then, in a truly disturbing episode, when Andy and his wife move to Portland, Oregon, to look for a job, Monica follows them. She pays cash for a Subaru station wagon, drives to Portland, enrolls in Lewis and Clark College, and introduces herself to his uncle, a math professor at nearby Portland State. She starts baby-sitting for the uncle's children and buys them expensive gifts. Then she makes friends with Andy's wife, Kate, and starts buying *their* children presents—all the while propositioning Andy when Kate and children are out of the room. Finally, Monica moves to Washington, but she still keeps in touch with the

Bleilers, even telling them about an affair she's having with a high-level White House official. She hints it's the president. The Bleilers joke to her that she needs "presidential knee pads." She also tells them of *another* affair she begins after transferring to the Pentagon; this leads to her having an abortion. As a final blow, eventually she tells Kate about her affair with Andy.

Even in the midst of her relationship with the president, she confesses to a close friend from high school days how she still yearns for Andy. "Oh, Cat," she e-mails Catherine Davis, "I want to get out of here so bad. You have no idea. I have been really sad about Andy lately, too. I keep having these dreams about Kate, him and the kid. It's really yucky. What really hurts is that I cared so much about someone who just threw me away so quickly. I miss having someone to be with and enjoy me. Ohh, woe is me, woe is me! . . . Well, I'm off to LA Tuesday night—the land of the skinny and fake boobs! Love and kisses, Monica."

At one point, under Linda's recorded urgings, Monica describes her sex life, listing the number of people she's had sex with—and it's a substantial list, including a number of men whose names she can't remember. But this, she tells Linda, is "pretty typical" of young women in the Nineties. She also has more than one "Schmucko" and "Big Creep," as she often calls the president, in her life. And she exhibits petulant bouts of jealousy: she gossips with Linda about the president's other supposed "girlfriends," sometimes reacting angrily at what she believes to be the more preferential job treatment they receive, derisively calling them "The Graduates." Similarly, she rails against the White House assistants—Evelyn Lieberman, Nancy Hernreich, Steve Goodin among them—whom she blames for keeping her from the president. To Monica, they're "The Meanies."

For all her experience and poise, at bottom Monica Lewinsky remains a troubled and unsure young girl.* All during her time in

---

*Once, while attending Santa Monica Community College, Monica "begins crying uncontrollably" after a motorist slips in ahead of her while she's waiting for a parking space. Her mother views this breakdown as so serious that

Washington she takes a number of prescription drugs, including anti-depressants: Prozac, Paxil, Zoloft, and Serzone, thereby adding another "typical of the times" note to the story. During the supposedly euphoric era of the boom, taking such drugs to reduce anxiety and tension—or ingesting the popular drug Ecstasy to produce synthetic emotional highs—is fairly common among young people like Monica.

There's a sad kind of childishness to her notes and the hearts and flowers she draws on them. Her handwriting has the look of a junior high school girl; it's not even cursive, just angular printing. Some of her notes contain huge hearts drawn in the middle, colored in blue, with the words "LUV U" above them. Hardly the work of a true adult. Many of the notes are so naive and childlike they make a reader think of them as both pathetic and heartbreaking. She also has a peculiar, but striking, way of always writing everything down, always taking notes. She doesn't just live like a normal person; she "creates a record."

The tapes:

Monica: I need to—hold on. I need to just—
Linda: What?
Monica: I'm trying to find, I need to write this. I need to write down all the stuff he said—
Linda: Yeah.
Monica: Before I forget it. Where the [redacted] did that damn notebook go? Shit . . . hold on—just don't talk to me 'cause I need to remember this. Hold on just a second.

This is the tough but vulnerable, naive but determined, self-assured but insecure child/woman who sets her eyes on the president, sees, as she recalls, her chance and seizes it. She fully knows what she's getting into, telling him in that very first sexual encounter after lifting her skirt and showing him her thong that she

she takes Monica to a psychologist who begins treating her and becomes one of two therapists whom Monica consults by phone even during the scandal in Washington. Yes, she tells them about the affair, too.

"understands the rules" about having a relationship with a married man. She understands, as she later explains, that they'll both lie to hide their relationship.

There's no doubt she's infatuated with the president, lovesick if you will. From the very first moment when they make "intense eye contact" as the president works his way down a rope line at a White House departure ceremony on the South Lawn, she becomes completely enamored with him, with his power, with his aura. She clearly wants to believe there's more to the relationship than just sex. At one point in trying to explain it to the grand jury, she gushes girlishly: "We would always tell jokes. We would talk about our childhoods [sic]. Talk about current events. I was always giving him my stupid ideas about what I thought should be done in the administration or different views on things. I think back on it and he always made me smile when I was with him. It was a lot of—he was sunshine."

Sunshine. Her naïveté is as touching as it is telling. She believes the president has "a beautiful soul. I just thought he was this incredible person and when I looked at him I saw a little boy." She goes to a Saturday night showing of *Titanic,* the soupy love story with its doomed affair, and returns home to write him what she herself later recognizes to be "a mushy and embarrassing" love note. At another point she brims with childish pride as she attempts to describe for the jurors why there was much more than sex in their relationship. They would hug each other, she says, and "sometimes" hold hands. She treasures an image that she thinks bespeaks love: "He always used to push the hair out of my face."

She also believes the expensive gifts she showers on him, and the few he gives her—including a T-shirt and coffee mug from the Black Dog restaurant and gift shop on Martha's Vineyard, where the president vacations with his wife and daughter; a New York skyline pin; a Rockettes blanket from the Radio City Music Hall—are a testament to their deep devotion. One of the most poignant moments comes when, near the end of their relationship, she spots a grocery bag under the president's desk in the back study where their sexual encounters occur. It's stuffed with her presents to him over many months. Trying not to seem hurt, she asks the president whether he likes her gifts. Yes, he replies, he just hasn't had time to take them upstairs.

So mesmerized is Monica she fantasizes that after the president leaves office he and the first lady will separate, enabling Bill and Monica to become a couple and live happily ever after. Once, she startles Vernon Jordan by asking him if he thinks "the president would leave the first lady at the end of his term." Jordan finds that "a both frightening and from my point of view unrealistic question about the president. So I just said that's a really crazy notion on your part that that would happen. And that I was confident that they would be together 'til death do them part. . . ."

But there's also no doubt that she sees their dangerous and intoxicating sexual sessions as elevating her own sense of self-worth. She bursts to tell people, especially wishing those earlier classmates who disparaged the fat unhappy Monica could see her now. How they'd react if they knew she was fooling around with the president! "God, you know if these people saw me," she tells Linda. "I mean I don't know. Like I'm fat and I was just . . . I just think I just see them seeing me and thinking to themselves, *this?*"

In all the torrent of testimony and deluge of documents that chronicle this story, there's no more poignant scene than the time when Monica, still the sad little fat girl, buys an "outrageously expensive" strapless red formal gown to wear at the president's second Inaugural Ball at the Kennedy Center. Then, after fancy salon styling with facial, manicure, permanent, and new jewelry perfectly in place, she positions herself at a rope line and stands for five hours pressed against that rope in hopes her beautiful gown will attract the president's attention. He never notices her.

Failure doesn't deter her, however. As she later candidly explains to the grand jurors, who wonder why she continually pops up at functions with the president: "I'm an insecure person and . . . I was insecure about the relationship at times and thought that he would come to forget me easily and if I hadn't heard from him, especially after I left the White House, it was very difficult for me and I always like to see him and usually when I'd see him it would kind of prompt him to call me. So I made an effort. I would go early and stand in the front so I could see him, blah blah blah." It's both sadly sweet and pathetic. As time passes, there are glimpses of increasing anguish in Monica, a sense that she's not only being tortured by the situation in which she has placed herself, but feeling cheapened by the experience. And for all her brashness and toughness, she seems to begin to realize she's hopelessly in over her head and yet still desperately battles to reclaim her former position with the president. Many of her notes to the president make you ache for her, for they are written to the point of being deeply embarrassing to herself: "It will have been two months since we last spoke. Please do not do this to me. I feel disposable, used and insignificant." (The underlining is as she wrote it.)

Some of her early friends, in whom she continues to confide about every step in her secret relationship, the girlish dreams, the adult hurt, could step straight from a script of a Nineties prime-time TV soap opera, *Beverly Hills 90210*. One of those "best friends," Neysa Erbland, has known Monica since Beverly Hills High, and Bel Air Prep. By the time her miniature portrait enters the official record, Neysa, twenty-five, lives in Sherman Oaks, has been married twice, attended four colleges, and graduated from none. She sounds very L.A., still very immature; even her grand jury testimony sounds childish, Valley Girl–like: "and I would say, 'Oh, my God. Could you imagine if, you know, you ended up having an affair with the President?'" Neysa recalls telling Monica after her White House internship begins. "And she [Monica] said, 'Yes, that would be wild and crazy.' And I said, 'Well, you have to tell me.' She said, 'No. Well I wouldn't be able to tell you. That would be just too much of a big deal. I couldn't tell anybody if that ever happened.' And it went back and forth like that for a few minutes and finally she said, 'Okay, it's happened.'" Monica says the affair started after she and the president "had been making eyes at each other and just flirty a little bit . . . and one afternoon she was dropping something off like that and he must have said something to her, you know, that she looked good in her suit or something to that effect, and she—she lifted up her skirt and showed him her underwear. And I guess it started from there." (She had flashed *exactly* the same way much earlier with Andy Bleiler.)

Neysa's first reaction on learning of the affair is "I thought, you know, atta girl on one hand. I thought, you know, I believe that she's actually, you know, having sexual relationships with, you know, the man who runs our country. You know, it was huge. It was just, I mean it was huge."

Monica implores the "best friend" not to tell anyone about the affair—and Neysa proceeds to tell her husband, her parents, her mother-in-law, and another friend who is a booking agent on a TV show.

In fact, everyone in this cheap melodrama promises to keep the secret—and immediately blabs to many others, as does Monica herself. She tells at least nine people about the affair, describing it in explicit detail in phone conversations and in e-mail messages. Pathetic and childish though that affair is, Monica clings to it as if her life depends upon it, and perhaps it does.

> Monica [on Clinton]: . . . what's hard for me is that and I know this is so stupid, but Linda I don't know why I have these feelings for him. Maybe I'm crazy. Maybe I don't have these feelings. Maybe I'm pretending it. . . . I never expected to feel this way about him.
> Linda: You protect him . . . every inch of the way.
> Monica: I didn't. And the first time I ever looked into his eyes close up and was with him alone I saw somebody totally different than I had expected to see. And that's the person I fell in love with.

And to Linda, far more than to any of her other "best friends," Monica pours out everything, over and over. Linda becomes the mother-figure Monica never had, and Linda encourages Monica to confide more, and be more consoled and comforted by, her.

> Monica: I can't take it any more.
> Linda: I know, I know.
> Monica: (crying) I just can't, I just can't (crying).
> Linda: Oh, my God.
> Monica: It's just too—it's too much for one person.

Linda: It is too much for one person. Yes it is . . .

Monica: It's just that I go to work every day (crying) and I just—I'm trying to keep it together and I just can't.

Linda: You've been a trooper [sic] through this, Monica, and you've been through a situation that most fully grown adult women couldn't handle, okay?

Even toward the end, when Monica's trust in Linda is rapidly eroding, she remains somewhat dependent upon and in need of the older woman's presence and approval. Her farewell e-mail note to Linda, on Monica's last day of work in the Pentagon in December 1997, as everything is about to collapse, is typical of Monica, and typically heartwrenching down to its effusive exclamation points and childishly gushing phrasing:

"LRT [for Tripp's initials]: I will miss working with you tremendously! Who will edit my letters? Who will tell me my grammar stinks??? Who will escape for coffee breaks with me? We'll only be a phone call away! I think the world of you and know everything will work out great!!! I can't wait to see how skinny you get! You go girl!!! All my love, MSL."

The Plotters

Of all the damaging pieces of information about Linda Tripp, the most arresting is that she tells Monica she's a witch and has psychic powers. The record isn't clear whether that's said in jest or not, but the description aptly fits the character Linda plays in the plot. It's impossible to read the massive outpouring of transcripts of tape recordings, affidavits, sworn testimony, and interrogations without coming to the same conclusion. She and her fellow plotter, Lucy Goldberg, seem like a witch's coven cackling and chortling and exchanging malicious gossip at the latest pathetic confession that comes their way unwittingly from Monica. Lucy even matches Linda's wicked witch role; posting on the Internet, Lucy actually refers to her fellow Clinton haters as "my pretties."

As for Lucy's self-portrait, perhaps the most revealing of her character are two remarks she makes to prosecutors. One: She's been offered $750,000 to sell the transcripts of the two early phone calls to her from Linda, raising the prospect of a scandal book on the White House. Fittingly, those recordings are made without Linda's knowledge. Second: She tells the prosecutors she "thinks all of Tripp's tapes are worth $2 million."

Of these two principal plotters, Linda seems the more embittered, Lucy the more cynical. Both are hard as nails.

Linda Tripp emerges in this tale as the bitter government secretary who never gets anywhere and thinks she's smarter and savvier than all her superiors. Life hasn't treated her fairly, she believes. She's determined to get even. Her résumé bears this out.

When she befriends Monica in the Pentagon she's approaching her fiftieth birthday. A graduate of the Katherine Gibbs secretarial school, Linda married Bruce Tripp, a military officer, in 1971 and began working in a support staff capacity for the Army Department in 1972, when Richard Nixon was seeking a second presidential term. After her children were born, Linda took time off from work to raise them. While her husband got posted from base to base, she found "a series of jobettes," nothing really challenging, nothing that advanced her. After twenty years, the marriage breaks up. Linda finds herself, an overweight single woman with two teenagers, increasingly bitter at her fate.* "I've spent several years in the federal government," she tells the grand jurors in her secret testimony, "clawing my way to the top."

The jurors, a number of whom are federal employees themselves in Washington, are curious why she uses the term "claw," asking her what she means by that. "Maybe that's a poor choice of words," Linda replies. "Climb up the career ladder. It was never easy for women back in the early Seventies, I can tell you in the federal government. It was not easy. I perceived it as a struggle."

Her entire summary of her earlier experience is laced with

---

*On one of her taped conversations with Monica, Linda confesses she has not had sex for seven years. How relevant, or revelatory, that may be is beyond my capacity to determine. So instead of practicing psychiatry without a license, I will leave it to the professionals—and the readers—to assess its significance, if any.

anger and resentment at the life of an Army wife; at not being "the one they're going to pick for the plum jobs"; at being "the family support group person... helping your husband's career along"; at being "the one who keeps the household running as you move from spot to spot"; at being left alone with the children for a year, as she was, while her husband was stationed in Korea.

Finally, Linda makes it as a secretary in the Bush White House and stays on, fatefully, as a holdover in the early Clinton years. From the beginning she never approves of the Clinton team; she dislikes their dress, their youth, their language, their method of working, and what she believes are their immoral personal conduct and their unethical actions in the Travel Office and Counsel's of-fice. She becomes a plotter, leaks the story of Deputy Attorney General Webb Hubbell's resignation to the media, shares informa-tion with and becomes a source for Mike Isikoff about Kathleen Willey and later Monica. Finally, she finds herself quoted as a source in a *Newsweek* article that brings White House disapproval, and, ultimately her removal to the Pentagon. This is the Linda who sits alone in her suburban Maryland home, eating chocolates, hunched over her secret tape-recording machine, strewing tapes about the room, and sometimes finding that Cleo, her cat, has stepped upon and stopped the recording of yet another treacher-ously recorded conversation.

The more Linda testifies about herself, the greater the sense of her bitterness and resentment becomes—and not just about the Clinton group, but about other government workers who rise through the ranks and eclipse her. Her jealousy toward Evelyn Lieberman, for instance, who becomes deputy chief of staff re-sponsible for getting Monica out of the White House, is palpable. When Monica talks about Evelyn, Linda grumps, "Hmpf, mmf, mmf, mmf, mmf, amazing. Haven't seen her in years. Remember when I knew her, she was a [redacted] secretary."

She's equally resentful of Monica, though she cloaks it under her surrogate mother role. Toward the end, in her final e-mail notes to Monica, her mask slips and she exposes a brutally cruel side: "From now on leave me alone," she writes. "Don't bother me with all your ranting and raving and analyzing of this situation. . . . I really am finished, Monica. Share this sick situation with one of your other friends, because, frankly, I'm past nauseated about the whole thing."

This from the mother-figure who has been coldly and calculatedly leading Monica on; who urges her to compile a written date-time-place record of all sexual contacts with the president; who asks her to repeat incriminating stories over the phone so she'll be able to give Monica better advice on what to do, explaining requests for the repetition by saying, "You know how I forget things"; who even warns the hapless intern to be cautious when speaking to other people on the phone because they might be recording her, which Linda herself is doing even as they talk; who revels in the confessions she elicits, and quickly passes them on to fellow plotter, Lucy. When Linda and Monica talk about whether to tell the truth about Monica's affair, Linda tells her young colleague how Monica has the administration on her side. "Me? I've got a house. I've got a mortgage. I've got two kids in college. They're not going to want to have anything to do with me after I perjure myself."

Self-pity aside, her true nature, and all its deep resentments and aggrieved airs, leaps out in an internal memo about her job performance written by her Pentagon supervisor, Cliff Bernath:

> Linda arrives for first day of work; doesn't like her office space; dashes off an email stating it isn't what was 'promised' her. . . . Linda ignores Susan Wallace when she tries to explain how things are done and turns her back. . . . Linda won't speak to coworkers or exchange greetings—unless she needs something, then all's fine. . . . Columbus Day weekend: Linda won't answer her phone, thereby requiring Susan to come in Sat/Sun/Mon to take care of Linda's as well as her own responsibilities. Essentially, Linda is a GS–15 public affairs officer who doesn't like her office space, thinks her responsibilities are only 8:30–5PM, isn't nice to coworkers, lends a disruptive manner to the office, doesn't want to be here.

In short, what a bitch.

However genuine Linda's sense of moral indignation, she's also interested in making money from the suspected scandals—first from the events of the first Clinton year ending in 1993, later with the Monica saga. She even presses Monica to demand more money when she's asking for a job, an act that arouses Monica's suspicions as to whether Linda thinks this will make the Clinton cover-up more damaging when Monica's job is secured. Even pushy Monica knows that some things have limits: "I just want you to look at this from [the president's] position for a minute," Monica tells Linda. "He's thinking I'm going to take this girl and I'm moving her...and I'm going to make her go from making $30,000 a year to making $90,000....I think that's gonna scare the [redacted] out of him."

Linda's hatred of the president is overpowering and all-consuming. Once, acting out of her obsession with what she believes are the president's numerous sexual dalliances, she "tests" him by sending a beautiful young woman, a friend of a friend, to shake the president's hand at a public event. Linda then stations herself with a camera and photographs the "meeting," later claiming the young woman describes the president's reaction to her as "instant letch." Another time, in discussing the president's affair with Monica, Linda says that "that's why I want to kick him in the nuts so that they flatten into little pancakes and he can never use them again...."

Moral: Don't get on the wrong side of a Linda Tripp.

Lucy Goldberg is no less a hater, but she comes over as more sardonic, worldly, and, if anything, more cynically calculating and sinister. It's Lucy who plays the leading behind-the-scenes role of anti-Clinton coordinator. It's her call to a friend, the conservative publisher Alfred Regnery, that leads her into the circles of Elves and other Clinton back-channel conspirators. Lucy asks Regnery if he knows how to approach the Paula Jones camp. He refers her to Peter Smith, the wealthy conservative Chicago funder of other efforts to get the president, who in turn recommends the lawyer who becomes the first of the Elves. Lucy then passes Linda's unlisted phone number to one of the Jones lawyers in October 1997; he calls Linda, initiating her debriefing with the Jones camp.

For years before Monica or Linda came to Washington, Lucy has been through the ideological political wars. Early in the Sixties, she was a fringe player in the Kennedy-Johnson years, boasting, as friends recall, of having affairs, and later actually operating as a Nixon secret surrogate and dirty trickster while traveling on George Mc-Govern's presidential campaign press plane.* She seems to have gloried in the role, confiding to a reporter in 1973 when her role was exposed, "They were looking for really dirty stuff.... Who was sleeping with who [sic], what the Secret Service men were doing with the stewardesses, who was smoking pot on the plane—that sort of thing." Her spy operative part pays well—a thousand dollars a week, from Murray Chotiner, the Nixon dirty trickster—and it shows that even then Lucy relies on a tape recorder for her work. She dictates her observations of doings on the press plane as often as five or six times a day, then replays them over the phone where they're typed up and rushed to the White House.

When she and Linda meet, Lucy Goldberg is solidly connected in the conservative world, and on the lookout for new publishing material.

Linda and Lucy's first dealing ends angrily when Linda drops the idea for a scandal book after deciding "it's not worth the risk, it's not worth the gamble." This comes after Lucy spells out what Linda might expect to earn after taxes, and ghostwriter fees, are deducted, a sum that, Linda concludes, "left me with not much more than two years' salary over the top."

Two and a half years pass before Linda, again through the intercession of Tony Snow, renews contact with Lucy about another, bigger scandal book involving the intern. Snow calls Lucy in New York on Linda's behalf in the summer of 1997. Linda tells Lucy about the "extraordinary" situation she's in with the intern. It will be "explosive" if this comes out, she adds. Tape-record your conversations with this Monica, Lucy tells Linda, even as she

---

*I knew her briefly when she was Lucy Cummings, a gravel-voiced, earthy, chain-smoking woman who loved trading political gossip and scandal stories.

surreptitiously records the early conversations with Linda. They begin almost daily conversations, reporting and assessing developments as they occur.

After Lucy's efforts lead to exposure of Monica's affair, she assiduously leaks salacious material to the press to keep the pot boiling. "I wanted to keep the story alive," she later admits to George Lardner of the *Washington Post*. "I wanted to give the story legs. I gave the *New York Post* [a notoriously anti-Clinton tabloid], which I love and work for, a story a day for eight straight days." Among them in that first week of the Monica bonfire, she tells Lardner, are the following *New York Post* trophy headlines which appear on successive days, January 24 and 25, 1998: "Monica Kept Sex Dress as Souvenir" and "Bill Had Hundreds [of Women]." She also acknowledges being the source for the January 29 story in the weekly scandal tabloid, the *Star*, which runs under the headline: "Monica's Own Story—Affair started on day we met—after I flashed my sexy underwear."

Not surprisingly, it's Lucy who confirms to Internet gossip Matt Drudge that *Newsweek* is holding back its planned intern exposé by Mike Isikoff after Drudge calls her at midnight on that momentous Saturday night, January 17.

They're a team for the times, Lucy and Linda—or for the ages. Transcripts of their taped conversations are repellant enough for the sheer level of duplicity and betrayal they reveal. Actually listening to them is even more chilling. They glory in exchanging gossip, in speculating on the size of the president's penis, on whether the president might be refraining from sexual intercourse with Monica because he has herpes. But always their focus is on what Monica might unwittingly reveal that will help the book and bring down Clinton, or both.

The tapes:

Linda: Things are hitting the fan over here with her.
Lucy: Really?
Linda: You know what she said on tape last night to me?
Lucy: What?
Linda: "I think he's on drugs."

Lucy: Wow. On tape—on tape, you got it on tape?

Linda: Yep.

Lucy: Good for you.

Linda: Yep.

Lucy: Well, I'll tell you, that justifies everything. That son-of-a [redacted].

Linda: Uh-huh. And I said, "Well, a lot of nice people are on drugs," like I wanted to keep her keep talking.

Lucy: Yeah.

Linda: And she said, "You know, it's because of the way he acts. I mean, he zones out. He just"—and you know. So I thought that was interesting.

This conversation concludes with Linda telling Lucy, "So, anyway, what I have on tape is very little. I mean, I'll get more."

To which Lucy replies, "All you need is a snippet."

"Oh, I got snippets," Linda happily says.

In time, the entire world gets to share them.

The Prosecutor

Public image should not be relevant in measuring character, but it contributes strongly to the picture people carry in their heads of the person being scrutinized. So it is with Kenneth Winston Starr.

The image of Ken Starr stamped in the public mind is of a neatly dressed, slightly chubby man with a smooth round face framed by rimless glasses. He stands in his driveway, holding a bag of garbage, while offering softly spoken homilies about truth, morality, and law to a phalanx of TV reporters and cameramen. They surround him each morning as he steps from his doorway, taking out his trash and preparing to go to his special prosecutor's office in downtown Washington halfway between the Capitol and the White House. There, he will continue his seemingly endless investigation of the president.

To the public, this Ken Starr seems destined to go down in history as the relentless, self-righteous Inspector Javert whose dogged pursuit of Bill Clinton year after year after year after year results in the second impeachment of a president of the United States in more than two centuries of American history, an outcome in which he, not the president, is seen as the ultimate loser. In the end, his personal standing with the public plummets while the president's actually rises. Ken Starr emerges in this public picture as a humorless moralist, a Puritan in Babylon, out of touch with contemporary attitudes. A zealous ideologue.

It's not an entirely false image, but, as in most such cases of intensely publicized public figures, it's more crude caricature than rounded portrait.

In yet another example of remarkable coincidences in this story where facts always seem more incredible than anything fiction might portray, the principal protagonists, the president and the prosecutor, grow up in the same humble circumstances, in the same section of the country, in precisely the same time. Both are classic baby boomers. They're born twenty-nine days apart in the first post–World War II year of 1946. The elder, by four weeks, becomes the prosecutor; the younger, the president. Neither comes from a family in which anyone before his generation has gone to college; both go on to forge distinguished academic and professional records as lawyers and prominent public officials. Both are seekers of the American Dream, and both achieve it in far greater measure than could possibly be expected considering the odds against them at birth.

Ken Starr's tiny hometown of Thalia lies in the barren, dusty windswept plains of East Texas near the Red River and the Oklahoma border. He's the last of three children born to Willie and Vannie Starr, both of whom come from farming families. Willie is a barber and substitute minister in the conservative Church of Christ denomination. In common with friends and neighbors in their southwestern area of the Bible Belt, religion plays a major role in their lives; Willie and Vannie begin taking their baby son to church when he's only two and a half weeks old.

Nothing in the record, or in later attempts to describe the soil from which the eventual prosecutor springs, suggests his parents to be anything other than loving people. They are not given to the extreme hatreds and bigoted views that afflict many fundamentalist denominations throughout the rural South and Southwest. But they strive to impart strong religious values of right and wrong in their children. Even the names of the small, struggling towns—Lone Oak, Buffalo—that dot the wheat fields and cow pastures of the prairies bespeak the importance religion and the hardscrabble pioneer West hold for them. His mother, Vannie, comes from nearby Palestine. By the time Kenneth enters elementary school they have moved to Centerville, population eight hundred, not far from Palestine. Later, they move to San Antonio.

Accounts of Ken Starr's boyhood and schooling, from early age on through college, uniformly depict him as a classic straight arrow—the good son who plays by the rules, faithfully attends Sunday school, studies hard, earns top grades, is voted "most likely to succeed" of his high school graduating class, and is regarded by all as an exemplary person of high moral character. He never smokes or drinks; no one seems to have heard him even curse. As for girls, not for him the sexual liberation and experimentation that sweeps much of American society throughout his formative teenage and college years during the Sixties. "He was not a bit girl crazy," Vannie later tells Sue Anne Pressley proudly. "He never did date girls to amount to anything. He was nice to all of them."

Nor does he seem to have been affected by the other strong national currents of social and political change then: civil rights, Vietnam, protests, legal or not, nonviolent or not, and ultimately challenges to all authority. Perhaps it's stretching the point, but Ken Starr can be said to represent one side of the political/cultural divide of those times: one of those whom Nixon would include in the "silent majority" of Americans who don't protest, don't experiment with drugs and sex, don't oppose authority.

Bill Clinton, growing up in the small town of Hope, Arkansas, and the gambling community of Hot Springs, can be placed in the other camp. Indeed, though their paths never cross, they are both in college in Arkansas at the same time—Starr, attending Harding University, a small conservative school numbering some eighteen hundred students affiliated with the Church of Christ in Searcy,

Arkansas;★ Clinton, the state university in Fayetteville some 150 miles to the north and west. Typically, during summers Ken Starr sells Bibles door-to-door; Clinton engages in political activities and enjoys the more freewheeling and active social life of his side of the Sixties generation.

However much they reflect distinctly different slices of their generational divide, in one important sense they are alike. Both avoid serving in Vietnam. But then, so do most of their fellow college contemporaries of the baby boom. That war is fought mainly by America's minorities and least-fortunate citizens.

Both of these ambitious young men leave their part of the country to advance themselves in the East; both attend colleges in the nation's capital, Starr at George Washington University, Clinton at Georgetown University; both go on to fine law schools— Starr, Duke; Clinton, Yale (after his Rhodes Scholarship in England)—and both find themselves rising rapidly as they pursue their careers and life dreams. Starr clerks for Chief Justice Warren Burger on the U.S. Supreme Court; he becomes the youngest judge ever named to the highly prestigious U.S. Court of Appeals for the District of Columbia Circuit and then is appointed by President George Bush as solicitor general of the United States, the "lawyer's lawyer" of the federal government.

Clinton, after having worked for his mentor Senator J. William Fulbright on Capitol Hill and earning his Yale law degree, returns to Arkansas, teaches constitutional law at the university, then, at the age of twenty-eight, becomes attorney general of his state and from there the youngest governor in the nation.

Still, despite their success, they haven't achieved their ultimate goals. From the beginning, Clinton's goal is the presidency. Starr's is the Supreme Court. Initially, Starr's prospects for achieving his goal

---

★A hint that Starr's portrait as a straitlaced goody-two-shoes is far too simplistic comes in the revelation by Sue Schmidt and Michael Weisskopf that Starr was reprimanded by Harding's president for criticial columns Starr wrote in the school paper, and that he left the school after concluding "its worldview was too narrow for him" on such things as its strict interpretation of the Bible and a campus ban on alcohol.

appear much brighter than Clinton's. When President Bush persuades him to leave his lifetime federal judgeship to become U.S. solicitor general, Ken Starr's name immediately leaps to the top of the list of those expected to be nominated to the highest court.

Friends and associates who worked with Starr when he was solicitor general say he was deeply disappointed when Bush failed to name him to the Supreme Court. "His career took a U-turn," says a conservative Republican lawyer who knows Starr well and admires him. "I suspect his willingness to take on the independent counsel investigation was prompted by the thought that it would continue his viability for that appointment."

By the time Ken Starr steps onto the stage of the scandal drama he has earned a reputation as a distinguished, painstakingly careful and cautious lawyer—no ideologue, his colleagues insist, and many of them are liberal Democrats. He's a conservative Republican, active in conservative political groups—anti-abortion among them—partisan when it comes to politics, but fair-minded and honorable when it comes to the application of the law. His private life reflects his public one, and reinforces the values with which he was raised years before in the small towns of East Texas. He's a dutiful father, devoted to his lifelong partner, his wife, Alice, and continues to contribute time and energy to his church in suburban Virginia where he teaches Sunday school.

At the same time, the Ken Starr who becomes the prosecutor and principal antagonist of the president clearly brings to his new position a number of jarring contradictions. He accepts the post even though the circumstances of his selection arouse instant—and justifiable—suspicion. His appointment comes from a three-judge federal panel known to be deeply conservative and closely allied to right-wing opponents of the president on Capitol Hill, and Starr accepts the post despite the fact that he himself was intimately involved with the anti-Clinton side of the Paula Jones case, having offered to work on a Supreme Court amicus brief in behalf of Paula. This alone raises serious questions about his judgment, particularly his seeming inability to see problems arising from the appearance of conflict of interest in a highly charged proceeding that is certain to generate the most intense controversy.

As the investigation proceeds, the same questions keep recurring. He continues to speak at conservative and pro-life functions; he brushes aside questions about when he became aware of a million-dollar lawsuit brought against his law partnership by the Resolution Trust Corporation, which Starr was investigating in connection with the Whitewater matter; he draws a chorus of criticism when he announces in the summer of 1999 that he intends to resign his prosecutor post to accept the deanship of conservative Pepperdine University Law School in Malibu, California, and then drops the idea after it becomes known that Richard Mellon Scaife, the most prominent funder of anti-Clinton efforts, was a benefactor of the law school.

Naturally, these and other actions draw heated criticism of Starr from the Clinton camp, from the media, and also from dispassionate leaders of the bar. Yet interestingly those who know him best, whether Democrat or Republican, dismiss the charges of unprincipled actions against him. The assessment of two prominent lawyers, one a conservative Republican, the other a liberal Democrat, is particularly useful in helping to unravel the enigma that Ken Starr presents.

The conservative is Douglas W. Kmiec, a distinguished law professor, public commentator, and Starr admirer. In the course of a long conversation about Starr at Pepperdine University overlooking Malibu and the Pacific Ocean in the summer of 1998, just as the Monica investigation is about to conclude, Kmiec says that to understand Starr's handling of the prosecutor's role you first have to understand the origins of the independent counsel statute under which he operates. Like so many well-intended reforms growing directly out of the Watergate scandal, the statute turns out in application to have far more negative than positive effects. In Kmiec's view, and that of countless others, the law represents "a naive supposition that ethical values could be written into a statute called The Ethics in Government Act, that...if we appointed someone outside of the normal government process, we would safeguard it from inherent conflicts of interest."

It sets an extraordinarily low legal standard for granting authority to undertake an investigation, stating that if "specific and credible information" about a crime exists then further investigation is warranted. That's far from the more rigorous standard of "probable cause" needed to grant power to undertake a search. Operating under such a lax legal standard for pursuing an investigation is only "one of the difficulties Ken Starr faces when invited into this make-believe world where the laws assume him to be the ultimate protector," Kmiec says. "He's asked to play out that role and his personality, it seems to me, is unfortunately the perfect one to play that out to its ultimate conclusion in every excruciating detail."

He explains, in words that go to the heart of an essential understanding of Ken Starr's strengths—and failings: "First of all, Ken is an extremely moral man who values integrity, but who is also extraordinarily cautious. He's a cautious non-prosecutor. His *real* background is not as a criminal prosecutor. You've probably encountered business lawyers in your life. One of the things you know about them is they're a bit like medical doctors. They order every test. They examine every nook and cranny of a legal idea. And that's Kenneth Starr's personality. So you have a statute that sets him on this trajectory and he's going to take that to its ultimate conclusion. The problem in his doing that is it has manifested to all of us the weakness of the idea [behind the independent counsel statute] in the first place. It has illustrated how we have unfortunately removed the responsibility of the Attorney General of the United States. She now has the capacity to execute her job merely by appointing independent counsels. She encounters hostile congressional questioning when she doesn't appoint one. That's not investigating a crime. That's not maintaining public integrity. That's not using public resources in a careful way. It's merely passing off responsibility to someone else."

Ken Starr faces two crucial decisions in the Clinton scandal investigations. Both involve questions of judgment. The first involves whether or not to accept the position of prosecutor when offered, especially if Starr thinks becoming prosecutor of the president will help him win a Supreme Court seat. On that score, there can be no question that history's verdict on his decision will be unanimous, and negative. The manner in which he handles the investigation dashes any hope of being nominated, and certainly confirmed, to the court.

Starr's second crucial decision comes when he's presented a choice on how to proceed in the Monica Lewinsky matter. Monica, mind you, comes three and a half years after Starr has been investigating the various other Clinton scandals involving Whitewater and the rest. When Monica arises, Starr could reasonably conclude that his best course lies in completing those investigations, which in fact are almost done, delivering his already far-too-long-delayed final report on them, then gracefully exiting the scene for private life, his reputation largely intact. That's the course that Doug Kmiec, for one, strongly believes to be in his best interest. To quote Kmiec again: "When the evidence came in of taped conversations from Linda Tripp, those conversations did suggest perjury on the part of the president. At that point, he had evidence of a crime and as an independent counsel would have statutory obligation to bring that to the attention of the Attorney General. In light of his personal commitments he might well have said to Janet Reno, 'This is serious. You have to handle it. That's why they pay you that big salary. And you may have to appoint a new independent counsel, but I don't want it because it's unrelated to the matters I was charged with originally investigating.' He could have been finished and be Dean of the Law School now."

In retrospect, clearly that would have been the wiser choice—and one that Starr himself comes to believe was the best course long after he leaves the scene.

So why doesn't he take it? Why doesn't he gracefully ditch Monica and sex and let others make those decisions?

"There's both a good interpretation and a less favorable one," Kmiec says. "The good interpretation is: 'Straight and narrow Ken Starr. I've got this job and I'm already doing it. I've got my staff assembled for it.' Combine that with the kind of persistence and determination and thoroughness that Ken Starr brings to everything and the message from Republican congressmen that we're not letting you out of town until you give us something, and he decides to see it through."

And the less favorable interpretation? That Starr "had started to perceive that he'd suffered a hit in terms of personal reputation. He needed to recover from that and [Monica] was a chance to do that. Not very many of us get a chance to make that up on a grand scale and I think that view coincides with the weakness of human nature. As strong as Ken Starr is as a lawyer and as a person, I think he readily realizes he's got human nature imperfections as well. Part of that is the fifteen minutes of fame syndrome."

The second prominent lawyer whose assessment on Ken Starr, prosecutor, is as arresting as it is credible is Sam Dash, the legendary chief counsel of the Watergate investigations.

Sam Dash, a liberal Democrat, forged such a towering reputation for integrity and fairness and competence during that national ordeal that his methods of operating became *the* standard for other prosecutors conducting highly sensitive investigations involving the highest officials of the country. In the generation from Watergate to—what?—Clintongate, it becomes almost obligatory for other chief counsels and lead prosecutors to consult Dash for guidance on how best they should proceed. To his credit, Ken Starr does just that after he's named the prosecutor in the summer of 1994, charged with investigating the alleged Clinton scandals.

Dash is a distinguished professor at the Georgetown University Law Center when Starr asks him if he is willing to come to Little Rock to lecture Starr and his entire prosecutorial staff, including the assigned FBI agents, on the role of the independent counsel. Dash agrees. As Dash recalls, he emphasizes to them "that not only must a prosecutor be fair, but he must be seen to be fair." By its very nature, the independent counsel occupies a central position in the nation's spotlight, Dash tells them. And because they are going to be investigating the president, they're going "to have to bend over backwards to be fair." He also tells them: "You've got to make judgments that the ordinary prosecutor wouldn't make." And he stresses the critical importance to the public of proceeding swiftly and efficiently; "get it done and close it up," he says, "because there's no great message for the public as we had in Watergate."

His comments are so well received that Starr asks Dash if he's willing to consult with them as their investigation proceeds. He does, becoming their outside, independent adviser on ethics and procedure. Actually, as Dash describes his role, he functions as their "hair shirt." As he finds himself often reminding Starr in the years that follow, "You made me the monkey on your back, and I scratch."

So he does, so much so that as the investigation grows more and more controversial Dash's internal criticisms in themselves become a source of contention among the staff, not infrequently producing heated shouting and arguments in their private meetings. But Dash perseveres. He has no illusions about the difficulties Starr and his team face, nor does he view Starr himself with anything other than hard-eyed realism. He's well aware that, as he remembers, "Ken Starr was brought into the middle of this because he was a very active Republican. He was a biased Republican. He had made the offer to write the amicus brief for Paula Jones. Then he gets appointed." At the same time, Dash does *not* believe Starr to be the ideologically prejudiced prosecutor depicted by Starr's critics, especially by his enemies in the Clinton White House. "What I think many people never knew," Dash says, "was that as active a Republican as he was, and as biased as he was against Clinton, his sense of responsibility and his sense of integrity made him believe he was going to put those things aside and do an honest job."

However well-meaning Starr's intentions, and however high his personal principles, the prosecutor never is able to rise above what becomes nothing less than an all-out war. It's a war that is waged brutally, viciously, and at times dishonorably on both sides. It takes place in a political climate poisoned beyond even that of Watergate, set against a backdrop of a white-hot, constant, all-consuming media spotlight that eclipses anything before it. The deep and embittered antagonism between the two political parties is also far worse than during Watergate. Watergate, for all its passions, in the end produced bracing political unity. A generation later, the scandals of the Nineties produce a near-total breakdown of political civility and comity. Bipartisanship is dead.

"It got out of hand not because Ken Starr or even his federal prosecutors wanted it that way," Sam Dash says. "It became such a battle, such a public fight. So many emotional things were happening. The White House response to the investigation became a very vicious spin operation, targeting Ken Starr. They were constantly being attacked and they can't respond back publicly. I'm sure Ken Starr had to go home every night and worry about what it's doing to his children and his wife. He took it very seriously, because Ken Starr's personal perception of himself is of an honorable person of integrity and ability. And he's reading about himself that he's the worst devil there is. Out of control. A nut. Therefore there's an effort to fight back. Unfortunately it created an atmosphere inside the office of 'Go get 'em.'"

The White House contributes to the increasingly hateful atmosphere. Starr and his prosecutors find themselves being investigated by private detectives hired by "friends of the president" seeking to dig up dirt and scandal on both their personal and professional lives. They also become subjects of negative leaks to the media, planted by the Clinton camp. Now, both sides are playing it dirty; each is out to destroy the other.★

Not surprisingly, the result, as Sam Dash views it, is to lead "Ken Starr and his staff—all very aggressive federal prosecutors, all borrowed assistant U.S. attorneys—to hate the president and the first lady. So you've got an atmosphere that was not conducive to objective, professional prosecution."

By the time Linda Tripp brings her tapes to the prosecutors, placing Monica Lewinsky at the center of the scandal, the prosecutor's operation has labored for years without successfully obtaining evidence to indict or impeach the president. By then, as Sam Dash sees it, the attitudes internally have changed. Hatred of the president is now endemic. "They were committed," Dash says of the

---

★How dirty becomes clearer with the news, in June 2000, that a former New York policeman turned private investigator had been working for more than a year on a projected book exposé of the sex lives, and sexual preferences, of members of Starr's prosecutorial team, Starr himself, the Elves, Lucianne Goldberg, and conservative commentators who had attacked Clinton's moral standards. The book project, tentatively titled, *Clown Posse,* was canceled by Disney's Talk/Miramax imprint after it stirred controversy because of supposed assistance from White House sources and because none of the allegations was proved.

prosecuting team. "They really were trying to get the president. They were trying to get the first lady. And they found nothing in Whitewater, and they thought maybe we didn't find it because we didn't have a confessor, someone who will tell us. Now [with Monica and the suspected role of Vernon Jordan] they really had a feeling they had to succeed. I always told them that you succeed if you do a fair investigation and find nothing. That's a success for a prosecutor. You can't measure success by indictment or conviction. If you do that, you'll be willing to indict anybody. That was my standard, and I always used it with them."

It's not the standard that prevails.

Ken Starr, in Dash's term, "was in a hurry to refer for impeachment—and way before he had Monica Lewinsky's testimony. All he had was her written proffer which was internally inconsistent. He was rushing because the Republicans in Congress were telling him that unless you get this here by June or July it will be too close to the election. It'll be dead. So he was rushing to get a referral. When I attended a staff meeting in which he said he's got overwhelming evidence, I said, 'Look, you've got nothing. You've got a sheet of paper that she gave you, a proffer which isn't even internally supportive. Operating on a standard of having 'substantial and credible evidence' you don't dare send that over."

Dash threatens to resign. Starr relents and doesn't send an impeachment referral to the Congress then. Dash succeeds in persuading Starr that "the only way you might meet that standard" for impeachment "is if you get the sworn testimony of Monica Lewinsky." Then Dash negotiates for Lewinsky's sworn, secret testimony.

In August 1998, eight months after her name first surfaces publicly, Monica Lewinsky testifies under oath. As part of her immunity agreement with the prosecutor, Monica agrees to turn over all relevant evidence bearing on the case. Among the material she gives them is the blue dress. Prosecutors had been frustrated for months by their failure to find it; a search of her Watergate apartment, under authority of a warrant, produced no such dress. That's because Monica removed it to her mother's New York apartment. As she later explains about the dress, she "rolls it up, covers it in plastic," takes it to New York, and throws it in a closet, where it remains until its fateful reappearance that same month of her testimony.

The long, long investigation is nearing its chaotic end, but not the final judgment on the principal players, and especially not on the prosecutor.

Let Sam Dash speak to that point.

"To understand Ken Starr," he says, "you have to understand prosecutors. I was district attorney for Philadelphia, I was a federal prosecutor for the Justice Department. Every prosecutor has tremendous power. Any U.S. attorney, and district attorney, has within his hands the ability to ruin people. One of the things that's absolutely essential is discretion, professional discretion, knowing how to use it and maintain at least some sense of humility as you have this power. Otherwise you can wreak havoc. But Ken Starr starts out not on a politically biased basis, but as someone who has a perception of law and justice seen through the eyes of a moralist. He's an extreme moralist. It's a religious moralism. He has the idea he has this appointment to do right by the law. It's almost as if he's an agent of God and that those who oppose him, even those asserting their constitutional privileges, are corrupt and are attempting to obstruct justice.

"There were times when I'd say to him on a particular case, that it seems to me you ought not to push for such a heavy charge. He could turn out to be a cooperative witness. And he'd say, 'I could never do that.' He believed he had a trust, a fiduciary, and 'It's my responsibility to see that he pays.' He called the courts 'Palaces of Justice.' It's a nice term, but it's completely unrealistic. They're not 'Palaces of Justice.' It was almost as though he had a naive, moralistic view of the law and the roles of the participants. Defense lawyers were conniving and dirty if they defended their clients. He didn't seem to perceive that he was in an adversary system."

Sam Dash's ultimate assessment of the prosecutor is likely to stand the test of time: "So I think, yes, he's a man of integrity. Yes, he's a man of honor. Yes, he's held these responsible positions. But he's never been a prosecutor. He's never had the responsibility to make these kinds of decisions. And he brought to it this moralistic view of the law and of crime and punishment which didn't permit him to use good judgment and discretion."

The President

Of them all, Bill Clinton best fits an age that simultaneously offers such promise and such degrading scandal. Talented, charming, possessor of political gifts far beyond the reach of most public figures—and presidents—he presides over a United States of America that enjoys the greatest advantages, and the greatest boom, in its history.

His public assets are formidable. Even his worst enemies concede he's one of the best informed, most knowledgeable public figures they've encountered, and one who demonstrates a capacity for seeing complex issues in larger perspective. Few presidents before him have been able to strike such positive chords among the public. His ability to communicate by connecting personally with people is extraordinary. He may well be the most empathetic of all presidents—and a president who, given the favorable conditions in which the nation finds itself, has an opportunity to fashion a record of progressive leadership that carries the United States into a new era in a new century. In reelecting him, the people hand him that very chance. Not since Franklin Roosevelt in 1936 at the apogee of the New Deal that reshaped the modern American political state has another Democratic president been reelected for two terms in the White House.

Historians will wrestle for years with a central question about Bill Clinton: Why would this superbly equipped political figure, ambitious, energetic, highly intelligent, substantive, risk all on a dalliance so clearly trivial in nature yet potentially so self-destructive? For in the end, it isn't his enemies, however conspiratorial, organized, and determined, who are responsible for the personal and political devastation the scandal brings. Only Bill Clinton can be held accountable for that. He, after all, knows better than anyone the nature of his enemies—who they are, how they operate—and of their determination to "get him" or "catch him" and destroy him.

He is well aware that once he enters the White House he will be scrutinized more intensely and more obsessively than any president before him. Yet he embarks on more breathtakingly reckless actions than any of his predecessors.

Here is no grand passion, no great love affair. No Tolstoy will find in it material to fashion the story of a doomed or tragic relationship. It is a self-indulgent, selfish, and, yes, consensual affair that viewed in context is both sad and pitiable on both sides. It hardly can even be considered to be "an affair." Its most striking aspect is its furtive, fleeting, fumbling nature. Sex, at its rawest and most mechanical, being performed in a bathroom, he standing, she kneeling, with no real fulfillment—or love, or tenderness—on either side. They never have intercourse, though the intern pleads with the president to do so, saying she deserves experiencing that with him at least once out of fairness to herself. Over a two-year span, Monica performs oral sex on the president nine times. On two of those occasions, he's on the phone.

Perhaps the most astonishing fact, the one most revealing of the president's breathtaking recklessness and that seems almost incredible given the location of their sexual encounters just off the Oval Office, is that in *every single one of them* the door is always ajar about a foot or less. Is the president so filled with hubris that he actually believes he could never be caught? Or is he actually courting detection in some internal psychodrama beyond comprehension?

Nor is this the end of the reckless acts. They exchange some fifty or more personal phone calls, of which logs exist to document their conversations, the hour, the place, and the date. Even more potentially destructive are the fifteen or so times they engage in phone sex, usually late at night or early in the morning. Devastating personally and politically as revelation of those conversations would be, the president's recklessness in making them raises an even grimmer prospect: that he could be subject to blackmail or other form of intimidation by enemies, whether personal enemies, or foreign ones. He himself clearly recognizes that risk. At one point, he warns Monica that he suspects his phone is being tapped by "an unnamed foreign embassy." Then he comes up with "the ruse" that if Monica is ever questioned about their conversations she could say they were friends and "just doing it to give people a run for their money."

The Intern can be excused for thinking the president's attention to her redeems her low sense of self-esteem and worth, even to dream that someday this might lead to a life with her Handsome. The president harbors no such thoughts; never does he tell her he loves her or needs her. Instead, he speaks of not wanting "to become addicted" to her sexually, or she to him. Nor does he permit himself to give himself physically to their meetings.

Only twice does he even allow himself to reach a sexual climax, and these occasions occur only after the insistent imploring, even begging, of the Intern that she be permitted to feel she has given him pleasure by completing the act. The first occasion, on their next-to-last sexual encounter, proves fateful. The semen-stained blue dress is the result.

Of all the scenes in this tawdry, pathetic story, the saddest comes when the Intern sees the president masturbating into a bathroom sink after he thinks she has left. Indeed, the Intern's accounts of their meetings show the president continually battling to distance himself, to hold himself true to the position he occupies and to the people who trust and love him most. He tells her, as he tells others, that he's struggling "to control" and "turn off" his body. He's trying to "be good." Twice he tries to end their relationship, telling her in April 1996 that he doesn't feel right about his conduct, that it's wrong, that he feels guilty. They can be friends, but nothing more. For eleven months, that resolve holds; then he succumbs, and at the end of February 1997, their relationship resumes. Nearly three months later, though, on May 24, in what Monica records on her calendar as "D Day"—for "Dump Day"—the president finally terminates their affair, telling her he wants "to do the right thing in God's eyes and do the right thing for his family."

One other conundrum: Does he, as the Intern so wants to believe, feel any real kinship with her? Or is he just cynically, selfishly using her as lords have always used vassals, she the kneeling supplicant, he the all-powerful master? As she tells numerous people, including Linda, the president relates to her "very touching" accounts of his troubled childhood. "I guess he had an unhappy childhood and an abusive stepfather," she recalls him saying. After hearing of her own unhappy upbringing, the president tells her he feels a connection between them because "they had similar backgrounds or similar childhoods [sic]." And Monica, for all her appalling selfishness and self-absorption, does show a touching sensitivity when speaking of other sides of the president's personality. She describes how he feels "awful" and cries after receiving news of the death of the first American soldier in Bosnia, how he's "very sad, lonely, and withdrawn" after learning of the shocking suicide of Admiral Michael "Mike" Boorda, the chief of naval operations, on May 16, 1996.★ How the president really feels about her, other than as a sexual vehicle, isn't clear from the admittedly incomplete record.

Three scenes, of numerous sordid ones that fill the public record, are particularly revealing.

The Intern tells her betraying friend, Linda, that she thinks the president has become more religious: "He has religious tapes and stuff in his office," she says. "I think he's turned to religion even more . . . I think he made some kind of a—in his head, like a pact with God . . . I think he made this pact, 'If you let me win again, you know I will behave.'"

On another occasion she adds to the silently whirring tape-recorded accumulation by quoting the president as saying: "I have an empty life except for my work, and it's a fucking obsession." "He said that?" Linda the Plotter asks. "Mm-hum," the Intern

---

★Boorda, the first enlisted man to rise through the ranks to become the Navy's commanding officer, became one of the most poignant victims of Scandal Times. He shot himself in the chest with a pistol outside his home in the Washington Navy Yard shortly before he was to meet with two *Newsweek* reporters who were going to question him about accusations they had received that Boorda undeservedly was wearing two Vietnam-era decorations for valor in combat. In two suicide notes, one to his wife, another intended to be distributed "to my sailors," the widely admired admiral said he was taking his life because his honor and integrity were about to be questioned. Two years later, after investigation, the Navy ruled that Boorda was entitled to wear the decorations on his uniform, describing their use and display as "appropriate, justified, and proper."

replies, quickly adding: "And then I said, I said, 'Well, don't you get any warmth and da da da from your wife?'" The Plotter: "You didn't!" The Intern: "I did! He said, 'Of course I do.'"

Finally, the Intern's belief that the president has a sexual addiction problem: "I think he has a problem," she tells the Plotter. "And I think he knows he has a problem. And I think he knows that faced with temptation with a willing partner that he finds sexy he would not be able to mend his ways." The president, she confides, tells her he's had "hundreds of women" before his fortieth birthday and quotes him as saying his life "was falling apart" when he turned forty. Since then, he's been trying to resist becoming involved with women, remarking to her that he "struggles with it" almost daily. She believes the president, a "very sensual man," has a "Saturday night personality" where he succumbs to his sexual desires and "a Sunday personality" where he's remorseful and goes to church.

Perhaps so, perhaps not. What really matters is why the president is so willing to risk all at the very time when he's struggling to forge the positive political legacy he so clearly wants to leave and which his reelection has handed him a historic opportunity to accomplish.

The answer, if there is one, lies more in the realm of pseudo-analysis and probably defies definitive determination; but there are clues to his character and makeup worth pondering.

One of those clues offers an insight into this most complex Dorian Gray kind of political leader. It comes from private comments by one of the president's closest associates, made long after the affair has entered history. This person is talking about the central question involving the president: Why would he have risked this? About the Intern he says, "None of us knew her. . . . She appeared to be smart. She was energetic. She was enthusiastic. And she was flirtatious. He caught her eye and she caught his. She put herself in his line of sight. There's no question about that. But he had a lot of sights that wanted her in it. Okay? Okay? . . .

"Now, did I think that he was taking her bra off and that she was sucking his cock in the back office? No. Did I think there might have been some mild flirtation? Yeah. Did I think it was more than that? No. I took his word for it. I didn't press him. Did he make a big mistake? Yes. I would have advised him to 'fess up right then and there. Would he have done it? Probably not. Because he thinks he's gotten away with these sorts of things all his life. Slick Willie didn't come out of the air magically. That's him. You have to accept it."

· · ·

Naturally, the global soap opera obscures more significant, and possibly lasting, aspects of the scandal. The endless revelations of yet more salacious tales involving the most powerful and identifiable people in the world provide titillating entertainment on a scale previously unknown, perfect fodder for late-night TV comedians, for stand-up nightclub acts, for incessant lewd gossip in homes and offices everywhere.

The characters, and the sexual acts revealed, quickly enter the cultural mainstream and provide script material for Hollywood films and TV sitcoms. Did the Intern really tell the president she wants to be "Assistant to the president for blow jobs?" Yes. Did the president really say "I'd like that?" Yes. Did the Intern boast repeatedly how she's earning her "presidential knee pads?" Yes. Did the president really tell her, after she says she has to return to work after their first nighttime White House sexual encounter, "Well, why don't you bring me some pizza?" and does she really promise to return shortly after asking him "if he wanted vegetable or meat"? Yes. Did the president really tell the Intern "Good Morning!" after they completed phone sex about 6:30 one morning, and then add, "What a way to start the day!"? Yes.

One lurid tabloid tale begets another.

Taken together, they cheapen public life, for they make it seem as if all politicians cavort across a huge burlesque stage. The already-strong view that officials can't be trusted to tell the truth is reinforced; the public is encouraged to believe the lives political leaders *really* lead behind their public relations screens are licentious and corrupt, thus unworthy of serious consideration. The disconnect between the people and their political system grows greater. The bedrock fundamental upon which the democratic experiment is based—the concept of truth, the essential belief that those invested by the people with power over their lives can be trusted to honor their oaths and responsibilities—becomes even more undermined. The cynical society becomes more so.

These consequences are self-evident. They are starkly, painfully, in public view. No one can avoid them. Others, not so evident, are at least as significant. Foremost among them is the concept of public service.

To a degree not appreciated by the public, the entire Scandal Times era wreaks extraordinary damage internally on those who work in the government, and particularly in the White House.

The point is not that the president and his administration should be exempt from investigation and prosecution for any offenses committed. The point is what legacy the rankly partisan tactics of the scandal era leave. Do the attack politics of personal destruction, of payback for past politically motivated investigations from those then in power, of leaks and spins and further erosion of the lines between public and private lives, become the norm in a political system already marked by excessive ill-will and hatred? If so, what effect will this have on those who serve future administrations—assuming anyone other than political hacks and sycophants would wish to serve the public under such conditions?

All the scandal frenzy during the Clinton years doesn't add up to answers, but it does suggest some consequences.

Long before Monica, basic governmental working relationships were being affected by the insistent pressures to respond to demands by investigators—in Congress, in the special prosecutors' offices, in the press—to produce records bearing on various phases of yet another inquiry. Many of these are nothing but fishing expeditions launched in hopes of discovering damaging information that will benefit political opponents. Example, only one of many: A relentless administration attacker, New York's Republican senator Alfonse D'Amato, has his congressional investigators subpoena the records of *every* telephone call made from the White House to the 501 area code—that is, to anywhere in the entire state of Arkansas. The Clintons themselves have to produce personal documents and records going back decades.

Frivolous and politically motivated though most of the investigations are, and fruitless as well, government employees are legally required to search for and produce literally millions of documents for investigators and keep searching and producing as more investigative demands are made. One obvious result is to place an increasing burden on already overworked employees, thus hampering the work they are charged to perform for the public.

More serious is the internal climate created by the constant assaults from scandal hunters. One of the most disturbing precedents comes when Secret Service agents sworn to give their lives to protect the president are compelled by the prosecutors against the vehement protests of their superiors, former presidents and senior White Houses officials of both parties, to testify about what they observe of the president's private behavior while on duty protecting him.

By the time Monica erupts, senior White House assistants no longer take notes even when meeting with the president for fear they will be subpoenaed by prosecutors or by a congressional panel. History, the ability to compile an accurate account to learn the lessons of the times, is the loser. Employees also learn to be wary of putting *anything* on paper, or in their electronic computer or voice-mail systems, that might possibly bear on, however innocently, a subject of an ongoing or potential investigation.

For the same reason, they no longer keep records of their phone calls. They're also much more cautious in saying anything that might be misinterpreted by investigators, or addressing taboo subjects, even if in jest. They even remove editorial cartoons or columns bearing on the investigations from their office bulletin boards. As the subpoena net keeps widening, senior presidential aides refrain from discussing various aspects under investigation out of concern they will be required to testify about such conversations.

Internally, suspicion increases: *Did he/she testify before the grand jury? Did he/she turn over documents that could be embarrassing to me or prove incriminating? Did he/she strike a bargain with the prosecutors for a grant of immunity in return for their testimony? Who else is he/she talking to, and about what?*

To an unprecedented degree, administration members during the Clinton years are subjected to a barrage of blanket lawsuits filed by well-financed opponents of the president. These represent much more than nuisance suits. Many of them are motivated by a desire to thwart administration policies and affect the legitimate processes of government. They are nothing less than an attempt to nullify the results of the last presidential election. Opponents use the legal or congressional investigatory process as an ideological club to attain political results other than those intended by the electorate.

These suits have another major negative impact on the workings of government and public service. They compel employees to testify, at their own expense, often on matters about which they have little— if *any*—pertinent knowledge. Witnesses are harassed and intimidated. They suffer not only personal but financial hardship.

The numerous suits filed by Judicial Watch, the Richard Mellon Scaife–funded project that targeted members of the Clinton administration holding both high or low positions, are a classic example of the tactics employed. A number of senior White House advisers, like Harold Ickes and Marsha Scott, amass as much as $400,000 in legal costs from being compelled to testify, often numerous times, in hectoring videotaped depositions such as those conducted by Judicial Watch's general counsel, Larry Klayman. Many leave government service deeply in debt and forever embittered by the experience.

It's not just the big political names that pay the price; often the most peripheral, poorly paid, and publicly unknown employees are forced to hire lawyers at their own expense. Then they're subjected to hours of relentless grilling about matters of which they have almost no knowledge.

To take a typically egregious example, consider the case of Stacy Parker.★ In 1993, while still a college student, she begins working part-time as a White House volunteer, then becomes an intern (at the same time as Monica) in the office headed by George

---

★When she first became a White House volunteer and then an intern, Stacy Parker was one of the students taking my university seminar on politics and the press. Later, she made available to me the disturbing transcript of her deposition before Klayman conducted on March 18, 1998, in Washington.

Stephanopoulos. In 1996, upon graduation, she again becomes a White House volunteer. Finally, in September of 1997 she gets a paid White House job as a clerical assistant to Paul Begala, one of the president's most visible defenders in public forums.

Merely because she works there, Stacy Parker, a black woman twenty-three years old, already in debt from college expenses, finds herself subpoenaed by Judicial Watch and ordered to give a deposition before Klayman. The ordeal she experiences throughout six hours and fifty-nine minutes of a bullying inquisition from Klayman is chillingly documented in the 373-page transcript of her deposition.

From the very beginning of the deposition, which is being videotaped, Klayman adopts a hectoring, threatening tone. The official transcript provides only a flavor; for its full impact you have to watch the videotape, and even that doesn't completely convey the intimidating atmosphere:

Q: Before your deposition today, did you talk to anybody about it other than the attorneys sitting at this table?

A: Yes.

Q: Who?

A: I told my mother.

Q: Who is your mother?

A: Must I give the name of my mother?

Q: Yes.

Her attorney: I object to the relevancy of her giving her mother's name.

Q: You can respond. What's her first name?

A: Her first name is Carol.

Q: Carol Parker?

A: No.

Q: And what's her name? What's her last name?

Her attorney: What's the need for this? Could you just explain for the record what the relevance is?

Klayman: It's obvious what the reason is. She may have talked to her mother and given her information.

Her attorney: That's not a proper inquiry.

At that point, Klayman begins a tactic he repeats constantly, threatening in a blustering manner to "certify it," that is, refer it to the judge for legal sanctions for refusal to answer. He continues in the same vein, asking to whom else she talked about the deposition. Her father, Stacy says. He demands *his* name. Again, she's reluctant to give it. "All right, certify it!" Klayman pronounces. Who else did she speak to? Her boyfriend.

Q: And what's his name.
A: Jonathan.
Q: And what's his last name?

At her lawyer's request, they take a quick break after which her lawyer complains that Klayman's line of question "is improper in that you haven't established what she might have talked about to these people. She's willing to tell you the substance of her conversations and generally who they were, her mother, her boyfriend, her father, but there's absolutely no purpose, other than an improper purpose, for her to be naming people when you haven't established any relevancy. . . . I view asking for these names as harassment unless you've established there is some kind of substance."

"It's not harassment," Klayman fires back, "and I'll certify this whole line of questioning and we'll go to court on it. I view your inappropriate objections as harassment."

On they go, hour after hour after hour, with Klayman often reminding her she's under oath and implying penalties that might be imposed. He threatens to certify this or that section of the transcript even though a fair-minded person would conclude Stacy's trying her best to be responsive to his questions. She isn't argumentative; she's not defensive; she's not disrespectful in any way. Repeatedly, she says she's trying her best to respond. He threatens to call her back for another session, saying, "This is one of the worst depositions I've seen defended." And the end result of all the grilling, all the abusive inquisitorial tone, is one clear fact: She has literally nothing of value to contribute—no pertinent evidence, no knowledge, no useful insight into anything being investigated. She becomes a victim of investigative zeal run amok. Soon, she leaves government disillusioned and in greater debt.

The grand jury inquiries launched by Starr further strike at the functioning of the administration. The Paula Jones civil suit, rather than not unduly affecting the functioning of the presidency, as the Supreme Court thinks will be the case, requires Clinton and his key advisers to devote inordinate time and attention to it and other investigative challenges. The marvel is that anything gets accomplished in the Clinton years.

Again, the point is not that investigations by themselves are wrong. They play an essential, indispensable role in a free society. The oversight role of the Congress is critical in maintaining a check on abuses of power, in examining potential wrongdoing, in correcting proved cases of corruption, mismanagement, or misjudgment. So is the watchdog role of the press, the very reason, in fact, the press is granted its extraordinary license, including its right to be wrong. The Clinton administration produced abundant enough examples of wrongdoing to require serious investigation of them. At the same time, the record of the Clinton years demonstrates how the investigatory power can be *misused,* and not only by official and private inquiries but also by an independent press that is supposedly an ultimate check on abuses of power, public and private. All this raises anew the enduring democratic dilemma: Who is going to watch the watchmen?

· · ·

Eventually, Bill Clinton may be seen as one of the loneliest of presidents. Historically, to be sure, the reasons for his increasing isolation differ from those of the few among his predecessors who became so engulfed in problems that they felt like captives in the lonely "prison of the White House." Four of those presidents—Lincoln, Harding, Lyndon Johnson, and Nixon—come to mind.

Lincoln, brooding and anguishing over the terrible decisions of the Civil War, knowing himself to be reviled publicly beyond any of *his* predecessors—the "ape baboon of the prairies," the figure of contempt among many of his political contemporaries who think him a crude and coarse teller of undignified jokes—battling what Churchill later calls "black dog" bouts of depression, suffers the death of his beloved eleven-year-old son, Willie, who falls sick with a cold and a fever after riding his pony in a chilly rain, then faces the deadly, incessant gossip that his wife, Mary Todd, is suspected of being a secret spy for the Confederacy. So virulent are these stories circulated by the president's enemies that they actually lead to the convening of a secret session of the Senate Committee on the Conduct of the War to address the suspicion that the president's wife is "a disloyalist."

This extraordinary, little-known session prompts Lincoln himself to appear before them unannounced and uninvited. With a look of "almost unhuman sadness" in his eyes and conveying "an indescribable sense of his complete isolation," as one senator remembers, he declares solemnly that: "I, Abraham Lincoln, President of the United States, appear of my own volition before this Committee of the Senate to say that I, of my own knowledge, know that it is untrue that any of my family hold treasonable communication with the enemy."

That's not the end of Lincoln's torment. His wife, becoming increasingly mad, continues to be the object of malicious gossip and whispered stories about her imperviousness, her extravagance, her still suspected disloyalty as a Southern agent. She turns to spiritualism, holds White House séances, sees apparitions of the dead, and leads one contemporary to write to a family member that "it is not all peace in Abraham's bosom" in the White House.

Harding, that "idol of the man in the street, the apotheosis of the Average American," at the peak of his personal popularity puts on a glowing front of exuberance and bonhomie as he conducts his official public business. This act continues until his sudden, still mysterious, death; but it masks his growing fear that the scandal net hovers over him, threatening to destroy his name and his presidency. He faces exposure of his longtime affair with Nan Britton, a young woman who traveled with the president as his "niece" and bore his child. He also discovers, to his hopeless horror, that old friends he appointed to high office—the "Ohio Gang"—have been engaged in widespread graft and, worse, that top members of his cabinet, among them the attorney general and interior secretary, are directly implicated in massive corruption involving some of the nation's most valuable oil reserves at Teapot Dome and Elk

Hills fields. As the end nears, with public exposure virtually certain, Harding muses bitterly that "it isn't my enemies that are causing me trouble, it's my friends," and complains that "the White House is a prison, I can't get away from it." Finally, he finds himself unable to speak candidly with anyone. He distrusts all his once closest cronies and advisers, and sinks into almost total isolation while still struggling to maintain a confident public front.

Lyndon Johnson's isolation is more complex, but no less emotionally crippling. Accusations of personal scandal and political corruption have plagued him from his earliest electoral ventures in Texas and continue to be raised against him in the White House. But the source of the historic torment this president of gargantuan appetites and abilities experiences lies not in charges of personal impropriety or corruption but in policies he pursues during the long escalation of the Vietnam War. It is Johnson's fate to preside over the most stunning turnabout in presidential history to that point. His Great Society domestic policies bring him levels of personal and political popularity unmatched since FDR's peak during the New Deal; then his Vietnam policies cause him to plummet to the lowest depths of recorded public approval. Accentuating his rise and fall are the sounds of loudly chanting protesters encircling the White House: *Hey, hey, LBJ/ how many kids did you kill today?* During sleepless nights when he pads from his White House bedroom in bathrobe and slippers to the basement Situation Room to get the latest battle reports from Saigon, his personal isolation rivals Lincoln's a hundred years before. In the end, Lyndon Johnson finds himself cut off from the country he leads and feels surrounded both by longtime political enemies and erstwhile nervous Nellie supporters whom he views as betraying him.

Like Johnson before him, Nixon soars to the heights of success during his first term before crashing to the depths of scandal and public disapproval. His entire political career has accustomed him to facing continual crises arising over accusations of scandal—indeed, he relishes them—but his growing isolation during the Watergate scandal turns out to be as psychically damaging as anything experienced by his predecessors. Nixon's final ordeal leaves him feeling even more besieged by enemies than LBJ. He prowls the

White House, staring at the portraits of other presidents on the walls, painting a presidential self-portrait as the most isolated of them all.

On the surface, Bill Clinton's presidency resembles none of these. Yet there are strong hints that the toll of the repeated investigations and the constant efforts to "get him" invest his presidency with at least a comparable sense of utter isolation.

Looking back on the scandal frenzy created by the Whitewater and related investigations from the vantage of the end of his first term, the president remarks revealingly that, "For three months there I thought I was lost in the funhouse. . . . I just thought it was bizarre. I had to fight hard to keep my mind and my spirit in the right frame. . . ." He adds: "It was maddening to me, frankly. I agreed to the special counsel even though I thought it was wrong, since no president had ever been subject to one before, for something unrelated to his service as president or to his campaign for president."★

In public, the president attempts to maintain his habitually sunny demeanor. In private, upstairs in the family quarters, he acknowledges that he and Hillary speak constantly about scandal developments. "We talked about it all the time," he says. "We thought it was just crazy."

For the most part, he succeeds in presenting a positive public face. But, again in private, he displays outbursts ranging from defensiveness and belligerence to anger and self-pity. It's during this time that a number of senators begin hearing the president pouring out his true feelings in late-night calls he makes to them. "It was the most gushing outpouring of rage, humiliation, frustration I've ever heard," one of those senators says shortly after the president calls him late one night. The president seeks advice, wonders how this can be happening to him, anguishes about how to combat it. He complains bitterly about the welter of rumors being fed to the

---

★These remarks of the president's and those immediately following were made to me and my colleague, David S. Broder, during an Oval Office conversation for our 1996 book, *The System*.

press about him by his enemies, and worries about how he and Hillary will ever be able to raise the required sums to pay the rapidly rising bills for their defense. Their legal bills are then approaching a million dollars; in the months and years ahead they multiply to ever-greater amounts. "They've become paranoid," a cabinet officer who admires them says during this period. "They think people are out to get them—this right-wing conspiracy stuff. They feel sorry for themselves. They talk about it all the time: 'There really is a conspiracy out there to get us. We don't have a chance. People don't understand how much good we've done. Our message isn't getting out because these people are beating us up.'"

And all this becomes a mere tepid prelude to the infinitely more inflamed feelings unleashed by Monica. Even then, the president manages to maintain his positive public stance. But as that scandal continues, creating greater and greater impact, there are glimpses of inner turmoil. Six months into the scandal someone who knows him extremely well, sees him regularly, and has been a confidant for years, notes a change in his private demeanor—more subdued, withdrawn, less focused, more anguished—to such a point that he leaves an impression of a man "whose great heart is breaking."

Whether or not that's true, it isn't the face the president shows in public. He displays a consistently optimistic manner, making him appear confident, largely untroubled, in control, comfortably at ease with himself, in supremely good humor. However else history will judge him, he will certainly have to be credited as being one of the rare presidents who possesses what can best be called an indestructible spirit. Underneath the smooth, jocular manner lies a tenacious brawler. In the twentieth century only three others possessed this kind of continually reassuring, criticism-deflecting, soft-seeming but tough public temperament: the two Roosevelts, Theodore and Franklin, and Ronald Reagan. Other presidents entrapped in lonely ordeals or facing destructive scandals leave images that linger in the public mind: the sad, brooding Lincoln; the ravaged, mournful LBJ; the combative, paranoid Richard Nixon proclaiming "I'm not a crook" when his furtive look and guilty manner betray him. Not so with Clinton. Unraveling this side of his character may prove even more difficult for historians.

A strong insight into that aspect of his character does emerge, however, out of the mass of recollections—most still private—of some of those who observe him closely in the White House.

This one comes at the very moment when the "affair" with the Intern is just beginning: It's November 1995, the government shutdown has occurred through the grand miscalculation of Republicans and Clinton foes who think him weak and destined to be voted from office. Each night leaders of the Republican congressional team huddle with the president and his key advisers in the Oval Office or the Cabinet Room. They're led by Newt Gingrich, the new Speaker of the House. His fellow Republicans have learned to be fearful of the president's ability to persuade, seduce if you will, even his most unregenerate critics. They're especially wary of seeing the president and the Speaker conferring alone. They fear their side will give away the store.

One night, in the Cabinet Room, the other Republicans see the sight they have been dreading. Off to the side, out of earshot, they see the president and the Speaker facing each other engaging in intensely animated conversation. Only when they return to the Capitol to regroup and take stock of their strategy do they learn what transpired.

What was that all about? they anxiously ask the Speaker.

The answer both fascinates and confounds them.

"Do you know who I am?" the president inexplicably asks as he leans forward into the Speaker's face. The Speaker is stunned by the question. "No," he says, "who are you?"

"I'm the big rubber clown doll you had as a kid," the president says, "and every time you hit it, it bounces back up."

Silence for a moment. Then the president applies the punch line. "That's me," he says. "The harder you hit me, the faster I come back up."

As those present will recall, that turns out to be perhaps the most accurate and revealing aspect of this president's character. It illustrates, if not explains, his amazing resilience in the face of exposures so disastrous and so humiliating that most public figures would have been crushed and destroyed by them.

The Public

As with "All O.J., All the Time," but with graver consequences, no one in "All Monica, All the Time" emerges unscathed. Not the president, not the attackers or defenders, not the opposing political players in either party, certainly not the media. All are diminished. It is a story without winners, but it does produce a singular hero: The American people.

From beginning to end, the American people display great maturity and sound judgment as they assess the scandal being reported so incessantly and excessively. And from the beginning, the overwhelming public reaction stands in stark contrast to the view of the scandal as reported from the political insiders of Washington.

Literally from the first day that Monica becomes *the* national story, a fascinating disconnect occurs between the prevailing opinion of the capital and that of the country. In the capital, the scandal is portrayed as historic. Clinton's demise is widely predicted; he can't possibly last. In the country, polls taken from virtually day one register strong approval for the job the president is doing. The scandal is not viewed as so serious as to force the president from office.

In all the weeks and months that follow, no matter what new revelations dominate headlines and telecasts, the same striking disconnect continues to be reflected in the opinion of Washington and the opinion of the people outside it.

People by no means condone Clinton's personal behavior. Uniformly, they condemn it. But they consistently exhibit a greater perspective on what's most important and what's not. They credit the president with doing his job, and in large measure believe him to be doing it well. They view the critics, whether in politics or in the press, as being driven by excessive partisanship and hypocrisy. Who among them could hurl the first stone? They draw a clear distinction between personal and public misbehavior, between a consensual sexual relationship, however demeaning, and criminality.

It is all seen as highly embarrassing, damaging to the country, unworthy of the United States. It is not seen as the kind of political criminal conspiracy that requires proceeding with the greater national ordeal involved in the ultimate political sanction of removal from office.

Part of their response derives from what they see and hear emanating from Washington. The worst elements of the media/political circus are on daily, nonstop display: Part sound and fury, part clamoring mob, part hypocritical political posturing. Part, too, stems from the fact that for months no proof is produced that anything other than consenting sex is involved, and even that continues to be loudly denied by the president and his daily defenders.

A critical factor in reinforcing public opinion comes in April when the judge dismisses the Paula Jones suit on grounds the Jones lawyers didn't offer sufficient evidence to present their case before a jury. This comes two months after the judge excludes Monica's testimony altogether, ruling that while her evidence "might be relevant to the issues in this case" it is "not essential to the core issues in this case." A civil suit that was all about sex turns out not even to be judged serious enough to go to trial, or even to hear the testimony of the central witness the federal prosecutors are relying upon to prove their case against the president. And even if the president is shown to have been lying about his sexual involvement with both Paula and Monica, people consistently assess his actions as being understandable while not admirable.

Implicit in this strong public judgment is a great shift in American attitudes toward such matters that has taken place over recent decades. Gone is the absolutism of puritanical American values about sex and personal behavior—attitudes held so strongly just a generation before that no divorced candidate seeking the presidency could win the highest office, or even be nominated for it.

Moralists in the Nineties could, and do, decry this change as a symptom of national moral and ethical decay—"moral and intellectual disarmament," as William J. Bennett, Reagan's former education secretary and author of the popular *The Death of Outrage,* terms it on another *Meet the Press* program devoted to the scandal. Pragmatists view it as evidence of more honest attitudes that re-

flect the realities of a nation in which six in every ten marriages in
the late Nineties are likely to end in divorce, in which 35 percent
of the more than 4 million households headed by unmarried
couples include children, in which sexual activity outside of marriage occurs for an average of between nine to eleven years before
marriage, in which grade school children are taught how to apply
condoms and practice "safe sex," in which openness about sexual
behavior and preference are the subject of endless public discussion beamed into every living room, in which infinitely greater
sophistication about personal behavior characterizes a society become both more tolerant and conflicted about such intimate
questions.

Rather than producing a public reaction against the president,
the "scandal" produces one against Washington and all its works.

The opinion polls consistently chart these attitudes even as the
investigation proceeds and its heated coverage continues for month
after month. Polls, of course, can be wrong—or, at the least, can
convey too simplified a view of a subject as complicated as *the* national opinion on anything. But in this case what the polls are reflecting is borne out to a remarkable degree in the conversations
being recorded for this book during all the months the scandal
dominates the national news. Those interviewed include Nobel
laureates at Caltech, senior Microsoft executives in Washington,
leading Silicon Valley entrepreneurs, top scientists racing to decode
the human genome or produce new forms of genetically modified
foods. They include conservatives and liberals, Democrats and Republicans, those who admire and those who feel contempt for the
president personally, but almost without exception their expressed
attitudes about the "scandal" reflect those being regularly reported
in the national opinion polls.

The remarks of one of the president's unhappy supporters in
Hollywood, speaking while the bonfire continues to rage and the
denouement remains far away, sum up the prevailing views.

"The worst part of this entire thing is that it's being treated as
if some terrible constitutional crisis is occurring because a guy had
sex in a bathroom and then lied about it to cover it up," says Gary
Ross, the Hollywood director and producer. "That is a natural,

human response for someone who acted that way. The only possible reason for optimism I see here is that the public, God bless them, somehow understands this. That's why you see this great distance between what the polls keep showing and what the scandal merchants keep shouting."

III. THE DRESS
AND THE RECKONING

On Thursday morning, July 30, 1998, nearly seven months after the name Monica Lewinsky entered the national consciousness, readers of the *Washington Post* awake to find a bold black headline emblazoned across their front pages:

CLINTON AGREES TO TESTIFY FOR GRAND JURY

Below that headline, they read:

Sources Say Lewinsky
Has Physical Evidence

It isn't the end, but it's the beginning of the end of the long period of lies. Even the most unsophisticated reader knows instantly that this latest news represents a major break, for the story also reveals something explosive about Monica and the investigators.

"As part of the pact granting her immunity from prosecution, sources close to the case said yesterday, Lewinsky agreed to turn over to Starr physical evidence that could help establish a close relationship with Clinton, including telephone message recordings with his voice on them and a dress she allegedly wore while with Clinton that could be tested for identifying DNA material."

The story continues: "The report that Lewinsky secretly withheld a dress from Starr's investigators despite their search of her Watergate apartment revives perhaps the most sensational allegation in the case—one that the White House had dismissed as discredited months ago. It also raises the possibility that the investigation may turn on more than the simple, he–said–she–said disagreement portrayed by Clinton advisers."

To Washington insiders, accustomed to the daily dance in which opposing sides send clear but anonymous signals to each other through the press, this story demonstrates a classic use of a leak to the press to influence political action. No one from the president down can fail to get the intended message—and threat. Now the prosecutors have the goods.

Indeed they do. For that very day, FBI labs in Washington receive what they label "Specimen Q3243," identified further as "Navy blue dress." Immediately, they begin a serological examination of the garment.

Now, events move with dramatic rapidity, though still out of public notice.

That next day, Friday, the thirty-first, David Kendall, the president's private attorney, receives a terse letter delivered by hand to his downtown Washington office. He opens it to read:

Dear David:
I telephoned you twice this morning but was unable to reach you. Investigative demands require that President Clinton provide this Office as soon as possible with a blood sample to be taken under our supervision. I assure you this information will be kept strictly confidential and will be restricted to a handful of persons on a need-to-know basis only. Your response to these requests will be greatly appreciated.
Sincerely,
Robert J. Bittman, Deputy Independent Counsel

There can be no mistaking the clearly implied threat in that note, with its blunt assertion that "investigative demands require" the president of the United States to undergo a legal process unique in the annals of the presidency. A further exchange of hand-delivered messages between the rival camps of the prosecutor and the president takes place that day, but there's no doubt which side now holds the upper hand. The prosecutor is dictating terms to the president.

Monday, the third of August, brings a rush of even more dramatic, but still secret, developments.

FBI labs report to the prosecutor that "Semen was identified on specimen Q3243," the navy blue dress. The report adds: "In order to conduct meaningful DNA analysis, known blood samples must be submitted from the victim, suspect or other individuals known to have contributed bodily fluids to specimen Q3243."

By that afternoon, the president capitulates to the prosecutor. In a final hand-delivered letter, marked CONFIDENTIAL and TO BE OPENED BY MR. BITTMAN ONLY, David Kendall notifies the prosecutor that the president agrees to provide the blood sample. Kendall's language sets out the best terms the president is in a position to negotiate: "The sample will be drawn by the White House physician, Dr. Connie Mariano, at the White House, in the presence of two representatives of the OIC, under medical procedures acceptable to the OIC."

At ten minutes past ten that night, in a scene unlike any other in the history of the White House, in a moment that will stand as one of the most humiliating in the history of the presidency, five people assemble solemnly in one of the most hallowed rooms in the Executive Mansion: the president of the United States; his lawyer, David Kendall; the prosecutor's representative, William Bittman; the White House physician, Dr. Connie Mariano. Standing aside, silently watching, is the fifth person present: an FBI agent.

They are gathering tensely in the Map Room where Roosevelt and Churchill met during the darkest days of World War II to chart their strategy and where FDR continued to receive daily battle reports from around the world for the remainder of the struggle. For the Clintons, and especially for Hillary, this is a favored meeting place, one where they gather with special guests. Hanging over the mantel in its original place is a map—found and placed there by Hillary Clinton—showing the European battlefront lines, drawn in red, as they stood on the last day FDR studied them in April of 1945 before leaving the White House for Warm Springs, Georgia, where he died.

Now, nearly half a century later, the president of the United States sits down in a chair in the center of that room and, after removing his suit jacket, quickly rolls up his right sleeve and stretches out his arm. His face is crimson.

Dr. Mariano steps forward, leans over, inserts a needle into the president's vein, and draws forth a flow of blood into a tube bearing a purple top. The business is done briskly, and silently. As soon as the tube is filled with four milliliters of the president's blood, she caps it, turns, and hands it to the FBI agent nearby. She writes the name "William Clinton" on a label, affixes it to the tube, adds her own initials, and, with the others, leaves. Exactly twenty minutes later the agent hands the purple-topped tube to a Unit 1 technician in the FBI's DNA analysis lab on Pennsylvania Avenue, halfway between the White House and the Capitol. Work immediately begins on analyzing the contents of the vial, now labeled officially "Specimen K39."

This time, there's no leak to the press.

Three days later, August 6, the fifty-third anniversary of the dropping of the atomic bomb on Hiroshima, the FBI submits preliminary findings of its DNA tests on Specimen K39, identified further as "Liquid blood sample from WILLIAM CLINTON." The test sample shows it to be "a potential contributor" to DNA found on the dress. Further tests are underway to reach a definitive determination.

None of that information leaks. Even if it had, the press focuses on what it collectively believes to be a greater story, certainly a vastly bigger spectacle that day: Monica begins her secret grand jury testimony in Washington. Naturally, that event is treated as epochal by the camera crews and correspondents who follow her car from her lawyer's office to the courthouse and the near-permanent blizzard of electronic gear called "Monica Beach" by journalists stationed there to provide live bulletins from the scandal front.

Two more weeks of All Monica pass; then, on Monday, August 17, the final veil falls, half in and half out of public light.

That day the FBI labs report to the prosecutor that "specimen K39 (Clinton) is the source of the DNA obtained from specimen Q3243-1 [the semen stain on the blue dress], to a reasonable degree of scientific certainty." How certain? Thanks to the scientific/technological advances of Technotimes, the probability of finding an unrelated individual chosen at random in the Caucasian population possessing the same DNA genetic profile is rated at being 1 in 7.87 trillion. Case closed.

The prosecutors have that knowledge when they assemble that afternoon in the White House Map Room where the president's blood was drawn exactly two weeks before. This time, starting at 1:03 P.M., they are going to take the president's sworn testimony before grand jurors who follow it via a closed-circuit TV connection between the White House and the courthouse. Now the number of people in the Map Room has increased threefold. For the White House side, the president and three of his lawyers, plus one Secret Service agent and a member of the White House technical staff are present. From the prosecutor's side, Starr brings six of his deputies, plus a senior consultant, and the court reporter.

Grand jury testimony is supposed to be secret, sacrosanct, its dissemination tightly held. Not this time. This time the entire world will eventually be able to witness everything that transpires there, for the president's testimony is also being recorded by video cameras at the insistence of the prosecutor's team, an action that further infuriates an already enraged and entrapped president, who suspects that the video will be released to the public in yet another demonstration of the prosecutor's determination to do everything possible to humiliate and defame him.

Assurances aside, the video, of course, is released to the public within a matter of days, and released in its entirety, unedited.

It shows a president facing his worst moment—and also shows the prosecutors at *their* worst.

The picture the president presents this day is not the charming, relaxed public figure so familiar throughout the long ordeal. It is a haggard, pale, tense president who sits uneasily in that chair responding to questions from prosecutors, a president who battles to control his emotions and doesn't always succeed.

He begins by reading a formal statement: "When I was alone with Ms. Lewinsky on certain occasions in early 1996 and once in early 1997, I engaged in conduct that was wrong. These encounters did not consist of sexual intercourse. They did not constitute sexual relations as I understood that term to be defined at my January 17, 1998 deposition. But they did involve inappropriate intimate conduct. These inappropriate encounters ended, at my insistence, in early 1997. I also had occasional telephone conversations with Ms. Lewinsky that included inappropriate sexual banter. I regret that what began as a friendship came to include this conduct, and I take full responsibility for my actions. While I will provide the grand jury whatever other information I can, because of privacy considerations affecting my family, myself and others, and in an effort to preserve the dignity of the office I hold, this is all I will say about the specifics of these particular matters."

Then begins what becomes a highly charged, highly embittered series of questions and exchanges.

The prosecutorial team, with Robert Bittman and Jackie Bennett leading the questioning while Ken Starr and his other aides intently observe the proceeding, is determined to press the president and get him to admit he lied during his Paula Jones deposition. The president keeps insisting he did not lie, but he was determined not to help his accusers in the Jones case.

At one point, after being hammered again and again, the president snaps back and speaks very to the point about his Jones testimony: "I did my best, sir, at this [that] time. I did not know what I now know about this. A lot of other things were going on in my life. Did I want this to come out? No. Was I embarrassed by it? Yes. Did I ask her to lie about it? No. Did I believe there could be a truthful affidavit? Absolutely. Now, that's all I know to say about this."

He goes on to say, "I will admit this": "My goal in this deposition was to be truthful, but not particularly helpful. I did not wish to do the work of the Jones lawyers. I deplored what they were doing. I deplored the innocent people they were tormenting and traumatizing. I deplored their illegal leaking. I deplored the fact that they knew . . . that this was a bogus lawsuit, and that because of the funding they had from my political enemies, they were [going] ahead. I deplored it. But I was determined to walk through the minefield of this deposition without violating the law, and I believe I did."

His bitterness, his candor at not feeling the need to flesh out the truth, his comment about being "determined to walk through the minefield"—that's probably as close a portrait of the true character of William Jefferson Clinton as emerges during all the assaults he faces over scandals, whether or not of his own making, in his years as president of the United States. He is a president determined to navigate carefully and cautiously all the dangers of this war, and a president determined to remain standing at the end.

In the one moment that makes him seem both a sympathetic and a sad figure, he is finally frank about the fact that he lied and tried to hide his secret sexual relationship with Monica from other people. "Well, I never said anything about it [to anyone], for one thing," he says. "And I did what people do when they do the wrong thing. I tried to do it where nobody else was looking at it."

Throughout, his bitterness and feeling of self-pity keep surfacing, especially at the tactics employed by the Jones lawyers and his political enemies. "I've not heard these tapes or anything," the president says at one point. "But they knew a lot more than I did. And instead of trying to trick me, what they should have done is to ask me specific questions, and I invited them on more than one occasion to ask me follow-up questions. This is the third or fourth time that you seem to be complaining that I did not do all their work for them.... Now, they'd been up all night with Linda Tripp, who had betrayed her friend, Monica Lewinsky, stabbed her in the back and given them all this information. They could have helped more. If they wanted to ask me follow-up questions, they could. They didn't."

At another point, he says: "By the time this case started they knew they had a bad case on the law and they knew what our evidence was. They knew they had a lousy case on the facts. And so their strategy, since they were being funded by my political opponents, was to have this dragnet of discovery.... What they wanted to do and what they did do and what they had done by the time I showed up here [as president] was to find any negative information they could on me whether it was true or not, get it in a deposition, then leak it, even though it was illegal to do so.... Their real goal was to hurt me. When they knew they couldn't win the lawsuit, they thought, well, maybe we can pummel him. . . . They just thought they would take a wrecking ball to me and see if they could do some damage."

That assessment, its self-serving nature aside, accurately describes the conditions that created the legal snare in which the president becomes entrapped. It doesn't, of course, answer the question about his own responsibility for the behavior that creates the opening for his enemies to trap him, actions that produce all the resulting consequences, personal and political, for him, his family, his supporters, and for the country he leads.

The other aspect of his personality, the "Slick Willie" side, also emerges strongly during the questions in the way he parses his words, splits hairs, engages in maddening semantic circumlocutions. A classic example of such sophistry involves his attempt to define the meaning of the word "is." This comes after prosecutor Bittman refers to an exchange during the Paula Jones deposition in January, specifically to a moment when the president's lawyer, Bob Bennett, vehemently denies to the judge that Clinton had a sexual relationship with Monica.

Bittman reads that passage to the president in the Map Room, triggering the following exchange:

> Prosecutor: That is a completely false statement, whether or not Mr. Bennett knew of your relationship with Ms. Lewinsky, the statement that there was "no sex of any kind in any manner, shape or form, with President Clinton," was an utterly false statement. Is that correct?
>
> President: It depends on what the meaning of "is" is. If the— if the—if "is" means is and never has been, that is not— that is one thing. If it means there is none, that was a completely true statement.

Similarly, the president offers a definition of "alone."

> Prosecutor: Do you agree with me that the statement, "I was never alone with her" is incorrect? You were alone with Monica Lewinsky, weren't you?

President: Well, again, it depends on how you define "alone." Yes, we were alone from time to time, even during 1997, even when there was absolutely no improper contact occurring. Yes, that is accurate. But there were also a lot times when, even though no one could see us, the doors were open to the halls, on both ends of the halls, people could hear. The Navy stewards could come in and out at will, if they were around, other things could be happening. So, there were a lot of times when we were alone, but I never really thought we were.

If the president's testimony is as painfully revealing as it is embarrassing, so the prosecution tactics and questions expose the truth of Sam Dash's comment that the prosecutors have long been motivated by hatred of the president and the first lady and are out to get them—if not legally or criminally, then by humiliating them personally in public.

There can be no other explanation, or justification, for the repeated gratuitous questions the president is asked about explicit sexual acts he may have engaged in with Monica and others. The president's opening statement clearly, if tacitly, answers the question of whether he engaged in sexual activity with Monica. The reasons he gives for not offering specific details of what he calls "inappropriate intimate contact"—to protect his family and preserve the dignity of his office—are understandable and should, it seems, be acceptable to disinterested, objective observers. But the prosecutors do not accept them; again and again, the president is asked to respond to the most explicit and embarrassing questions about sex.

On the blue dress:

Prosecutor: Mr. President, if there is a semen stain belonging to you on a dress of Ms. Lewinsky's how would you explain that?

President: Mr. Bittman, I, I don't—first of all when you asked me for a blood test, I gave you one promptly. You came over here and got it. That's—we met that night and talked. So that's a question you already know the answer to. Not if, but you know whether. And the main thing I can tell you is that doesn't affect the opening statement I made. The opening statement I made is that I had inappropriate intimate contact.

The prosecutors are not willing to leave it at that. The Kathleen Willey case, never proved, is introduced as if it's a fact the president sexually fondled and groped her.

"Mr. President, you did make sexual advances on Kathleen Willey, is that not correct?" Jackie Bennett asks.

"That's false," the president replies.

Not content with that unambiguous answer, Bennett aggressively asks more and more explicit questions.

"You did grab her breast, as she said?"

"I did not."

"You did place your hand on her groin area, as she said?"

"No, I didn't."

Still not satisfied, Bennett presses on.

"And you placed her hand on your genitals, did you not?

The president, struggling to control his temper, replies: "I didn't do any of that, and the questions you're asking, I think, betray the bias of this operation that has troubled me for a long time."

That doesn't stop the prosecution's clear intent to get it all in, as explicitly, as embarrassingly, as humiliatingly as possible.

> Prosecutor: If Monica Lewinsky says that while you were in the Oval Office area you touched her breasts, would she be lying?
>
> President: That is not my recollection. My recollection is that I did not have sexual relations with Ms. Lewinsky, and I'm staying on my former statement about that.
>
> Prosecutor: If she said—
>
> President: My, my statement is that I did not have sexual relations as defined by that.
>
> Prosecutor: If she says you kissed her breasts, would she be lying?
>
> President: I'm going to revert to my former statement.
>
> Prosecutor: Okay, if Monica Lewinsky says that while you were in the Oval Office area you touched her genitalia,

would she be lying? And that calls for a yes, no, or revert-
ing to your former statement.

President: I will revert to my statement on that.

Prosecutor: If Monica Lewinsky says that you used a cigar as a
sexual aid with her in the Oval Office area, would she be
lying? Yes, no, or won't answer.

President: I will revert to my former statement.

Prosecutor: If Monica Lewinsky says that you had phone sex
with her, would she be lying?

President: Well, that is, at least in general terms, I think, is covered by my statement. I addressed that in my statement. . . .

Prosecutor: Let me define phone sex for purposes of my question. Phone sex occurs when a party to a phone conversation masturbates while the other party is talking in a
sexually explicit manner. And the question is, if Monica
Lewinsky says that you had phone sex with her, would she
be lying?

President: I think that is covered by my statement.

· · ·

Walter Lippmann, in a celebrated remark, once said that we are all
captives of the pictures in our heads. We think the world we believe in is the world that exists, and the world that will endure. So
it is when it comes to the pictures people carry of their public
leaders.

For Americans, an indelible mental picture of William Jefferson Clinton begins to form at eighteen minutes after ten o'clock,
eastern daylight time, Monday night, August 17, 1998. That's when
their forty-first president stares into a television camera and, after
receiving a signal from a technician, begins speaking to them.
"Good evening," he says. "This afternoon, from this chair, I testified before the Office of Independent Counsel and the grand
jury. I answered their questions truthfully, including questions about
my private life, questions no American citizen would ever want to
answer."

Presidents have had to speak to their fellow citizens in times of
crisis and celebration, sadness and joy—and, occasionally, of scan-

dal—but never has one found himself forced to make such a public report as this president faces this night. His address is part excruciatingly public confessional, part apology, part self-defense, part desperate attempt to save his office, and all delivered after completing perhaps the single most humiliating episode in the history of the presidency. Only two and a half hours have elapsed since he completed nearly five and a half hours of draining, demeaning testimony. Since then, he and a handful of trusted aides have been huddling inside the White House, thrashing out what he will say in a live, nationally televised address they all know is guaranteed to attract a vast audience and will strongly affect how his fellow citizens feel about him.

It's possible future Americans, reading the words he utters this night, will find them politically persuasive. It's doubtful those actually seeing the president on videotape will react as sympathetically. His angry appearance, his defensive manner, his underlying tone of defiance, all portray a president who fails to come to terms with the problems his conduct has created, and who can't bring himself to render a simple public apology for them. Indeed, he never uses the word *apology*. He never asks for forgiveness. He conveys no sense of repentance. He acts without a hint of humility.

For the most part, his words are carefully chosen. They say, at a minimum, what has to be said.

He admits he "did have a relationship with Ms. Lewinsky that was not appropriate," and adds: "In fact, it was wrong. It constituted a critical lapse in judgment and a personal failure on my part for which I am solely and completely responsible." He defends himself against the charge that his conduct was unlawful, and says "at no time did I ask anyone to lie, to hide or destroy evidence or to take any other unlawful action." He doesn't concede that he lied or otherwise failed to tell the truth, but acknowledges that "my public comments and my silence about this matter gave a false impression." In the closest he comes to an apology, he admits that "I misled people, including even my wife," and adds: "I deeply regret that." He explains his actions by saying he was motivated by "many factors: First, by a desire to protect myself from the embarrassment of my own conduct. I was also very concerned about protecting my family. The fact that these questions were being asked in a politically inspired lawsuit, which has since been dismissed, was a consideration, too."

Then he offers what amounts to a justification of his behavior: "In addition, I had real and serious concerns about an independent counsel investigation that began with private business dealings twenty years ago, dealings, I might add, about which an independent federal agency found no evidence of any wrongdoing by me or my wife over two years ago. The independent counsel investigation moved on to my staff and friends, then into my private life. And now the investigation itself is under investigation. This has gone on too long, cost too much, and hurt too many innocent people."*

Until that point the president's remarks, and the way he looks making them, are coolly, if not coldly, delivered. He's not rising to oratorical heights or striking positive emotional sparks, but then he's not groveling or indulging in false or mawkish appeals, either. Given the embarrassing nature of the subject, most of those watching can understand how in this painful moment their most empathetic president falls flat.

Had he left it there, after a few final grace notes of peroration, doubtless the overall public response would have been more favorable. But he can't seem to leave it there, for now the president, nearing the end of his remarks, becomes notably more tight-lipped and pale. His very manner breathes belligerence. "Now, this matter is between me, the two people I love most—my wife and our daughter—and our God," he says tensely. "I must put it right, and I am prepared to do whatever it takes to do so. Nothing is more important to me personally."

Here he seems to spit out the words, "But it is private, and I intend to reclaim my family life for my family." Then he says something that, however intended, comes over so defiantly that it sounds shocking: "It's nobody's business but ours."

---

*He refers to the action by a U.S. judge in Washington, D.C., ordering an inquiry into charges brought by Clinton's lawyers that Starr's office is guilty of leaking information to the press that is both illegal and damaging to the president's defense.

At that point, some of those intently watching their president find themselves asking, "Fine. Then why not resign your public duties and attend to your private responsibilities?" For, after all, it's *his* private behavior that places so heavy, and so unwanted, a burden on the public, a burden that directly affects how the *public's* business is being performed.

The president continues speaking, now more in an undertone of self-pity: "Even presidents have private lives. It is time to stop the pursuit of personal destruction and the prying into private lives and get on with our national life. Our country has been distracted by this matter for too long, and I take my responsibility for all of this. That is all I can do."

He then calls for the country to move on, tackle its important unfinished work, seize its opportunities, solve its problems, face its security challenges—all easy to say, difficult to achieve, considering the circumstances. "And so tonight," he concludes, offering a string of bromides, "I ask you to turn away from the spectacle of the past seven months, to repair the fabric of our national discourse, and to return our attention to all the challenges and all the promise of the next American century."

Of course, the people cannot turn away from this spectacle, however much they would like to, for now the bonfire roars almost out of control, threatening to consume everything in its path. And the pictures people carry in their heads of this personally disgraced president will be joined only hours later by ones that burn even more into the public mind. They leave perhaps the most lasting images of the entire long scandal story.

· · ·

There have been several fragmentary secondhand accounts of what Hillary really says to Bill when she finally learns the truth, how Bill attempts to explain his betrayal to her, and how they try, separately, to deal with their beloved daughter, Chelsea, after she realizes the lies her father has told everyone. Undoubtedly, there will be many more to add to the briefly reported snippets about shouts, slaps, cold consuming rage upstairs in the White House family quarters, Hillary telling him, "You've done more to give adultery a bad name than anyone since Moses," and, eventually, postmortem memoir accounts by the first family members themselves. They will all be superfluous. Nothing more is needed to paint an accurate portrait of how the characters caught in this personally tormenting psychodrama are really reacting. The entire world sees that clearly and unmistakably, and all on live TV.

As fate would have it, the day after the president finishes his national address, the Clintons are scheduled to leave for a twelve-day vacation to Martha's Vineyard, where they have enjoyed three previous relaxing breaks from the demands of the White House. Now, in late afternoon of a bright summer day, the presidential helicopter, *Marine One,* stands ready on the White House South Lawn to take them to Andrews Air Force Base in nearby Maryland for their flight to the island in the Atlantic just off the Massachusetts coast.

It takes only a few minutes for them to walk from the South Portico entrance of the mansion to the helicopter, but those moments recorded by TV are almost too painful to watch. It is a scene of a family in maximum emotional distress, a family breaking apart. Only the attempt of the daughter to hold her parents together by grasping their hands firmly keeps them from proceeding individually and alone on their stiff, awkward stroll toward the helicopter. Hillary's face is a study in sorrow and suppressed anger: She never smiles, never even glances toward her husband flanking Chelsea on the other side. At times, she veers away as if wanting to break off right then; Chelsea's grip keeps her from doing so. It's Chelsea, displaying cool poise, who keeps them moving forward. Bill attempts a forced smile that succeeds only in making him look more embarrassed, like a guilty little boy. In his free hand, he holds the leash attached to Buddy, his Labrador retriever. Only prancing Buddy acts without a care.

The emotional distance between the president and first lady appears infinite as they cluster momentarily around the helicopter steps before boarding. Neither looks at the other. They stand stiffly apart. Later, upon their arrival on Martha's Vineyard, the same scene repeats itself for the TV cameras: The president studiously turns away from his wife as he talks to greeters, including Vernon

Jordan; the first lady consciously stands off to the side, chatting with others. Then they depart. For the remainder of this most difficult vacation period, in a marriage marked by continual personal crises and disappointments, they will barely be seen together and almost never appear in public.

Cynics, hardened by the continuing scandalous spectacles of Teletimes, express a jaundiced view about the seeming disintegration of the presidential marriage, especially the impression these scenes leave of a first lady who appears to have been shocked and surprised by proof of her husband's betrayal. *She had to know,* they say. After all, Bill Clinton's history of philandering is something she, better than anyone, knows all too well.

Perhaps she did suspect, but the evidence strongly suggests she didn't know, perhaps again because she doesn't want to know. One private glimpse during the months in which the Monica scandal builds to its conclusion provides persuasive evidence of this belief: The first lady is having her nails done inside the White House while watching the news reported on CNN. When the broadcaster mentions an update on Monica, Hillary instantly and almost by second nature lifts her hand to the control panel and immediately changes the channel. She doesn't say a word, just continues watching something else.

There seems no doubt that Hillary Rodham Clinton had convinced herself that her husband could never be so recklessly stupid as to risk his presidency, and their White House standing in history, by engaging in such a relationship—especially one that represents such a massive betrayal. She learns not only that he did, but, even more destructively, that he carried it out over a two-year interval with a young woman close to her daughter's age right in their own home—even once after returning to their home from Easter Sunday services, when, most wounding of all, he had oral sex in the Oval Study just after she called him there and heard him tell her "I love you" before he hung up and turned to Monica.

The revelation that this turns out not to be some single failure, a transitory fling, not a concoction of their enemies, but in fact an ongoing liaison in which the intern tracked Hillary's Washington schedule daily to learn of *her* time away from the White House and maintained logs of *her* travels away from the capital in order to arrange clandestine meetings in *her* house, is a betrayal on such a scale as to be unimaginable.

Terrible as these disclosures are, others are no less humiliating and infuriating. The record shows the intern, having been invited to a White House Christmas party in 1997, made a point of introducing herself to the first lady and telling her she's a friend of Walter Kaye's. It also shows that the first time the intern engages in sex with the president, back on that November night in 1995, the president interrupts himself to take a phone call from Hillary. Then he resumes the sexual encounter.

Cynics also are quick to dismiss this obvious anguish of the first lady's by saying it is part of a bargain of her own making, one she struck with her husband long ago, a bargain motivated by her own ambition. She'll overlook his indiscretions so long as they are kept private; he'll support her ambitions to play a major public role in the nation's life, perhaps someday even a presidential role.

Again, there's no question that Bill and Hillary Clinton form a political partnership of singular drive and ambition. In many respects, theirs is a partnership unique in the presidency. Even more than Franklin and Eleanor, with whom they are most often compared, Bill and Hillary come close to being coequals in the White House. In the making of policy, in the selection of personnel, in the crafting of strategy, in the dual public roles they perform, theirs is a more consequential political relationship than any other before them. It's also, obviously, a most complicated relationship, even beyond the difficulties that mark virtually every marriage and particularly political ones of highest public visibility. That they have needed, and depended upon, each other in the past is unquestionable; neither would have been able to reach their high public positions without the total backing of the other.

They are the first baby boomer couple to occupy the White House, the first to exercise power after the post–World War II Cold War era has ended, the first to reflect so many of the cultural changes and ambiguities—in lifestyle, in career choices, in personal values—that have affected national life. They are representative of their times in numerous respects, Hillary no less than Bill. She is the first professional career woman to become first lady, a fact that some enemies and traditionalists hold against her. She attracts a level of hatred unmatched since that leveled against Eleanor half a century before. Yet without her strength and support, it's inconceivable he would have fulfilled his ambition to become president. Standing by her man and defending him in all their crises past, she has repeatedly saved his political life.

Now, in their most critical moment of all, the question is whether she can do so—or wants to do so—one more time. The pictures of them as they painfully make their way across the White House lawn suggest not. But the words Hillary authorizes her office to issue for her on the day of her greatest public humiliation suggest otherwise.

"Clearly, this is not the best day in Mrs. Clinton's life," the formal statement from the first lady's office reads before she departs for Martha's Vineyard. "But her love for him is compassionate and steadfast."

Asked if she forgives him, her press spokesperson tells reporters, "Yes. She believes in this marriage."

* * *

Two days after the Clintons begin wrestling with their personal problems in the privacy of their vacation retreat on Martha's Vineyard, Monica makes a notable official appearance in Washington. That August 20 she completes her formal grand jury testimony. After this, she will appear only briefly again as another of the fleeting celebrities claiming their fifteen minutes of infamy in book or other commercially arranged appearances—most memorably during the impeachment trial and many months later when 70 million Americans watch her during a breathless Barbara Walters prime-time TV interview—but this time, in Washington, her words are still uttered in private. They're still supposed to be secret, although as it quickly develops, not for long as far the prosecutor is concerned.

Whether because she's now become accustomed to the embarrassment of giving testimony about intimate acts, or because she's even begun to enjoy it, it's a far more relaxed, even jaunty, Monica who appears behind closed and guarded doors for her last grand jury appearance. She gets friendly with the jurors: "Can you guys call me Monica? Are they allowed to call me Monica instead of Ms. Lewinsky?...I'm just 25. Please." She makes wisecracks to the foreperson when they return from jury breaks, repeatedly interrupting teasingly to say before the prescribed legal admonition can be given, "Yes, and I know I'm still under oath." She even acts out with apparent relish a scene that previously would have produced sobs or sighs from her.

Asked by one of the prosecutors to describe yet another embarrassing intimate encounter with the president at a rope line inside a public ballroom, she cheerily asks the prosecutor: "Can I stand up and show you?"

Prosecutor: Sure, sure.
Monica: Okay, if this is the rope line and here are all the people and the President is standing here...I had my back to him and I just kind of put my hand behind me and touched him.
Prosecutor: Touched him in the crotch area?
Monica: Yes.
Juror: I didn't hear that.
Prosecutor: Touched him in the crotch area.
Juror: Oh.

By now, the jurors are clearly sympathetic with her. Their final Q & A period with her almost turns into a parody of an encounter session, a *Donahue, Oprah, Sally Jesse* daytime-TV talk show. They are solicitous, understanding. They offer their Dutch uncle advice. One says: "Monica, none of us in this room are perfect. We all fall and we fall several times a day. The only difference between my age and when I was your age is now I get up faster. If I make a mistake and fall, I get up and brush myself off. I used to stay there awhile after a mistake. That's all I have to say."

At the end, Monica volunteers a final statement: "I think because of the public nature of how this investigation has been and what the charges aired [have been], that I would like to say no one ever asked me to lie and I was never promised a job for my silence. And that I'm sorry. I'm really very sorry for everything that happened. [She begins to cry, then continues speaking.] And I hate Linda Tripp."

This triggers an emotional outburst from the jury, transferring the somber jury room into something approaching a religious revival meeting.

> A juror: Can I just say . . . we sin every day. I don't care whether it's murder, whether it's affairs or whatever. And we get over that. You ask forgiveness and you go on. There's some that say that they don't forgive you, but 'he whose sin,' you know, that's how I feel about that. So to let you know from here, you have my forgiveness. Because we all fall short.
>
> Juror: That's what I was trying to say.
>
> Juror: That's what it's all about.
>
> Monica: Thank you.
>
> Juror: And I also want to say that even though right now you feel a lot of hate for Linda Tripp, but you need to move on and leave her where she is because whatever goes around comes around.
>
> Juror: It comes around.
>
> Juror: It does.
>
> Juror: And she is definitely going to have to give an account for what she did, so you need to just go past her and don't keep her because that's going to keep you out.
>
> Juror: That's right.
>
> Juror: And keep going to keep you from moving on.
>
> Juror: Allowing you to move on.

It's become group therapy, Nineties style.

At the very end, the jury foreperson gives what becomes a benediction.

> Foreperson: Basically, what we want to leave you with . . . we wanted to offer you a bouquet of good wishes that includes luck, success, happiness, and blessings.

Monica: Thank you. [She starts crying.] I appreciate all of your understanding for this situation and your—your ability to open your heart and mind and—and your soul. I appreciate that.

What an ending, the perfect closing for one aspect of Scandal Times.

Except it's still not *the* end.

· · ·

Of Bill Clinton, more than *any* American president, it can be said that he's lucky in the nature of his enemies. Nothing better demonstrates this than the way his political enemies in Washington handle the information they've gathered about him. From the beginning, they misread the American public—and do so again once the investigatory process led by the prosecutor nears its end with the testimony of the president and the intern.

The president is historically fortunate, too, in the backdrop of the times. As the scandal process wends to its conclusion, the long boom continues to set economic record after record. Americans may not credit the president for creating the boom, but they certainly don't think negatively of him because of its benefits. And however damaging to the president's *personal* reputation the scandal revelations are, the public renders a different judgment on how they affect his *political* standing. Instead of producing a public groundswell for removal of the president, as many of his critics hope or expect, his job approval ratings *rise* in the immediate days following his admission of the affair with Monica. A poll by the Pew Research Center for the People & the Press four days later finds 62 percent of the people approve his presidential performance. Only 32 percent disapprove. Two days later an ABC poll puts the president's job approval rating at 64 percent; 68 percent of its sample credit him with being a "strong leader" and 79 percent believe he's doing "a good job keeping the economy strong."

A strong clue about why the public registers such stunning support for the president after such a humiliating spectacle—and after a speech the polls also show does *not* rate well with the people—can be found in the public's reaction to the first lady. Her popularity soars. Two-thirds of people polled say they admire her for standing by her husband; this represents an almost complete reversal of public attitudes about her from a year before. Once again, it seems Hillary has saved her husband—at least for the moment.

Polls, of course, can prove to be fickle instruments in measuring public opinion. But the attitudes expressed about the state of the country, about the job the president is doing, and about the performance of both the news media and the politicians in the scandal are remarkably consistent. As the Pew Research Center concludes, in assessing the findings of its various public opinion samples throughout the endless year of scandal: "A full year of the Clinton-Lewinsky scandal has had little negative impact on how Americans judge the country or the president's job performance and overall record. In fact, more people are satisfied with the way things are going in the country than was true in early 1998 before the scandal broke. Further, as many Americans say Clinton's accomplishments will outweigh his failures as said that about Ronald Reagan at a comparable time in his second term."

The negative feelings come when people assess the press and the Republican leaders of Congress. From beginning to end, their views on scandal coverage are scathing. In February 1998, one month into the scandal, for instance, a Media Studies Center survey funded by the Freedom Forum finds wide majorities of the public characterizing the Lewinsky coverage as "excessive" (80%), "embarrassing" (71%), "biased" (67%), "disappointing" (66%), and "disgusting" (57%). Two-thirds of those surveyed say the Clinton-Lewinsky coverage is "not important enough to deserve the level of coverage it has received."

As for attitudes about Republican leaders, the public judgment grows more critical the longer the scandal continues.

In late August, after Clinton's grand jury testimony and national address, 48 percent of people polled by Pew say they approve of the job the Republican leaders are doing as opposed to 36 percent who disapprove. Two months later, in November, a complete shift takes place. Then 48 percent disapprove while 41 percent approve.

The views of women are particularly telling. In February, in the same Media Studies survey about scandal coverage, women by far are the most critical segment of society when it comes to judging the press, and that's true regardless of party affiliation or political ideology. By more than a two-to-one margin, women say the media has "gone too far in disclosing details of President Clinton's private life." Men are split down the middle. Half think the media has gone too far; half think it hasn't.

Women are uniformly more likely than men to describe the coverage in negative terms—*excessive, embarrassing, biased, disappointing, irresponsible*—and far more likely than men to say the story is being blown out of proportion by the news media. Only 28 percent of women believe the story is important enough to deserve the coverage it receives. By comparison, 43 percent of men think it deserves such coverage—still a minority, but nearly twice as many as women.

That finding comes before the president admits he hasn't been telling the truth about his relationship with Monica—but even after he does, women *still* view the scandal as being overblown and unworthy of the kind of obsessive attention it receives.

A liberal Democratic supporter in Hollywood, active in the president's two national campaigns, who knows key administration people, puts into words the feelings of many of those women polled throughout the year. She's speaking just after the president's national address and confesses to holding intensely conflicting feelings about him. "I went through an incredibly huge range of emotions," she says. "I went from being incredibly angry at him basically for lying in the first place and for being so stupid to lie in the first place and putting the people who were the closest to him and who worked for him through it, all for no reason. That was the biggest betrayal for me. Then I saw the Republicans, and I saw those swarmy looks on their faces and I started to turn and say, 'You know what? What he did wasn't so bad. I don't want it to go that far. I can't have them smirking at him.' And I said to myself, 'Reagan lied, and Franklin Roosevelt sent off a boatload of Jews and they all died.' Where is the perspective on what he [Clinton] did, for the stupidity of it, and the fact that it's really about sex?"

Then she adds, "Women believe ultimately that men do this and also women believe that it's between a husband and a wife. If it's okay with Hillary, it's okay with them. The saddest part for me is the notion of what public service has become, the aspiration to become a public servant. Nobody in their right mind would do it anymore."

Given the evidence about the relative lack of significance the public attaches to the scandal, the president's high job-approval ratings, and similarly strong public opposition to impeachment, the failure of his opponents to change public views after the president's lies and irresponsible behavior are exposed would seem to send a strong signal for them to cease their efforts against him. Instead, his foes in Congress and in the prosecutor's office redouble their attempts to bring him down. They do so out of a conviction they will win their case against the president and force him from office, either by resignation or by impeachment. Nor is their reasoning implausible, however much it represents another strong disconnect between the capital and the country.

· · ·

By the summer of 1998, a kind of deadly parity exists between the two opposing political parties. For two decades, both parties have been locked in a mortal struggle for supremacy—a struggle that becomes more vicious with each passing congressional session. Ultimate political control is completely up for grabs.

The Clinton years represent a reversal of the political patterns of power from previous decades. In seven of the nine presidential elections from 1932 to 1968, Democrats control the White House. For five of the next six presidential elections, from 1968 to 1992, Republicans have the White House. Over the sixty-year period from 1932 to 1992 Democrats have a solid lock on the Congress, failing to control it only twice. In the Nineties, the position of the two parties shifts dramatically. Under Clinton, the Democrats hold the presidency for two successive elections while the Republicans break through to claim both houses of Congress in back-to-back elections for the first time since the 1920s.

Intensifying that situation is another factor: In neither of his two winning presidential campaigns does Bill Clinton receive a majority of the votes cast.

Mere recitation of this political history fails to capture the virulent partisanship that characterizes political life in the Nineties. Part of the reason for the poisonous climate lies in the actions of Newt Gingrich, whose goal for years is nothing less than to demolish the old order and create a new one out of the wreckage.

By 1998, Gingrich has already wrought a revolution. In the congressional midterm elections of 1994 he leads Republicans to one of the greatest triumphs in modern political history. The Gingrich-led forces do more than win the House for the first time in forty years, gaining 52 House seats and reducing Democratic numbers from 256 to 204 while Republican ranks swell from 178 to 230. They also sweep out of power many of the Democratic barons who have exercised power on Capitol Hill for decades. Gingrich doesn't achieve this victory by employing gentle tactics. He applies a blowtorch; his continuing assaults on the private lives and character of his opponents exemplify "the politics of personal destruction." In the summer of 1998, with the president reeling from the long months of scandal, Gingrich and his Republican strategists set their sights on the upcoming congressional elections as a crucial first target on their way to achieving full political power.

Notwithstanding the findings of the polls on Clinton's still-favorable public standing, Gingrich believes the scandal will energize the Republican ranks and motivate them to vote while discouraging deeply disillusioned Democrats from going to the polls. Nervous Democrats, anxiously assessing the political aftershocks of the president's exposure and public humiliation, see *their* chances of regaining power rapidly declining, and not just for one election cycle but potentially for years to come.

By hammering on the "Clinton scandals" around the clock in their speeches and public hearings, Gingrich and GOP strategists plan to undermine the president's personal standing. They count on continuing mass media coverage of the scandals and further damaging revelations to be made public by the prosecutor during the fall campaign to erode the president's popularity. His voters will desert him. Then, as in 1974 when the aftermath of the Watergate scandal and Nixon's resignation resulted in Republican losses of forty-seven House and five Senate seats, the Clinton scandal will produce big Republican electoral gains.

Privately, Gingrich gets daily reports that summer from Republican pollsters tracking the likely outcome of the elections just two months away. These convince Gingrich the Republicans are likely to pick up *at least* thirty House and five or more Senate seats. Capturing the White House two years later in the millennial year of 2000 will achieve their long-term goal of majority political power, a dream unfulfilled even during the Reagan era.

* * *

When the president returns to Washington at the end of August from his most difficult vacation on Martha's Vineyard, from what the public can see, the vacation hasn't resolved his personal problems with his wife. She's seen walking up the steps to the plane alone. No cheering waves and smiles from the first couple to be flashed via TV to the American people on this occasion, as is normally the case in their arrival and departure photo-ops.

Once they return to the always-suffocating heat and humidity of Washington in the summer, the political climate becomes even more inhospitable.

To a degree unmatched even during Nixon's final days before his forced resignation twenty-four years before in August, a swirl of vicious rumors envelops the capital. Nearly all of them find their way into the public discourse through the media. Leaks from "those close to the investigation" promise devastating revelations of yet-undisclosed sexual details about the president and the intern; these will shock the nation, it's said, in the forthcoming Starr report to Congress. As *Newsweek* reports, these revelations will make Americans "want to throw up." *Fox News* tells viewers the prosecutor will describe sexual "activities that most Americans would describe as unusual." *Newsday* tells readers that Monica provides "specifics about a half-dozen sex acts" during a secret two-hour session with Starr's deputies and her lawyers in a final sworn deposition taken before them August 26 after her grand jury testimony is completed. "In deference to Lewinsky and the explicit nature of the testimony," *Newsday* further reports, "all the prosecutors, defense lawyers and stenographers in the room during the session were women."

As Joan Didion later writes of these and other leaks being reported then, "Since the 'explicit nature of the testimony,' the 'unusual activity,' the 'throw-up details' everyone seemed to know about (presumably because they had been leaked by the Office of the Independent Counsel) turned out to involve masturbation, it was hard not to wonder if those in the know might not be suffering some sort of rhetorical autointoxication, a kind of rapture of the feed." As she adds: "Seventh graders in some schools in this country were as early as the late 1970s reading the Boston Women's Health Book Collective's *Our Bodies, Ourselves,* which explained the role of masturbation in sexuality and the use of foreign objects in masturbation. The notion that Americans apparently willing to forgive a dalliance in the Oval Office would go pale at its rather commonplace details seemed puzzling in the extreme, as did the professed inability to understand why these Americans might favor the person who had engaged in a common sexual act over the person who had elicited the details of that act as evidence for a public stoning."

No matter, expectation of even more degrading—and titillating—disclosures drive the coverage and heighten the sense of a capital city about to explode.

Democrats, fearful they face political disaster at the polls little more than eight weeks away, grow even more nervous. Something close to a stampede begins to develop as they seek to distance themselves from the president. Already, 140 newspapers are calling for his resignation. Republican congressional candidates, in a campaign tactic authorized by Gingrich, are beginning to air commercials denouncing the immorality of the president and urging his ouster. Democrats, watching helplessly, know the worst is yet to come.

Never have they been so panic-stricken, never so enraged at a president they have never really admired, except for his political skills and ability to connect with the people. Increasing numbers of them are speaking out, urging the president to accept, at the least, a congressional resolution of censure expressing strong condemnation for his acts. Some want him to resign, and actually say so to his face behind closed doors in a tense and never reported White House meeting between the president and party leaders early in September.

In the midst of that inflammatory atmosphere, Gingrich fans the flames even higher. The Speaker, according to a believable account given Sue Schmidt and Michael Weisskopf by a presidential adviser, phones the president at the White House ostensibly for a private discussion about terrorism. Before hanging up however, the Speaker, "with his well-known instinct for the jugular," offers the president some gratuitous political advice. He tells the president that his private Republican polls on the election outcome in November are showing the GOP riding high and with the likelihood of picking up dozens of seats. Then, applying greater pressure, Gingrich warns the president that he'll be hearing soon from worried Democrats who will want him to resign.

From all sides, the political news can't be worse. A lead editorial in the *New York Times* titled "The Gloom around Bill Clinton" perfectly captures the atmosphere.

"Bill Clinton returned to work this week in a Washington unmistakably transformed since his hedged, much ridiculed confession to the nation last month about his relationship with Monica Lewinsky. With Democrats increasingly allergic to their leader, and Republicans no longer treating impeachment proceedings as unthinkable, Mr. Clinton faces a rapid erosion of support that imperils his Presidency."

If all this isn't enough, Washington simmers amid further evidence that the scandal fever is spreading beyond the president to claim victims among his most conservative opponents.

In a matter of weeks that summer, while the nation absorbs the disclosures about its president, the people are fed exposés of the sex lives of three House Republicans who have been condemning the president's immorality.

On the Internet, *Salon* magazine reveals that Henry Hyde, the courtly rotund white-haired chairman of the House Judiciary Committee who will preside over an impeachment inquiry, had an extramarital affair thirty years ago. Now old pictures of a slender "Hank" and his long-ago paramour posing happily together are widely published, and even more widely discussed.

About the same time Helen Chenoweth, a hard-edged Republican conservative from Idaho, a member of the Gingrich earthquake Class of '94 who denounces the president's lack of family values, is forced to acknowledge her six-year affair with a married man.

A leading Republican congressional investigator of the President's alleged scandals, Dan Burton of Indiana, who has publicly called the President "a scumbag" and conducted reenactments for the media of Vince Foster's death in his backyard to "prove" Foster was murdered, confirms he has a son born out of wedlock from an extramarital affair. This comes after *his* sex life is exposed publicly in the press, and by who knows what private investigatory process paid for by whom.★

Now the fear leaps beyond Democratic ranks. Who's next? politicians anxiously whisper among themselves. Talk of "sexual McCarthyism" and "sexual Armageddon" begins to be heard as the fear of scandal ripples through the capital in those last days of summer. Meanwhile, the mother of all scandals continues to unfold.

· · ·

On Wednesday, September 9, at a quarter to four in the afternoon, the prosecutor's top deputy, Jackie Bennett, places calls to the House sergeant of arms and the majority and minority staffs of the House Judiciary Committee. The report is on the way, he tells them. No one needs to be told *what* report. It will be at the front door of the Capitol in fifteen minutes, Bennett says tersely. Even as

---

★Later, in another press exposé, Burton was shown to have approved nearly $.5 million in payments and salary to a woman acting as his campaign manager, who also worked as a part-time clown. The disclosure came after it was discovered that her name appeared simultaneously on Burton's political and official House payrolls. The payments, including monthly amounts for her to rent office space in her home outside Burton's congressional district boundaries, were made over a nine-year period.

he speaks, two FBI vans are preparing to leave the prosecutor's office on Pennsylvania Avenue between Tenth and Eleventh Streets.

Whether by coincidence or design, this sudden and unexpected news comes just forty-five minutes after television carries live reports of a now-penitent President Clinton that day in Florida pleading for forgiveness and another chance in a dimly lit hotel ballroom in Orlando. "I have no one to blame but myself for my self-inflicted wounds," the president says, offering words he was unwilling to say previously. Earlier, before embarking on that long-standing Florida fund-raising trip, the president offers a much the same, but more emotionally charged, apology before a private White House meeting of House Democrats.

Newsworthy though the president's remarks are, they are quickly overtaken by the more dramatic news surrounding delivery of the long-awaited prosecutor's report to the Congress. Instantly, the TV cameras shift away from the president to Capitol Hill, where a far greater public drama forms in the parking plaza in front of the Capitol dome.

Promptly at four o'clock the vans pull up before the front steps of the Capitol's East Front where every president from John Adams to Jimmy Carter took their inaugural oaths before the ceremonies shifted, under Ronald Reagan in 1981, to the West Portico and the more telegenic vista of the Capitol Mall, the distant Lincoln Memorial, and beyond, unseen but always noted, the American nation stretching west for three thousand miles. Gathered to greet them are somber-looking congressional staff members, Capitol policemen, and a hastily formed press phalanx kept yards away.

As FBI agents keep watch, members of the prosecutor's staff begin handing over thirty-six boxes containing copies of the prosecutor's 445-page report and more than three thousand pages of supporting material the prosecutor terms "other evidence." Twenty minutes later, after Capitol police load the boxes into two sport utility vehicles, those symbols of the Nineties, the SUVs proceed slowly down Capitol Hill to the Ford House Office Building below. In a scene telecast live around the world, policemen are shown carrying the heavy boxes inside. There they are deposited in a special suite where the press is breathlessly told the locks have been changed just the night before in preparation for the moment when the historic report is delivered. Police immediately take up their posts around the suite to safeguard the material.

Peering out a window at the Ford Building, a young Capitol employee watching the scene offers a comment appropriate to the moment—and the times. "It wasn't a big deal," he tells a reporter, "but it's our building's fifteen minutes of fame."

There it stands, an impeachment report which no member of Congress has read; a report which the president's own lawyers have been denied access to despite repeated requests to receive a summary of its findings in order to prepare a preliminary response; a report that represents the product of more than four years of investigation by the prosecutor into a wide range of suspected wrongdoings from Whitewater to Monica but about which not one case has yet been resolved; but a report which the watching world is told by the prosecutor's official spokesman contains "substantial and credible evidence" that may be grounds for impeachment of the president on charges of perjury, obstruction of justice, and abuse of power.

No one seems clear exactly what's to be done with it; no one doubts that it initiates a special chapter in the nation's history. One of the congressional chairmen responsible for acting upon the report, the New York Republican Gerald Solomon of the Rules Committee, tells reporters he expects the full House will vote the next day on a resolution drafted by his panel. If passed, that will clear the way for each representative to leave for the weekend with a copy of the report and some accompanying documents. Thus, everyone will have an opportunity to study it before deliberating on the next step.

A congressional resolution *is* passed the next day, Thursday, by Solomon's Rules Committee, but far from the one suggested.

Acting under fevered political pressures, with the urgings of leaders like Newt Gingrich, the committee's Republicans rebuff attempts by Democrats and the president's lawyers to give the White House just one hour to review the report before its release. A plan is formed for its distribution in its entirety not only to the elected representatives of the people, but to the American public by releasing it, unedited for graphic material, over the Internet. By a strict party-line vote the committee passes that resolution and sends it on to the full House for final action the next day.

On Friday, September 11, every single Republican votes to make public a document whose contents none of them has read. Democrats, even more panic-stricken and fearful of appearing to defend a disgraced president, break ranks in massive numbers. Only sixty-three of them oppose releasing the report in such fashion.

The argument that nothing like this has been done before— not in Nixon's impeachment process, not with other fateful public documents—carries little weight. Neither do the arguments that it's a travesty, grossly prejudicial, and sets a terrible precedent for future investigations of all kinds to release such documents before assessing them and deliberating how best to handle them. It's all dumped into the ether of cyberspace where it quickly speeds around the world.

The hard news of the report—that the prosecutor says the president betrayed his "constitutional duty to faithfully execute the laws" by engaging in a pattern of "abundant and calculating lies" making him liable for eleven possible grounds for impeachment— is almost lost amid the search to discover the most salacious details ever made public about any political leader at any time in history.

Nearly lost, too, amid the massive dump of graphic sexual material is the portrait of a humbled president finally, belatedly, attempting to reveal what he calls "the rock-bottom truth" of his actions and his responsibility for them.

The setting that same morning is the White House. In still another scene that would seem unbelievable if rendered in a Washington potboiler, the president happens to be scheduled to speak before a gathering of the nation's religious leaders. The occasion is the annual White House Prayer Breakfast.

When the president stands to address them, he clearly shows signs of emotional strain and weariness. As well he should, for after months of trying ceaselessly to fight this scandal by lying and covering up, the president now decides to try and address it definitively and personally all at the last possible moment. He's been up until four o'clock that morning writing out on three small sheets of paper what he wants to say to the religious leaders. When aides ask to see his remarks before he delivers them, he refuses their request. This time there will be no soundbite scriptwriter.

Now, his voice husky, his eyes glistening, his face marked with new lines of fatigue, he finally says what he should have said so many months before, "I am sorry." He also says: "I don't think there is a fancy way to say that I have sinned."

He tells them, "I have been on quite a journey these last few weeks to get to the end of this, to the rock-bottom truth of where I am and where we all are." He agrees with his critics who say he was "not contrite enough" when he addressed the country after testifying before the grand jury. "It is important to me that everybody who has been hurt know that the sorrow I feel is genuine." He enumerates them: "first and most important, my family, also my friends, my staff, my Cabinet, Monica Lewinsky and her family, and finally, the American people. I have asked all for their forgiveness," he says in a low, soft voice, then adds: "But I believe that to be forgiven, more than sorrow is required." That requires "at least two more things: genuine repentance, a determination to change and to repair breaches of my own making." He *has* repented, he says, and he intends to repair "what my Bible calls a broken spirit: An understanding that I must have God's help to be the person that I want to be. A willingness to give the very forgiveness I seek. A renunciation of the pride and the anger, which cloud judgment, lead people to excuse and compare and to blame and complain."

He's been speaking nearly ten minutes now, biting his lips repeatedly, head often bowed, face tightly drawn. If he succeeds in doing all this, he continues, "if my repentance is genuine and sustained, and if I can then maintain both a broken spirit and a strong heart, then good can come of this for our country, as well as for me and my family."

Sitting in the audience with the religious leaders, intently watching, is his wife, offering mute testimony that she stands with him. When he finishes his remarks, one of the ministers, a woman, approaches the first lady. You've been "wonderfully strong," the minister tells the first lady. Hillary replies: "Not always."

Later, toward the end of one of the most extraordinary days in the presidency, a day filled with scenes of pathos and bathos, solemnity and soap opera bouffé, the first lady accompanies the president to a dinner meeting of the Democratic Business Council at a downtown hotel. There, after sitting side by side with her husband acknowledging applause, she proceeds to the lectern and delivers a rousing introduction in which she expresses her pride and admiration for him. She praises his "unrelenting determination to do what is right for America" and says he has "given faith again to people all over our country that the American dream can mean what it should mean."

The president steps forward and hugs her. She pats his back. As if the public doesn't catch this signal intended to show that now all's well between them, her office releases a statement in which the first lady "stresses her support, her love, and her forgiveness" for her husband.

As the nation scrambles to discover what's in the ultimate scandal report calling for the first impeachment of an elected president of the United States, Hillary stands by her man.*

• • •

Even in an age of excess, even after O.J. and Diana and JonBenet and the Bobbitts and all the rest, Americans have never experienced anything quite like this mad rush to participate in the most scandalous of all spectacles. All over the nation, indeed all over the globe, people with Internet access are logging on to the now-posted Starr report in a frenzied rush to search for its most prurient contents.

They aren't hard to find. They appear on almost every page, presented not in the dry legalisms of a brief or even an indictment, but in language that reads like a novel. It even includes a 280-page centerpiece containing "a narrative" of the president's relationship with the intern. Lest readers have difficulty finding the "hot" stuff, the report helpfully provides section headings that read "Sexual

---

*Andrew Johnson, the only other impeached president, moved from vice president to president after Lincoln's assassination in 1865.

Contacts," and "Initial Sexual Encounters," and "Continued Sexual Encounters." Even more helpfully, it breaks them down into subheadings that chronicle exactly what happened on what date and where: "November 15 Sexual Encounter"; "November 17 Sexual Encounter"; "December 31 Sexual Encounter"; "January 7 Sexual Encounter"; "February 4 Sexual Encounter and Subsequent Phone Calls"; "Continuing Contacts"; "March 31 Sexual Encounter" . . .

Nor is anything left to the imagination. Punch in the word *cigar* and up comes the most explicit account of how the president, with the encouragement of the intern, employs a cigar as a sexual device, and then describes what he does and says after completing that act.

As the *Washington Post* notes, "In the brief history of news on the Net, yesterday broke all records." So it does. Dot-com sites report record "hits." At the *Washington Post* site, a manager reports in awe-struck tones that they've tripled their previous record of "visitors." "It's unprecedented," Erin O'Shea says. "We were almost maxing out."

Across the country, newspaper editors wrestle with decisions never before faced about the propriety of publishing such explicit sexual details. Most decide to publish the entire report verbatim, adding, as does the *Post* in cautionary words of explanation to readers: "We recognize that the independent counsel's report contains extensive sexually explicit material that normally would be unacceptable for publication in the *Post*. However, we have decided not to edit the text of the report because of the circumstances of its release by the U.S. Congress, on government Internet sites and by other means, in the midst of a public controversy over whether the president has committed impeachable offenses."

Still, the paper feels constrained to warn, daintily, that "some material in these unedited texts is inappropriate for children and younger readers, and some of the material will be offensive to some adults."

Inside, it publishes an article supposedly aimed at parents, under the headline:

HOW TO (NOT) TALK TO KIDS
ABOUT THE CLINTON AFFAIR

Outside, on page one, it informs readers about the madcap nature of this national scandal moment:

"The oddest mass political act in the history of American democracy began with an electronic storm, a blizzard of attempts to find the Starr report on the Internet. By late afternoon, the details, the explicit stuff—the sex—was everywhere: in solemn readings on television, in breathless recitations on talk radio, in gossip that sizzled across workplaces and shopping malls" as a stunned nation attempted to come to terms with such an "explicit recitation of the President's alleged sexual behavior." All this produced "an afternoon of giggles, whistles, and sighs of disgust. But was this the nervous laughter of a nation writhing under an avalanche of sexual detail, or a first, emotional exhalation at the beginning of a constitutional crisis?"

In other words, stay tuned—or keep reading.

Television, lacking the luxury of time to prepare stories and forced competitively to go live with the material even as it scrolls sentence by sentence across computer screens courtesy of the Internet, finds itself in uncharted journalistic waters.

On CBS, reporter Sharyl Attkisson begins reading a passage about Monica's navy blue dress; then Dan Rather interrupts her to caution: "This is daytime television and there are children in the house. . . ."

On CNN, correspondent Candy Crowley warns viewers "who are the least bit squeamish" either to turn off their set or switch to another channel; then she reads: "According to Ms. Lewinsky, the president touched her breasts and genitalia. . . ."

From the *Wall Street Journal,* under the headlines:

STARR REPORT: THE FALLOUT

A Wildfire Transforms the Global Village

comes a sprightly story that explains the linkage between Teletimes and Technotimes making possible this cyberspace moment. "It began with a single 3½-inch floppy disk with WordPerfect files, sealed in a white envelope buried in the vanloads of cartons Independent Counsel Kenneth Starr sent Congress. With dizzying speed, it turned into an information wildfire that sent the Starr report almost instantaneously onto millions of computer and TV screens. And as it raced around the world, the Starr report made communications history just as dramatically as it was making political history."

The political history is well analyzed by R. W. (Johnnie) Apple's page-one *New York Times* account: "A political process has begun, and in political processes, as the Founding Fathers foresaw, the people have the ultimate power. It is by no means clear how they will use it, or what the result will be."

After reporting that "one prominent Democratic political consultant told his clients this week that he expected Mr. Clinton to be out of the White House within weeks," and noting that "only one president in the last fifty years, it is true, has been in trouble this deep, and that one, Richard M. Nixon, was of course forced from office," Apple presciently comments: "But many clear-eyed people still think that Mr. Clinton will hold on."

In his most telling observation, Apple says of the mood of the capital then: "The fact is, Washington is lost. There is no precedent for the present bizarre spectacle—lurid tales of sexual high jinks in and near the Oval Office, poker-faced lawyers in serious suits debating the constitutional implications thereof and rowdy legislators trying to behave themselves."

Rowdy high jinks aplenty are apparent across America, with an outpouring of bawdy jokes to accompany them.

On NBC's *Tonight Show,* the comedian Jay Leno begins his broadcast that Friday night by asking his studio audience: "How many of you are here because you gave up trying to get on the Internet?" After a raucous reply, the comedian goes on to say: "I guess this answers the question of whether a gentleman should offer a lady a Tiparillo." Leno notes that Ken Starr has sent Congress thirty-six boxes of evidence, then quips: "Thirty-six boxes! That's a lot of evidence. Even O.J.'s saying, 'Give it up!'"

Amid the massive amount of material Americans are offered to read and assess, one memorable account stands out. It comes not from a political analyst, but from the book critic of the *New York Times.*

Writing on deadline, and appearing for perhaps the first time amid the welter of other daily news stories about the Starr report, Michiko Kakutani produces as fine an overview of the meaning of

Scandal Times as anyone has written. She begins by reacting to what she, and countless others, have been witnessing that day: "It was a surreal and unnerving spectacle sadly translated from the world of tabloid television to the world of governance: the White House issuing a press release yesterday about the nature of President Clinton's sexual activities, a reporter asking Michael McCurry whether the President suffered from 'satyriasis' and the entire globe becoming privy, via the Internet, to Mr. Clinton's intimate, sexual proclivities—the sort of thing people do not know about their best friends, much less the most powerful man on the planet."

The president, Monica, and the supporting cast of Linda & Co. "have become cartoon figures in the popular imagination," Kakutani writes.

And what has brought the nation to this miserable moment? she asks. The answers she suggests are plausible: "In a day when deconstructionists, feminists, and radical Afrocentrists have declared truth a relative matter, we have a President who has suggested that sex does not mean sex, that smoking dope does not mean inhaling, that sidestepping the draft does not mean trying to avoid it. And in an age of confessional talk shows and twelve-step programs, we have a President happy to discuss his underwear (boxers or briefs) on television, a President who has now followed the likes of Jim Bakker and Jimmy Swaggart in confessing to sins and asking to be forgiven."

With nice historical memory, and appropriate historical irony, she concludes:

"Marx once observed that history repeats itself, the first time as tragedy, the second time as farce. It is the peculiar nature of Mr. Clinton's predicament—and other contemporary phenomena, like the O. J. Simpson trial—that tragedy and farce co-exist, that the horrifying and the absurd now frequently dwell side by side."

* * *

For one long weekend, aptly described by the *Times* editorially as "one of the most tumultuous weekends in modern political history," the American people wrestle with the revelations about their president.

No definitive answers quickly emerge to resolve the critical question of whether the people think the president should stay or go, but once again there are strong signs the public continues to support him. Four days after the Internet posting, and after endless public hours of discussion about it, one of those antique multiple-deck, one-column headlines that would have been familiar in Lincoln's day captures the essence of the lead story in the *New York Times*:

MOST IN POLL SAY
PRESIDENT SHOULD
REMAIN IN OFFICE

Job Rating Is Still High

But Public View of His Moral
Character Sinks to a Low
—Reprimand Backed

With his job approval rating still standing at a remarkably high 62 percent, the *NYT/CBS* News Poll also finds the public holding strongly negative views of the prosecutor and the contents of his report.

The following days bring more severe criticism of the Starr report from scholars and opinion leaders. A respected Yale law professor and author, Bruce Ackerman, writes: "Mr. Starr's report contains thousands of words about sex, but not one word explains why the impeachment clause or the basic principles of the Constitution justify removing the President.... Millions of Americans have read the dirty details without being invited to take the larger constitutional view. It's the O. J. Simpson trial all over again, but the stakes are much greater."

Others warn that the offense—lying about consensual sex—does not merit the ultimate political sanction of impeachment under the constitutionally prescribed grounds of a president having committed "treason, bribery and other high crimes and misdemeanors." For the nation to proceed down this path, in such a case, raises the danger of establishing a new impeachment standard—impeachment as a political tool.

However much merit such arguments hold, and they correctly

raise serious long-term implications for society, it's fair to say that most Americans are not following them closely or are not persuaded by them one way or another. At this point, and on through to the end, Americans are grappling with more elemental emotions, what the public opinion experts call "the public revulsion factor." As the *NYT/CBS* News Poll and other new polls make clear, the public has a "plague on both your houses" attitude. People are disgusted with the president's private conduct, but even more disapproving of the actions of the prosecutor and the presidential opponents who are pressing for impeachment.

In the face of this latest, seemingly definitive, public reaction to the scandal, one would think the opponents in Congress and in the prosecutor's office would proceed far more cautiously. Yet, again, they respond by intensifying the attack. Now, they prepare to launch a second wave by making public, through television, the video of the president's grand jury testimony plus an additional twenty-eight hundred pages of "evidence" that supposedly contain even more personally damaging material.

Nothing better demonstrates their massive political miscalculation than this act.

For days after the Starr report is released on the Internet, Republicans and allies of the prosecutor spread the word, widely reported in the press, that when people actually see the way the president appears while testifying they will be appalled; the president's appearance will be devastating for him. The word is passed, and immediately reported on TV and in the papers, that the president will be seen blowing up, purple-faced, erupting in anger, losing control, even storming out of the Map Room in fury.★ Not only this, it is whispered, the additional "evidence" will show the president to have been guilty of even more egregious sexual misconduct, and not just with Monica Lewinsky.

To rally the troops for this second-wave assault, Gingrich urges his Republican colleagues to release all this damaging material to

---

★Bob Schieffer of CBS, one of the best and most respected reporters in Washington, becomes embarrassed after reporting over that network that sources tell him the president will be seen "storming out of the room."

the public. He still holds the belief that further negative material will depress the Democratic vote and further motivate Republicans in the critical elections now just six weeks away.

Release of the grand jury video itself provides the president and *his* allies another weapon in the battle for public opinion. They argue, persuasively, that the *real* reason for taping the president's testimony is to make it public later, contrary to long-established grand jury practice that holds such testimony sacrosanct and secret. It is being released for the sole purpose of further humiliating and damaging the president.

From the beginning of negotiations over his testimony, the White House takes the position the president will voluntarily appear before the grand jury and invites the jurors to come to the White House for that purpose. In fact, nearly five and a half hours after he begins his testimony, in the closing minutes of his long day before the grand jury, the president refers to that wish to have the jurors present.

After a brief moment of wrangling, one of the prosecutor's deputies, Solomon L. Wisenberg, says: "Just for the record, the invitation to the grand jury was contingent upon us not videotaping, and we had to videotape [today] because we had an absent grand juror."

"Is that the only reason, Mr. Wisenberg, you have to videotape?" asks the president's clearly suspicious lawyer, David Kendall.

Before an answer is given, the president interjects: "Well, yes," he asks Wisenberg. "Do you want to answer that?"

At which point Jackie Bennett cuts off all further discussion of that subject. "Thank you, Mr. President," he says, abruptly terminating the grand jury session.

"Whereupon, at 6:25 P.M." as the court reporter notes, "the proceedings were concluded."

On Monday, September 22, eleven days after the Starr report release, on the vote of every House Republican, the president's grand jury video is released, immediately igniting another media storm and another exposure of explicit sexual descriptions—the cigar, masturbation, oral sex, phone sex—about which the president is asked to respond and explain.

This time, the networks are better prepared for the latest dump. Labels warning that explicit sexual material is about to be shown are superimposed on the video pictures.

Once again, the belief that the president will be severely damaged is proved wrong. Even some of his strong critics grudgingly concede the pictures aren't as bad as they were led to believe. Alan K. Simpson, the former conservative Republican senator from Wyoming, who earned a reputation as a skilled soundbite artist noted for his caustic humor, best expresses that view. "That's his life," Simpson says of the president. "His life is bounding from precipice to precipice like a huge mountain goat, bridging crevasse after crevasse with people shooting at him with high-powered rifles and the other side crumbling as he lands, just like in the movies."

The prosecutor's case is further weakened by the discovery that he omits one of Monica's most significant—legally, perhaps *the* most significant—sworn grand jury remarks in his impeachment report to Congress. This is her closing statement to the jury, in the final minutes of her secret testimony, when she says she wants to make something clear: "No one ever asked me to lie and I was never promised a job for my silence." This comment undercuts the most serious case of criminal subornation of perjury or obstruction of justice against the president. Not until journalists examine the new mass of papers the prosecutor releases does this statement of Monica's become known. Naturally, its omission from the Starr report generates criticism of prosecutorial prejudice.

Again, the polls are not favorable to the prosecutor, and far from devastating for the president.

In one overnight poll, seven of ten people who watch the president's testimony think he seems evasive. The identical number responding to the same poll say the president is right not to talk about sexual details. His job approval holds steady, even rises, after he scores political victories on the budget, on a Middle East peace agreement, and on the apparent withdrawal of Serbian forces from the bloody ethnic conflicts that have ravaged Kosovo in the Balkans.

Now the congressional campaign enters its critical closing phase, and as it does the politics of impeachment becomes even more venomous.

Far from being deterred by the polls, Gingrich continues to reassure his Republican colleagues that in the end the president's popularity will fall, disillusioned Democrats will not vote, and energized Republicans will turn out in greater numbers. Besides, as Gingrich the historian reminds them, they have history on their side. Traditionally, the party that controls the White House loses seats in the last midterm political cycle before an eight-year presidential term ends. In fact, the Republicans actually will benefit from keeping a wounded, discredited President Clinton in office for now. He symbolizes immorality, corruption, and controversy. His public presence stands as a daily reminder to voters that they have a disgraced Democratic leader in the White House.

So they plunge forward. In the first week of October, once again with the backing of every Republican member, the House votes to begin a formal impeachment investigation after the election just weeks away.

Something else happens in those early days of October 1998 that will figure prominently in the history of Scandal Times.

On Sunday morning, October 4, readers of the *Washington Post* open their papers to find a full-page ad paid for by Larry Flynt, sleazy sex-trade "overlord of porn" and publisher of *Hustler,* one of the most offensive men's "skin" magazines. The ad is addressed to women. "Have you had an adulterous sexual encounter with a current member of the U.S. Congress or a high-ranking government official?" it reads. If so, *Hustler* will pay as much as $1 million to women willing to name names and go public about their affairs. Before being paid the money, they must have "documentary evidence"—snapshots, home videos, taped phone and answering machine conversations and messages, dinner and drink receipts, phone bills, witnesses, divorce papers—to prove the truth of their claimed affairs.

The ad attracts little public notice; just another publicity stunt from another Scandal Times hustler. Privately, it sets in motion a chain of events that will explode with great impact as the impeachment process reaches a climax. For before dawn that Sunday morning, separated by a three-hour time-zone difference from Washington, phones begin ringing in the darkened Los Angeles offices of *Hustler* as women respond to the phone number listed in the ad. In a short time, two thousand women claiming to have evidence of their affairs with Washington public figures have answered the ad. When *Hustler* employees return to work there Monday morning they find, as an editor tells a reporter, that "every voicemail the calls were routed to was full, and every time we took down the numbers and deleted the messages, the system would immediately fill up again." Their names, audiotapes, photos, and other material are turned over to the Washington private investigator Flynt has retained, who in turn shares the information with two investigative reporters working anonymously for Flynt. In short order, they narrow down the list to, first, forty-eight, then twelve, women whose stories merit a closer look. Immediately, out of public view, a sordid search for scandal begins.

On the eve of the election, Newt Gingrich plays his last card. He authorizes the expenditure of $10 million to air campaign commercials specifically injecting the president's affair with Monica into the election to boost Republican turnout in selected battleground congressional districts.

To the end, Gingrich remains certain that his strategy will prevail. By cruel timing, he permits CNN cameras to trail him and his political operatives in the weeks leading up to the election as they plan for the coming votes. Election night itself, at a Gingrich war room in an Atlanta hotel, Gingrich and his aides are heard expressing optimism that they are on the verge of a great victory as polls begin closing around the country. Then, even as bad news begins trickling in to their command center, they still aren't able to accept the electoral facts being spelled out by voters. Things still look good, Gingrich assures uneasy Republicans around the country by phone. Minute by minute the news grows grimmer; maybe they'll only be able to pick up seven seats, the cameras catch his aides saying to each other. Then, as the clock nears seven that night, with closing of more polls in the East, nervousness begins creeping into their discussions. Maybe they'll only pick up one seat. Then they may lose two, three, four five, seven seats, maybe even control of the House itself. Panic and consternation. As Gingrich leaves for an early evening news conference, he's still expressing cautious optimism that when all the votes are counted his attack strategy will have been rewarded by the voters, not by as great a margin as expected, but the election won't be a disaster, either. By midnight, it's clear all these hopes have been dashed.

Election day, Tuesday, November 3, has produced another electoral earthquake—but the reverse of the one Gingrich engineered four years before.

Instead of Democrats losing massive numbers of seats, giving the Republicans a veto-proof Congress to face a falling president in whatever months he has left of his term, the voters confound history and political expectations and deliver a stunning upset. Republicans still hold the House, but Democrats gain five House seats. This narrows the majority in the House to the smallest margin in forty years and places effective control of the House in the hands of moderate Republicans and newly energized Democrats. A shift of only six Republican votes will signal defeat on any GOP initiative. Instead of losing five or more Senate seats, as widely projected, to give the Republicans the sixty votes necessary to override any presidential veto, Democrats hold their own, further imperiling future Republican legislative success in the forthcoming battles between White House and Congress leading toward the year 2000 presidential election.

The lessons are clear to every political player in each party. Impeachment politics are a failure. It is the Democrats whose voters are energized to turn out, the Republicans who are turned off by the scandalmongering in which their party leaders have played so prominent a role.

On Capitol Hill, where Republicans are planning how to conduct forthcoming impeachment hearings, the reaction is swift. Within twenty hours of the election results, Judiciary Committee Chairman Henry Hyde announces tentative plans for a scaled-down impeachment proceeding, possibly calling only one witness—the prosecutor, Ken Starr. Shocked party members anxiously begin quarreling among themselves. Many condemn the leadership strategy that so badly miscalculated, leading to a midterm debacle that places them all in future electoral jeopardy. The central target is Newt Gingrich.

That Friday, November 6, just two days after the final vote tabulations are in, confirmation of the scale of this upset comes in the form of an astonishing announcement by the Speaker himself.

In the wake of the election, Gingrich resigns as Speaker, a stunning event that creates new political turmoil in Republican ranks. The White House is jubilant at this dramatic fall of the once all-powerful Speaker whose ambition to be the first president in a new century is well known; Republicans are thrown into spasms of confusion and concern.

Gingrich's decision is completely unexpected. Nothing has so surprised official Washington in decades, if ever. The politics of destruction have claimed not the president—at least not yet—but the most aggressive exponent of those divisive tactics.

The Speaker's explanation for voluntarily giving up his power is not entirely persuasive, but generally he's credited with recognizing he faces rebellion in the ranks, making his continued tenure impossible. By January, he will be gone, not only as Speaker, but as a member of the House itself, for he also resigns his congressional seat.

What isn't known until later is that Gingrich has been engaged in a longtime sexual affair with a young congressional aide that began even before the 1994 electoral earthquake that propelled him into the Speaker's chair, and that exposure of his affair and the breakup of his marriage looms. With no small irony, Newt Gingrich himself becomes a victim of Scandal Times. He will not be the last.★

Republicans, facing both a leadership void and fateful decisions

★A continuing mystery about Gingrich's resignation is the later claim by the executive editor of Larry Flynt's *Hustler* that the pornographer's search for scandal had documented Gingrich's six-year-long affair with the young congressional aide, that in fact a "good dossier" had been compiled on the Speaker including the affair. Asked why this information wasn't made public as Flynt had threatened in exposing Washington officials, his editor claimed "we were holding back" and that they "wanted to see what Gingrich was going to do next as far as his career was concerned." This clearly implies the threat of blackmail, but it has not been determined whether *Hustler's* information was made known to Gingrich, and, if so, in what manner.

about impeachment, huddle and announce that a veteran congressman from Louisiana, Robert Livingston, scion of signers of the Declaration of Independence, is their new Speaker-designate of the House of Representatives. Livingston first steps forward immediately after the disastrous election results are known to announce before the cameras assembled in front of the Capitol that he's challenging Gingrich for the Speakership. When a reporter asks him if this public challenge might not be interpreted as an *et tu, Brute* stab in the back, he merely replies, "That's a good question," and walks away. As for impeachment, the inside wisdom of Washington now holds that the Republicans, given the new electoral realities and clear signal of continued public opposition to removing the president, will settle for a truncated impeachment process. Then, ultimately, they will agree to some sort of compromise in which the president is censured but not removed by impeachment. Once again, the wisdom of Washington is wrong.

· · ·

Before exiting the scene to await history's judgment on his performance, the prosecutor makes his most heralded public appearance in widely anticipated testimony before the House Judiciary Committee Thursday, November 19. It occurs under difficult circumstances, for even before Starr raises his hand to be a witness, he knows that his ethics adviser, Sam Dash, intends to resign in public protest if Starr testifies that day as, in Dash's words, "an aggressive advocate" for the proposition the president had committed impeachable offenses. In the two weeks preceding the scheduled Starr appearance before the Judiciary Committee, Dash had repeatedly warned Starr of the dramatic step he intended to take. Members of the press also had learned of his resignation plans—an action, as one Democratic operative put it later, that would be "a fairly devastating indictment of Mr. Starr," one that demonstrated that Starr's ethics adviser, whom the prosecutor had "trotted out a several critical juncture of his investigation," resigned out of concern over Starr's ethical conduct.

Dash plans his extraordinary public protest not because of any personal views he holds—never expressed publicly—about the president's guilt or innocence, or about whether the record justifies impeachment. He does so because he believes the intended testimony of the prosecutor injects Starr squarely into the impeachment process as a strong advocate in favor of impeaching the president. This, in Dash's mind, exceeds his legal mandate under the independent counsel statute and goes beyond Starr's "narrow duty" under the law to "objectively" give the House substantial and credible information that *might* constitute grounds for impeachment.

"By your willingness to serve in this improper role you have seriously harmed the public confidence in the independence and objectivity of your office," Dash writes in a letter to Starr that he makes public on the day after the prosecutor's testimony. This represents "an abuse of power."

Hardly an auspicious sign for the prosecutor as he takes his case against the president public under the full glare of the media spotlight.

His twelve hours of testimony that begin in the morning and last into the night take place in a rancorous hearing in which the opposing political sides trade repeated insults. Except for the times when he begins to lose his cool under a relentless nighttime hour of cross-examination by the president's lawyer, David Kendall, Starr's courtly, careful, dignified demeanor that his admirers know well from his previous judicial service is on display. But temperament and personality aside, the prosecutor also makes what turn out to be a number of damaging disclosures that unleash even greater criticism of his role and his motives.

He acknowledges that after four years of relentless investigation, and the expenditure of more than $40 million, he has not found any grounds to bring charges against the president and the first lady in connection with their roles, actual or suspected, in Whitewater, Travelgate, or Filegate—the very matters that first launched him down his long investigatory path. Yet, he also acknowledges, he never reported his exculpatory findings on those cases to the public and never mentioned them in his vaunted impeachment report to Congress. Nor is he prepared to close the books on those matters even now. Then he raises the disturbing hint that he might still want to bring criminal indictments of the president *after* the congressional impeachment process ends. The investigation cycle promises to be never-ending.

In its lead editorial the next morning, the *New York Times,* a severe critic of the president's behavior throughout the years of Scandal Times, strikes an eminently fair balance on the prosecutor's appearance, but delivers a devastating final judgment on him: "Mr. Starr hoped to use his testimony to salvage his legal reputation while also proving his case that lying in a sexual harassment suit was in and of itself grounds for impeachment. His long opening presentation was at once an orderly and useful summation of the evidence to date, and a devastating admission that he had no additional evidence that would strengthen his case. . . . Indeed, Mr. Starr testified that he could not prove his suspicions about President and Mrs. Clinton on Whitewater and that he found no offenses by them on Travelgate and Filegate. This deck-clearing exercise was useful, but Mr. Starr failed to explain why he could not have settled those issues months or even years ago. . . . As for the final resolution of these matters, the mess does not look so large today as it did yesterday."

Then the paper delivers its final negative judgment: "Mr. Starr's reputation for legal acuity and balanced judgment has been seriously damaged."

With the prosecutor's testimony drawing as much criticism as praise and clearly failing to change opinion on the case against the president, the old puzzle about public opinion rises again. In the aftermath of all the disclosures, all the damaging and embarrassing reports and videos and personal congressional appearances, not a single mind seems to have been swayed when it comes to the question about impeaching the president. All the soundings continue to send the same signal: the public remains even more opposed than before. The president's approval rating now stands at a high of between 68 and 70 percent, higher even than Reagan's at a comparable point in his second term.

The explanation, if there is a good one, lies in murkier factors that go far beyond any questions of logic, facts, or even legal findings. They are rooted in emotions—raw, charged, passionate emotions. Many of the emotions that surface so explosively in public are genuine; even more of them are driven by cynical calculations of self-interest.

How else to explain the actions that swiftly follow the prosecutor's final bow? Even as other Republican members of Congress are telling reporters, and themselves, that they now need to search for an exit strategy; that they have no enthusiasm for a protracted impeachment battle with the president; that they fear to do so will propel their party over the cliff, their Judiciary Chairman Henry Hyde immediately acts to *expand* the grounds for impeachment beyond that even suggested by the prosecutor. Hyde issues new subpoenas for new witnesses; he signals that his inquiry is now heading into matters not previously cited by the official impeachment investigators: into campaign finance irregularities and their possible criminality; into the contents of even more foul sexual allegations about the president involving unproved, uncharged accusations of rape decades before and other whispered abusive treatment of women.

On its face, these actions seem so politically destructive as to be incomprehensible. Except they aren't.

Filling the immense void left by Gingrich's fall, with their new Speaker slated to begin exercising power in the next Congress to be installed in January 1999, effective power in the House is now being exercised by a short, feisty politician, Tom DeLay of Texas.

Even before Gingrich's fall, DeLay has become a formidable power among House Republicans, a power that goes beyond his official position as party whip, or third in the ranks of GOP command. DeLay, a former pest exterminator from Sugar Land, Texas, is a perfect example of the minority of the minority that in fact dominates the House—sternly ideological, fiercely combative, from a "safe" Republican district whose constituents reward efforts of representatives who oppose "liberal excesses" in Washington.

Throughout the Nineties, DeLay is one of the president's strongest opponents; he consistently attacks him and congressional Democrats for being, in the deprecating Vietnam-era term, "chicken hawks"—politicians who weaken America's defense through budget cuts. DeLay himself, as is typical of many other politicians who raise the antimilitary, draft-dodging arguments against their ideological opponents, avoided Vietnam service through student deferments.★ After the Monica story breaks, he becomes one of the first to call for the president's resignation. In the following months, he intensifies his efforts to bring down the president, operating for the most part behind the scenes in private sessions with his House Republican colleagues.

How intensely DeLay and his political aides feel about the need to drive Clinton from office is revealed, two years after the fact, by Peter Baker, the *Washington Post*'s White House correspondent, in his telling book, *The Breach: Inside the Impeachment and Trial of William Jefferson Clinton*. Nothing better captures the venomous hatred then coursing through Republican ranks than an e-mail exchange Baker quotes between DeLay's deputy chief of staff and his press secretary just hours before Clinton goes on national TV that August to acknowledge his relationship with Monica. "He's going to admit it," the press secretary writes the other in typical e-mail unpunctuated stream-of-consciousness style. "the big q is on what level—I still say we need to attack!" His fellow GOP operative agrees, saying, "we need to force dems to distance themselves from the liar." Back flashes the answering e-mail: "This whole thing about not kicking someone when they are down is BS—Not only do you kick him—you kick him until he passes out—then beat him over the head with a baseball bat—then roll him up in an old rug—and throw him off a cliff into the pounding surf below!!!!!"

Within hours after those messages between top DeLay aides, DeLay himself places a conference call to vacationing staffers and summons them back to Washington. From that time on, DeLay tells them, according to Baker, they are going to make it their mission to drive Clinton from office—not just impeach him, but force

★In the 1988 Republican convention, when Vice President–nominee Dan Quayle comes under attack for having avoided Vietnam service through the National Guard, DeLay is quoted as explaining his own failure to serve by saying he and Quayle were victims of an unusual phenomenon during the Vietnam era: "that so many minority youths had volunteered for the well-paying military positions to escape poverty and the ghetto that there was literally no room for patriotic folks like himself."

him to resign. "This is going to be the most important thing I do in my political career," Baker quotes DeLay telling his aides, "and I want all of you to dedicate yourselves to it or leave.... As of today, I want a war room. I want a communications strategy. I want a political strategy. I want you to work day and night."

They do just that and are undeterred months later when political reality suggests the impeachment drive is dead and nervous Republicans are publicly searching for an "exit strategy." Not DeLay. He pushes the anti-Clinton effort even harder behind the scenes. To his conservative colleagues, he argues that by impeaching the president they consign Bill Clinton to the historic disgrace he deserves, and their constituents will reward them for an action that places morality above politics.

This isn't the message being sent to the press by many politicians, Republicans as well as those inside the White House. They see signs of incoherence and desperation on the part of Republicans, a party stumbling out of control.

Three days after the prosecutor testifies, the *Times* refers to the disarray afflicting the Capitol Hill Republicans and comments editorially: "The Republican leadership, and to some degree all of political Washington, are dashing around blindly. The leaders cannot seem to absorb the week's main political development. Mr. Clinton, barring the emergence of dramatic new evidence, has won the political battle over impeachment. Having heard Mr. Starr's evidence presented in its most damning form, the American people do not want Mr. Clinton removed from office, and the Republicans cannot muster the votes to accomplish it."

Wrong again. In less than a month, the House of Representatives, in almost a straight-party-line vote, impeaches the president of the United States.

· · ·

Only once amid the wretched, petty wrangling that sweeps across the floor of the House of Representatives, prompting rowdy legislators to erupt in boos, catcalls, and jeers, does the political rhetoric approach the historic nature of the moment. That comes after Dick Gephardt of Missouri, the painstaking, straitlaced leader of the Democrats, regarded not always kindly by some there as a Boy Scout in politics (he was, in fact, an Eagle Scout as a youth), steps to the well of the chamber and addresses his congressional colleagues on the question of whether to impeach the president. "Our Founding Fathers created a system of government of men, not of angels," he says. "No one standing in this House today can pass a puritanical test of purity that some are demanding that our elected leaders take."

For this time only, silence descends over the otherwise raucous chamber. They actually are listening. "If we demand that mere mortals live up to this standard," Gephardt goes on, "we will see our seats of government lie empty, and we will see the best, most able, people unfairly cast out of public service."

Then he says, in a rare burst of passion: "We need to stop destroying imperfect people at the altar of unobtainable morality. The politics of smear and slash-and-burn must end!"

Suddenly, the sound of cheers and applause reverberate around the old chamber, erupting spontaneously from both sides of the aisles dividing bitterly partisan opponents. In that briefest of moments, as Francis X. Clines of the *New York Times* later writes, Republicans and Democrats unite in expressing mutual anguish over the self-destructive tactics that have brought them to this point and find "common ground over the haunting pathology of character attack that has become text and subtext of the impeachment debate and the political beyond."

It doesn't last long.

Soon after, a furious Patrick Kennedy, the young Democrat from Rhode Island whose family has been haunted by tragedy, including his younger brother who lost a leg to cancer, angrily upbraids Bob Barr of Georgia in the corridor outside the House floor for a speech Barr has just made on the floor. Barr, one of the most unregenerate of the Republican conservatives, has been calling for the president's impeachment long before Monica. Now, in pressing that case during the impeachment debate, Barr recalls to his congressional colleagues a speech John F. Kennedy, Patrick's uncle, made on the need for obedience to the law during the racial integration crisis in the Sixties.

"How dare you!" says young Kennedy, whose father, Ted, has led liberals in the Senate for a generation. "Anybody who has been to a racist group has no right invoking my uncle's memory." He's referring to a controversy over a recent disclosure that Barr made a speech before a southern group espousing white supremacist views.

"Young man," Barr begins patronizingly.

Kennedy, even more furious, snaps back, "I'm a duly elected representative of my state."

With deep and cutting sarcasm, Barr replies, "I'm impressed. I'm *duly* impressed."

It's like that all throughout the hours of the impeachment debate on this Friday, December 18, as the House prepares to vote the following day. It's raw and rancorous, unworthy of a great democracy deliberating over a fateful decision. Not a trace remains of the solemnity and deep sense of gravity that characterized the Watergate impeachment hearings twenty-five years before. Then, Republicans and Democrats visibly struggled with the conflict between their consciences and their political interests. In the end, they set an example of serious, dignified deliberation that became an exemplar of the meaning of bipartisanship. Now, a generation later, it isn't even a debate that's taking place; what the nation and the world are witnessing is a debasement of the meaning of that term. This isn't a debate. It's a series of set ideological speeches aimed at the cameras and the audience beyond, delivered to appeal to individual constituencies, not to sway minds or votes.

Before a word is spoken, everyone knows that all minds already are made up, and long have been; the divisive hearings that preceded this day demonstrated that fact beyond dispute. Nor is it a debate that is going to be influenced by public opinion. This morning the latest *New York Times*/CBS News Poll shows the standing of the Republican Party with the public dropping to its lowest point in the last fourteen years. Nothing will now stop the process racing toward its conclusion, nor turn Republicans from the path they have chosen.

Speakers await their turn to make their remarks, then deliver them before colleagues who for the most part pay little attention. The rhetoric is familiar.

Hyde of Illinois: "The matter before the House is a question of the willful, premeditated, deliberate corruption of the nation's system of justice. Perjury and obstruction of justice cannot be reconciled with the Office of the President of the United States."

DeLay of Texas: "What the defenders want to do is lower the standards by which we hold this president and lower the standards for our society."

Democrats speak of a coup d'état taking place, of the House launching itself on "an historically tragic case of selective moralizing."

Occasionally, a flash of genuine personal anguish sounds, as when Dave Obey of Wisconsin, one of the most thoughtful of the Democrats, says: "I honestly believe this is the worst day for this institution in this century."

Another that transcends the conventional party-line rhetoric being expressed comes when Charles Schumer of New York links both parties' responsibilities for what is taking place with the unintended legacy of Watergate that produced a generation of scandal seeking: "What began twenty-five years ago with Watergate as a solemn and necessary process to force a president to adhere to the rule of law, has grown beyond our control so that we are now routinely using criminal accusations and scandal to win the political battles and ideological differences we cannot settle at the ballot box."

But none of these words, or the scenes the television cameras are portraying comes close to capturing the extraordinary backdrop against which this demeaning spectacle plays. For it takes place in the midst of a major military strike launched by the United States against Iraq aimed at demolishing the foundation of Saddam Hussein's power, a sudden strike that once again arouses deep public cynicism about the timing of the attack and the president's motives. As happened when the president ordered air strikes on a presumed terrorist base during his miserable Martha's Vineyard vacation, critics murmur that he's playing out a real-life *Wag the Dog* scenario, a reference to a new film about a president who orders military action to divert the public from a sex scandal in which he is involved. Similar dark suspicions are raised anew with this strike against Iraq in the midst of Clinton's impeachment.

Throughout the previous two days American missiles have been raining down on Baghdad, setting off massive explosions that turn the night skies red.

The juxtaposition of these events, the impeachment of a president and the actions of that same president in sending American men and matériel into "harm's way" as their commander in chief, are so confounding, so disturbing, that they leave virtually everyone straining to find ways to express them. *Bizarre, unbelievable, surreal*—these are the adjectives of choice. Now the television news departments are showing split-screen views of the military and the political battles. So, too, the people read of "surreal split-screen frenzy" taking place in the capital even as TV screens are "aglow with ghostly green images of bombs and missiles exploding in Baghdad."

There's no way of escaping the connections. Readers that morning in the capital open their *Washington Post* to find the rare, two decks of headlines spread across the entire eight columns of their front page:

IMPEACHMENT DEBATE TO BEGIN
AS AIRSTRIKES ON IRAQ CONTINUE

They read further and find Dan Balz, the paper's political writer, describing how the capital struggles to disconnect the parallel events but cannot. He sketches the hard new political realities that confront politicians and policymakers: "A decade of destructive partisanship, personal attack and win-at-all-cost politics has crystallized in Washington this week, and the question no one can begin to answer is where it will end. The extraordinary events of the past 48 hours suggest that the simple civilities that once helped smooth the rough game of politics are being swept away."

Then Balz quotes Ken Duberstein, chief of staff in the Reagan White House. After saying that Washington is beginning to resemble the bombed-out Beirut of the 1980s, Duberstein shifts the metaphor: "It now reminds people of the napalm-bombed Vietnam: total scorched earth. It is very sad. . . ."

Just when it seems no more shocks can be sustained, a new one occurs that eclipses all the others. For that same morning, readers of the *Post* see a third eight-column page-one headline that says:

SPEAKER-DESIGNATE'S ADMISSION ON INFIDELITY CAPS A DAY OF DRAMA

The night before, after a day of increasing rancor over whether to put off the impeachment debate and vote until the military action is completed, Republicans decide to push ahead; then they convene privately in the Capitol basement where their new Speaker-designate, Bob Livingston of Louisiana, has summoned all House Republicans to discuss the impeachment schedule.

Livingston is three-quarters of the way through his briefing when, suddenly, he detonates his own bomb. He reaches inside his suit jacket, pulls out a piece of paper, and begins to read. It's a confession. "On occasion," he says, during his thirty-three years of marriage he has had extramarital affairs. As his colleagues sit, slack-jawed and shocked into silence, their Speaker-designate tells them these affairs have caused problems in his marriage. But, he reads, "I sought marriage and spiritual counseling, and have received forgiveness from my wife and family, for which I am eternally grateful."

His reason for making this extraordinary confession, at this moment, on this of all occasions, he explains, is that only in the last few days has he learned that reporters are investigating his private life and he faces the threat of exposure. He and his wife, Bonnie, have decided that "now was the time" to say something about his situation. Although he doesn't tell them, the unnamed reporters are working for Larry Flynt, the *Hustler* publisher; Livingston is a victim, and spectacularly, of Flynt's scandal hunt that has turned up evidence of his affairs; plans are being made to publish the exposé shortly.

As Livingston finishes his statement, he says this is the last word he will have on the matter. His fate is in their hands, he tells them.

After a heavy, uneasy silence, Livingston's colleagues rise and give him a standing ovation. They're standing by him—at least, that's what the press is later told. The reality is something else. Many feel a sense of betrayal, and a fear that this unbelievable second fall of a Speaker in little more than a month is sure to present devastating political problems for a party about to impeach a president for moral, if not criminal, misconduct.

Younger members especially immediately begin pressing for him to step aside as their new Speaker. "We have so many bombshells," says Michael N. Castle, a Delaware Republican. "We have bombshells in Baghdad and we have bombshells in the House. You can't turn your back for ten seconds."

Yet another incredible sexual scandal hangs over Washington, but this time it produces little public shock, and, headlines aside, even less public attention—and for excellent reason. By now, a thoroughly disgusted American public is completely turned off by the workings of Washington, *all* of Washington. The latest sex scandal merely confirms the wisdom of Michiko Kakutani's observation that in the America of Teletimes and Scandal Times, of synthetic celebrities and fallen leaders, tragedy and farce co-exist side by side.

· · ·

At 1:22 P.M., on Saturday, December 19, 1998, Bill Clinton—the first elected president of the United States to be impeached—suffers a near-party-line loss in the House of Representatives. By 228 to 206, the House approves an article accusing him of perjury for misleading the federal grand jury in August about his relationship with the intern. A second impeachment article, charging the president with obstruction of justice, passes by a narrower margin of 221 to 212. His impeachment trial will begin before the Senate after the new Congress convenes in January.

Historic as this is, it seems so anticlimactic that one network, CBS, switches away before the vote to carry a Saturday football game. The game easily wins the day's ratings, attracting far greater numbers of viewers than those who watch the impeachment of an American president. Indeed, the entire process diminishes public interest in the fate of the person who leads the nation and in the political system he represents. During the Watergate impeachment trauma in 1974, some 47.3 million viewers saw some or all of the Nixon impeachment hearings on network TV. In 1998, some 1.6

million viewers watch the Clinton impeachment hearings on cable TV, which alone covers them in their entirety.

The only unexpected development during the actual impeachment involves Bob Livingston. When Livingston, still the Speaker-designate and thus third in constitutional line to the presidency, stands to explain why he will vote to impeach the president, he draws a torrent of boos and catcalls when he calls upon the president to resign. *He* is the one who should resign, some Democrats shout.

Livingston proceeds to do just that. "I was prepared to lead our narrow majority as Speaker," he says, "and I believe I had it in me to do a fine job. But I cannot do that job . . . under the current circumstances. So I must set the example that I hope President Clinton will follow. I will not stand for Speaker of the House on January 6."

Thus he follows Gingrich in resigning, and also in vacating his seat in the House.

When *Hustler*'s Larry Flynt learns of Livingston's resignation, he expresses pleasure. He's still investigating claims of other women who say they have had affairs with public figures in Washington, most of them Republicans, including four members of the House supposedly closely associated with the impeachment proceedings; he also claims to have an expanded scandal list he threatens to make public of "people who go on TV and keep attacking Clinton." Flynt promises there's more to come. Asked by the *Wall Street Journal* whether he has any regrets or compunction about offering money for stories, Flynt says no; then, expressing still another epitaph for the times, he adds: "There's nothing that changes people's moral outlook like money."

And who is to say otherwise? At this point, with a president impeached, his trial looming, and more scandals on the horizon, the American political system appears to be coming apart. The entire country is becoming a casualty of Scandal Times.

Trial of the Century—
Part Two

*"Pick your own adjective to describe the president's conduct.
Here are some that I would use: indefensible, outrageous,
unforgivable, shameless. I promise you the president would
not contest any of those or any others. . . ."*

W hen Dale Bumpers returns from the grocery store that
Monday afternoon, his wife is waiting for him as
soon as he opens the door. "The president is desperately trying to reach you," she tells him. *Oh, shit,* Bumpers thinks;
he knows exactly what it's about.

The night before, after he returned from a lecture in Arkansas,
Bumpers's old friend and former Senate colleague, Tom Harkin of
Iowa, phoned to urge him to think about delivering the closing argument on behalf of the president in the impeachment trial later
that week. "Tom," Bumpers said, "I really am flattered you'd call
me, but let me give you a piece of advice: George Mitchell [the
former Senate majority leader and federal judge] is the guy who
ought to make this speech. George is much more cerebral than I

am, and he's a very good speaker. He'll tie it together better than
anybody. And, second, you know there isn't anybody in this town
who doesn't know that Bill Clinton and I have been friends for
twenty-five years and come from the same state. That would diminish anything I might have to say."

"Well, just don't say no," Harkin said.

"I'm not gonna tell you categorically no," Bumpers answered,
"but, Tom, I'm really not inclined to do it. It's not something I relish doing, and it's not something I think I can do."

They hung up.

What is all that about? his wife, Betty, asked. Bumpers told her,
adding, "I think it's just something Tom conjured up on his own."

Now, less than twenty-four hours later, Bumpers instantly understands that Harkin was not acting on his own. He'd called on
behalf of the president. Bumpers feels a deep sense of ambivalence.
He waits for an hour, thinking about what to say to the president,
before calling him back.

It's Monday, January 18, 1999, Martin Luther King's birthday,
a holiday in Washington. On Capitol Hill, the second impeachment trial of an American president, then in holiday recess, is approaching its climax. Tomorrow, when the trial resumes, promises
to be the most melodramatic moment in the near-endless spectacle
that already has provided one unbelievable scene after another, for
tomorrow the president's lawyers are scheduled to begin their final
defense of him. Then, just hours later, the president will travel to
the Capitol to deliver his prime-time State of the Union address.
The address will be broadcast live around the world. Countless
millions will be watching the impeached president. Gathered before him in the congressional chamber will be the entire Congress
of the United States: the senators, who are acting as jurors in his
trial; the Republican House managers, who are acting as prosecutors after presenting their case for convicting and removing him
from office in strong, often contemptuous, language; and every
other member of the House—those who voted to impeach him
and those who did not. Also attending will be the cabinet, the joint
chiefs, the judiciary, and leaders of the diplomatic corps. Looking
down from the visitor's gallery will be his wife. She knows, of

course, that the cameras will be trained on her face to catch every flicker of reaction. It will be a scandal voyeur's dream, a dramatic Teletimes moment—news as entertainment raised to even new heights.

Bumpers thinks about all this before placing his call to Clinton in the White House.

The president *really* wants him to do it, Bumpers now hears him say. Bumpers demurs; he repeats what he said to Tom Harkin the night before. "I think George Mitchell would be a much better person," he says. "Tell you the truth, Bill, I don't know what I'd say that would be meaningful considering our relationship. I've not immersed myself in this thing; I don't know what time Betty Currie called, and who hid the gifts, and where they were hidden. In fact, I've found this whole thing so God-damned appalling I haven't even listened to it. The repetition of those House managers has driven me crazy. I really think you, or someone else, should contact George and see if he's interested in doing it."

The president is solicitous. He doesn't press, but before hanging up he says, "I'll understand if you don't want to do it, but, Dale, I sure wish you'd help me."

In the next few hours, the president calls back twice. Each time they have essentially the same conversation; each time, the president ends their conversation by saying: "If you don't want to do it, I'll understand, but I'd really appreciate it if you could help me." His persistence, as Bumpers knows, is typical of his character—and so, he's learned through long years of close observation, is another aspect of this president: "He's the most resilient man I've ever known, absolutely incredible."

Finally, about nine o'clock that night, the president calls one more time. This time Bumpers relents. "Let me just get a little solitude here," Bumpers tells the president. "No television, no nothing. Let me scribble off some notes and see if I can develop an outline for something that I might want to say that might be meaningful and helpful."

For the next three hours, until shortly after midnight, Dale Bumpers sits in his study hunched over a legal pad making notes as he always does for his speeches, the very lessons learned from an

English teacher when he was an Arkansas schoolboy: Roman numeral I, followed by big letter *A,* then little number 1, followed by little number 2. In all his quarter century in the United States Senate, and before that his years as governor of Arkansas and, earlier, as a small-town lawyer, he has never deviated from this practice.

In a political world where speechwriters craft the words to be spoken by the great public figure, where politicians study how to deliver them over teleprompters and read from uppercase texts carefully marked by notations to PAUSE for dramatic effect and to place EMPHASIS on certain words and phrases, Dale Bumpers is a throwback. He's an original: the stump speaker as orator, the nearly vanished breed who shuns prepared texts, who speaks from notes, whose words and thoughts are his own.

Though it's true that he and Clinton have formed a close friendship, the reality of their relationship is, as is often the case with political figures, much more complicated. It's Bumpers who wins his Senate seat by defeating Clinton's mentor, J. William Fulbright, while Clinton passionately stumps the state for Fulbright; Bumpers, who thinks of himself as becoming the first southern president since Lyndon Johnson, and who might well have, except for a characteristic caution that keeps him from risking all when he thinks the Democratic nominee faces certain defeat★; Bumpers, who remembers seeing the young Bill Clinton deliver a masterful short campaign speech in Clinton's race for attorney general, a first impression that led Bumpers to think "I hope that son-of-a-bitch never runs against me"; Bumpers who faces the prospect years later that Clinton, in fact, plans to run against him for the Senate, though Clinton eventually decides against it.

★Bumpers thought seriously about becoming a presidential candidate in 1976 and stepped back at the last moment only to see a fellow southerner and governor, Jimmy Carter of Georgia, run and win. In 1983, he again anguished over running but concluded Reagan was the almost certain 1984 winner. At that time, Bumpers remembers how Clinton "did everything in the world to get me to run for president. I didn't have a stronger supporter for the presidency than Bill Clinton."

In his years in the Senate, Bumpers earns a reputation for independence and eloquence. He becomes a political nonconformist. Stately yet folksy, anecdotal as well as analytical, his lean, tall figure and droll but piercing manner of speaking have led more than one commentator to call him Lincolnesque, a Gregory Peck–type come to Washington.

But over those same years Bumpers has grown increasingly disenchanted with the new lone-wolf media-age slash-and-burn politics. By the Nineties, he finds himself reacting strongly to the rising meanness in political life, a destructive partisanship beyond anything remembered, accompanied by what he comes to feel is "a serious deterioration in the quality of the Senate."

Gone is the bipartisanship that marked the Senate when such Republicans as Howard Baker of Tennessee, Jake Javits of New York, Ev Dirksen of Illinois, Mac Mathias of Maryland, Clifford Case of New Jersey, Hugh Scott of Pennsylvania, Ed Brooke of Massachusetts, and Lowell Weicker of Connecticut played prominent roles. They were the kind of people, as Bumpers would say in the privacy of his Senate office, who could be depended on to rise above party when the interest of the country was involved. "These people were born and trained to govern," Bumpers would say. "They understood the public interest. The intellectual quality, to say nothing of the understanding of the public good, is just totally different now."

He also feels a deepening frustration about the lack of understanding and appreciation for what in recent years has become his new passion, and his subject in numerous Senate speeches—the workings of the United States Constitution and the intent of the Founders.

Finally, contemplating whether to seek an almost certain fifth term in the Senate in the November elections of 1998, the Year of Monica, Bumpers, at the age of seventy-three, decided to step down after the new Senate convened that January 6.

Before announcing his plans not to seek reelection, Bumpers quietly approached several people he thinks "were classy, well-educated," people whom he regarded as "suitable replacements" for his Senate seat. "I didn't even get a sniffle," he says, in an implicit negative comment on the way the public now views political life. "None of them would even think of it."

Now, just twelve days after vacating his Senate seat, Dale Bumpers continues scratching out the outline for what he might say to his erstwhile colleagues in a speech that can define his public career and affect the presidency long beyond the tenure of Bill Clinton of Arkansas.

The next morning he phones the president, who is preparing for his critical State of the Union address that night. "I think I'll do it," Bumpers says. "Are you sure you still want me to do this thing?" "Yes," the president says, he's sure.

For the rest of that Tuesday, while the president's lawyers begin their defense of him in the Senate, and on through the night while the world watches the president over live TV from the Hill, Bumpers keeps scratching away on the notes for his speech. By Wednesday morning, he's halfway through the second version of his edited, and reedited, notes, when two White House aides visit him to discuss what he plans to say before the Senate the next day, Thursday, January 21.

Bumpers, feeling apprehensive, uncertain whether he's on the right track, welcomes the chance to see how others might react to his remarks. But, he tells them, once they take seats opposite his desk, "I'm gonna tell you fellows something. Number one: I'm halfway through the second half of this thing so I'll deliver the first half of what I'm gonna do on the Senate floor tomorrow, and then I'll tell you what the second half is. But don't stop me and don't interrupt me when I'm talking, because I'm not inclined to change anything. If you have any suggestions, I'll be glad to entertain them, but I'm not likely to accept them."

With that, he begins speaking in his clear, slow Arkansas drawl. Halfway through, he knows from the look on their faces that it's going to be all right.

· · ·

Flashback: Two hundred eleven years earlier, on Friday, March 7, 1788, literate citizens of the small city of New York, population

25,000, on the lower half of Manhattan Island, pick up a copy of the *New York Packet* and read the latest—No. 65—of a series of essays under the standing headline, THE FEDERALIST, and addressed "To the People of the State of New York." The author, identified at the end of the article, is "Publius."

The topic this Friday is the difficulty of creating "a well-constituted court for the trial of impeachments" that Publius and the other two anonymous authors of the articles, all using the same pseudonym, are considering in their ongoing public debate about fashioning and adopting a Constitution to govern the new United States of America. Such an impeachment court, the citizens read in their newspaper, would be convened to determine the guilt or innocence of those charged with "offenses which proceed from the misconduct of public men, or, in other words, from the abuse or violation of some public trust. They are of a nature which may with peculiar propriety be denominated POLITICAL, as they relate chiefly to injuries done immediately to the society itself."

For that reason, the author goes on, prosecuting public figures for such political offenses against society through impeachment "will seldom fail to agitate the passions of the whole community, and to divide it into parties more or less friendly or inimical to the accused. In many cases it will connect itself with the preëxisting factions, and will enlist all their animosities, partialities, influence, and interest on one side or the other; and in such cases there will always be the greatest danger that the decision will be regulated more by the comparative strength of the parties, than by the real demonstration of innocence or guilt."

This will be especially delicate and difficult since it goes to the heart of a government "resting entirely on the basis of periodical elections." Thus, society faces the inevitable danger that leaders "of the most numerous [political] faction" will be inclined to employ the tools of impeachment to remove political opponents lacking the power of the majority.

This fear of "faction" haunts the young authors—Alexander Hamilton, thirty; James Madison, thirty-six; John Jay, forty-two—of those eighty-five newspaper articles published in New York during 1787 and 1788. Later, in book form, they come to be known as *The Federalist Papers,* the founding charters of the American Constitution.

Madison, Hamilton, and Jay all agree that in creating an impeachment process they will borrow directly from the British example where the House of Commons, representing those most directly close to the people, brings the impeachment charges, and the House of Lords, the body most independent and removed from the pressures of daily politics, acts as the jurors in the trial.

Still, they argue, the danger exists of an abusive majority assuming dictatorial powers. By giving the legislative body, the Congress, the power to impeach the leader of the executive branch, the president, an obvious threat arises. Congress possesses the power to make the president subservient to its whims and wishes, serving at its pleasure, thus irreparably weakening the office and the ability to govern a diverse democracy. As a remedy, they give the third independent branch of the planned new government, the judiciary, a central role in the court of impeachment. While the Senate sits as the court of impeachment, the chief justice of the United States makes the rulings on law and procedure as the presiding judge. This, they hope, will prevent one independent branch of the government from acting alone in trying and finding the guilt or innocence of the other, the executive, or, in the case of other impeachments, trials of federal judges, including those on the highest court.

Animating their analysis and their proposals for drafting the constitutional rules under which the new American system will operate is one overriding desire: to protect liberty and freedom in the face of the inherent human instinct to naturally divide "into different interests and parties" whose differences often "split the community into the most violent and irreconcilable factions."

From that view comes their conclusion: The causes of factions can't be removed; therefore, efforts to control their deleterious effects must be made. So the rules have to be established to protect the people against the abuses of a dictator. Impeachment is the remedy. At the same time, the executive, in the person of the president, has to be protected against the abuses of a majority party bent on assuming absolute power for itself. The remedy is to make the act of impeachment extremely difficult. A simple majority vote of the House will be sufficient to impeach the president. But it will take a two-thirds vote of the Senate to convict and remove him from office.

They address, at length, the grounds for impeaching a president. Writing in the same New York paper exactly a week later that March of 1788, Hamilton spells them out: "The President of the United States would be liable to be impeached, tried, and, upon conviction of treason, bribery, or other high crimes and misdemeanors removed from office; and would afterwards be liable to prosecution and punishment in the ordinary course of law." Hamilton has already established, just a week before, that the president's "high crimes and misdemeanors" are POLITICAL (as he emphasizes through applying uppercase letters) offenses against society. They are not *personal* offenses.

During the four-month-long convention in Philadelphia that leads to the adoption of the Constitution, the delegates take that language and incorporate it into the Constitution. But before doing so, they debate long and hard and ultimately further narrow the grounds for impeachment to make it even more difficult to accomplish. To the term "high crimes and misdemeanors," George Mason, that architect of liberty in Virginia, adds the phrase "against the United States." He does so to ensure that Hamilton's language about "political offenses against society" is specifically spelled out as political offenses against the state. The constitutional convention delegates vote to approve Mason's addition, and send it along to the Committee on Style and Arrangements, charged with drafting the new Constitution of the United States in language that everyone can understand. In that drafting process, Mason's phrase "against the United States" is dropped on the grounds it's redundant; the meaning, the committee members assume, is clear, understood by all.

So it seems then, and for more than two centuries to come, including the only other impeachment trial of a president in 1868. It is not understood in 1999 when the Senate weighs the evidence and the impeachment grounds that bring Bill Clinton to judgment before it.

· · ·

In the closing years of the twentieth century, political rhetoric no longer holds the power to sway the citizenry that it once did. Even the art of political phrase-making evokes a curl of cynicism from people conditioned to distrust emotional appeals: just more political bull, more promises, easily expressed, hardly persuasive or believable. In the decades leading up to the president's impeachment, Americans can count on the fingers of one hand those times when they were truly moved by memorable public addresses—John Kennedy's inaugural address; Martin Luther King's "I have a dream" speech; Lyndon Johnson's "we shall overcome" remarks in the wake of the civil rights violence and murders in Selma, Alabama; Ronald Reagan's words after the *Challenger* explosion.

Add to these the impeachment trial of the forty-first president, which provides two such memorable moments on the public stage within the space of little more than twenty-four hours.

First comes the president.

Bill Clinton's State of the Union address Tuesday night, January 19, will not rank as great oratory or be remembered for a single phrase, but it certainly rates as great political theater.

That Tuesday becomes what one commentator calls a "looking-glass day," a day in which everyone turning on the TV set can see the president's "glaring personal weaknesses and manifest political strengths" take the stage in quick succession. All eyes are fixed on the Capitol, watching as the drama unfolds in two acts: the setting remains unchanged and many of the players are the same, but the central role of the president is up for grabs. Throughout the day, the impeachment trial has focused a relentless light on the president's character flaws and sordid behavior. That night, the president returns to the Capitol and enters the spotlight alone. His performance will be subject to the most intense scrutiny that any president has ever experienced. The stakes are insurmountably high. Everyone watching understands this; it heightens the suspense and induces speculation about how this latest televised spectacle will turn out.

The answer comes quickly: Once again the master political escape artist, the Houdini of his times, puts on a spectacular performance. And this time, he surpasses all of his previous acts.

From the moment he walks into the packed chamber, ablaze with lights and rent by the sound of partisan cheers from *his* side and stony silence from the other, until nearly an hour and a half later when he walks back through that aisle with partisans on *both* sides clamoring to shake his hand and be acknowledged by him, impeachment judges and managers among them, he revels in his leading role.

As he stands at the podium against the backdrop of a huge American flag, never has he seemed so confident, so ebullient, so at ease. He smiles broadly, salutes his wife and the baseball slugger Sammy Sosa sitting beside her in the balcony, a sports icon chosen this night by the cynical political stagecrafters as the ritual celebrity prop designed to enhance the president. Then, turning, the president faces his friends and foes and delivers one of the most ambitious political agendas the nation has been given in years.

His agenda is all-encompassing, something for everyone: strengthening and saving Social Security by letting citizens for the first time invest their benefit savings in the stock market. Providing better schools, better health care, stronger Medicare provisions. More defense spending, greater federal aid for crime prevention, continuing to press government suits against the predatory tobacco companies, stopping employers from discriminating against parents, expanding federal assistance to welfare-to-work programs. Who can be against this marvelous bag of goodies? Look, says the wizard, at all the benefits I have brought you in this greatest of all booms. "America is working again," he says. "The promise of our future is limitless. But we cannot realize that promise if we allow the hum of our prosperity to lull us into complacency."

He cloaks his appeals in a sunny veil of bipartisanship and pauses pointedly to praise the new Republican Speaker sitting behind him, the pedestrian Denny Hastert. He assures Gingrich's successor that together they and the Congress can work together in a spirit of mutual cooperation for the good of the country throughout the remaining 730 days of his presidency.

Not once during the seventy-seven minutes that he speaks does he mention the word impeachment, or even obliquely refer to his responsibility in contributing to the embittered partisanship that brought it about. Instead, he focuses entirely on his agenda for the future. His message is hardly subtle: Put all this miserable business behind us and get back to the real unfinished work for the people. Thirty-nine times he exhorts Congress that it "must" or "should" act on some urgent national problem—the obvious implication being that they're not doing their *real* job in the public interest. He appears to be a president not lost in a state of paranoia or one who seeks vengeance for offenses committed against him. But throughout he wields the political scalpel with the skill of a surgeon. Repeatedly, but in apparent good humor, he jabs the Republican majority about initiatives they abhor—the minimum wage, on campaign finance reform. Turning the television cameras against his foes, he forces sullen Republican ranks to stand and applaud his call for equal pay for men and women. How can they be seen sitting silently and sullenly when he evokes such popular goals?

It's more than a tour de force; it's a political triumph, and everyone in that chamber knows it, most especially his political enemies who want to remove him from office.

The reviews next morning are uniformly glowing. In the *New York Times,* John M. Broder describes him "towering above the three branches of government assembled before him," while giving a "virtuoso" performance. "A lesser politician, or a more timid soul, might have balked at facing his House accusers and Senate jurors at a moment of political peril greater than that faced by any other President for many years," he writes. "But Mr. Clinton seems to thrive on such adversity." Mary McGrory, in the *Washington Post,* with her customary felicitous touch, writes: "Depending on the time of day and which side of the Capitol you were on, the President was either a man on the lam or the lord of the earth." She adds: "The State of the Union was made for an extrovert like Bill Clinton. It is a collection of overstatements and extravagant promises delivered with great force and frequent pauses for applause." And: "Was there any question in his mind about his survival in office? Apparently not. What the Senate is grinding through is the past. He is about the future."

His bravura performance rallies his restive supporters and

leaves his many detractors choking with frustration, a situation that recalls the begrudging tribute a lifelong Republican opponent, William Allen White, paid the mesmerizing FDR at the peak of that confounding president's political mastery: "We who hate your gaudy guts salute you."

Much as they dislike it, some Republican bitter-end opponents are now forced to say something similar, most notably Pat Robertson, the conservative televangelist and power in the Republican party, who has been pressing for the president's impeachment and removal from office. On his national TV program, Robertson says the president "hit a home run." "From a public relations standpoint, he's won," Robertson tells millions of viewers of his TV program, *The 700 Club*. "They might as well dismiss this impeachment hearing and get on with something else, because it's over as far as I'm concerned."

The polls reinforce this judgment. In a January *USA Today/CNN/Gallup* Poll, 81 percent of the people judge his presidency a success. Sixty-nine percent approve of his overall job performance. A solid two-thirds want him to remain in office. But it's not over as far as others are concerned, especially since the Senate has yet to begin its crucial deliberations on whether to convict or acquit him in its ongoing trial.

That sets the stage for Dale Bumpers, who by peculiar coincidence returns to the Senate for his closing defense of the president a year to the day after the Monica story first appeared in the *Post*.

· · ·

Dale Bumpers's impeachment speech before the Senate becomes the only time those proceedings rise to a level worthy of their gravity. Until that moment, the public debate and discourse leading to Bill Clinton's impeachment and trial are debased by shameless tactics and false claims. And all the excesses are gleefully exploited as devices to raise ratings by a twenty-four-hour electronic media bent on capitalizing commercially from a sex scandal. Bumpers's speech doesn't redeem the dismal state of the American political system, but it does demonstrate the political world can still produce figures capable of eloquence and historical perspective.

Bumpers strides into the Senate in his gray pinstripe suit and chooses not to speak from behind a lectern before a stationary microphone. Instead, he adopts the manner most comfortable to his style and personality—strolling up and down a Senate aisle, referring only occasionally to handheld notes, and directly addressing those he keeps referring to as "colleagues" in either party. He appears entirely at ease; even what could be a major embarrassment—when his lapel microphone is pulled off in the midst of his remarks because Senate aides make his cord more than a foot shorter than needed—doesn't affect him. He merely reaches down, casually picks it up, uses it as a hand-held mike, and continues what he's saying.

He is part historian, recalling in personal terms how Hamilton, Madison, Mason, and the rest arrived at their definition of impeachable offenses and the standards that should be applied. He's part country lawyer, telling his colleagues to appreciative smiles, that "you're being addressed by the entire South Franklin County, Arkansas, Bar Association!"

Easygoing manner and homely observations aside, his core message is deadly serious. "Ah, colleagues, you have such an awesome responsibility," he tells them, after remarking how "the weight of history" lies "on all of us."

So many words have been used to describe these proceedings, Bumpers says, "historic, memorable, unprecedented, awesome." All are apt. To these he adds *dangerous*—"dangerous to the political process, dangerous to the unique mix of pure democracy and republican government Madison and all his colleagues so brilliantly crafted, and which has sustained us for 210 years."

That word isn't his, he says. It's Alexander Hamilton's, who warns that the greatest danger in impeachment is that the verdict will "be based on the comparative strength of the parties rather than the innocence or guilt of the president."

How did we come to be here? Bumpers asks, before supplying his answer: "We are here because of a five-year, relentless, unending investigation of the president; $50 million, hundreds of FBI agents fanning across the nation examining in detail the microscopic lives of people. Maybe the most intense investigation not only of a president, but of anybody, ever."

He describes the innocent people hurt financially and emotionally by all the inquiries in the attempt to "get" the president, then says:

"Javert's pursuit of Jean Valjean in *Les Misérables* pales by comparison."

He halts, glances around the Senate chamber, and says, "I doubt that there are few people—maybe nobody—in this body who could withstand such scrutiny."

And for what? After all those years of investigations, after those $50 million expended "on Whitewater, Travelgate, Filegate, you name it—nothing. Nothing. The president was found guilty of nothing, official or personal."

So, he goes on, returning to his theme, why are we here?

"We are here today because the president suffered a terrible moral lapse, a marital infidelity; not a breach of the public trust, not a crime against society, the two things handled and talked about in Federalist Paper No. 65. . . . It was a breach of his marriage vows. It was a breach of his family trust. It is a sex scandal."

"H. L. Mencken," he recalls, "one time said, 'When you hear somebody say, "This is not about money," it's about money.' And when you hear somebody say, 'This is not about sex,'" Bumpers reminds his colleagues, "it's about sex."

Then Bumpers begins to articulate the main points he wants to make, the points he labored over in all those hours in his study, points he believes haven't been mentioned so far in the entire impeachment process. "Pick your own adjective to describe the president's conduct," he says. "Here are some that I would use: indefensible, outrageous, unforgivable, shameless. I promise you the president would not contest any of those or any others.

"But there is a human element in this case that has not even been mentioned. That is, the president and Hillary and Chelsea are human beings. . . . But when I talk about the human element, I talk about what I thought was, on occasion, an unnecessarily harsh, pejorative description of the president. I thought that the language should have been tempered somewhat to acknowledge that he is the president. To say constantly that the president lied about this and lied about that—as I say, I thought that was too much for a family that has already been about as decimated as a family can get."

You can see from the utter stillness in the chamber and the intent expressions of the senators that he's captured their full attention.

"The relationship between husband and wife, father and child, has been incredibly strained, if not destroyed," he says. "There's been nothing but sleepless nights, mental agony, for this family for almost five years, day after day, from accusations of having Vince Foster assassinated, on down. It has been bizarre. "I didn't sense any compassion, and perhaps none is deserved.

The president has said for all to hear that he misled, he deceived, he did not want to be helpful to the prosecution, and he did all of those things to his family, to his friends, to his staff, to his Cabinet, and to the American people. Why would he do that? Well, he knew this whole affair was about to bring unspeakable embarrassment and humiliation on himself, his wife whom he adored, and a child that he worshiped with every fiber in his body and for whom he would happily have died to spare her this or to ameliorate her shame and her grief."

Having described the Clintons' ordeal, Bumpers shifts to discussing an impeachment case that in his opinion comes down to lying about a sexual affair.

"The House managers have said shame and embarrassment is no excuse for lying," he says. "Well, the question about lying, that's your decision. But I can tell you, you put yourself in his position, and you've already had this big moral lapse, as to what you would do. We are, none of us, perfect. Sure, you say, he should have thought it all out beforehand, and indeed he should, just as Adam and Eve should have. Just as you and you and you and you"—he sweeps his arm around the chamber, emphasizing the personal allusions—"and millions of other people who have been caught in similar circumstances should have thought of it before."

In his law career, Bumpers tells his colleagues, he's tried maybe five hundred divorce cases in his small Arkansas town and its surrounding counties. In about 80 percent of those contested cases perjury was committed. "And do you know what it was about? Sex. Extramarital affairs. But there is a very big difference in perjury about a marital infidelity in a divorce case and perjury about whether I bought the murder weapon or whether I concealed the murder weapon or not. And to charge somebody with the first and punish him as though it were the second stands justice—our sense of justice—on its head."

Once more, he revisits the constitutional convention and the long debates about framing a remedy for presidential abuses that impeachment is designed to address. The key players in drafting the impeachment provision were in complete agreement about one great concern: "They did not want any kings. They had lived under despots. They had lived under despots, under kings and under autocrats, and they didn't want any more of that. And they succeeded very admirably. We have had [forty-one] presidents and no kings."

The term "high crimes and misdemeanors," he says, where did that come from? It comes from English law which defines it as meaning "distinctly political offenses against the state."

"So, colleagues, please for just one moment forget the complexities of the facts and the tortured legalisms. And we have heard them brilliantly presented. . . . But ponder this: that 'high crimes and misdemeanors' was taken by George Mason . . . as 'political offenses against the state.' What are we doing here? If, as Hamilton said it had to be a 'crime against society' or a 'breach of the public trust,' what are we doing here? Even perjury, concealing or deceiving an unfaithful relationship, does not even come close to being an impeachable offense. Nobody has suggested that Bill Clinton has committed a political crime against the state. . . . You cannot indulge yourselves the luxury or the right to ignore this history."

Turning toward the House Republican impeachment managers, sitting silently to his left behind a small table, Dale Bumpers addresses the remarks that their leader, Henry Hyde, made in a florid, flag-waving opening speech telling the senators why they should find the president guilty and remove him from office.

"There has been a suggestion that a vote to acquit would be something of a breach of faith with those who lie in Flanders Field and Anzio and Bunker Hill and Gettysburg and wherever," Bumpers says. "Chairman Hyde alluded to that and said those men fought and died for the rule of law."

Glancing around the Senate, searching the faces before him, Bumpers cites the wartime example of a number of those present, both Democrats and Republicans.

"If you want to know what men fought for in World War II or in Vietnam ask Senator Inouye. He left an arm in Italy.... A certified war hero. I think his relatives were in an internment camp. So ask him, what was he fighting for? Or ask Bob Kerrey, a certified Medal of Honor winner, what was he fighting for? You'll probably get a quite different answer. Or Senator Chafee, one of the finest men ever to grace this body. And a certified Marine hero of Guadalcanal. Ask him. And Senator McCain, a genuine hero. Ask him. You don't have to guess. They're with us, and they're living. And they can tell you. And one who is not with us in the Senate anymore. Robert Dole. Ask Senator Dole what he was fighting for."

He's nearing the ending now, approaching his peroration. Employing a familiar metaphor from Watergate, he says the American people are asking for an end to this nightmare. "Colleagues," he says, "this is easily the most important vote you will ever cast. If you have difficulty because of an intense dislike of the president— and that's understandable—rise above it. He is not the issue. He will be gone. You won't. So don't leave a precedent from which we may never recover and almost surely will regret.... But if you vote to convict, you can't be sure what's going to happen."

He recalls the only other impeachment trial of a president and the regret of one of those Senators who voted to find Andrew Johnson guilty. Twenty years later that senator, James G. Blaine, recanted, saying, "I made a bad mistake." Blaine came to believe, as Bumpers says, that "having convicted Andrew Johnson would have caused much more chaos and confusion in this country than Andrew Johnson could ever conceivably have created." The same is so with William Jefferson Clinton: "If you vote to convict, in my opinion, you're going to be creating more havoc than he could ever possibly create.... So don't, for God's sakes, heighten people's alienation, which is at an all-time high, toward their government. The people have a right, and they are calling on you to rise above politics, rise above partisanship. They're calling on you to do your solemn duty. And I pray you will."

No sooner does he finish speaking than his colleagues from both parties surround him. They press to shake his hand, pat him on the back, congratulate and praise him. No one watching this closing scene can have any doubts about what it signals. However they feel about the character of their president, and there's bipartisan condemnation of the president and deep anger toward him in that chamber, there's no way that two-thirds of them will convict and remove him from office.

Two weeks later they vote to acquit. It isn't even close. On the first article of impeachment, charging the president with committing perjury, ten Republican senators join all forty-five Democrats to vote for acquittal. On the second, charging obstruction of justice, five Republicans join all the Democrats in voting for acquittal. Thus, at 12:39 P.M., Thursday, February 11, with the pronouncement of the presiding chief justice of the United States that William Jefferson Clinton is "acquitted of the charges in the said articles," the trial ends. Impeachment proponents don't win a majority on either charge, to say nothing of coming close to achieving the necessary two-thirds of the Senate to remove the president.

*   *   *

On the day after it finally ends, Saturday, February 13, 1999, the *Economist* offers a brilliant postmortem on the entire sorry episode. Throughout the year the magazine has been harshly critical of the president's "grave and shameful" behavior, more than once calling for his resignation and saying that if he possessed the necessary moral character "the proper response of a man of honour would have been to resign his office."

Now, sketching the debris littering a national capital "mocked by the rest of the country," a capital "sunk in political and physical exhaustion," the magazine says, "It is hard to remember where it all began; only that, while it lasted, almost everyone did things and said things they wish they had not. As at the end of a drunken orgy, there is a need to clean up: sort through the wreckage, sweep the floor, throw open the windows, and repent."

It's not enough that none of this should ever have happened—or never should again, for "The end of this awful tale still leaves in place the elements that spawned it: a diminished presidency, a bitterly divided Congress, an over-mighty prosecutor, and a media pack that is proud to seek out scandal wherever it can. . . . Since Mr. Clinton would not go, his political enemies determined to root him out, using impeachment to try to overturn, by a sort of constitutional coup d'état, the result of two elections. This is not impeachment's purpose, and no one has emerged from the process untarnished."

As the magazine also points out, "Impeachment once devalued may be used this way again, not least by those who feel that vengeance and partisanship are more neatly countered by more of the same.

"Morning-after Washington also bears the scars of other excesses. The scandal saw a sort of Faustian pact between the special prosecutor, Ken Starr, and the press, whereby each seemed to feed and encourage the prurient appetites of the other. . . . Where the Internet dared, the old press followed. Rumour was published before it had been verified; the prosecutor's office sprang leaks for which no apology was offered; and Congress, when the time came, pushed titillating material immediately and unthinkingly into the public domain. This howling after sex stoked the Republican moralists, obscured the valid reasons for condemning this president, and made the public think one thing: this prosecution was unfair."

Brilliant as this analysis is, and it should stand as a definitive assessment, it doesn't examine (nor was it intended to) other large questions that loom after the formal impeachment stage is struck. Nor does it address their consequences for the future.

Is the Clinton impeachment debacle merely a momentary diversion for a society aglow in the best of times or does it signal something more seriously amiss at the core of the nation's character? Why does one great scandal, Watergate, produce a uniform feeling of pride in the way the nation's public institutions—representing prosecutors and politicians, judges and journalists—deal with it while a generation later the second great scandal leaves only profound disgust in its wake? Why, in short, is one celebrated for displaying the strength of the society, the other condemned for showing its fault lines, one for its heroic public figures, the other for its pygmies? What's responsible for so sharp a change in so relatively short a period?

The writer Neal Gabler finds that at least a partial, and plausible, answer lies in the intermingling of the cult of celebrity with the ever-expanding mass entertainment industry. "After decades of steady inundation," he writes, "we have come to think of nearly everything as entertainment, and we demand more and more outlets for it." He believes that "all of this makes for a marvelous symbiosis. Viewers think of these blockbuster news stories as entertainment, the cable networks think of them as inexpensive programming, and the talking heads seem to think of them as vehicles for their own celebrity. This last factor is the only way to explain why lawyers and pundits allow themselves to be recycled from O.J. to JonBenet to Monica."

Clearly, he's correct. In the Nineties, everything is being recycled from old films to old scandals to old performers. That's especially so when it comes to the talk-news operations. Instead of spending additional money on new programming, they use the same facilities to recycle the same news around the clock. If holding viewers means continually investing the newest story with a hothouse crisis atmosphere—well, it sells, it's cheaper, and it's entertainment. If attracting a figure associated with a great scandal— an Oliver North, a G. Gordon Liddy, a Johnnie Cochran, a Marcia Clark, a spinmeister from the Republican or Democratic PR attack camps—lends spice and attracts viewers, well, it's show business, not news. There's no such thing as shame, or shamelessness; infamy equates with success. Darva pleads virtue in marrying for love, then takes off her clothes for *Playboy* for half a million or more and hustles for her own TV show.

Who cares? It's all entertainment—and so, in this instance, is the impeachment of the president of the United States. Just another O.J. moment, on a grander scale, complete with crude cartoon characters—the chief justice presides at the trial wearing a robe he designed bedecked with bright orange chevrons on his sleeves that makes him look like the Lord High Executioner out of Gilbert and Sullivan and the House managers strut and strike theatrical airs as they deliver turgid rhetoric during their fifteen minutes of televised fame on the Senate floor. Impeachment becomes the latest Scandal Times spectacle, a momentary diversion, if that, for a public turned off by all things political.

That's appropriate for a post-Watergate era where each political disappointment breeds new disillusionment and each broken political promise generates new feelings of betrayal. No wonder Americans are becoming conditioned to disbelieve and are voting less and less and less. Leaders are killed, or self-destruct. There are no heroes. Everyone sells out, from sports stars to sports franchises. Loyalty, from corporate to collegiate, is a one-way street: Mine, not yours. Cash in and cash out—that's how they do it in the dot-com world of Silicon Valley. It's nonsense to stick around building a company, or a career with this or that firm; everyone knows no one stays in place any more. Beware of committing yourself, personally or professionally. Take care of yourself first, because no one else will. Selfishness becomes a necessary trait.

It's the David Letterman syndrome writ large: Make fun of everything; identify with nothing. It's cool and hip to feel that way. And acting cool, hip—detached, apart, disconnected—are the characteristics of the times. Nothing distinguishes fact from fiction. That's all right, for there is no such thing as "truth." It's all relative. Your myth is my fact. Your right is my wrong. The media nudges the public consensus in one direction, then in another. Twenty years earlier most Americans could define themselves as Democrats or Republicans, liberals or conservatives, with increasing numbers declaring themselves, understandably, Independents, ideologically and politically. A generation later fewer and fewer people can fix themselves anywhere in the political spectrum—or want to. From a time when most could apply their own labels, now almost no one does. That's especially so when it comes to politics. Just a game. Meaningless. As for government, it's irrelevant. Besides, the country's doing well, thank you. Witness the endless boom. It's the political system that keeps failing.

Out of this, inevitably, arises the Cynical Society.

Of course, cynicism is nothing but a defense against dashed hopes. It's armor, protection against feeling greater disappointment. So America becomes a nation that dons thick armor. In a time when idealism appears dormant, we protect ourselves from feeling further hurt through cynicism—we protect against it politically, we protect against it culturally, we protect against it in the family. We *have* to be cool, detached, apart. In what becomes a sick cycle, the more we react cynically, the more the media feeds what it perceives to be an appetite for more cynical material.

Impeachment provides only the latest example of the disconnect between politics and real life, between the machinations of the capital and the interests of the country. It reinforces the already-prevailing public belief that you can't expect anything better out of the political world, even though people still cling to the hope that they'll find it. As one of the people interviewed for this book put it, when the impeachment process was reaching its final act, "Everyone is debating is this the best we can get when we have this media madness going on? What normal person is ever going to run? Are there only going to be sociopaths, abused children, people fundamentally broken in some way who run? Who would take this on?"

· · ·

Nothing more powerfully validates the wisdom of the *Economist*'s assessment that impeachment's destructive elements are still in place—and still capable of sowing even more public discord and disgust—than the immediate aftermath of the president's acquittal. Far from putting the scandal behind it, the public finds it erupting in more lurid and disturbing fashion.

Even in the final days before the Senate vote, Republicans, facing defeat, are proposing new subjects for investigation of the president through whispers about "other women" and other supposed sexual transgressions by the president vastly more serious than the clumsy consensual Monica affair. Only days after his acquittal, the first of these long-rumored sexual offenses bursts into the mainstream press, courtesy of a *Wall Street Journal* editorial page article telling of "Jane Doe # 5," Juanita Broadderick, whom allegedly Clinton raped and beat a generation ago in an Arkansas hotel room.

The allegation was never reported to police, charges were never preferred, there are no witnesses to corroborate her claims, and she earlier signed an affidavit denying it. It's gathered by scandal hunters in the Paula Jones camp and, later, the prosecutor's, but never included in the formal impeachment referral to the Congress. There it now stands for all the world to see, another scandal first: a president accused of rape in an account that cannot now—and never will be—proved. But it leaves more damage and opens even more vicious rumor-mongering.

Nor is this the end of what becomes a new media/political rerun of the old scandal. Monica's moment on the public stage is far from over. To the intense beating of public relations tom-toms, new scandal tidbits are leaked about her forthcoming two-hour prime-time network interview with Barbara Walters, timed to appear simultaneously with publication of her hastily produced "memoir" written by a British biographer of the royals. All this triggers more cable-TV scenes of the intern, the beret, the rope line, and the wagging finger denial of sex with "that woman." Scandal heaven revisited. Now Monica gets her very own fifteen minutes, or more accurately, gets to add yet another fifteen-minute segment to her accumulating tally of celebrity fame.

At the same time, the endless investigation lumbers on, attracting further press and public attention. From the prosecutor's side are strong hints that both the president and the first lady might yet be indicted, and this while the final formal public accounting has yet to be made on the results of years of investigations into Whitewater, Travelgate, Filegate. If this catalog isn't enough to feed the voracious scandal appetite, the attorney general is now investigating the prosecutor for possible abuses of the power of his office and the prosecutor's chief spokesman has been charged with criminal contempt of court, accused of leaking secret information to the press.

The scandal lives on. As Francis X. Clines writes in the *New York Times,* "The very idea of remilking the mass entertainment aspect of the debilitating Clinton-Lewinsky-Starr saga seemed the year's ultimate obscenity," but within two weeks of the impeachment acquittal it keeps flaring up. "For sheer durability," he writes, "the story is beginning to resemble a toxic waste dump of fetid ingredients and methane energies. The news media cannot afford not to see the story through. The President's enemies, naturally, stand ready to continue feeding the pile. And then there are the sorry deposits offered by President Clinton himself, whose search for historic legacy seems increasingly overwhelmed by vapors from the past."

To this are added embittered public comments of conservative political leaders enraged and baffled by the failure of the Republican-controlled Senate to find the president guilty. In their frustration and fury, and their search for scapegoats, they begin turning on each other—and blaming the American public.

In an extraordinary mailing to hundreds of thousands of members of his influential New Right organization, the Free Congress Foundation, Paul M. Weyrich writes that the president's acquittal demonstrates American "moral standards have collapsed." Weyrich, who coined the term "Moral Majority" and suggested that the Reverend Jerry Falwell adopt it for Falwell's politically powerful conservative religious organization, now says, "I no longer believe that there is a moral majority" and "I do not believe that a majority of Americans actually shares our values." He's "not suggesting that we all become Amish and move to Idaho," Weyrich tells his supporters, but he believes "we have to look at what we can do to separate ourselves from this hostile culture."

In similar dispirited vein, he adds: "What steps can we take to make sure that we and our children are not infected? We need some sort of quarantine." He concedes that the conservative cause has "probably lost the culture war," saying the cultural collapse implicit in the president's survival was "so great that it simply overwhelms politics."

William J. Bennett, the former Reagan aide who has made a reputation—and money—from his writings and lectures about morality and virtue, places the blame for the president's acquittal on the American people. Ordinary Americans, he tells Richard L. Berke of the *New York Times,* "are complicit in his [Clinton's] corruption." "I will not defend the public," Bennett says. "Absolutely not. If people want to pander to the public and say they're right, they can. But they're not right on this one." He also rejects "the canard that conservatives are putting out that we didn't make our case." That accusation, perhaps more than any other factor, infuriates the conservatives. They are convinced they made their case against an immoral president. As Bennett remarks: "A lot of conservatives said, 'Wait, there's another woman. There's another affidavit. It will all turn around.' . . . But at each point they found out more about Clinton, his approvals went up."

Berke also revealingly quotes New Hampshire's staunchly conservative Republican senator Robert C. Smith lashing out at the public: "The President's acquittal is a sad commentary on the prevailing values in America today." Berke describes the senator as sneering at polls that showed the public consistently supporting the president and opposing his impeachment for week after week, month after month. "My wife likes to say they must be polling people coming out of Hooters [where waitresses are topless] on Saturday night," the senator says.

Blame the mess in Washington on the people, not the politicians.

· · ·

Years will pass before all the consequences of the collective "Clinton scandals" and his impeachment are known, but some of the effects are already evident.

Foremost is what this episode says about the American people. Their response, as already noted, provides the only heartening aspect in what otherwise is a dark chapter in the American story. They are its only heroes.

Alone among all the players the people display a reassuring maturity and measured common sense in their judgments. Throughout, they manage to balance revulsion at the near-daily revelations of their president's personal conduct and their appreciation of his ability to perform his public job. In so doing, they demonstrate an appreciation for maintaining a proper line between the public and private aspects of an official's life, for distinguishing between when private conduct becomes dangerous to the public good and when personal frailty doesn't cross the line into dangerous criminal behavior. Despite the sustained barrage of intimate, demeaning details about a president's personal life, they reject the offerings of the scandal merchants as more threatening to national stability than the deplorable misconduct of their president. They see the president's principal opponents for what they are: at best, mean, small-minded, vengeful, true believers; at worst, gleeful character assassins determined to gain power at any cost. If the president demonstrates an astonishing resilience under endless assault, even more so do the American people. They may not achieve what F. Scott Fitzgerald calls a "wise and tragic sense of life," but their tempered reaction comes close to that state.

Only one other positive aspect emerges from this tale, and it's the most important of all. That is the reaffirmation of the wisdom of the framers of the American Constitution. Those distant young men, seeming so antique in their powdered wigs, breeches, and buckle shoes, so removed from the lives of millennial Americans, turn out to have been remarkably prescient in fashioning a political system that performs *exactly* as they intended. Their great concerns over abuses of power led them to create a structure that permits the political factions to clash in heated ideological combat but also a structure that provides adequate checks and balances against overturning the will of the people and imposing the rule of a zealous minority.

Both the reaction of the people and the reaffirmation of the enduring structural strength of the political system are promising for the future. But these virtues don't minimize other problems—and questions—that keep resurfacing amid the aftershocks of scandal times. Here are some of them:

Have the repeated attacks weakened the presidency to a point where it stands in danger of being unable to provide necessary strong leadership for an increasingly fragmented society? Have the people, for all their good sense and maturity, been so disillusioned by the falsehoods and deceits emanating from a succession of presidents and members of Congress that they no longer believe in their leaders, their political process, and their government? Have they become so alienated, if not so cynical, that they will withdraw even more from participating in their nation's public life? Have the negative campaign operatives been emboldened by their ability to win electoral victories by raising the greatest sums from private interests and then attacking the character of their opponents? Have the forces that produced the zest for launching endless investigations been checked, or will they continue even more to divide and divert the society? Have the leaders of the news media been able to withstand the pressure to provide less serious information and capitulate to the apparent commercial success generated by those who offer more scandalous spectacles in the guise of news-as-entertainment?

Finally, questions arise that become even more difficult as America enters what appears to be the post-boom era. During the great economic expansion, it was understandable, if foolish, for Americans to be so lulled by the material successes of the age, so convinced that they were living amid endless prosperity and opportunity, that none of these questions seemed important, especially questions about the workings of government and the clash of political forces. In those diverting best of times, many became so comfortable that they began thinking they alone among the peoples of the world were crisis-free. They alone were immune to the problems afflicting others. So entranced were they with the undeniable successes of the private sector that they no longer saw the relevance of the public sector or appreciated how both private and public sector need each other for a successful society. How will they deal with the aftermath of the already sharp decline—or worse—that America was experiencing as the Clinton age ended? What will their response be to a crisis, global or domestic, that suddenly erupts and requires mobilization of public and private resources of the nation?

Clearly, the scandal era has turned people away from public life. One of the most evident consequences of the scandals that mark the Clinton years is the almost infinite widening of the chasm between the capital and the country. If the purposes of politics and government are really no longer relevant to increasing numbers of Americans, none of this matters. But that, of course, is not the case. If anything, Americans of the new millennium face a host of ever-more-complex issues that are central to the future. Every single one of these will require a public solution, or, at least, a public-private consensus and partnership.

Some of the issues that must be addressed are the consequences of our newest technologies, others are the legacy of older technologies, and still others are as old as life itself. Among those needing to be addressed are the risks and benefits of genetic engineering; the limits society might choose to place on what science is able to do; the availability and allocation of health care; the threat of new plagues like AIDS spreading fearsomely through Africa; the protection of privacy; the role of the media in a free society; the dangers of weapons of mass destruction, disease, overpopulation, and the degradation of the environment; the inequitable distribution of wealth and resources; the tensions between corporate and public interests; the allocation of dwindling and/or renewable natural resources; the role (and quality) of education in a democracy...

Every one of these challenges existed before the scandal era. Consideration of every one was diverted by the obsession with scandal. It may well be that historians will look back on those times and find the greatest legacy was one of needless neglect of major issues, of drift and diversion and division that postponed serious consideration of serious questions. Certainly, the president's actions contributed strongly to the public's failure to address these and other questions. By his conduct Clinton weakened the office of the president, created greater public disillusionment with public officials, and wasted a critical year that might have been spent dealing with problems. Even worse, this neglect occurred at a time of unparalleled peace and prosperity when a second-term president had a rare historical opportunity to provide significant long-term leadership. That opportunity was squandered.

BOOK
FOUR

Millennial
Times

The People

*"What is missing in people's lives? Why do they respond to
a Jerry Springer? To a Rick and Darva? Why do people get
more excited about baseball and football than they do their
own lives? Why are people passionate about conflict? Why?
Why war? War! Why do people need conflict artificially created?
Isn't there enough conflict in their lives? Conflict sells. Why?
Is it because we're angry? Helpless?"*

In the fall of 1998, even as the Monica/impeachment scandal
was nearing its critical moments, the *Wall Street Journal* dispatched an enterprising reporter, Douglas A. Blackmon, to explore a different kind of question about the character and tastes of
the American people far away from their festering capital. The results of his search were disclosed in a bank of page-one headlines:

Metamorphosis

Forget the Stereotype:
America is Becoming
A Nation of Culture

The U.S., Wealthy, Worldly
And Wired, Undergoes
'An Explosive Change'

'Madame Butterfly' in Waco

These were followed by a sparkling lead written from the dateline of Alpena, Arkansas.

> At a crossroads Texaco station, where a strip of pavement curving through the Ozark Mountains meets Highway 62 on the way to Yellville, a flashing yellow sign beckons weary travelers.
>
> "Two hot dogs $1," reads the first line. "Propane Gas," the second.
>
> And the third, all in red capital letters: "CAPPUCCINO."

The reporter then sketched what he described as "a revolutionary transformation of American culture," a transformation in which "Americans are buying serious books in dizzying numbers," scores of new regional theaters and opera companies are proliferating, symphony orchestras are being founded at a historic rate, and citing figures from the National Endowment for the Arts, the percentage of Americans attending the performing arts is rising dramatically. Nor is this all of the transformation: cinemagoers are displaying heightened tastes for classic drama and offbeat storylines, elegant restaurants featuring a wide range of tastes are flourishing in areas dominated in past years by fast-food outlets, small-town grocery stores stock the finest imported wines and beers, and sprouting in hamlets as well as cities are coffee bars offering a staggering array of choices.

While the reporter conceded that "some of these new American tastes are merely signs of conspicuous consumption in a gluttonous era," his larger conclusion correctly captured the positive side of the American people and their culture at the millennium: "Historic levels of wealth, educational attainment and cultural exposure have converged over the past decade in such a way that the lowest common denominator of American culture is rising rapidly. Hardly any place is remote as it once was. Contrary to the wails of many cultural critics, middle-class, mainstream Americans have become, simply put, sophisticated."

This analysis clashes with much of the perceived wisdom about the culture and stands in sharp contrast to many negative critiques offered by intellectuals and critics. They despair at what they see as clear signs of the decline of American culture.

One of the most telling of those critiques came from Daniel Bell, the distinguished sociologist. In a celebrated lament early in the Nineties, Bell said he believed the burst of creativity that enriched all fields of American culture—in literature, music, painting, the sciences—following the huge influx of Europeans during the Nazi period had dissipated. That outpouring of talent helped America develop a cosmopolitanism it lacked. More than half a century later, as Bell saw it, the impulses sparking that cultural renaissance were largely spent. In their place was an American culture marked by increasing provincialism and intellectual exhaustion, producing an America "changing in far different ways than it has before." His analysis led Bell to conclude: "We may be at the end of old ideologies and old history, but there are no unified sets of belief to take their place, only the splintering of cultures and political fragmentation."

It was a scathing indictment and uncomfortably correct in delineating the splintering of the culture and the increasing political fragmentation, but it was far from the whole story.

True, no one looking closely at the contemporary American scene could fail to observe a society in the process of further fragmentation. Long past was the America of the idealized melting pot: *e pluribus unum,* "out of many, one." The days of talking about *the* American people, *the* American mind, *the* American mood as easily identifiable and representing one people, indivisible, united for the purpose of achieving liberty and justice for all, were also long gone—if those days ever existed outside of grade school myths about Washington and his cherry tree and political leaders never telling lies. There was no one America, but many Americas. Hardly a new condition, either.

America has always been divided among its many parts—between urban and rural, rich and poor, master and slave, aristocrat and indentured servant, WASP and immigrant, owner and worker, farmer and mechanic—and now, at the millennium, it was even more so. Divisions separating society along lines of race, ethnicity, gender, sexual orientation, religion, and ideology were solidifying. It required no special vision to see an America drawing farther apart and a society so immersed in materialism and individual comfort that it elevated selfishness into a national virtue, valuing private over public, profit over service. All these surely were among society's shortcomings, but the depiction of them was, at best, incomplete and in one sense wrong. Whatever the nation's many failings, it was not an intellectually exhausted America that raced pell-mell into the new century, nor an America whose creative talents were declining while its provincialism increased.

By any measure, Americans were less insular, more directly connected to the currents of the greater world through travel, electronic contacts, and higher education. Though others still disparaged American society, partly with justice and partly with envy, as loud and crude, Americans were infinitely more cosmopolitan in their tastes and their lifestyles. Whatever the elusive, probably indefinable "American culture" was, there could be no doubt that a distinct American culture existed—and a mass culture that represented more than mere materialism or such easily mocked symbols as Disney World, shopping malls, Big Macs, ubiquitous fast-food chains, and raucous rock concerts.

At century's end, Americans displayed vitality and creativity. In music and the arts and letters, and especially in film, their influence was transcendent.

Nor was this litany of success all that characterized the condition of the American people. Their recent record of righting wrongs provided one of the nation's most prideful chapters. Legal segregation, the last vestige of the shame of slavery, had been eliminated. The civil rights revolution that struck down racial barriers in voting, education, and public accommodations in the mid-1960s spread to other elements of the society.* In only a generation, women, gays, ethnic Americans, Native Americans, the handicapped, the mentally ill, all of whom suffered from historic

---

*But it didn't eliminate such barriers, as the results of the 2000 presidential election made shockingly clear.

patterns of discrimination, won gains in securing greater rights and opportunities. Along with other legal victories that further expanded individual liberties and rights, came a new public openness and frankness in discussing such hitherto taboo subjects as sexual behavior and orientation, abusive families, alcoholism, and mental illness. The result was a more tolerant, self-correcting America, and an America whose people enjoyed numerous advantages over the generations that came before them.

The change in life expectancy alone was extraordinary: The average life span had increased from forty-seven in 1900 to seventy-six in 1999. If those trends continue, demographers predict that within another half century life expectancy will increase so greatly that numbers of people can expect to live to the astonishing age of 150. Already it's no longer remarkable for people to live to a hundred and beyond. In the first decade of the twentieth century, only one in every five Americans lived to celebrate a sixty-fifth birthday. At century's end, at least seven out of ten were achieving that milestone. Put another way, in 1900, no matter how rich you were, in the United States you could not expect to live longer than fifty-five. In 1999, no matter how poor you were—unless you were an intravenous drug user—you could reasonably expect to live at least to the age of seventy, and probably longer.

Older Americans were also remarkably healthier than those of past generations. Surveys conducted by the MacArthur Foundation at the end of the Nineties revealed these statistics: 89 percent of those between the ages of sixty-five and seventy-four reported no significant physical or mental disabilities. Seventy-three percent of those between the ages of seventy-five and eighty-four experienced the same good health. Even 40 percent of those older than eighty-five said they enjoyed the same.

Behind these figures documenting markedly greater longevity for Americans lie other best-and-worst-of-times kinds of questions. While increasing longevity holds the prospect for people to live a full life and remain healthy until very old age, at the same time society faces immense, largely unaddressed challenges about how to accommodate and care for more and more of its older citizens.

American culture, its values and its attitudes, are being profoundly shaped by these demographic changes. Only one among the many challenges leaps out as obviously critical. As people routinely live longer, they know they will need to be on Social Security retirement income longer, and will also require medical benefits like Medicare for an equally long period. These realities explain in great part why those two government benefit programs are so controversial and generate such doubts about whether there will be enough money to sustain an increasingly aging population.

In the early years of Social Security, large numbers of workers paid payroll taxes to support each person receiving benefits—in 1945, for example, there were approximately 42 workers per beneficiary. By 1995, that number had shrunk to 8.6 workers per beneficiary. Three years later, the number had been sliced by almost a third to 3.4 workers per beneficiary. In just another generation, in the year 2030, it's estimated there will be only 2 workers per beneficiary.

These are explosive demographic/economic figures. Given these estimates and current tax and benefit provisions, revenues will fall short of covering benefits by around the year 2012. Social Security trust fund reserves will make up the difference until the year 2029; then, current revenues are projected to cover only about 75 percent of benefits. This shortfall would be hugely helped by removing the income cap on Social Security taxes and taxing capital gains at a higher rate, but the prospect of that reform being adopted is remote given the political climate that currently exists.

Another stark statistic underscores the problems facing an aging America: As many as half of all women over sixty-five in the United States and a third of all men will spend some of their remaining years in a nursing home. The situation will grow worse, as the *New York Times* comments, as new assisted living communities draw off the more able and affluent retirees, and as increased federal funding allows others to be supported by health care visits at home. That leaves nursing homes tending to the sickest and most dependent of old citizens.

Of that concentration of 1.6 million older Americans in nursing homes at century's end, less than 8 percent can bathe, dress, go to the bathroom, move about, and feed themselves without help. As the *New York Times* comments editorially, "Bedded down usually two to a room, they are cared for at an average charge of about $50,000 per year—$94,000 in New York City—enough to impoverish most of them. About two-thirds of them end up being welfare cases." These grim statistics underscore the problems facing an aging population.

The paper adds, with the kind of proper indignation that should—but does not—produce a public outcry: "It is a disgraceful dose of misery and death for a generation that survive[d] the Great Depression and World War II to be treated like this."

All of these demographic changes will place tremendous new pressures on government and promise to strain the entire social fabric. There's no way of escaping these challenges; they are inevitable, and they grow more serious and more complicated as those curves on the life expectancy graphs keep rising and as the costs of sustaining life through new developments of medical science and technology keep multiplying.

So, as greater physical well-being enhances people's lives, it also engenders new concern that greater struggles lie ahead, struggles that will pit one generation against another in a bitter battle over obtaining scarce public funds. Together, increased longevity and age-old parsimony add up to a picture of an American people simultaneously exhibiting their best and worst traits—a confounding condition befitting a society both great and gross, a society in which new tolerance and new sophistication coexist with self-centered tastelessness and vulgar exhibitionism.

· · ·

In America, everyone is a cultural critic, everyone thinks he or she is entitled to an opinion—and to express it, however well- or ill-informed it may be. The quintessential American expression is the Bronx cheer: loud, spontaneous, judgmental, absolutely assured in the correctness of its derisive verdict.

By the Nineties, the two central competing pillars that defined the society of a generation ago, the Establishment and the Counterculture, had splintered. Out of them emerged a new pluralistic, multicultural, and ever-more-fragmented American culture. Diversity, not commonality, became the new cultural ideal. Along with these changes came heated cultural debates that further divided the society. On one side were the critics who feared the dangers of "defining deviancy down," "dumbing down the culture," and "pushing the envelope" beyond all prior norms of accepted private and public behavior. On the other were those who argued for greater individualism and cultural separation. "Let it all hang out" and "do your own thing" became their cultural mantras.

"There are no dark corners left in America," said Don McLean, singer and songwriter, at decade's end. In 1971 his hit, "American Pie," brilliantly captured the public disillusionment and alienation in the wake of the assassinations and riots erupting against the backdrop of the civil rights revolution and the Vietnam War. "Every square inch is spoken for and spoken to, and as a result the American public is psyched out and homogenous," McLean said. "We've been simplified. Today, self-esteem is more important than Shakespeare. Ideas have been substituted with slogans and information manipulation. As technology moves us forward technologically, it moves us backward culturally."

Throughout the Nineties, two of the sharpest cultural critics were a pair of *New York Times* columnists, Maureen Dowd and Frank Rich, who provided a stream of biting commentary about the excesses of the period. Historians seeking to understand America during the boom will find their writings illuminating. In one of her columns, Dowd offers as good a description of the era as exists: "The 90's are the 80's without moral disgust."

So they were. As she also wrote, "In our pampered, narcissistic culture, vices are affectations, or advertisements for oneself, served up with pretentious argot about blending, aging and flavoring, about lineages of seed and vine and bean.

"People used to be sinful as a way of having fun. Now they are sinful as a way to accessorize. Cigar candles. Wine throw pillows. Martini bathrobes. Humidor ties. And for golfers, a stainless steel tee with cigar rest attached for only $750."

No one wrote more delightfully and insightfully about the single most significant change in American society—the role and status of women—than Dowd. In one column, titled "Cowboy

Feminism," written during the Year of Monica, she addressed the old Freudian question of what women want, then provided an answer: "When we were gatherers, we wanted hunters. Then we wanted clean-cut doctors and lawyers instead of hirsute alphas in loin cloths. Then came the 60's. We wanted equal pay for equal work, co-partners and zipless adventures. Then we wanted Mr. Moms, male nurturers to share the burden of food-shopping and baby-diapering (alpha Alan Aldas). And now that we have feminized society and domesticated our men, now that we have forced them to help pick Martha Stewart stencils and watch Olympic ice dancing, guess what we *really* want. Cowboys. Yup. Brawny, rippling, rawhide hunks who don't know a Flaubert from a flambé, or King Lear from King Kong. The Cowboy Option is the perfect coda for a year when feminists committed hara-kiri to defend their President, and when Hillary Clinton was unmasked as a counterfeit feminist after she let her man step all over her."

All this and more was taking place against a background, as her colleague Frank Rich put it, of an America waging cultural wars in a time of a cultureless culture. In a number of incisive commentaries, Rich wrote of the excesses—and the entertainment—being produced by what he called "the mediathon" constantly at work during the Clinton years. No sooner had Clinton left the stage, than Rich, tongue-in-cheek, reminded readers how much they would miss the profligate president. "Rational—and irrational—people may differ over the merits of Bill Clinton's presidency," he wrote, "but few can dispute that he was without peer as our entertainer in chief. He turned the whole citizenry, regardless of ideology or demographic, into drama fiends. When he left office, the cliché had it that late-night comics, after eight years of sure-fire punch lines, would miss him the most. Now it turns out we're all junkies for the fast-paced, round-the-clock theatrics that defined the Clinton years, and we're desperate for a fix, any fix, to fill the cultural vacuum he left behind." It was, as Rich said, an age in which Americans had become "more addicted to combative entertainment than ever."

Certainly for displays of overindulgence and tasteless materialism, the Clinton years had few equals in previous decades. Everywhere you looked, yet more signs of extravagance and wealth appeared. Open your Sunday *New York Times* and you find a three-quarter-page Saks Fifth Avenue ad promoting a Bill Blass collection showing a model wearing an embroidered silk gown that to the untutored eye looks ordinary. Price: $11,800.

Thumb through a summer issue of *Vogue,* and you spot an article tucked far back amid the lavish advertisements that begins: "In an era when Jackie O's sister can get $500,000 for a summer rental in the Hamptons, when a plate of risotto in a restaurant costs $100, and John D. Rockefeller Jr.'s old apartment goes for $35 million, it was bound to happen. May as well tell the sun not to shine, the stock market not to gyrate, champagne not to bubble. Somebody, somewhere, was going to break the $500 price barrier for a face cream."

Of course they did: In the basement of Fifth Avenue's Bergdorf Goodman (now called a "new level of beauty," since "basement" no longer is politically correct), Kanebo Sensai's La Crème goes on sale for $500 for 1.36 ounces (plus $41.25 for tax). As *Vogue* adds, in a bit of writing that in itself reflects the prevailing emphasis on overstatement: "In the world of skin care, the arrival of what might be history's most expensive cream sent ripples of envy, outrage, and self-questioning throughout the land." Sure it did.

Shattered were boundaries of good taste—the term itself had virtually ceased to exist, and with good reason. Cashing in commercially on celebrity, on tragedy, on infamy were so commonplace, so expected, that they barely attracted any notice.

No sooner had Princess Di died in that high-speed car crash in a Paris tunnel than hucksters were racing to capitalize on the event. Immediately offered for sale was a new line of Princess Di lingerie, featuring "Lady Di" undies and perfumes to reflect "her life and beauty." For the right price, you, too, can possess panties for a princess. If that doesn't satisfy your longing for status, there's "The Princess Beanie Baby," a purple teddy bear with a white rose emblem.

Naturally, the death of John F. Kennedy Jr. in that airplane crash off Martha's Vineyard produced a similar wave of hucksterism. During the weeklong reporting about his flight and death, an advertisement appeared on the Internet under the title GOODBYE JOHN-JOHN!

Offered for auction was Item #133123483, with the commercial come-on: "What better way to show your affection to this great American than by displaying this beautiful tribute in your home! Produced in 1964—just weeks after the death of John F. Kennedy (our martyred President)—this touching ceramic replica of John-John (as we affectionately knew him) was issued in a limited edition. . . . It seems appropriate that this loving tribute now serve another important purpose—that of helping to celebrate the life of this much-loved lad! The figure is white-glazed ceramic (unmarked) and stands approximately 7″ tall (base is about 3″)."

What price it fetched is unknown, and unimportant.

When Monica appeared on her prime-time interview with Barbara Walters, the corporate sponsors demonstrated their taste by airing the following commercials: Victoria's Secret lingerie; Oral-B Deluxe toothbrush; Burger King ad with musical number, "It's My Party and I'll Cry If I Want To"; a promo for the movie *Cleopatra* that included the voice-over: "When she was only twenty, she seduced the most powerful leader in the world"; a Maytag washing machine ad that boasted of its product, "It actually has the power to remove stains."

In popular music, music videos, TV sitcoms, soap operas, and talk shows the operative sales pitch was anything goes—and that means *anything*. In an aging America already obsessed with youth and sex, American mass culture, and its cultural hucksters, smashed what few remaining taboos and barriers were left.

On TV, both network and cable, on radio, and especially on the Internet, the notion that anything should be censored for reasons of taste became ludicrous in view of the fare being offered. A Rip Van Winkle reawakening from a slumber not of twenty but of only a few years into the America of the new millennium would be astounded at what he saw and heard. From Jerry Springer's catfights, as the *New York Times* put it in an article addressing the remarkable changes, to Howard Stern's genitalia gags, *South Park*'s scatological cartoon children, *Dawson Creek*'s libidinous children, and the anorexic Ally McBeal salivating over nude male models, there were no limits to what could be portrayed—or said.

In the rush to capitalize on sex in its most explicit form, no one could escape the reach of the sex marketers. Some cloaked their message in pseudo-psychological jargon. The Erospirit Research Institute in Oakland, California, "which explores and teaches the connection between sex and spirituality through video and other media," claimed to employ sex therapists trained in the art of "erotic genital massage" who also taught the benefits of "ritual masturbation."

Presumably not so pleasurable, but no less exciting, were the proliferating computer games, many of them celebrating violence and appealing to young Americans. One "cool" game, targeted at white suburban youths, was called Postal. The player portrays a deranged psychotic loser who wanders around shooting innocent people at random. Great fun.

The erosion of standards in language was equally astonishing, considering the more recent prim past. Kurt Andersen, one of the sharpest chroniclers of the culture, who wrote a fine novel about the changes, *Turn of the Century,* traced the evolution of what at first was the shocking use of the word "suck" into the American cultural mainstream. That occurred in 1993 when the cartoon characters, Beavis and Butthead, premiered on MTV and began saying, "This sucks," over and over.

"What was too coarse for polite conversation a few years ago has become the mainstream's vernacular," he wrote in the *New Yorker* toward decade's end, describing how the term "everything sucks" now turns up in the pages of such "respectable" magazines as *Fortune* and *GQ* and is used regularly by DJs across the country. "The word 'fuck' recently appeared in a piece of fiction excerpted on the Web site of the *New York Times*—the *New York Times!*" he went on. "The *Wall Street Journal* reported that, according to Howard Kurtz's new book on the White House propaganda machine, the president's press secretary described the first lady to reporters as 'the mummy [Clinton's] been fucking.'" The most talked-about new plays, he said, were *Art,* "which is about art, and '*Shopping and Fucking,*' which is about what you'd expect. On *60 Minutes* a few weeks ago, Mike Wallace, interviewing Sally Quinn about the Clinton scandals, used the word 'pussy,' but the show bleeped him—an interesting, awkward, realtime moment in the transition to the new explicitness."

Andersen wrestles with what all this means for an America experiencing bewildering change in the midst of a rapid cultural transition, but transition from what to what? His answer strikes just the right note of concern for aberrant behavior and tolerant amusement for wretched excess. "What is the appropriate response to the epidemic of stark obviousness, high-concept literalism, rude candor?" he asks. "Is it thumbs up or thumbs down? Neither. Some of it is refreshing, some of it is appalling; a lot of it is amusing; most of it is just the American way. By definition, therefore, it doesn't totally suck."

· · ·

Our millennial Rip Van Winkle would not be surprised at one aspect of the American people. More than ever, they are distinguished by the casualness of their dress and the informality of their manner.

In a sense, America is a copy-cat culture, with its people striving to look and act like the people they see on their screens. This was always the case in the past when Americans emulated the dress and the manner of the Hollywood stars. It seems even more so in the era of mass media entertainment and celebrity culture. Now everyone emulates everyone else. Stars dress down to look like "the people." The people try to look like stars in their studiedly casual, and expensive, fashion. If that means splurging for designer jeans and blouses and sports shirts—well, hey, it's the fashion.

Look at the people that come on *Letterman* and *Leno,* says Norman Lear, the TV producer who has as sharp an eye for American cultural changes as anyone. The stars and the starlets used to dress up for those appearances in shirts and ties and dresses. Now they dress down to be one of the audience, as does the president and the presidential candidates, thereby diminishing the office—a condition the new millennial president, "Dubya," attempted to change when he issued White House dress codes mandating men to wear ties whenever visiting the Oval Office. They all want to be seen as one of the group, one of the crowd. Their conversation is all about *their* good life, *their* media celebrity lives: Where they're eating. What they're doing. To whom they're married.

The people watching, who are not rising on the economic tide, are also influenced by them. "As they go to sleep," Lear says, "they've seen three dozen commercials in the same hour showing how people dress, how they smell, how they bathe. They see commercials filled with happy people running down the beach with a beautiful girl, dancing in the moonlight in a seven-hundred-dollar tuxedo, patting each other on the back in the bar, having such a good time. The guy watching, who rolls over and goes to sleep, wants to be like them, but he feels more frustrated, resentful, because he can't afford to send his kid to college, he's caught in a squeeze, he's overextended with debt, his safety margin is razor thin, he's worried about his company's merger."

A similar kind of hunger to emulate cultural models accompanied by a similar feeling of disconnection from society affects the young and their music.

David Geffen, who helped create the modern music industry, makes the point that the protest music, the music of the young, had changed radically by the end of the twentieth century. Now, for the most part, he says, people who buy rap records and gangsta rap records are white suburban teenagers. "White kids have always tried to look at what black kids are doing," he says. "They may not have wanted to go to school with them, or sit at the lunch counter with them, but they were interested in their culture. Now all these white kids in Beverly Hills have pants that sit on their hips and hats on backwards and listen to very hard violent gangsta rap music. They have a great affinity for the black culture, but it doesn't create a great affinity for black people. In fact, I think it does almost the opposite. It's frightening. So instead of bringing people together, it may in fact bring them farther apart."

· · ·

If a commonality to the culture exists it's in the homogenizing influence of the TV set, and that, of course, as so many critics agree, produces a culture appealing to the lowest common denominator. Yet even the tube no longer holds the power to bring together so many diverse elements of the society. As we've seen, with cable, with the explosion of new channels, with the Internet, with the sharp decline of network viewers, the mass television audience, too, is as fragmented as the society it seeks to reach.

Although television's imprint on the people may not be as pervasive as in the recent past, it still disproportionately influences the American people—how they act, how they dress, how they talk, perhaps even how they think.

As the veteran TV producer Ted Harbert says, "People are learning about how we treat each other as human beings from Jerry Springer instead of from our parents. Yes, there are shows that are very positive models. But, boy, America is famous for its ability to be suckered into low-brow entertainment—and there's so much of it in prime time, in daytime, all over the dial. The ratings of WWF Wrestling, for instance. I grew up with wrestling on TV. Now it's a scripted form. They have writers printing out the story lines for each of these wrestlers. It's like daytime soap opera. The WWF wrestling people sued one of their wrestlers because he refused to go along with the story line. He wanted to go out there and win, and they said, No, no, no, no. The plan is you're supposed to *lose* here. And he said, Forget it. They sued him because he'd signed a contract saying he'd do what he was told. I guess deep in their hearts people know that it's fake, that it's just entertainment. I think Jerry Springer has become wrestling with words. It's more acceptable to watch Jerry Springer and those other shows than be white trailer trash and go to the wrestling arena."

The more difficult question is why people are so taken with the Jerry Springer kind of programs—and whether these programs signal real changes in the American character and culture.

There's no definitive answer, but for a provocative insight into the colliding cultural forces at the millennium listen in on a conversation between one of the greatest cultural icons of recent decades, the songwriter and performer Carole King, and her husband, the film producer, Phil Robinson.

In the early Seventies, as one writer said of her, you couldn't turn on the radio without hearing a Carole King song. By then she had achieved international fame as songwriter, singer, Broadway musical performer. She was regarded by her admiring public as an "archetypal earth mother dressed in tattered jeans and gauzy shirt, perched atop a thoroughbred, riding through a field of wildflowers and prairie grass," her face as common "as macramé plant holders and shag carpeting." King was a true megastar. Such luminaries as the Beatles, the Monkees, and the Everly Brothers helped popularize her songs, while her own albums, recorded in her distinct singing style, became instantly popular; one of them sold more than 20 million copies worldwide.

Now, during an animated conversation one night, King and Robinson are talking about the daily dose of violence and aberrant behavior filling the television screens. "What are people missing in life that they must come to this?" King asks. When no immediate answer is forthcoming, she asks again: "What is the answer to that?"

Her husband replies. "We've gotten away from the basics of life, family and community and values," Phil Robinson says. "That sounds very sort of right wing. Well, anyway, the right wing has co-opted it."

Still not satisfied, Carole King says, in a burst of emotion, "What *is* missing in people's lives? Why do they need this violence? Why do they respond to a Jerry Springer? To a Rick and Darva? Why do people get more excited about baseball and football than they do their own lives. Why are people passionate about conflict? Why? Why war? War! Why do people need conflict artificially created? Isn't there enough conflict in their lives? Conflict sells. Why? Is it because we're angry? Helpless?"

Phil picks up her question about the disgraceful Jerry Springer TV spectacles in which guests are chosen to reveal their most sordid personal affairs, their betrayals, their embarrassing personal actions, and their abusive behavior before supposedly unsuspecting family members or lovers sitting beside them on the stage. They are subjected to obscene insults and physical assaults until show bouncers rush to stop the fighting. As the scene erupts in violent displays—hurling of chairs, throwing of punches, even guests tackling each other—the audience, as if on cue, cheers loudly, "*Jerr-y, Jerr-y, Jerr-y.*"

This deliberate pandering to violence and the worst instincts in people is not limited to the Springer show. It is a staple of Nineties talk-shows. In the mid-Nineties, TV's pursuit of sensational revelations led to a highly publicized murder of a gay man, Scott Amedure, who was killed after he revealed he had a secret homosexual crush on another show guest during the taping of the *Jenny Jones Show.* Five years later a Jerry Springer guest was murdered after appearing along with her former husband and his new wife on an episode entitled "Secret Mistresses Confronted." The ex-husband and his wife fled after the crime and were sought by the police on homicide charges.

Part of the reason people watch these programs, Robinson says, is because it makes them think, "'Bad as my life is, it's not as fucked-up as their lives.' You get to feel better about yourself and laugh. The question to me isn't why people watch that show. The question is, Why do people go on that show?"

"That's easy," Carole King says. "It's attention. They get attention to themselves."

It's also a prime example, as we've earlier seen, how market-driven forces dictate programming—what King calls "corporocracy." With an eye only on the bottom line, corporations are not personally accountable for anything they do, good or bad. "If Jerry Springer doesn't make money," she says, "he'd be off the air. If he does make money, those corporations will insist he be on the air."

· · ·

Millennial Headlines:

In the U.S. Nearly 1 in 10 Is Foreign Born

For First Time, Nuclear Families Drop Below 25% of Households

Vast Majority of Teen Mothers Are Unmarried

Black Teen Suicide Rate Increases Dramatically

Nation's Prison Population Climbs to Over 2 Million

In Prosperous Idaho, Social Spending Declines as Prison Spending Grows

Gap Between Rich and Poor Found
Substantially Wider

PHILANTHROPY: During the 1990s stock-market boom, proportionately fewer households gave to charity

New Machine Spies on Everyone.
The Snoopmaster 6000

. . .

As those headlines suggest, the United States enters the new century in the midst of perhaps the most explosive and complex change in its history, change that will test the very nature of its experiment in democracy, and change that will require greater public and private efforts to address the many imbalances and inequalities that are now reshaping the society. At the heart of these challenges lie fundamental changes in the composition of the citizenry, socially, racially, ethnically, and an accompanying dramatic acceleration in the rapid rate of change affecting the evolving structure of the American family. Taken together, these all pose fundamental questions about the ability—or willingness—of the society to both educate and provide economic and social opportunity for this new, immensely more diverse, American people. In the larger sense, these structural societal changes are infinitely more important than the public preoccupation with scandal, entertainment, and amassing of wealth, however temporary, that characterized much of the Nineties. No one of these elements stands alone; each affects the other, but here in admittedly too brief form are some of the most significant areas that require public attention and understanding.

Immigration. One of the clearest signs that America remains a dynamic society is the government's demographic data emerging out of the 2000 census that describes an explosion of new immigrants to the United States. New immigrants have always provided an infusion of energy and talent, but the rate of their entry into American society in the Nineties reveals one of the most significant societal changes in a century.

During the Nineties, the percentage of the nation's foreign-born population increased nearly four times as fast as that of the native-born population. Throughout that decade the foreign-born population grew by 27.1 percent; at century's end, it amounted to 10 percent of the overall U.S. population. In America's cities, the rise of the foreign-born population was even more dramatic. In New York at the millennium, for instance, 40 percent of the population was foreign born, and the rate of increase was accelerating. How great a demographic change was taking place can be seen in census figures of just a generation ago when 4.7 percent of the total U.S. population was foreign born. Nor was the rate of increase in the Nineties the most striking aspect of this demographic shift. The *nature* of America's foreign-born population was changing rapidly, bringing with it new challenges in how to breach the vast differences in educational levels of new, less affluent immigrants, increasing numbers of whom are Spanish-speaking, and other highly skilled immigrants like those drawn to such high-tech centers as Silicon Valley where they immediately assume a higher position on the American social ladder.

Unlike the flood of Europeans who, in the closing years of the nineteenth century, formed the classic "nation of nations" immigrant story, the greatest number of new immigrants in the Nineties were Hispanics arriving from South and Central America and what the Census Bureau defines as "Asian-Pacific Islanders." From mid-1990 through 1998, for example, the number of foreign-born Hispanics grew by 34 percent—in itself, a demographic revolution that is reshaping American culture and will affect the nation even more profoundly in years to come. At very least, this means the America of the new millennium will have to demonstrate that its democratic ideals of tolerance and equality of opportunity work as well in assimilating and educating this new polyglot culture as it did for the previous waves of immigrants.

The Family. The cheery, comforting image of the "typical American family," a Norman Rockwell portrait of smiling children gathered in safe and secure suburban living rooms with Mom and Dad before their Fifties TV sets, long since has passed into the realm of American myths. In its place is a dramatically new American family supplanting the much-studied old "nuclear" one. That old "traditional American family" consisted of married couples with children younger than eighteen. In 1960, such families represented 45 percent of all U.S. households. Twenty years later, in 1980, the figure had dropped sharply to 30.2 percent of all households. When the new Census 2000 demographic survey data about family composition was made public in mid-May 2001, an even greater shift was shown to have occurred. Only 23.5 percent of all U.S. households then consisted of married couples with children under eighteen. Cited by demographers as among causes for this truly dramatic change were such factors as both men and women delaying marriage and having children; and single-parent families increasing much more rapidly than those with married couples. As a result, and often popularly cited, America was experiencing a "growth in empty nests." Marriage itself was becoming rarer. During the Nineties, the number of unmarried couples in the United States had nearly doubled to 5.5 million couples in 2000 from 3.2 million in 1990.

The implications of these changes were profound. As Eric Schmitt noted in reporting on the latest census figures that spring in the *New York Times,* "the number of families headed by women who have children, which are typically poorer than two-parent families, grew nearly five times faster in the 1990s than the number of married couples with children, a trend that some family experts and demographers described today as disturbing."

Another problem he cited involved the stress these changes placed on the nation's already grievously burdened public school system. "With more communities having fewer households with families," he wrote, "public schools often face an increasingly difficult time gathering support for renovating aging buildings and investing in education all over."

These latest demographic statistics immediately produced a strong ideological reaction. Leading conservatives, who already were blaming the continued decline of the traditional American family on a moral breakdown in society, seized on the new data as evidence they were correct in their assessment. "We're losing;

there isn't any question about it," Bill Bennett, the prominent conservative commentator and former secretary of education in the Reagan administration, told Nicholas Kulish in the *Wall Street Journal.*

Yet another looming problem presented by the census data involved the nation's housing structure. "Just a few years ago," the *Wall Street Journal's* Patrick Barta wrote in his report on the new census findings, "economists and housing analysts were predicting a glut of homes as baby boomers began to settle into retirement and a much smaller generation took over the U.S. housing market." But now, he added, "in one of the unforeseen findings to emerge from the U.S. Census Bureau this year, new data suggest that demand for housing is actually rising faster than expected and could lead to shortages in some parts of the country. Housing advocates have been complaining about the amount of affordable housing for low-income Americans for years, but now the problem appears to be spreading to the middle class."

Teen Mothers, Black Youths. Here stands a classic good-news/bad-news story. Throughout the Nineties, the birthrate of babies born to teen mothers fell sharply, a positive development that eased some of the inevitable familial pressures on young mothers. The bad news is that the number of unwed mothers is rising alarmingly. Only a generation ago, unwed motherhood was still the exception in America. In 1980, most teen mothers were married. By the end of the Nineties, two-thirds of Hispanic and white teen mothers were unmarried. Figures for black teen mothers were both astonishing and appalling: *Ninety-five percent of black teen mothers were unmarried.* As the *Wall Street Journal* commented on these overall figures, "These bare facts have huge and horrific implications for society. Many of our seemingly entrenched social pathologies can be directly correlated to the marital status of the young mother. A teenager with a baby but no husband is three times more likely to end up on welfare than a teen mother who is married. One of the best predictors of whether a child will grow up in poverty, use drugs and turn to crime is whether or not his parents are married. (This is true no matter the age of the mother.)"

A related figure about infant mortality rates speaks volumes about the continuing disparity facing blacks in America: White infants are surviving at a rate twice that of black infants.

Yet another indication of increasing problems afflicting black Americans was a report by the Centers for Disease Control and Prevention made public in 1998 that showed the suicide rate for black teen-agers had risen sharply since the 1980s—and it was increasing, as the *Washington Post* reported in a page one report then, "at a pace much faster than that of white teen-agers." In 1980, for instance, the suicide rate for young white Americans was 157 percent greater than it was for young blacks. By 1998, the white suicide rate was 42 percent greater. No clear cause was cited for this change, though some sociologists and African-American scholars cited by the *Post* suggested a best-and-worst scenario: That the "rising prosperity and social integration for blacks over the last few decades" might also have caused blacks to distance themselves from family and church, thus leaving them "much more isolated now in times of crisis."

Drug Use; Crime; Prison. A surprising, and disturbing, statistic involves drug usage among high school seniors. Despite all the educational efforts warning of problems from drug usage, in the aftermath of the pervasive experimentation that began in the Sixties, by 1997 high school seniors were actually using marijuana and LSD at a greater rate than they were in 1975. Use of heroin had not increased significantly.

As for violent crime rates, they increased from 475.9 per 100,000 inhabitants in 1977 to 634.1 in 1996.

The figures for America's prison population, long the world's largest, grew even worse during the boom, rising to over 2 million at the end of 1999. In the decade of the Nineties, the nation's prison population swelled by a depressing 77 percent. By then, one in every 110 men in the U.S. had been sentenced to at least a year's confinement.

Once again, the statistics for blacks were grim. In the United States at the millennium, one in every eleven black males in his late twenties was serving a sentence of a year or more in either a federal or a state prison. Of all inmates serving prison sentences of at least a year, blacks accounted for nearly half—46 percent—of those in jail. The comparable figure for whites was 33 percent and for Hispanics, 18 percent. Because of harsher sentencing provisions, all of those incarcerated were serving longer time in jail.

These figures speak for themselves; they make a mockery of the "law-and-order" adherents who in numerous jurisdictions succeeded in spending more for prisons than for social or educational programs aimed at alleviating the causes of crime and violent behavior.

Rich and Poor and Philanthropy. As we have already seen in these pages, the great boom of the Nineties vastly widened the income gap between America's richest and poorest families. In figures compiled by two respected non-profit and nonpartisan Washington think tanks, the Center on Budget and Policy Priorities and the Economic Policy Institute, and made public early in 2000, earnings for the poorest fifth of American families rose less than 1 percent during the decade. At the same time, income for the richest fifth of U.S. families soared 15 percent. It hardly needs to be said that this provides more hard evidence that for all its benefits, the boom was leaving the nation's poor farther behind economically.

Nor, ideological commentary about the benefits of volunteerism to the contrary, did Americans in the Nineties bridge this wealth gap by contributing more to charitable organizations. "The sobering lesson of the 1990s," as a telling special report in the *Wall Street Journal* made clear after examining the state of philanthropy during the boom, was that proportionately fewer American households gave to charity. "During the unprecedented era of wealth creation of the past few years," Eileen Daspin wrote in the fall of 1998, "fund-raisers figured they had a platinum opportunity to persuade more people to give to their causes. But it has been a promise unfulfilled." As she reported, "in this period of prosperity, the proportion of households making charitable contributions actually has dropped, to 68.5% in 1995 from 71.1% in 1987, according to Independent Sector, a Washington, D.C. advocacy group.

What's more, those who do give have been giving proportionately less: an average of 1.6% of personal income in 1997, compared with 2.1% in 1967. Worse, now that that stock market is gyrating, and the economic and political outlook is uncertain, fund-raisers feel the opportunity has been irretrievably lost."

The Nineties, she quoted Georgetown University's director of development, Michael Goodwin, as saying, have been "the best and worse of times in fund-raising." Best, because of such creation of wealth in so short a time span. Worst, because those who bene-fited the most from the great expansion had not shared as much of their wealth with those less fortunate.

It was, sadly, a hallmark of the times—and a phenomenon that could be seen in the public sector's response to inequality as well as that of the private sector. No better example of that public sector failure could be found than in that previously cited eight-column *New York Times* headline toward the end of the Nineties that told a dispiriting tale: "In Prospering Idaho, Social Spending Declines as Prison Spending Grows."

Privacy and Pornography. The Internet and the wired world of Technotimes connect people as never before, bringing many ben-efits to individuals and to businesses. The Net also opens opportu-nities for invasions of privacy, voyeurism, and proliferation of pornography beyond anything that could have been imagined by earlier generations of Americans.

As evidence of what can be called the new Snooper Society, Americans who can afford the $3,900 price could hop on down to the Counter Spy Shop on New York's Madison Avenue and pur-chase a phone device that detects changes in a person's voice when he is lying. If that's not enough of a hook to lure you, The Sharper Image offers the Truth Quest Phone supposedly able "to detect subtle and suspicious modulations in the voice of the person you are talking to." But to fulfill your longing for spying on people, for prurient or other reasons, not least checking up secretly on a spouse or lover, you can acquire, for a price, a host of new software that will do the trick: WinWhatWhere Investigator; Desktop Surveil-lance; Cyber Snoop; and 007 Stealth Activity Monitor are among the computer sleuthing programs being produced for home—or workplace—use.

As for sex in all forms, privacy can be invaded via the Internet in a number of ways. You might wish to log on to "Intrusive Eye," which boasts that it's the "home of the uninvited guest." There, you'll find "The Dark Room," promising material designed "for the sole purpose of fulfilling your voyeuristic rights!" and "invasion privacy at its very best!"

"You can't drop film like this off at a drugstore!" the viewer reads. "Upskirts, flashers, accidental exposures, sleeping beauties, unsuspecting neighbors, parked cars . . . and everything else you can dream up."

If that's not enough to entice you, the commercial come-on adds, "Our guerrilla photographers have hidden beneath sewer grates, in air vents, under beds, in bushes, and even in airplane bathroom towel dispensers. We've put over 50 spy cams in women's change rooms, washrooms, tanning salons, under desks and even in strip joint private booths."

And how about this for delicious invasion of privacy: "Watch the incredible toilet cam in action! . . . See sex at work . . . dads banging babysitters . . . is that your girlfriend sucking off your boss under the table?"

It informs you that "there are over 200 million security, closed circuit and peek cameras in use right now." Which ought to make every American feel reassured that his or her inalienable right to privacy is still secure in the new millennium. Best of times, for sure.

Chapter

11

The Markets

"How do we know when irrational exuberance has unduly escalated asset values, which then become subject to unexpected and prolonged contractions as they have in Japan over the past decade?"

—ALAN GREENSPAN

Throughout the Nineties, as the long boom gathered greater force, smashing records and surpassing forecasts about the economy's capacity to generate new wealth, one influential voice was raised in quiet warning year after year after year. Each time Alan Greenspan's words were different; his message was always the same: Be cautious. Be aware of history, of past cycles of boom and bust, soaring hopes and dashed dreams.

Greenspan was not alone in urging investors to be cautious, but no one, not even the president, spoke with more authority or commanded such instant attention from the investing public.

As chairman of the Federal Reserve Board, given statutory independence from both the president and the Congress, in Alan Greenspan's hands rested the immense power of deciding whether to raise or lower interest rates—or leave them alone—and thus directly affect the functioning not only of America's markets but the world's. This role alone gave Greenspan extraordinary power over global fortunes. But during the boom, he assumed a position that exceeded even his legal authority.

He became the Oracle, an almost cultlike figure whose very deliberate distance from the public stage invested him with a certain air of mystery. His occasional pronouncements, always circumspect, always carefully crafted, always maddeningly open to varying interpretations, were instantly combed word by word for clues about which direction he would take in presiding over America's money supply. Would he ease or would he tighten? Did he think inflation was being contained or did it pose a looming threat? Did he see stability or turbulence as he surveyed the global economic scene?

His most famous remark about "irrational exuberance" came early in December 1996, at a time when visions of instant new dot-com wealth fueled investor hunger to get in quick and ride the market upward. "Irrational exuberance" perfectly captured the essence of the mentality that fired the boom, just as Herbert Hoover's confident prediction that America stood on the verge of achieving "one of the oldest and perhaps the noblest of human aspirations," the "abolition of poverty," became the quintessential remark that typified mass thinking during the great bull market of the Roaring Twenties. Already, Hoover said in famously accepting the Republican presidential nomination that summer of 1928, "The poorhouse is vanishing from among us"; he promised his fellow citizens that "given a chance to go forward with the policies of the last eight years, we shall soon, with the help of God, be in sight of the day when poverty will be banished from this nation." Little more than a year later came the crash and the Great Depression.

Greenspan's comments, whether or not out of memory of that earlier example of misplaced economic optimism, consistently counseled caution, a message he repeated like a Greek chorus.

In 1999, three years after his "irrational exuberance" remark, the Fed chairman warned that, "History tells us that sharp reversals in confidence happen abruptly, most often with little advance notice.... Claims on far-distant future values are discounted to insignificance. What is so intriguing is that this type of behavior has characterized human interaction with little appreciable difference over the generations. Whether Dutch tulip bulbs or Russian equities, the market price pattern remains much the same."

A year later, speaking against a backdrop of ever-soaring market prices, he warned that "a set of imbalances, unless contained, threaten our continuing prosperity."

Later, during that same year of 2000, he wondered aloud whether the boom would be remembered as the best of times or as "just one of the many euphoric speculative bubbles that have dotted human history."

* * *

It isn't science, but as Wall Street calculates the equation, a bull market begins when the Dow Jones Industrial Average rises 20 percent, and ends by turning into a bear when it falls 20 percent. By that formulation, the great bull market of the Nineties began October 11, 1990, after the Dow had closed the previous day at 2,635.10.★

One of numerous statistics demonstrates how historic a boom ensued. At the end of 1990, the stock market was valued at $3 trillion. It took Americans two hundred years to build that value. At the end of the Nineties, the stock market value had increased five-fold to $15 trillion. It took investors only a decade to quintuple that overall value. Nothing like that rapid rise had ever occurred before.

In one sense, the timing of the beginning of the long boom is misleading, for the Nineties began with America locked in a recession, burdened by a mountain of debt driven by soaring federal deficits, and experiencing a wave of major bankruptcies. Those deficits were officially projected to continue rising long into the future, further imperiling the economy.

★Sidney L. Jones, who served in important capacities at the U.S. Treasury Department, as senior economist at the Council of Economic Advisors, and as special adviser to the chairman of the Board of Governors of the Federal Reserve System, actually dates the great expansion as beginning in November 1982 and producing 209 out of 217 months of economic growth from then to February 2001.

At that point, unemployment had just risen to 5.5 percent, the largest increase in four years. Over the next year, the deficits continued to accumulate, the bankruptcies increased, and so did the unemployment rate, rising to 6.5 percent that spring. Those conditions worsened in the 1992 presidential election year, creating a disconnect between the slowly rising market values and the day-to-day economic worries that plagued millions of Americans.

When Bill Clinton became the Democratic presidential nominee that July, the unemployment rate had jumped again to an eight-year high of 7.8 percent and the national debt was rising at a rate of $13,000 a second, at an annual interest cost to the nation of nearly $300 billion. When Clinton took his presidential inaugural oath on January 20, 1993, the national debt stood at more than $4 trillion; it had risen nearly $400 billion in just a year.

There was no talk of a boom.

· · ·

However it's defined, and however slowly it gathered accelerating force, the bull market of the Nineties became much more than a visible symbol of prosperity, an astonishing accumulation of great new wealth. It became part of the American culture, a centerpiece of conversations in living rooms and offices across the country, prime fodder for best-selling books about finance, for TV series— *Bull* and *The $treet*—for movies, even for sports stars showing their supposed knowledge of the arcane workings of the financial markets in TV commercials crafted for them by investment houses.

Not since the Twenties had Americans been so captivated, if not obsessed, by the stock markets. And not just obsessed; more and more of them became players in the markets. Indeed, the role of the stock markets in the U.S. economy was the greatest and widest since the Twenties. At the beginning of the Nineties, just a third of all American households owned stocks; at decade's end, more than half of all households did.

As the markets began to move steadily up, public fixation on returns grew more intense. With good reason. It soon became apparent that something historic was taking place.

On April 17, 1991, the Dow closed above 3,000 for the first

time. Four years later on November 21, 1995, propelled by the new Internet–dot-com craze, it closed over 5,000. Now the real Technotimes boom was in full swing. It roared ahead so rapidly that by the next December Alan Greenspan felt impelled to issue his warning about the dangers of irrational exuberance roiling the markets. No matter, investors were undeterred. The markets shot up at an even more remarkable rate.

Up, up, up, the prices went. The Dow crossed the 6,000, the 7,000, the 8,000, then the 9,000 mark, each time entering unexplored financial territory, each time producing more ebullient predictions about greater gains to come.

Early in 1999, the Dow breached a seemingly unbreakable barrier and closed over 10,000. Here, experts said, was a truly symbolic breakthrough—and far from the end, either. Just twenty-four days later it closed over 11,000. On January 14, 2000, the Dow hit its peak of 11,722.98. One book then promoted the idea of a 36,000 stock market in the not-distant American future. Still another published about the same time foresaw a Dow of 100,000. And why not? By then, the bull market of the Nineties had become the longest in United States history, and the second strongest in terms of percentage gains. From its close at 2,365.10 on that October 10, 1990, the Dow rose over 400 percent during the decade. Only the fabled bull market of the Twenties gained more, rising nearly 500 percent before the crash; but its gains were achieved over a shorter period, during eight years from August 24, 1921, to August 3, 1929, while the boom of the Nineties was moving beyond the ten-year mark into a new century and a new millennium.

The boom of the Nineties had rewritten financial history. It had repealed the business cycle; from then on, there was no way to go but up. Or so "they" said, "they" being the tough and savvy new investors who flourished in these best of times. After all, as one of the prophets of the era had written, the world had turned a historic corner and the old patterns and habits were being replaced by dramatically differing ones. In his *The End of History and the Last Man,* Francis Fukuyama proclaimed the end of the ideological conflicts that had dominated the last half of the twentieth century.

Henceforth, the whole world would be embracing democratic free-market capitalism.

While the Dow was recording its great gains, the true winner in showering new wealth on increasing numbers of investors came in the Nasdaq market, home of the record-shattering technology sector of the U.S. exchanges. There, the gains were breathtaking. From its close of 325.61 on October 11, 1990, the Nasdaq soared beyond the 5,000 mark before decade's end and hit its peak closing of 5,049 on March 10, 2000. In the last year of the Nineties alone, the Nasdaq rose by 86 percent; for the decade, it increased tenfold.

The Nasdaq was the darling of the Technotimes savants, the place where the smart young investors placed their bets on the future and were rewarded beyond their dreams by the vast returns on their IPOs. Theirs was the home of the new economy, the real wave of the future, not the dull, plodding financial world of the old economy lumbering along on mere 10, 15, or 20 percent returns. And the records reinforced their exuberance.

As the *Wall Street Journal* reported on the tenth anniversary of the bull market on October 10, 2000, the ten best-performing stocks of companies that went public during that great run were all from the tech sector—and what returns! EMC, up 83,841 percent. Cisco, up 75,324 percent. Emulex, up 39,057 percent. Dell Computer, up 29,718 percent. The same almost unbelievable percentage gains were recorded by the ten best-performing IPOs during the bull run. America Online, up 61,506 percent after having gone public. Veritas Software, up 24,662 percent. Network Appliances, up 14,159 percent.

In truth, these figures were—or should have been—a sign of danger of a vastly overinflated market. But warning signals weren't flashing. These figures stirred even more feelings of avarice—and contained even greater danger signs. Consider the price-to-earnings ratio achieved by the golden new economy tech stocks as opposed to the leading old economy ones. At the end of 1999, Yahoo!'s price-to-earnings ratio was 735; America Online's, 186; Cisco System's, 184; and Sun Microsystems, 110. By comparison, such still profitable old standards showed the following P/E ratio: General Electric, 49; Exxon Mobil, 39; General Motors, 9; K-mart, 8. Despite these disparities, an avalanche of new companies were issuing stock in the Nineties—5,371 of them compared to 3,665 in the Eighties—and the greatest number of new issues were in the tech sector. Despite the disparities in P/E ratios between these new economy issues and the less glamorous old economy standbys, investors were eager to snap up tech stocks. Even though most of them never produced a profit, that didn't stop investors from seeking to catch a new winner, another Microsoft, another Cisco, and become another billionaire, another Gates, another Chambers.

Not surprisingly, this explosion of wealth triggered familiar human reactions. In the new economy, opportunities to get rich quick seemed greater than ever. Perhaps they were, but they also produced new examples of something age-old: greed and speculation.

· · ·

In the new wired world of instant online access to markets and trading desks in the Nineties, new techniques—and new traders— emerged to garner the greatest profits in the proliferating dot-com business world. The methods and the instruments of investing were new, but the rush—mania? madness? lunacy?— to gain quick and easy profit was all too familiar.

Seventy years before, when the bull market of the Twenties was nearing its pinnacle, firing dreams of easy riches and endless prosperity, speculative fever caused more and more "ordinary" people, from clerks to street sweepers, to invest their meager savings into purchase of stocks on margin, borrowing from their brokers to buy even more. They did so in the belief they could profit quickly; they were equally certain they could always get out before the fall. Even if prices *did* drop, the seemingly endless boom made them think any decline was only temporary. As Frederick Lewis Allen described the prevailing mind-set in his great narrative history of the Twenties, *Only Yesterday,* people thought: "Two steps up, one step down, two steps up again—that was how the market went."

A similar form of speculative fever seized the nation in the Nineties. As stories of overnight wealth won by investors smart enough to cash in early on the latest Silicon Valley wonder to go public were widely popularized, people who had never owned a single share of stock and knew nothing of the intricacies of markets began trading from their homes, thanks to their new ability to connect with markets via their personal computers. From there, they climbed higher on the speculative ladder to enter the even riskier new online world of day trading.

Here was the perfect marriage of old and new: Old dreams for instant riches linked to new technology that enabled the individual to manage his or her own portfolio and initiate frenzied, instantaneous transactions through electronic stock trading.

In the Nineties, countless numbers of people, ranging from young professionals to housewives to retirees, plunged into this dangerous new investment venture. Suddenly, they were swept up in the excitement of quick in-and-out buying and selling, all before the markets closed each day. Many gave up jobs to trade stocks daily from new brokerage offices formed to accommodate them and allow them to bypass traditional market firms. Whether operating from their homes or from the new trading casinos—which included cocktail bars and coffee bars like Starbucks (so very Nineties!) that permitted their clientele to dabble in deals while socializing over mixed drinks or lattes—day traders engaged in the pulse-racing daily game of matching wits with the masters. The masters, of course, being the pros.

It was most democratic, the ultimate in people's capitalism, a reinforcement of the naive belief that in American democracy one person's opinion is as good as another's—and to hell with experience, knowledge, training, or, dare say it, wisdom. Every man a king, every "expert" a fraud waiting to be exposed or overthrown.

· · ·

One of the hottest by-products of the new electronic economy, a free online financial advice service started in 1994 called The Motley Fool, spelled out this creed on its electronic home page by quoting from its own best-selling investment guide book, *The Motley Fool Investment Guide:* "The wise would have you believe that 'A Fool and his money are soon parted.' But in a world where three quarters of all professional money managers lose to the market averages, year in and year out, how Wise should one aspire to be?"

The answer, they told the more than a million monthly visitors to their online site, was, You have wisdom enough to beat the markets. Let us teach you how to invest on your own and then go for it! They added: "The Wise have prevailed in the money world for long enough. Now it's finally time that some Fools showed up and leveled the playing field."

Fools indeed. Many enjoyed playing that role, and the heady sense of risk it brought; this, too, was familiar. "Hello, suckers!" the brassy Texas Guinan would cry each night to her speakeasy nightclub customers joyfully breaking Prohibition laws during the Twenties. And each night her New York customers would roar back their greeting, reveling in the risks they were taking. Then and now a bit of Barnum beats in the heart of every American.

In the Nineties, more and more inexperienced first-time investors-as-suckers joined in the game. A flood of how-to books further enticed the uninitiated into believing they, too, could find fulfillment of their oldest dreams and become one of the instant rich. And how much could one expect to gain after learning how to play this real-life Monopoly game? How about "ten thousand dollars a minute!" gushed one such investment guide come-on.

They would have been better advised to listen to the cautionary words of Burton G. Malkiel, a Princeton economics professor and respected author of *A Random Walk Down Wall Street.* "Day trading in the stock market," he wrote near the end of the decade, "is akin to playing a computer video game each day while gambling lots of money at each click of the mouse."

No matter that most day traders, as he said, were likely to fail with frightening suddenness (some studies showed between 80 and 90 percent of such investors lost money), the intoxicating prospect of those Internet stocks kept them in the game. "Just as tulip trading became truly insane in seventeenth-century Holland when ordinary people started trading bulbs in Dutch taverns," Professor

Malkiel observed, "so have the valuations of Net stocks as cyber-armies of day traders have eagerly snapped up the scores of new is-sues immediately after their offering. . . . The spreading philosophy of the day traders that 'fundamentals don't matter' may well have contributed to valuations of Internet stocks that can only be de-scribed in terms of a financial bubble."★

The day traders of the Nineties in effect became day specula-tors and as such reflected a phenomenon as old as America itself. The land of the Puritans, home of sober moralists searching for the higher purpose of life, has also always been the home of the gam-blers, believers in big casinos and winning the lottery, an inherent contradiction that exists in the American character. The earliest American immigrants possessed the same blend of piety and cau-tion combined with a gambler's instinct to pull up stakes overnight and risk all by heading west into the unknown. In the Nineties, amid the boom, the lulling sense of the best of times, hazy memo-ries of hard times, and the belief that the old rules no longer ap-plied led risk takers to triumph over cautious natures.

Years of prosperity had created, as Bernard Wysocki Jr. of the *Wall Street Journal* rightly described it, a "culture of risk-takers." The sense of living on the edge in the Nineties had been trans-formed into a mainstream American virtue. "From cautious saver to citizen speculator in just a decade—that's quite a trek across the spectrum of financial risk," Wysocki wrote, in an apt characteriza-tion of much of the mind-set of the decade. He was citing the lat-est official data showing America's rate of consumer consumption rising rapidly while its rate of savings was falling equally quickly. Once again, Americans were demonstrating an old trait. They were spenders, not savers.

Stories of how some day traders profited greatly stimulated others to take the plunge. Many more followed the path of the margin plungers of the Roaring Twenties; it was a path that ended in bankruptcy and shattered lives. Some ended violently, like the

---

★A popular Silicon Valley book in the late Nineties was a reissue of Charles McKay's 1841 *Extraordinary Popular Delusions and the Madness of Crowds* de-scribing that seventeenth-century Dutch tulip speculative craze.

crazed day trader in the summer of 1999 whose overnight hundred-thousand-dollar losses drove him to murder nine others in the Atlanta firm where they worked.

Tragic endings didn't stop the craze to ride the rising markets. So young whizzes continued the rush to form start-up companies designed to attract giant high-tech firms eager to take them over, and not incidentally make overnight millionaires of the start-up entrepreneurs. Some founders accomplished exactly those ends; in their cases, the fables about new success were well founded.★ They were the great exceptions. Still, the failure of the many didn't halt the stampede to capture a corner of success, and in the Nineties, success stories abounded, especially success stories involving twenty-somethings entering the dot-com world.

· · ·

Not all the winners were plungers. Some shrewd investors capitalized handsomely on the boom by the time-tested method of knowing when to get in and—more important—when to get out.

David Geffen, the Hollywood billionaire, was one of these, and his experience was telling. "I sold my company to MCA in 1990," he recalled to me late in 1999, "and I ended up with $700 million—and in cash—at a time when everything had collapsed. At that time the Japanese seemed to be on the top of the pyramid and America was no longer on the top of the pyramid. Everything Japan did was right, and everything we did was wrong. And I ended up coincidentally, and not out of any great genius on my part, with a lot of money. So I was able to ride the stock market from 1991 until I got out of it last August [1998]."

How he made that decision was equally revealing. Watching

★Princeton's Professor Malkiel on the absurdity of Net valuations: "When the market capitalization of Priceline.com, a dot-com auction company whose site sells empty airline seats, exceeds the combined market capitalization of the other major carriers Delta, U.S. Airways and United, one can only shake one's head in disbelief." Priceline.com became a metaphor for the dot-com craze of the Nineties. From a peak of $104 a share, by the end of 2000 Priceline.com stock had plummeted to just over $6 a share.

the markets soar beyond all reasonable expectations, and seeing the vastly inflated valuations of so many hot new market entities, Geffen began to be wary about what might be coming. He remembered the classic old Wall Street story from the Roaring Twenties when Joseph P. Kennedy, the multimillionaire father of the future president, supposedly stopped to get a shoe shine from a boy stationed across from the stock exchange. This was at the peak of that boom, just before the crash. When Kennedy heard the bootblack describe how he and everyone he knew were borrowing money to invest in the market on margin, the hard-eyed investor decided it was time to get out of the market—and did. It was bound to fall, Kennedy rightly thought.

Seventy years later, Geffen had something of the same feeling. "I don't think it's like the Twenties," he said, "but I did remember something I'd read, that bulls win and bears win but pigs always lose. And I thought that's it. I'm getting out of the stock market. I've made enough money. Zillions of people have enough money that they've given their lives to making and they don't have to be taking risks. What's happening in America at the end of the century is that people have started to believe that mutual funds are like savings banks. Instead of saving in their banks, they'd save in mutual funds. And mutual funds, of course, give you much more interest. That's not saving. That's gambling."

Geffen's reference to Japan's rise and fall in the Nineties echoed that of Alan Greenspan when the Fed chairman cited the Japanese as a prime example of what can happen when irrational exuberance leads to unduly inflated stock values which turns into "unexpected and prolonged contractions as they have in Japan over the past decade."

What happened in Japan should have provided, as Greenspan intended it to, a cautionary warning to American investors. It didn't.

On the last day of 1989, on the eve of the decade of the boom, the United States was locked in a recession and awash in pessimism about its economic future. Japan was in the ascendancy. On that day the Nikkei, the Japanese stock market, was streaking upward and closed at a record 38,915.9—up nearly 500 percent for the decade. It had then recorded twelve straight years of upward gains. This was seen as a sign of a historic change in world fortunes. Now, or so it was said, the United States was declining while Japan was taking its place as the world's economic superpower. Japan's work ethic, its educational system, its method of production, its energy, its products from cars to Walkmen, and its financial acuity were hailed as the way of the future. Huge Japanese majority investments in American firms and Japanese purchase of celebrated American symbols such as Rockefeller Center were all cited as signs of what one best-selling book called "Japan Inc." And not far behind Japan, following in its wake, were other rising economic powers—the so-called Asian Tigers who were then said to present another severe economic challenge for the United States.

No sooner had these gloomy assumptions begun to take hold in the United States than the perceived wisdom of the moment once again proved wrong. In less than three months into the Nineties the values of Japanese stocks had fallen by nearly a quarter. Instead of bouncing back, as widely predicted then, they kept falling. By the end of 1990, the Nikkei was off 39 percent to 23,848.7. It continued plummeting until it closed at 14,309 in August of 1992, off 63 percent from its peak. Amid sighs of relief from American investors, who had bought heavily in Japan's market, experts predicted the worst was over. Not so. By October of 1998 the Nikkei had dropped to 12,880, and by decade's end the Japanese market had lost more than half of its value since the beginning of the Nineties.

Greatly compounding the Japanese fall was another, even greater, economic crisis that shook the world's capital markets throughout 1997 and 1998. What came to be known as the Asian debt crisis began when Thailand devalued its currency, the baht, an event that triggered currency devaluations among the other Tigers leading to their virtual economic collapse. That frightening situation was stabilized in large part through the patient, effective behind-the-scenes efforts of America's Alan Greenspan and the U.S. Treasury secretary, Bob Rubin, a former major player on Wall Street. Together, as two largely unsung heroes of the Nineties, they kept what was a genuine financial crisis from becoming not just a debacle but an unmitigated international disaster threatening a worldwide depression.

During that same period, another financial crisis sent ripples of fear around the globe. That was when Russia devalued its ruble, in effect defaulting on its debt. Amid fears of a global financial melt-down, foreign investors poured more and more money into the presumably safe harbor provided by the booming U.S. markets, further accelerating the long boom. Though warnings were sounded by people like Greenspan about America's own irrational exuber-ance, and its accumulation of debt and overextension of credit, what counted most for most people were the benefits of the boom. In the aftermath of the Japanese, Asian, Russian (and to some ex-tent, Latin) economic crises, the U.S. markets responded with the greatest surge of all.

It was, after all, a new economy, and leading it ever upward into new record profitability was the U.S. tech sector. Techno-times and its boom were unstoppable.

· · ·

October 10, 2000, the tenth anniversary of the bull market, was not a time for celebration.

For half a year, since the spring, enormous volatility had hit the markets, with almost unprecedented swings that hurtled stocks up and down. In the process, both stock values and investor be-liefs were battered. Hardest hit were the golden tech stocks and, most of all, the Internet stocks. The Dow was down, but nothing like the 40 percent loss recorded on the Nasdaq. The wild swings continued through the summer and into the fall, but the overall result remained the same: the markets, and the values, were down, substantially.

By the tenth anniversary of the great market run, numerous analysts began to try to put the long boom into historic perspec-tive. For obvious reasons, assessing the long-term effect of the In-ternet stocks and the high-tech revolution—if, indeed, that term *really* applied—was a central focus. One of the best analyses came in a lengthy summary of events that shaped the boom by the *Wall Street Journal*'s Greg Ip, who described the promise of the Internet sector in this provocative fashion: "The most-exciting business to come along in a century had sales of $30 billion last year, a bit less than Kmart's. But it lost billions and will lose twice as much this year. For this, the price tag is a cool $1 trillion—more than the gross domestic product of Canada. That, in essence, is what investors in the Internet sector had bought: an incomparable business opportunity at an inexplicable price."

Another, far more ambitious, effort was a special section surveying the myths and realities of the new economy by the *Economist*. History, as the publication properly observed, is littered with foolish predictions about the wonders of technology, and so it is with the Internet and the "IT," or information technology, sector of the economy: "Boosters have boldly proclaimed it as the greatest invention since the wheel, transforming the world so radically that the old economics textbooks need ripping up. At the other extreme, skeptics say that computers and the Internet are not remotely as important as steam power, the telegraph or electricity. In their view, IT stands for 'insignificant toys,' and when the technology bubble bursts, its economic benefit will turn out to be no greater than that of the 17th-century tulip bubble."

The *Economist*'s view, and mine, falls between those extremes: Yes, the Internet is a revolutionary device that will continue to affect "the way we communicate, work, shop, and play." No, it hasn't eliminated the business cycle and made inoperative "the old rules of economics and traditional ways of valuing shares."

As the *Economist* says, correctly assessing the contradiction between best-of-times forecasts and worst-of-times outlook: "The trouble is that IT commentators go over the top at both extremes. Either they deny that anything has changed, or they insist that everything has changed. This survey will argue that both are wrong, and that the truth—as so often—lies somewhere in the middle. The economic benefits of the IT revolution could well be big, perhaps as big as those from electricity. But the gains will be nowhere near enough to justify current share prices on Wall Street.

America is experiencing a speculative bubble—as it has done during most technological revolutions in the past two centuries."

. . .

Bubbles burst.

More than a few believed they had witnessed just that at the tenth anniversary of the bull market, October 10, 2000, a time when the markets were experiencing even greater volatility, with prices plunging and investors fleeing. The comments of one respected investor then seemed to sum up part of the awareness of new realities. "Investing was never meant to be exciting," Edward Kerschner, chief investment strategist for PaineWebber, told a *Wall Street Journal* reporter. "It was never meant to be recreational. It was never meant to be the stuff of talk shows. It was meant to be where you put your money for future needs. So it won't be the stuff of cocktail party conversations any more. And that's fine."

The wreckage was not fine. It was devastating, and it was spreading. Overnight, it seemed, talk of recession—or worse—was in the air. Most savagely hit were the dazzling dot-coms.

As the Nasdaq kept sinking—by the end of February 2001 it had dropped by more than 55 percent since the previous March high—a wave of layoffs swept the Internet companies. They were toppling like a row of dominoes. Beginning slowly early in 2000, then accelerating rapidly, by that year's end more than 50,000 Internet jobs had been cut. In December alone nearly 13,000 Internet job cuts were recorded.

Nor did the new year, or a new administration, stem the losses. The layoffs came with a swiftness and a brutality that caught the young dot-com workers unprepared. In company after company, scores of workers were told they had an hour to clean up their desks and head for the door, often with no severance pay or medical benefits to sustain them in the suddenly bleak job market. So much for stock options, which had become virtually worthless. So much for dreams of ever-rising prices and easy, endless prosperity.

Among the hardest hit were vaunted companies that had typified the cocky new economy ethos. At the Motley Fool offices, a third of the workforce found themselves on the street. At Amazon.com, by February 2001 the value of Amazon's stock had dropped 88 percent from its high fourteen months earlier. Losses at other highly publicized dot-coms were even greater. Many Internet company shares sank by 90 percent or more. Salon.com, the edgy, irreverent media online operation, saw its shares fall from $14.25 to $.75. At The Street.com, which had attracted major investors like the *New York Times,* the stock dropped from $71 to $2.72. Those were the fortunate ones; they were still clinging to life. WebMergers, a statistical firm in San Francisco, reported that 210 dot-coms had failed in 2000, with more carnage to come. The Gartner Group, another respected firm, surveyed the field and predicted that 95 percent of all dot-coms will fail.

Cruelly, but fittingly, the first anniversary of the peak of the Nasdaq market on March 10, 2001 brought even more disastrous news. From its high of 5048 just a year before, the Nasdaq had sunk to around 2000. It had taken two years for that golden high-tech market to soar from 2000 to 5000. Now, in just a year, it was hovering around 2000, a precipitous fall that came with even greater speed than its meteoric rise. Worse news followed. On Monday, March 12, the papers were filled with stories about the reemergence of the dreaded bear after a decade of hibernation. This bear sighting came after the Standard & Poor's 500-stock index registered a 22.7 percent decline market on Wall Street, eclipsing the standard definition of a bear market—a 20 percent decline from a record high. As the *New York Times* commented in its lead editorial, under the headline "A Moment of Economic Suspense," "Wall Street's bull market formally expired Monday. Will the nation's longest-ever expansion follow suit?" That development triggered further slides. Two days after entering a bear market, the Dow Jones industrial average plunged 3 percent as the 10,000-point stock floor collapsed, sending greater ripples of fear through traders and investors. This, in turn, sparked even greater losses—6.3 percent—on the Nasdaq, causing the tech market to sink below 2000 for the first time in two years. When the brutal week ended, Wall Street had recorded the worst week for blue-chip stocks in eleven years. Inevitably, a number of highly critical post-mortem articles appeared examining how this debacle had occurred, and pointing embarrassingly at some of the boom years' greatest cheerleaders.

The *Wall Street Journal,* in a devastating article, took dead aim at one of the era's hottest Internet analysts. "Where is Mary Meeker?" Randall Smith and Mylene Mangalindan wrote. "A year ago, the Morgan Stanley Dean Witter & Co. Internet analyst was basking in adulation. Dubbed the 'Queen of the Net,' by Barron's magazine in 1998, subject of a glowing profile in the *New Yorker* in 1999, Ms. Meeker graced the cover of *Fortune* as the third most powerful woman in business later that year, and addressed the World Economic Forum in Davos, Switzerland, in early 2000. Then the bubble burst. The Nasdaq Stock Market is now down 56 percent since its peak last March 10—and Ms. Meeker has fallen off the radar screen." Their article included a number of Meeker's economic cheerleading quotes from years past, including these two: April 1999: "There is no doubt in my mind that the aggregate market value of the Internet sector will be a lot higher in three years than it is today." January 24, 2001: "We love the idea that there appears to be a universal perception that Internet-related investing is dead, as we believe it gives us the opportunity to be 'crazy like foxes' and come up with what should be some good money making ideas in 2001."

The falling markets inspired other painful critical assessments. "It has now become painfully clear, and last week's gyrations only added more proof, that Wall Street is a different place than it was even a short time ago," Gretchen Morgenson wrote in that Sunday's *New York Times* Money & Business section under the biting headline: "How Did They Value Stocks? Count the Absurd Ways." As she observed, "A brutish, bearish market has brought with it deflated hopes and new realities. Among the harsh realities one stands out. Of all the hot air generated during the great bull market of the late 1990's, none propelled stock prices further than the notion that new economy stocks were a breed apart...."

Perhaps the most trenchant commentary came, appropriately, from Ron Chernow, author of *Titan,* the splendid biography of that original Robber Baron, John D. Rockefeller. Writing on the

New York Times's op-ed page, Chernow put into historic perspective "the magnitude of the Nasdaq collapse" and pointed to some of its long-term consequences. "Think of the stock market in recent years as a lunatic control tower that directed most incoming planes to a bustling, congested airport known as the New Economy while another, depressed airport, the Old Economy, stagnated with empty runways. The market has functioned as a vast, erratic mechanism for misallocating capital across America. In such an atmosphere, the little people of America, egged on by Wall Street's hired optimists, wrote blank checks to indulge the giddy fantasies of high tech entrepreneurs. In the early stages, this sparked robust expansion and innovation among software, computer, telecommunications, bio-technology, semiconductor and fiber optic companies. To many New Economy gurus, the pot of gold from Wall Street seemed fully justified and certain proof of their own genius."

Of all the mighty that had fallen, none provided a more powerful example than the most fabulously successful high-tech firm of the late Nineties, Cisco Systems Inc. By decade's end, the Silicon Valley company, renowned as the leading maker of Internet-switching gear, had supplanted Microsoft as the world's largest company, and had exceeded published earnings estimates for fourteen straight quarters in a row. When other high-tech companies began slicing employee rolls after the steep market fall that began in the spring of 2000, Cisco reacted to changing market conditions by taking the opposite, bullish, corporate approach. They continued to expand aggressively, confident their explosive growth would continue into 2001. It did not. By that spring, Cisco was forced to post its first quarterly loss in its eleven years as a publicly traded company—a gargantuan quarterly loss of $2.69 billion. At the same time, it announced plans to slash 8,500 workers, and grappled with the disappearance of more than $400 billion of its market value. How fast a fall that represented could be seen in the public comments that Cisco's chief executive, John Chambers, made to stock analysts in those early months of 2001. On February 6, Chambers, one of the notable high-tech leaders of the Nineties, told the analysts that despite the falling prices, "We remain confident about the market opportunity ahead of us over the next three to five years." Two months later, on April 16, he was forced to acknowledge that ". . . this may be the fastest any industry our size has ever decelerated, which has required us to make difficult business decisions at an unprecedented speed." By May 9, he told analysts, "We've clearly had an understanding that the peaks are much higher and valleys much lower than many of us anticipated. We are now in a valley that is much deeper than many of us anticipated."

For the young techies, these blows were as shocking as they were unexpected. Depending on personality, the suddenly jobless employees reacted to their changed employment condition with bravado or bitterness. Pink-slip parties proliferated. So did mordant jokes and gallows humor. But nothing could hide the new reality they all faced as they saw their dreams fade, their hopes dashed.

The national mood shifted abruptly from euphoric to pessimistic. The media, ever ready to focus on the new and the troubling, began producing major stories on what the high-flying young dot-commers faced as they came crashing to earth. A *Newsweek* cover story after a new president began his term captured the new focus on their grim plight. Superimposed over a picture of three glum young dot-commers were the bold headlines:

LAID

How Safe Is Your Job?

OFF

Under those words were the subheads:

Surviving the Slowdown
Who's Hiring Now

The Federal Reserve, reacting to the sudden change in fortunes—or, more accurately, reflecting the back-to-reality sense of investors—cut interest rates twice in January 2001 and three more times in the next four months. Each time the markets rose temporarily, and then dropped even more. Whatever the long-term outlook, no swift economic revival seemed in store. The losses were phenomenal; by spring, as prices continued to fall, more than $4 trillion in value had disappeared.

Alan Greenspan, the Oracle, again traveled to Capitol Hill to deliver yet another assessment of where the economy was headed. He was, as ever, cautious, but now, instead of warning about "irrational exuberance," he struck a tone that attempted to restore eroding consumer confidence even while suggesting more bad times might lie ahead.

Speaking a month after the presidential inauguration, before newly elected members of Congress, Greenspan gave them the best news he had to offer: The economy was not then in recession, and in fact had improved slightly in the first month of the new year. At the same time, he acknowledged that "because the extent of the slowdown was not anticipated by businesses, it induced some backup in inventories." Still, he said, "The exceptional weakness so evident in a number of economic indicators toward the end of last year, perhaps in part the consequence of adverse weather, apparently did not continue in January. But with signs of softness still patently in evidence at the time of its January meeting the F.O.M.C. [Federal Open Market Committee] retained its sense that the risks are weighted toward conditions that may generate economic weakness in the foreseeable future."

This time, it required no soothsayer to understand the meaning of his carefully crafted words about the future. "It is difficult for economic policy to deal with the abruptness of a break in confidence," the Oracle said. "There may not be a seamless transition from high to moderate to low confidence on the part of businesses, investors and consumers. Looking back at recent cyclical episodes, we see that the change in attitudes has often been sudden."

He recalled that in previous testimony he had "likened this process to water backing up against a dam that is finally breached. The torrent carries with it most remnants of certainty and euphoria that built up in earlier periods. This unpredictable rending of confidence is one reason recessions are so difficult to forecast. They may not be just changes in degree from a period of economic expansion, but a different process engendered by fear. Our economic models have never been particularly successful in capturing a process driven in large part by nonrational behavior."

For once, Greenspan's magic did not appear to be working. The Oracle's halo was slipping. Angry investors, thrashing about to affix blame for the sudden market collapse, began sharply criticizing Greenspan for supposedly being too hesitant to act in attempting to stop the decline. "Impeach Greenspan," some Wall Street traders were heard to say. Even the sternly conservative *Wall Street Journal's* editorial page published a surprisingly condemnatory article about Greenspan's alleged mishandling of the economy that April. "A few weeks ago," Bruce Bartlett, a conservative Washington policy analyst began, "a reporter called and asked me if I thought the current economic slowdown was Bill Clinton's fault. Knowing my opinion of the former president, he clearly expected an affirmative answer. Reluctantly, I replied that as much as I would love to blame anything bad that happens on Mr. Clinton, I was forced to say that the slowdown is 100% a monetary phenomenon. If anyone is to blame, I said, it is Alan Greenspan."

The economic expansion wasn't over—yet—and neither was the promise of Technotimes. But the dam had been breached, the dot-com bubble had burst. That party was over. It took its place along with the other speculative bubbles that littered economic and social history—the seventeenth-century Dutch tulip bulb craze, the eighteenth-century British railroad investment frenzy, the twentieth-century margin mania that drove the great bull market of the Twenties. It was time to step back and reassess. Events at home and abroad suggested a more sober approach was in order: for reason not euphoria, realism not irrationality were the order of the day both for the functioning of the markets and for the role of America in the new century.

12

The Millennials

"We're a very career-oriented, a very chasing-the-good-life kind of generation. Maybe that's because of the wonderful economy and the fact that we've never seen a war, we've never seen a huge economic depression. We're a pretty comfortable generation."

In 1997, as the then-seemingly unstoppable economic expansion surged to higher and higher historic levels, a national survey of American college students by a Big Five international accounting firm produced astonishing findings. Of all college students questioned, 77 percent said they expected to become millionaires in their lifetimes. Stunning as those findings were, they perfectly reflected the attitudes being created in a new generation by the benefits of a boom that more and more young Americans believed had become a permanent part of their lives.

· · ·

Human nature and basic attitudes don't *really* change from generation to generation, but that doesn't stop cultural popularizers and commercial mass marketers of our aging but youth-obsessed society from trying to give each wave of Americans distinctive charac-

teristics, the better to label and profit from them—the Silent Generation, the Me Generation, the Baby-Boomer Generation. Most
of these attempts to give each generation a name and its dominant
attitudes a common value are of little use in determining fundamental generational change. Most of the opinion surveys sample
only an infinitesimal fragment of the target group, providing at
best transitory snapshots of American attitudes; and those attitudes
are affected by constantly changing events.

One national survey, however, stands as a notable exception in
tracking real long-term changes in attitudes and values of generations of young Americans from the Sixties through the Nineties.
That is the annual survey of entering college freshmen attending
two- and four-year colleges and universities conducted each fall by
UCLA in association with the American Council on Education.
As the millennium approached, the results of this survey documented a kind of best-and-worst-case scenario about the submil-
lennial generation who are destined to lead America through the
first half of the twenty-first century and beyond: Best—more sophisticated, more tolerant, more talented technologically, more
concerned and careful about their health and physical and sexual
habits; Worst—more cynical, more self-centered, more insensitive,
more driven to succeed at any cost.

Since 1966, when the annual survey began, more than 9 million students attending some fifteen hundred institutions of higher
learning have participated in it. Over the decades, the results track
a steady turning away from public life, public service, political and
social activism. In place of the idealism expressed by students in the
Sixties, the survey shows young Americans placing increasing emphasis on making money and materialism.

By 1998, as the *New York Times* noted in reporting on the results of the annual student survey, the findings confirmed what college professors and administrators had long been sensing: "students
are increasingly disengaged and view higher education less as an
opportunity to expand their minds and more as a means to increase
their incomes." Thus, UCLA researchers concluded, two suggested goals of education—"to be very well off financially" and
"to develop a meaningful philosophy of life"—had switched

places. In 1968, for example, 82.5 percent of the students surveyed cited the importance of developing a philosophy as an essential life goal as opposed to the 40.8 percent who believed achieving financial security most important. Three decades later those numbers were reversed: 74.9 percent of the students said being well off was their essential goal while 40.8 percent chose developing a philosophy of life.

This means, the survey director, Linda J. Sax, told the *Times,* that students of the late Clinton years were "using education more as a means to an end, rather than valuing what is being learned." Students' strong interest in high incomes has been rising steadily over the years, according to Sax, who noted that "the trend took on more significance when added to the fact that incoming students showed unprecedented levels of academic and political disengagement."

These trends continued as the boom accelerated and the social and cultural forces of the Nineties became even more distracting.

"Business executive (management administrator)" again topped the list of students' probable career choices in the survey of 364,546 entering freshmen at 683 of the nation's two- and four-year colleges and universities conducted in the fall of 1999 and made public early in the year 2000. At the bottom, picked by only six-tenths of a percent of the students, were "policymaker/government" and "military service."

Student interest in politics continued to plummet. A record low of 25.9 percent of the millennial freshmen believed that "keeping up to date with political affairs" was either a very important or essential life goal. Thirty-four years earlier, in the first freshmen survey conducted in the mid-Sixties, 57.8 percent of the students embraced this goal. By 1999 only 14 percent of freshmen frequently discussed politics. More than twice as many did in 1968.

Further evidence of disinterest in civic affairs and declining commitment to social activism emerges strongly in that century-end survey data. Only 20.6 percent had voted in a student election. While a record number of freshmen—75.3 percent—performed some form of volunteer work in their last year in high school, much of it required and thought to be essential to gain a spot at a top college in today's pressure-filled college application process, only 18.9 percent expected to continue their community service work in college. And their long-term goals for social activism continued to drop.

The percentage of freshmen who thought it "very important or essential" to "influence social values" fell to 35.8 percent, its lowest point since the mid-Eighties. Their desire to participate in community action programs also reached its lowest point in over a decade (21.3 percent), while interest in becoming a community leader dropped four percentage points in the last three years—from 32.1 percent in 1996 to 28 percent in 1999.

Even more disturbing for what it says about the desire or willingness of the millennial generation to become involved in improving the quality of public life, the survey documents their declining social consciousness. Their commitment to "help clean up the environment" declined for the seventh straight year, reaching its lowest point (17.9 percent) in more than a decade. In 1992, as the boom was just getting underway, nearly twice as many wanted to become involved in environmental clean-up programs. Their commitment to "helping others who are in difficulty" declined for the third straight year, and also fell to its lowest point in more than a decade (59 percent). Similarly, commitment to "help promote racial understanding" declined for the third straight year, dropping to 28.4 percent, its lowest point in fourteen years. That represents a falloff of more than a third from the number of students who previously expressed a commitment to work for better racial understanding.

In another revealing commentary about attitudes on race, the percentage of college freshmen who agreed that "racial discrimination is no longer a major problem in America" increased to 23 percent. That marks the highest number recorded in the ten years since survey officials began asking students that specific question about the state of American race relations.

Not surprisingly for a generation greatly influenced by the technological revolution that sparked the boom and its good times, and a generation whose models for success are the young techies amassing new wealth, the millennials are highly proficient technologically.

The survey results show them becoming even more dependent on their new electronic tools and toys.

By century's end, more than four out of five students used the Internet for research or homework. Nearly two-thirds communicated via e-mail. More than half participated in Internet chat rooms. Nearly three-quarters engaged in "other Internet use." More than 80 percent of the students played computer games at least occasionally.

All these findings naturally led Linda Sax, the survey director, to conclude that the Internet had become a way of life for the majority of America's college students. "What remains to be seen, however," she says, "is whether proficiency on the Internet enhances student learning during college."

The answer to that question won't be known for years, but there's no doubt how the millennial generation is being affected by another critical societal aspect of the technological revolution. Technotimes is exposing the inherent inequality that still divides America racially and economically, publicly and privately.

More than 80 percent of all freshmen attending private universities communicated with e-mail during their last year in high school. Yet only half that number, representing 41.4 percent of students at all the nation's public black colleges, used e-mail in high school.

In large part, this dramatic disparity reflects the woeful lack of access minorities, particularly blacks, have to computers in public schools from the earliest grade up, to say nothing of the availability of home computers. As a result, these young Americans are being culturally shortchanged by their inadequate training in the new technology that will be even more crucial for their success in the future. As I described earlier in examining the miserable public-school facilities and the dangerous world of minority inner-city children in South Central Los Angeles, these children are technologically handicapped, often hopelessly, by the circumstances of their birth.* Even if they are able to attend a black pub-

*See Chapter 7.

lic college, the millennial survey documents how they continue to operate at a disadvantage when compared to their more favored contemporaries.

And although the majority of freshmen at all institutions have used the Internet for research or homework, this figure ranges from a high of 90.2 percent among students entering private universities to a relative low of 77.6 percent among freshmen attending black public colleges. This finding provides further evidence of how blacks and whites begin their college years with markedly different levels of exposure to, familiarity with, and dependence on the computer.

What part racial, social, and economic inequities play in other striking disparities revealed by the latest survey data is not so easily determined, but the results illuminate a number of growing problems.

Rising numbers of college freshmen—a record high of nearly 40 percent of all entering college students in the millennial survey—frequently feel "bored in class." An all-time high record, 62.6 percent, "frequently or occasionally" come late to class, and another 36.2 percent, "overslept and missed class or (an) appointment"—nearly double the rate reported in 1968.

The hours students spend doing homework each week in their last year of high school continue to decline. In 1987, when researchers began asking students about their studying habits, 43.7 percent spent six or more hours a week on homework. In 1999 a record low of 31.5 percent spent that much time on homework. At the same time, more and more are studying less than three hours a week, 40.2 percent, while 17.1 percent devote less than one hour a week to homework in their last high school year.

Growing numbers of students have to take remedial courses in high school before being admitted to college. The number of students who took remedial courses in mathematics and foreign languages rose to an all-time high, while those having to take remedial work in science soared to a twenty-year high. Moreover, the percentage of college freshmen taking at least one remedial course in high school, 18.3 percent, had increased by more than half since 1982.

These statistics all reveal what the survey analysts inadequately call evidence of the continuing rise of "academic disengagement." Whatever term best describes that phenomenon, the long-term societal implications are unnerving. Although the root causes are not easily determinable, one factor, a most familiar one in our story, is the impact of television on student attitudes and behavior.

Alexander W. Astin, who founded the annual college survey and heads UCLA's Higher Education Research Institute that conducts it, links the increasing amount of time young Americans watch TV to the evidence of an increase in student materialism, boredom, and disengagement. "Kids who started college in the late 60's had much less television," he says. "Today's kids never didn't have it. We tracked freshmen of 1985 for four years to see how much TV they watched during college. The more TV they watched, the more their materialistic tendencies were strengthened."

That's not surprising given the bombardment of commercial inducements flitting across their screens and the breathless tales of new fortunes being made by both their contemporaries and their idols in sports and entertainment.

That America is the greatest consumer society in history is not news; the ability of marketers to reach every element of the mass electronic culture will only stroke society's consumerism. Neither is it surprising that teachers find growing numbers of students adopting a consumerist view of education. Attending college is viewed as a means to a job, a practical way, as Professor Mark W. Edmundson of the University of Virginia put it in a *Harper's* article, "to learn a skill, a capacity that you can convert into dollars later on."

Sobering as these results are, they should *not* be interpreted as conclusive evidence the millennial generation is so obsessed with material gain and so driven by the dream of making money that these values eclipse all others. Nor do they prove that the soon-to-be leaders are basically selfish, lazy, or academically inattentive. In fact, the evidence strongly suggests that a wide gap exists between the rising number of bored, ill-prepared, disadvantaged underachievers and those who come to college with greater advantages, better prepared, and more motivated to succeed. In that latter group are students who have never worked harder on their studies. They may well prove to be the most technologically accomplished, innovative, and talented group of young Americans ever.

One of the most telling of all the millennial survey findings involves the increasing pressures many students feel as they enter college. Record numbers of incoming college students, or 30.2 percent, say they experience such high levels of stress that they frequently feel "overwhelmed by all I have to do." That's nearly twice as many as reported feeling "stressed-out" in 1985.

Obviously, the ever-rising cost of college, forcing increasing numbers of students to work full-time and assume heavy long-term debt to pay off student loans, looms as a major factor in creating greater feelings of stress. One out of every four entering freshmen—another record—reports facing either "some" or a "very good" likelihood of having to work full-time to meet collegiate and personal expenses. Survey director Linda Sax sees the rise in levels of student stress as "a reflection of an increasingly fast-paced society, made more so by computers and other media. . . . Students feel more competition, they're applying to more colleges than ever before, they're worried about having to work during college. That can be overwhelming."

Most striking is how differently these pressures affect women and men.

Twice as many women students as men, or 38.8 percent, say they frequently are overwhelmed by multiple demands. Buried in the survey results are other findings that help explain the notable differences between women and men. More women than men, 69.6 percent vs. 57.2 percent, are worried they may not have enough money to finish college. Significantly larger numbers of women than men, or 44.1 percent vs. 33.3 percent, find they will need to get a job to pay for college expenses.

Another reason why women experience greater stress than men is that they spend their time differently. Women study more, volunteer more, participate more in student clubs and groups. They also bear the greatest burden of being the ones who, if married, must tend to child-care or housework responsibilities.

Men, on the other hand, spend more time than women exercising, playing sports and video games, watching television, partying. According to Linda Sax, "These findings suggest that women spend more time than men on goal-oriented and potentially stress-producing activities."

If so, and the evidence for this is overwhelming, it is only more evidence that changes in gender roles are profoundly affecting American women, especially young American women, and most specifically the new millennial young women—and, beyond them, greatly affecting the nature of American society.

· · ·

When Amanda Jones was in grade school, her first-grade teacher asked her class to draw pictures of themselves when they grew up. Amanda, then six, the daughter of a successful lawyer, drew herself as a lawyer, briefcase in hand. As she presented her self-portrait to the class, she heard a girl across the aisle comment, "Just like her daddy."

The boy behind her drew himself as a race-car driver, which is what his father did, either as a living or for a hobby. When he presented his picture to the class, the same girl said, "Just like his daddy."

Looking back from the vantage point of a twenty-three-year-old graduate student, that memory strikes Amanda as remarkable. A generation before, the boys would have drawn careers, just like their daddies, and the girls would have drawn kitchens, just like their mommies. A generation from now, if and when her child is in first grade, *everyone* will draw careers just like all their mommies and daddies.

As Amanda reflects on this astonishing societal change, she says: "My generation of women is the only one without a visible image of womanhood to follow. The rules wouldn't let me draw a picture of myself baking cookies and singing the baby to sleep. Even in first grade I somehow knew that girls are expected to have career goals from early childhood on, exactly as boys are. This is wonderful. Still, women my age are in a difficult spot. We need careers, but we never saw how that works. I don't mind admitting it's left me a little muddled."

It makes her feel, as she puts it, "as though I'm caught in a crack between one generation and another."

When she thinks about what defines her generation, she says, the first idea that forms in her mind is about technology. "We've all been inundated with images of five-year-olds pushing mice," she says. "You know, 'The youth of today are computer wizards. They'll leave their parents in the dust.' That's the conventional wisdom. As I think about this, though, I realize it's just a pat response, drilled into my mind and everyone else's mostly by—who else?—the media. Actually, our technological revolution has yet to affect people's lives to anywhere near the extent of the automobile, or the electric light, or even the telephone. The changes we're going through now are just the same old changes. Those who learn how to use it the fastest will have an edge for a while, while people who cling to their manual typewriters will lag behind. But before too long a leveling will occur. In another twenty years there won't be anyone in America who doesn't know how to turn on a computer and put it through its paces, any more than there is anyone now who doesn't know how to dial a telephone."

What sets her generation apart is not its ability to maneuver through the electronic intricacies of Windows 2000. What makes her generation truly remarkable is that it is the first to grow up under what she aptly calls "the new gender rules."

For women, the new rules are not only exhilarating but confusing. Like Amanda, women today arrive at young adulthood acutely aware of how different their lives are from those of their mothers and grandmothers. As Amanda says, "There never was a day during my childhood when I felt I couldn't do something because I was a girl. My father played baseball in the yard with my brother and me. He bought us both packs of baseball cards and gave us both tools for Christmas. When I was little, I played with Star Wars *and* with Barbies. My brother only played Star Wars. I felt lucky; as a girl, I was allowed to have twice as many kinds of toys, twice as many kinds of clothes. As an adult, I see things differently: I'm required to be twice as many things. Right now I have to figure out whether I'm going to be a lawyer or a journalist or a teacher or I'm not a productive member of society. That's the same pressure that's placed on my brother. At the same time, my TV tells me I had better have something hot on the table when my hypothetical kids get home for dinner. And I had better spend an hour a day on my hair and makeup and accessories, or I'll die lonely. My brother gets off the hook on that. Somehow the 'man works' versus 'woman stays at home' dichotomy has died, while the 'swinging bachelor' versus 'bitter old maid' is alive and kicking."

Her mother was on the cusp of great generational change, too, but didn't have the pressure that Amanda and her contemporaries now experience.

Amanda's mother was in the first class of women to graduate from Yale, but Ivy League degree notwithstanding, she chose to be a housewife and a homemaker. When wrestling over *her* ultimate career path, Amanda finds herself conflicted. "There are so many career choices available to me that I have never been able to make up my mind," she says.

Reassuring, if not enticing, though her prospects are, Amanda also finds an ambivalence, even resentment, about emulating her mother. "It's odd," she says, "that the one thing I feel I am not per-mitted to choose is the thing my mother actually does. Secretly, a very big part of me does want to spend my life raising kids. But I can't imagine the response I would get if I told a casual acquain-tance I aspire to be a housewife. The amazing thing is that I used to think I was the only person who had this guilty wish, but over the years I've had a number of close female friends confide the same thing to me. We are probably the only generation of women in history who have not been allowed to aspire to imitate our mothers. What it comes down to is that a woman is supposed to fill a bunch of roles now that she was never supposed to before, and that is miraculously good, because the new roles are better than the old roles. Unfortunately, though, she has to keep filling the old ones at the same time, and she's not allowed to just choose one or the other."

This, she thinks, will change. "My daughters, if I have them, will never know what it's like to live with a woman who cooks breakfast every morning and doesn't collect a paycheck. What worries me, though, is this: my mother's was the first generation of middle-class American women to truly have a choice between staying at home or going to work. As a group, they have been successful, because the only women in the professional world now are the ones who really wanted to be there. My generation, on the other hand, is the first truly *not* to have a choice. If you are a twenty-three-year-old American woman from a middle- or upper-income family you *must* go to college and get onto some kind of a career path, at least until you have children. Nothing else is socially acceptable."

At the time this conversation took place, Amanda was agonizing over that career choice. After earning her undergraduate degree, she had worked as a public school teacher in inner-city New York. Then she was attracted to journalism and was one of my most talented graduate students. She was, however, torn, and feeling great personal pressure to become a lawyer like her father.

"The final irony for me is that I've spent today working on my law school applications," she says. "My father tries to act nonchalant about it, but he's practically drooling in anticipation of me following in his footsteps. It occurs to me that I'm feeling the same kind of paternal pressure that men have felt since time immemorial. It's a strange situation to be in. Would I trade it in for the chance to wear an apron and rubber gloves? I don't think so, but every once in a while I'm not sure."*

· · ·

No amount of surveying and statistical analysis adequately defines the young millennials. Yes, they appear more materialistic, more pressured to succeed, more enamored—seduced?—by the visions of great wealth many of their contemporaries enjoyed before dot-coms began crashing and thousands of young people lost jobs. Yes, they do not seem as concerned with social activism and certainly not with politics. Yes, they are probably the most economically favored of all generations, but that doesn't mean they all lack values

*Amanda Jones was admitted to Yale Law School, where she is now (2001) studying.

or concern for others. Indeed, many who observe them most closely cite numerous positive signs about them. Robert McKinn, for example. He's a distinguished Stanford professor who teaches a justly famous course there on the impact of science and technology on society. McKinn concedes he relies on anecdotal evidence to assess the new millennials he teaches, but what he sees makes him optimistic about their values.

"Except for the gung-ho, full-speed-ahead, damn-the-torpedoes engineering students who are coming here just to make the better microprocessor—the better nanoprocessors, the better nano-this, nano-that—who aim to grab on to the brass ring of the spinning merry-go-round," he says, "I'm finding there are a lot of kids here who are searching for a certain set of core values in terms of community, in terms of family, in terms of interpersonal relationships—character, beauty, whatever. We have people from a lot of different cultural backgrounds—ethnic, religious, moral—and yet I find a hunger among them for those core values, for achieving a way of life with qualities that make life worth living."

But neither McKinn nor anyone else watching them can fail to be struck by how many conflicting emotional and intellectual crosscurrents are affecting what I have come to call—for reasons that will be quickly apparent—the Good Life Generation.

I say that by way of introducing a group of Stanford seniors, young men and women of varying backgrounds, who I gathered together to hear express what *they* think about themselves and their generation in a lengthy freewheeling and wonderfully candid conversation.

Obviously, these students can't be described as "typical"; their very admittance to such a prestigious university, with all the advantages it brings, separates them from many of their age group. But the more they talked, the greater was the sense that they reflect basic attitudes and values held by their generational contemporaries across the country.

They are, all of them, children of the boom, young adults shaped by all the best and worst roller-coaster events of the Clinton years.

They remember where they were as children when the *Challenger* blew up. They remember where they were in high school when the O.J. verdict was handed down. "We were in biology class," one recalls, "doing our genetic studies and the verdict was Not Guilty, and everyone said: 'He got off. This DNA must be a sham!'"

They remember growing up just before the video-game generation. Though they themselves starting playing those games while young, they didn't devote nearly the time to them that their younger brothers and sisters do. As one says, "The younger you are, more the prevalent it [video-game playing] is."

They remember, to their chagrin, being hooked by the Jerry Springer daily outrage-and-insult TV offerings. "We all watch it together," one of the women says, "and laugh at it and make fun of it and go, 'Oh, my God!' It's a mockery. It's not our world, I guess. I usually watch it with feigned disgust. You know, 'I can't be watching this; it's horrible.' I hate that I watch it, but I'm definitely hooked." The women remember how the TV character, Ally McBeal, quickly became a symbol of young females in their age group entering the intensely competitive career track. Not all of the Stanford women regard Ally favorably. Quite the opposite. As one of the students said, "I'm sick of how she's supposed to represent women everywhere. Wearing really short skirts and having emotional problems, that's not Everywoman to me."

They remember the advent of MTV, and how it influenced their lives by making everything seem to move far more rapidly, projecting shortened scenes and encouraging shortened attention spans. "I didn't watch TV in the Sixties and Seventies," a male student says, "so I don't know how it's different, but now we have MTV where they flash these huge bright lights at us and with our three-second attention span, there's a beautiful model here and then a beautiful model there who grabs our attention away. You know, everything today is immediate gratification; playing video games is an immediate gratification. Now the books kids of my brothers' and sisters' ages read are always abridged. They read a forty-seven page version of *A Tale of Two Cities* in their school!"

And they vividly remember many scenes revolving around Bill Clinton and the scandal years, and they report being repulsed by most of them. "I hated it," one of them says. "It makes people more disgusted with the political system. I hate that all that's been a focal point for our country, that that's what's been talked about more than anything else in the world. If I could ever raise any political discussion with my friends at home, it was always all about Monica. They all read the Starr report. They made jokes about it. Sex jokes. You know, 'Can you believe he did that!'"

As for the Clinton years, these young men and women are, like the public at large, strikingly ambivalent about them, and nowhere more so than in their feelings about the central public figure of their lives, Bill Clinton himself—a president for whom they express admiration and sympathy, as well as disgust and pity.

"There's this unique American sensibility that demands its heroes be absolutely flawless," one senior says. "Classically defined, a hero has a flaw. Bill Clinton wouldn't necessarily be my hero, but I do have a lot of respect for his mind. I also think he has an unbelievably large ego, which I think you need to be president. We're attracted to a personality who could be a womanizer. We elect people who are like that. Look at the extramarital histories of a lot of our presidents."

His remarks spark a flurry of reactions. "He hasn't left me with anything to identify with," a young woman says. "I respected his mind and his intelligence, but I think he let himself down and let us down in a lot of ways. He let his family down. I don't agree with what he's done. It's reprehensible."

Another woman says, "Lots of people other than Bill Clinton accomplish things. I don't think we should ask perfection of our leaders, but I also have a moral issue with what he's done. I find it kind of nasty. People had JFK and lost JFK, and even though he had his womanizing tendencies, he conveyed a sense of wisdom and vision, and he represents something to people. People still have posters of him on their walls."

One of the men interjects another view, again triggering a round of reactions: "The question is how Kennedy would be regarded now because we know all these things."

"Why do we set ourselves up?" one of them asks. "Why do we demand good-looking charismatic men to be our presidents? Why do we demand that?"

One of the women raises the subject of Monica, and says, "I hope she doesn't become an icon of our generation. That would be brutally misrepresentative."

All the women quickly nod in agreement.

Curious, I ask them to react to the following: Ken Starr, Linda Tripp, Lucianne Goldberg, the press.

"Vultures," one of the women snaps.

The rest, men included, all signal their agreement, except for one who adds, "And you can put the president and Monica Lewinsky in that group. They're all controlled in some sense by ego and personal ambition."

"Do you have a hero?" I ask them.

Immediately, one asks, "Alive?"

A young man responds, "For most kids, it'd be Michael Jordan, but you don't have a Martin Luther King figure. No Kennedy."

"National heroes?" another says. "I think we're at a loss now. You look at those big white buildings in Washington and the monuments to Washington, Jefferson, Lincoln, Roosevelt. We did a monument tour with one of our professors and I was thinking that all those we've erected monuments to followed in the wake of something huge. They all operated in some kind of context where they could be heroes. Now there's no cause for anyone to be at the forefront of. No Depression, no war to make you heroic."

Listening carefully, one of the women waits until her classmate has finished, then says, "Maybe we're living in a time that's not conducive to heroes. We don't even think of having a hero. There's a cynicism."

"I'm falling into the same trap," a young man says. "You know, 'Oh, yeah, Lincoln was great,' but 'Oh, yeah, he didn't really want to free the slaves. He wanted a Union.' So scratch him off the list."

When I ask about the main reason for our gathering, to hear them define their generation, the answers become even more revealing.

One of the young men speaks first. "I don't think there's a tangible definition for our generation," he says. "In the Sixties, they had a political movement across the country, a rebellious, anti-Vietnam movement. They had that to unite over. We don't have anything like that. A lot of our generation feels we're doing so well now, and America's succeeding in so many ways, that there's not anything that you feel charged to change at this point."

His answer is quickly followed by the views of one of the women. "We're a very career-oriented, a very chasing-the-good-life kind of generation," she says. "We're entering into a faster-paced lifestyle. We'll bounce around from career to career and nobody's going to stay anywhere for a long time. We're going to take the fast track to the top as fast as possible. That's what we are: career-oriented, good-life-oriented, fast-paced."

Another woman responds. "I definitely agree," she says. "There's nothing that's uniting us, or defining a movement for us. Except maybe we're the information age generation, the generation that started using the Internet. But I don't think that really binds us together."

Another female student says: "In terms of not having a cause to unite over, maybe that's explained by the fact that there's no political idea that seeks to unite people like me and people like lower-class kids and minorities in the inner cities. I'm not gonna march anywhere holding hands with somebody that's entirely different because the cause doesn't exist. Maybe if it did, things would be different."

The conversation becomes more introspective and searching. A certain kind of yearning creeps into their remarks, a sense that for all the good times they have experienced, they've missed out on some larger public purpose and cause. "I wish we had something to fight for, something that brings you together," one of them says. "I love hearing those stories about the Sixties. We met with a couple of people who were [student] presidents of Stanford in the Sixties and they were talking about how they'd unite and hold protests and sit-ins and do all these things. But for us, there's nothing to unite about. They say you grow through strife, it builds character. So maybe it's been too easy for us."

There's general agreement and a tone of skepticism, too. "I think we're creating an image of the Sixties that's a little too ideal," a young woman says. "Everything wasn't good about all the protests then. Maybe the fact that a common cause doesn't exist now offers an opportunity, which is that somebody from the political world can step into the void and create something affirmative. It's easy to talk about and difficult to do. Maybe it isn't happening because there hasn't been another time when we've been so at peace and so prosperous."

One of her friends, also a woman, says: "I do feel like I missed out on something in the Sixties. Not that I would have wanted to live in that time, but there was something that so united everyone."

When asked why they think there has been what one describes as "a loss of spirit and a common cause," no one has an answer that satisfies them. One suggests that "the anti-tax, anti-government trends of the Eighties contributed to stealing the fire from those common causes." Another wonders if the "opportunity was lost because potential leaders didn't take over and change things." Finally one says, inconclusively, "A lot of different trend lines have been coming together to make things different now."

I return to the striking theme of their chasing the good life. "What is the good life?" I wonder. "What are you chasing?"

The young woman who first used that phrase describes it this way: "From the people I've spoken with, and the people I know, we're all chasing the same thing: Quintessentially the American Dream. We're looking for economic stability, economic security, but we're also looking to feel good about what we do. There's a lot that's different from pure, raw capitalism. I know a lot of people who want to go into business or medicine but also want to do something good with their lives—volunteer a lot, go into community service, be on the board of a foundation. Things like that. We're looking to give our children the same wide range of opportunities that we have right now. We don't want that ever to close down."

"She's right," a female friend says. "There's this weird trend now, especially at Stanford, with Internet start-ups and management consulting, all these things. People want to make a lot of money quickly. They don't just want economic stability, they want economic prosperity, and they also want it without being beholden to anyone. You don't have one boss. You're hopping around from place to place. In an Internet start-up, your only boss is yourself."

"That's desirable?" I ask.

"That's definitely desirable," one says.

"It's appealing," another adds. "Being your own boss, getting everything you want—sounds pretty good to me."

"You mean no one aspires to go to corporate America, to have a job—" I start to say before being interrupted.

"And grind at it for forty years?"

They all laugh.

"No, really, you don't want to work your way up the ranks," one of the male students explains. "You want to become a millionaire quickly. I don't think that's a reflection of a different generational mind-set. It's just that the opportunities are out there. Any generation would have thought it pretty okay to be able to become self-made millionaires very quickly. We have those opportunities with these Internet and software industries that change so very quickly."

They talk about how they differ from their parents, especially their grandparents who were affected forever by awareness of the hardships of the Great Depression and the war. "For them," one of the women says, "economic stability was good, and eight-hour-a-day jobs were fantastic. You were lucky if you got to be in corporate America and ground away for forty years. I don't think our parents want us to backpedal and sort of suffer for ourselves and learn about causes. My parents would say, 'You're prosperous, you have prosperity. Why wouldn't you take that opportunity?'"

A young man says, "I see my parents, the generation that we've been discussing that grew up in the Sixties, I see them wanting me to go and accomplish things. But this is what concerns me about *my* generation. They [his parents' generation] keep an eye on what's going on in politics and the government. So do their friends; they know what's going on out there. But even at Stanford there's little talk about politics, and within my circle of friends at home I'm hard-pressed to get into any kind of political discussion.

No one's interested. Members of my generation have no interest in politics. It has no affect on their lives. They see everything through private businesses and private industry. Government doesn't even play a role. That's a little naive, but that's the sense I get of my generation.

"When I was in high school, I used to say that I wanted to be president of the United States. There was even an article in the *Orange County Register* about me, saying how this kid wanted to be president of the United States. And friends couldn't believe it. 'Why would you ever want to be that?' they'd say. 'Who would want to do that?'

"We're not trained to look at the political issues and keep watch. Things are going all right now, but I'm worried that they may not in the future, that they may slip away. I'm worried that we're not keeping our eyes peeled to make sure that everything's working twenty years down the road, like I still think my parents are. I worry about that for my generation."

"How about the rest of you," I wonder. "Do you agree with what he's saying?"

"It would be an unfortunate statement to say that what the government does doesn't affect me, but it would be a fair statement to say that what I see the government do doesn't affect me that much," one of the women answers. "I know every time I get on the highway I have the federal government to thank. But I'd say that things like the new economy, the new global economy, the things we see most tangibly affecting our socioeconomic future, are not [caused by the government]."

The conversation delves further into their attitudes about government and society. "Maybe that's the big disconnect between we [sic], as Stanford students—people who are well-educated, who have upper-middle-class resources, who don't need the government, who plan to strike out on our own—and people we might not come into contact with who are affected by government programs and federal funding," one of the group says. "I'd be interested to hear what someone in South Central L.A. would say about that."

"Maybe we're a microcosm of how people feel about government," another woman suggests, "but, um, we, I, really find myself pretty impatient with bureaucracy and the system. It's too bad that I'm going to say that government services don't affect me, because they do, but there is a political impatience now. People don't want to wait around for the government to fix things, and they're doing it in the private sector instead. Maybe that's where the government's getting left behind, because structurally it's huge and massive and cumbersome and the pace of business gets faster and faster and fasters."

The young man who wanted to become president when in high school speaks next. "We don't have our best and brightest going into government by any stretch of the imagination," he says. "Why *would* the best and the brightest, people with all these prestigious degrees, go into government? That's sad, especially from the way I think. That's why software companies do so well. They're structured on an incentive basis. They play to human nature: 'If I work, if I do this, I will get there.' There's no sense of that within the government. You don't even get fired in the government. If you're not doing well, they just move you to another office."

He then expresses what turns out to be an inherent conflict they all feel between chasing the good life and pursuing less materialistic callings and losing out materially if they stray from the money track.

"As I say, I personally always wanted to go into politics," he begins. "Now I think, after school I want to try to be an actor. I'm dead serious about that; I think it would be a lot of fun. But I'm used to a middle-class lifestyle and all these things that cost money. My dad has a lot of friends who didn't graduate from high school, and some who dropped out of college, but economically they're all doing fantastically. Very wealthy, live in the biggest houses. But it doesn't seem that that opportunity exists for our generation anymore. Because we *are* the good life generation, because there's so many people trying to snatch up the good life, I think if I spend years trying to do acting and fall on my face I could find myself behind all these kids who have gone forward. I could have kept pace with them, and done all these things, and I could be—who knows? So many people now are taking steps to reach the good life and there's only a finite amount of spaces to go forward. It's a system that discourages taking risks. You almost can't afford it."

"Pressure to stay on the track?" I say. Immediately, a chorus of voices responds.

"Yeah."

"Right. Right."

"And that scares the shit out of me."

"I definitely feel the same thing," says the young woman who first employed the "chasing-the-good-life" metaphor. "I'm a senior and I'm applying for management consulting positions right now [some of her colleagues hiss], and here it's not a popular thing." Then, she adds surprisingly, "I'm not a traditional candidate in that at some point I anticipate going into the public sector. But it's really interesting to talk to people about how any minor step deviating from the career path makes you a more difficult sell later on. It's more difficult to come back in. So there's a lot of pressure to stay on the path."

The pressure, it quickly becomes apparent, comes not only from themselves and the actions of their contemporaries; it stems, as well, from parents pushing them to equate material gain with a successful life.

"Ten to fifteen years from now I hope that I'm doing work that I enjoy," one of the women says. "My parents always told me, 'Why do you think you can find a job that you'll like? Work is just something you do so you can support your family and earn money and be stable. We don't understand why you're so concerned with figuring out what your calling in life is and worrying about having an impact on the community.' They're sort of 'Look out for nmber one and for any children you might have.' You shouldn't worry about anything else. Another thing I'm fighting against is this notion [her parents express] that if you find one thing to do, do it for the rest of your life. I don't think I'm going to find one thing and do it for the rest of my life. I can see myself starting off somewhere and then something else will come along."

"We're in the age of extreme specialization," a male student says.

"In ten or fifteen years I'll probably have changed jobs at least five times," a female contemporary adds.

When I ask about their greatest concerns for the future, their responses quickly disprove the idea they're so self-absorbed, so riven with selfishness, that they either are callously insensitive to problems of others or oblivious to greater economic and social challenges confronting America. They talk about growing disparities between have and have-not worlds, particularly the vast differences between America and much of the planet. They express fears that at some point these conditions of widening inequality could ignite an uprising, trigger a global conflict. They raise the threats posed by terrorism, global warming, depleted resources, rising population. "To me, it's incomprehensible that people will start killing each other in the next fifty years because they don't have fresh water," one of the young men says. "And I take a fifteen-minute shower and go on a run and drink a gallon of water every day. But there are people who are killing each other over that."

To which, a fellow student replies that Americans are out of touch with the rest of world—and so are they. "It just seems with the good life that you pay attention to yourself and not to taking care of other people," he says. "It's inevitable. The rich get richer and they get more educated. And that separates us from the rest of the world. What concerns me is that that separation will become insurmountable."

Yes, one of the women says, and that kind of societal separation is increasing. "More and more people will be more and more oblivious to the have-nots," she says. "Thinking that racism doesn't exist anymore and gay marriages are a terrible thing— those types of things will become more accepted while we'll be moving more on autopilot at the same time these other groups on the side are floundering."

While this conversation is taking place the original "chase-the-good-life" young woman has been intently listening. Finally, she speaks up. "It's telling that our largest individual concerns are about *our* jobs and *our* careers whereas our largest concerns for the future for the greater good are about social problems and things like income distribution problems."

To which one of her friends says, "I don't mean to sound so

cynical, but the directions of the global economy make us more 'Every man for himself.'"

In articulating their concerns, a number have mentioned the fear of the outbreak of a new epidemic like AIDS or the introduction of a fatal virus. When I ask them to address these concerns, one of the men says, "This might be a dumb thing to say, but I have a certain amount of faith in science that if a new disease arises we'll be taken care of. It's hard to think of a problem in those areas that somebody isn't already on the path to solving it or has solved it. I have a great deal of faith in those people."

This prompts me to recall for them a recent dinner conversation I had in Manhattan with two young college students, both of whom smoke, and are longtime close friends of my family. When I asked them why they continued smoking despite their knowledge of its long-term health implications, they both quickly dismissed those concerns. The reason, they told me, was their belief that the new wonders of science and technology would save them. As one of those young students said, "In a few years, they'll make new lungs."

"Does that kind of thinking resonate with you?" I ask the Stanford students.

"If you put it that way, yeah, it's a foolish statement that with all the education people have they still smoke," one replies.

"But you understand what I'm saying," I say, "this idea that science and technology can solve anything."

A young man seems to speak for them all when he says, "Technology in a way scares me. Cloning and those kinds of issues are very scary to me. I want to make sure there are mechanisms in place to control those kinds of things, and I don't know that there will be. Now we have economic distribution inequities that separate classes. Maybe, with the way things work, these genetic make-ups will separate classes. I fear that. Another thing: We have so many conveniences now, things can be done so quickly. We have technology to send things around the world. We exchange information so much more quickly. But we don't have any more time to enjoy. We have less time. The more we can do, the more capabilities we have, the less leisure time we have to enjoy them. I have no idea what it will be like in however many years, but the more we can do, the more we want to do. I'd like to sit back a little more. But I can't afford to."

All of them agree, all confess to feeling greater demands on their time and energies, greater pressures to perform and to succeed. That's why, one of the women explains, we spend so much of our spare time on e-mail. "Your time is chopped up into tiny little pieces," she says, "so you want to get in touch with someone whenever you have a spare moment."

Maybe, another woman suggests, these increasing cumulative pressures of time and careers help explain one of the demographic facts about their generation, particularly about the well-educated career-oriented young women like her Stanford classmates. That is their desire to have fewer, or no, children.

"I think the reason they wouldn't bring children into their world," this female student says, in a fittingly sardonic and cynical final remark, "is that it might interfere with their career paths."

·   ·   ·

Throughout this long and fascinating conversation, I could not help contrasting these favored and attractive young people with John and his fellow gang-bangers in South Central L.A. They expressed many of the same aspirations, but the Stanford students were confident they would achieve them—indeed, they were *certain* that no matter how difficult the conditions facing the nation became, *they* would continue to enjoy a good life. With John, that kind of optimism was notably lacking. The gulf between them seemed hopelessly wide.

·   ·   ·

It's not surprising, if ironic, that these millennials, so scrutinized by elders seeking clues to their behavioral values and career goals, express concern about how different the next wave of young Americans will be from their generation. As one of the Stanford students said, after the discussion turned toward the present high school age group and its supposedly shorter attention span, obsession with video games, even greater disinterest in larger public questions,

"What are these kids gonna be like? How are they gonna interact with society if they spend so much time in front of machines?"

One thing is certain: America's preoccupation with youth will continue to accelerate, and at even faster rates than during the rapid pace of change that marked the Clinton years. Already, the greatest wave of entertainment beamed at Americans through their TVs and movie screens is aimed at capitalizing on an aging America's hunger to emulate teens, who increasingly are seen as setting the cultural trends.

"There's been a big sea change because of the success of *Titanic* and all these teen-girl TV shows," says Ted Harbert, the Hollywood producer who spent much of his career fashioning popular prime-time network entertainment programs and now heads TV productions for DreamWorks, the Spielberg/Geffen/Katzenberg studio. "The reason that teen shows have made a comeback and are so popular is because of the changing mores and behavior of young people. The way seventeen-year-old girls talk about sex is very similar to the way twenty-seven-year-old women talk about sex and relationships. There used to be a big difference. College was that big dividing line. If you were younger, you were in a totally different world, older in a different world. Now they're all blended together. They have the same relationship problems, the same issues; they're having sex the same amount. So the twenty-seven-year-old woman can watch *Dawson's Creek* and see it as entertainment for them, which would never have happened five or ten years ago. Never would have happened."

Harbert recalls seeing rows of limos filled with young girls and their mothers pulling up before the Forum in Los Angeles for a Spice Girls concert. "There were these little girls, from six to sixteen, dressed like hookers," he says, "and more importantly, their moms were dressed like their daughters. The moms and the young women had figured out the teens are setting the trends, so if they wanted to stay hip, they had to dress the same way. It was just amazing to see little girls up on their chairs dancing like they were in go-go cages."

As for *Titanic,* it became the biggest box-office hit in history because it strongly appealed to younger Americans, especially teen girls, and captured their fatalistic, disbelieving, if not cynical, view of life. "Let's not forget," as Harbert says, "that the biggest movie in history, even if it was preachy and sappy, was still ultimately a love story. But it was about *doomed* young love. What speaks to kids today is that even if they open up, even if they fall in love, they're gonna die and their love will not be realized. *Titanic* became an accidental but brilliant metaphor for how these kids feel today. It's almost easier for them to contemplate hitting that iceberg, having that kind of disaster and that kind of doom, than it is for them to contemplate hope, which is that much scarier to them. So in a bizarre sense what *Titanic* did was rip down the barriers we're talking about and allow teens to experience something uncynically for a moment, to be kids and have idealism. But it also appealed so much to them because they all *knew* these two lovers weren't going to be together forever. They *knew* that frigging ship was going to sink, and if it sinks, well, that explains why their young love wouldn't survive in the end."

No happy endings in the new America? No prospect for changing the cultural messages celebrating youthful styles and attitudes, however deceitful those commercial come-ons are? No, no way, Harbert says. And why? "There's too much money in it," he says. "The guys that control the messages to the folks today aren't the government leaders or the business leaders. Now, they're mostly entertainment/media leaders. What we have now are people whose primary goal is to make money by deciding what those messages are and then selling them."

Millennial Generation Update. As I write, results of the latest fall survey of entering college freshmen—members of the Class of '04—have just been published in the *Chronicle of Higher Education.* Under a headline reading LOOKING INWARD, FRESHMEN CARE LESS ABOUT POLITICS AND MORE ABOUT MONEY, the lead sentence in the article says: "If this fall's perplexing presidential race provided a lesson in politics, college freshmen may have slept through the class."

Student interest in politics normally rises during a presidential election year. This time, political engagement "reached an all-time low."

The overwhelming disinterest could not be blamed on an election whose outcome was thought to be a foregone conclusion. In September 2000, when nearly 300,000 incoming freshmen responded to the survey, the most tumultuous election in American history was entering its climactic closing weeks, and it was an election whose outcome at the time was widely—and accurately, as later events proved—expected to be extremely close. While students' interest in politics continued to plummet, their desire to be "very well off" was still rising; three-fourths of them cited this as their primary life-time goal, thus continuing the materialistic trend of recent years.

Scholars examining these latest dismal findings found fresh evidence in them of disturbing national trends.

To Kenneth S. Sherrill, chairman of the political science department at New York's Hunter College, for students to be so disengaged in the face of such a momentous election signaled "a classic danger sign for any democratic political system." To John Gardner, who directs the Policy Center on the First Year of College at North Carolina's Brevard College, the documentation of even greater disinterest in public affairs and increasing interest in personal gain seemed an opportunity to fashion an epitaph for a remarkable era. "We've just gone through a decade—perhaps more than any in history—where the whole country has worshiped the pursuit of wealth," he said. "Many of these students want to cash in on it, too."

All of which led the directors of the annual college survey to conclude, as their findings were reported in the *Chronicle of Higher Education,* that "the wealth and prosperity of recent years may have lulled many young people into thinking that money is what's most important and that it will never be difficult to find."

Yet even as that was written, history, ever impatient with conventional wisdom, forged a new course. Suddenly, a sharp shift in national mood and outlook occurred. The luster of the long boom faded. In America's markets, speculative bubbles burst. Prices plunged, bankruptcies rose, layoffs increased. Fear of looming recession cast shadows over the nation. For the Good Life Generation, illusions of endless prosperity and easy accumulation of wealth were dashed. All this took place against a backdrop of an election that resulted in a historic breakdown of the democratic process, reflecting adversely on the nation's press, its courts, and its local, state, and national governmental functions. It also coincided with the melodramatic, but entirely fitting, best-and-worst events that marked the end of the Clinton years.

Chapter

13

The Fiasco

*"Although we may never know with complete certainty the identity
of the winner of this year's presidential election, the identity of the
loser is perfectly clear. It is the nation's confidence in the judge as
an impartial guardian of the rule of law."*

—JUSTICE JOHN PAUL STEVENS

The Capitol, January 20, 2001. An inauguration of an American president is both an occasion for hope and for America to demonstrate anew to the world that, no matter how bitterly fought an election contest, power always passes peacefully in the democracy. So it did again on this fifty-fourth inaugural day, but this was no ordinary celebration of the peaceful electoral process.

From the inaugural platform facing west from the Capitol, dark clouds and heavy mists nearly obliterated the historic monuments to Washington and Lincoln and Jefferson. Huddled together on the grounds below, braving lashing rain and numbing cold, the inaugural crowds were more subdued than celebratory. They greeted the words of their new president with a smattering of applause. Only once did they voice a genuine cheer—that, when the president repeated a campaign pledge to cut their taxes.

The parade route was rife with signs of unprecedented security measures. For the first time, citizens seeking to witness the procession had to pass through numerous checkpoints where their possessions were inspected. Behind the grandstands, sodden with rain, sullen masses of demonstrators stood, their presence watched by wary police and military forces stationed to keep them from moving forward to Pennsylvania Avenue where the new president would soon pass.

Even before the parade began, the demonstrators began waving their placards: HAIL TO THE THIEF ... CRIME SCENE ... THE PEOPLE HAVE SPOKEN—ALL FIVE OF THEM ... ELECTION FOR SALE ... FRAUD. Sounds of their angry chants echoed off the marble buildings housing the executive departments and agencies of the United States government—the Justice Department, the FBI, the Interior Department: *Hey, Dubya, What Do You Say, How Many Votes Did You Steal Today?* Some demonstrators, wearing black masks, succeeded in forcibly evicting people with parade tickets from grandstands. Then they occupied the stands and waited, ready to pounce on the procession. Others, their faces also hidden by masks, climbed a flagpole at the Navy Memorial across the street from the National Archives, which houses the Declaration of Independence, the Constitution, and the Bill of Rights. They tore down and burned the American flag and other military banners.

The icy rain, the cold, the swirling clouds, the palpable feelings of anger and recrimination, perfectly reflected the extraordinary nature of the election that had been concluded only so recently, and as many believed concluded illegitimately.

*   *   *

On election night, November 7, 2000, television screens best captured the results of the millennial contest by showing an America deeply divided. Swaths of red marked territory captured by the Republican candidate. They encompassed most of the areas from the Appalachians to the Rockies, from the Canadian border to the Gulf of Mexico—an area, in its totality, more sparsely populated than bi-coastal America, more rural than urban. Swaths of blue marked territory captured by the Democratic candidate. They encompassed narrow bands of territory running the length of the West Coast and from Maine to the Mid-Atlantic states on the East Coast, with a band of upper midwestern industrial states—far less land mass, but far more heavily populated, far more urban than rural.

Out of this emerged a portrait of an America almost perfectly split along geographic, cultural, ethnic, racial, gender, political, ideological lines. The way the votes were cast highlighted an almost total division of the nation.

More women voted for Al Gore, the Democratic vice president and presidential candidate; more men voted for George W. Bush, the Republican son of a former president. Gore won more black, union, urban, and non-gun-owning voters; more new immigrants or recent descendants of the old ethnic melting-pot population; more non-churchgoing families; more pro-choice women. Bush won more white men, especially in the South, the old stronghold of the Democratic party; more gun owners; more men and women opposed to abortion; more rural inhabitants; more religious conservatives. As the *Economist* commented on the returns, the election mirrored the portrait of a starkly divided America sketched by the conservative social historian Gertrude Himmelfarb in her *One Nation, Two Cultures*. One America was hedonistic, secular, socially tolerant, rooted in the values of the Sixties; the other was religious, puritanical, insular, grounded in the mores of the Fifties. The Republican pollster Bill McInturff, analyzing the election results for the *Washington Post,* came to a similar conclusion. "We have two massive, colliding forces. One is rural, Christian, religiously conservative with guns at home, terribly unhappy with Clinton's behavior... And we have a second America that is socially tolerant, pro-choice, secular, living in New England and the Pacific coast, and in the affluent suburbs." The more often voters went to church, the more likely they were to vote Republican. As the *Economist* noted, Gore won the godless coasts and the industrial Midwest. Bush won the rural vote, capturing much of the Midwest and the South, though in the South he needed the overwhelming support of southern white men to offset the votes of blacks and women.

Bush lost the popular count to Gore by over half a million votes, or half of one percentage point of more than the 100 million ballots cast.★ He became the first president in 112 years to win the White House while losing the popular vote. He eked out an electoral college victory only after thirty-six emotion-laden days of intense uncertainty marked by recount after recount and court challenge after court challenge. The election was finally decided by historic but dubious rulings of a United States Supreme Court that for the first time interceded in and determined the outcome of a presidential election. Bush won by one more electoral vote than he needed to gain the presidency.

The makeup of the new Congress, too, reflected the razor-thin margin between victor and vanquished. In the Senate, Democrats gained four seats to produce a fifty-fifty tie with Republicans, the first such equal division of power there in more than a century; but the Republicans still ruled, since the new vice president, Dick Cheney, would cast tie-breaking votes as the constitutionally prescribed presiding officer of the Senate.★★ In the House, Democrats

---

★The 2.7 million votes cast for the liberal/progressive candidacy of Ralph Nader were the decisive factor in determining the winner. Nearly a hundred thousand of those Nader votes were cast in the ultimately crucial electoral battleground of Florida, which Gore lost by only a hundred or so votes. By siphoning off Gore votes, Nader's candidacy clearly made *the* difference in electing Bush. The final national vote tabulation by the Federal Election Commission showed: Gore: 50,996,039, Bush: 50,456,141. Gore's popular vote margin was 539,898.

★★Four months after Bush's inauguration, in a stunning move that turned the political world upside down, control of the Senate went back to the Democrats after Senator James M. Jeffords, a Republican moderate from Vermont, defected from the GOP, became an Independent, aligning with the Democrats to give them the power to set the Senate's political agenda and head its key committees. Jeffords made his extraordinary move out of what he described as his deep concern over the strongly conservative stance Bush had taken in pressing his new agenda and the increasingly more conservative cast of the Republican Party. "Looking ahead," he said, in an emotional news conference May 24 in

made a slight gain of one seat to bring the margin between the two parties even closer to parity than it had been in the last Congress; a shift of only five votes would give Democrats control. Although Republicans ended up controlling both houses of Congress and the White House for the first time in nearly half a century, neither the new president nor his party on Capitol Hill could claim any kind of mandate. Seldom in all of American history had political divisions been so great and power held so evenly.

Not since 1876, in the embittered aftermath of the post–Civil War Reconstruction period when the Democratic candidate Samuel J. Tilden won the popular vote but lost the electoral college vote to Rutherford B. Hayes under clearly fraudulent actions by the Republican-controlled Congress, had America experienced anything like this presidential contest. The anger stirred by that ancient contest rivaled the rancor arising out of the millennial one a century and a quarter later.

Election 2000 was a fiasco of many dimensions with possible long-term consequences.

The television networks disgraced themselves by a series of inaccurate election night projections. First they seemed to give the election to Gore. Then they reversed themselves by saying it was too close to call. Hours later they ignited a brief celebration among Republicans across the country by falsely declaring Bush to have been elected president of the United States.★

Vermont, "I can see more and more instances where I will disagree with the president on very fundamental issues—the issues of choice, the direction of the judiciary, tax and spending decisions, missile defense, energy and the environment and a host of other issues, large and small." His greatest disagreement, he added, was with his party and his president on the issue which Bush had made his centerpiece during his election campaign: education.

★In a scathing report commissioned after the election by CNN, three journalism experts excoriated election night TV as a "news disaster" and blamed the networks for creating a political climate of "rancor and bitterness" that created a premature impression Bush had won, a characterization that carried through the postelection period in which "Gore was perceived as the challenger and labeled a 'sore loser' for trying to steal the election." The report called CNN's coverage a "debacle," but leveled much broader criticism at all

In an intensely divisive action, the Supreme Court damaged its reputation for impartiality and judicial discretion by intervening for the first time in a state court's proceeding in a presidential election. On December 9, the Supreme Court halted a crucial recount of presidential ballots just one day after the Florida Supreme Court had ordered it to proceed. Not since the Dred Scott case, which fanned the flames that were to ignite the country in Civil War, had a Supreme Court decision been so controversial.

Gore was gaining when the high court stopped the recount. At that point, on December 9, a mere 154 votes out of 6 million Florida presidential ballots separated the candidates—an astonishing margin of less than one ten-thousandth of the votes cast. Had the recount been concluded, it could well have provided enough votes for Gore to win. One early independent analysis later suggested that an accurate recount would have given Gore the victory by some twenty thousand votes. In fact, at every stage of the byzantine legal and political wrangles that marked the attempts to accurately recount Florida's presidential votes, Gore had consistently gained on Bush. The night the election ended, the official tally showed Gore trailing Bush by 1,784 votes, a margin so thin that an automatic recount was mandated by Florida law. Within twenty-four hours, after that mandatory recount ended, the margin between Gore and Bush stood at only 327 votes, less than a fifth of what it had been.

Despite the favorable vote trend, it is far from clear that Gore would have won if the ballots had been recounted. In the months after the election, attempts to answer that ultimate question produced only more confusion—and, in some cases, the surprising verdict that Bush, not Gore, would have carried the Florida vote. Five months after Bush was declared the winner, an extensive survey of

network and cable coverage and called for, among other remedies, an end to the use of exit polls to project winners and a ban on calling a winner while polls were still open. The inaccurate TV calls declaring Bush the winner and subsequent retractions "played an important part in creating the ensuing climate of rancor and bitterness" that wracked the nation, the report concluded, and constituted "a news disaster that damaged democracy and journalism."

Florida's undervotes and overvotes by the *Miami Herald,* Knight Ridder newspapers in Florida, and *USA Today,* joined by five other Florida papers, produced findings that added to the uncertainty. After examining 61,000 undervotes and 110,000 overvotes throughout the state, on May 11, *USA Today* reported on the groups' conclusions in a page one headline that told a classic story of ambiguity:

Florida Voter errors cost Gore the election

But just beneath, readers were informed that the survey showed another result:

Bush still prevails in recount of all disputed ballots, using two most common standards.

The accompanying story was, if anything, even murkier. "George Bush would have won a hand recount of all disputed ballots in Florida's presidential election if the most widely accepted standards for judging votes had been applied," Dennis Cauchon and Jim Drinkard reported. Then they said: "However, the review of 171,908 ballots also reveals that voting mistakes by thousands of Democratic voters—errors that legally disqualified their ballots— probably cost Al Gore 15,000 to 20,000 votes. That's enough to have decisively won Florida and the White House." The paper summarized the group findings by giving what it called "answers" to four main questions about the election:

1. Who would have won if Gore got the manual vote counts he sought in four counties? Answer: Bush.
2. Who would have won if the Supreme Court had not stopped the hand recount of undervotes? Answer: Bush, under three of four voter counting standards.
3. Who would have won if all disputed ballots had been re-counted by hand? Answer: a split verdict: Bush, under the two most widely used voter recount standards, Gore, under the least two.
4. Who did it appear most voters intended to vote for? Answer: Unequivocally, Gore.

Far from resolving the election controversies, these kinds of findings only intensified them. And they and other post-mortem voter analyses did not address the greatest of the controversies surrounding the election—the manner in which it was finally decided, and the decisive role played by the Supreme Court.

In first intervening and then deciding the election in such stunning and controversial fashion, the high court showed it was as divided as the rest of the nation, and at least as bitterly polarized ideologically. Not only was it polarized; the rationale some of the justices offered for their decisions reeked of political considerations.

Justice Antonin Scalia, the court's strongest, most articulate, and conservative member, came close to acknowledging that the majority was acting to ensure Bush's election. In a statement accompanying the court's narrowest possible five-to-four decision to halt the Florida recount on December 9, setting the stage for the final arguments on whether to permit them to proceed, Scalia said, "The counting of votes that are of questionable legality... threaten[s] irreparable harm...by casting a cloud upon what [Bush] claims to be the legitimacy of his election."

Scalia added: "It suffices to say that the issuance of the stay suggests that a majority of the court, while not deciding the issues presented, believe that the petitioner [George W. Bush] has a substantial probability of success." While his legal rationale for issuing the stay—that the petitioner had "a substantial probability of success"—is one often employed by the court in its rulings, to many in the public at large Scalia's statement created an impression that the court's conservative majority already had made up its mind before hearing the final appeal.

Three days later, when the court ruled on Gore's appeal challenging its decision in *Bush v. Gore* to halt the recounts, the public had evidence of even greater depths to the court's ideological divisions. Its final decision again showed the court's five most conservative members—Scalia, backed by Chief Justice William H. Rehnquist and Justices Clarence Thomas, Anthony Kennedy, and Sandra Day O'Connor—were implacably aligned against its four more center-to-liberal ones, Justices John Paul Stevens, David H. Souter, Ruth Bader Ginsburg, and Stephen G. Breyer.

That decision effectively terminated the presidential contest and handed Bush the presidency.

Dissents to the narrow-but-decisive ruling by the high court were remarkably sharp, and tinged with unprecedented tones of anger. In her dissent, as the *Washington Post* reported, Justice Ginsburg protested bitterly that the court lacked any good reason to second-guess the interpretation of Florida state law by that state's highest court. As the *Post* put it, "in a thinly veiled accusation of hypocrisy aimed at the court's majority who have championed states' rights in past cases," Justice Ginsburg said, "Were the other members of this court as mindful as they generally are of our system of dual sovereignty, they would affirm the judgment of the Florida Supreme Court" ordering the vote recount.

She concluded by saying: "I dissent," pointedly omitting the customary court modifier, "respectfully."★

Joining her in dissent, Justices Souter and Breyer argued that problems of fashioning remedies to inexact standards on how to recount Florida votes could be overcome. Souter blamed the majority for stopping the recount in midstream, thus placing even greater time constraints on the process, virtually ensuring it could not be completed by the constitutionally designated deadline for members of the electoral college to cast their respective state's presidential ballots. "To recount . . . manually would be a tall order," Souter wrote, "but before this Court stayed the effort to do that, the courts of Florida were ready to do their best to get that job done. There is no justification for denying the State the opportunity to try to count all disputed ballots now." Justice Breyer, in his dissent, warned his colleagues of the potential long-term consequences of the court's decision. "In this highly publicized matter, the appearance of a split decision runs the risk of undermining the public's confidence in the Court itself," he wrote. "That confidence is a public treasure."

★Nearly two months later, after the new president was in office, in a law school speech in Australia, Justice Ginsburg referred to the court's historic election decisions as "this breathtaking episode" and as a "December storm over the U.S. Supreme Court." She said that pressure for a resolution had "pushed the unanimity of the United States Supreme Court past its breaking point."

He added sharply, "We do risk a self-inflicted wound, a wound that may harm not just the Court, but the Nation."

By far the most extraordinary of the dissents, and the one that most clearly indicated the depth of the emotions dividing the court, came from Justice Stevens, at eighty its oldest member, a Republican who had been appointed to the bench in 1975 by President Gerald R. Ford. Justice Stevens, joined by Justice Ginsburg, said the court's decision "can only lend credence to the most cynical appraisal of the work of judges throughout the land. It is confidence in the men and women who administer the judicial system that is the true backbone of the law."

Then he wrote, in words that will always hang over the 2000 presidential election: "Time will one day heal the wound to that confidence that will be inflicted by today's decision. One thing, however, is certain. Although we may never know with complete certainty the identity of the winner of this year's Presidential election, the identity of the loser is perfectly clear. It is the Nation's confidence in the judge as an impartial guardian of the rule of law."

The ruling ignited a storm of outraged reaction, and not least coming from well-known members of the bar. "The court's decision a week ago to stay the hand count of undervote ballots was the most overtly politicized action by a court that I have seen in 22 years of practicing law," Scott Turow, the Chicago lawyer and successful prosecutor turned novelist, wrote in the *Washington Post*'s Sunday editorial section. "It was an act of judicial lawlessness that effectively terminated Gore's chance to win the presidency. It deviated so far from governing legal principles that Terrance Sandalow, a conservative legal scholar and former dean of the University of Michigan Law School, was quoted calling the decision 'incomprehensible' and 'an unmistakably partisan decision without any foundation in law.'" Turow added: "The court ignored the lower federal courts, which had four times rejected similar stay requests from the Bush campaign, because it could not prove that Bush would be irreparably harmed by the recounts." While Turow doubted the *Bush v. Gore* decision will seriously erode the authority of the court with average Americans, he feared "its impact on those who live the law—lawyers and legislators—is likely to be far-reaching. . . . Unless there is a sustained outcry from the legal profession calling for the restoration of the boundaries we have known for a century, the reliability of our courts will remain uncertain for a long time to come."

Ronald Dworkin, the distinguished legal scholar writing from a liberal-left perspective in the *New York Review of Books,* said: "The conservatives stopped the democratic process in its tracks, with thousands of votes yet uncounted, first by ordering an unjustified stay of the statewide recount of the Florida vote that was already in progress, and then declaring, in one of the least persuasive Supreme Court opinions that I have ever read, that there was no time left for the recount to continue."

Far more intemperate—even "savage" doesn't do justice to the feelings expressed—but no less revealing of the deep emotions being stirred, was the attack from another prominent lawyer, Vincent Bugliosi, who had successfully prosecuted 105 out of 106 felony jury trials as a Los Angeles deputy district attorney and 21 murder convictions without a loss, including that of mass-murderer Charles Manson. In a lengthy postmortem article in the *Nation,* the journal of opinion of the left, Bugliosi erupted furiously: "The stark reality, and I say this with every fiber of my being, is that the institution Americans trust the most to protect its freedoms and principles committed one of the biggest and most serious crimes this nation has ever seen—pure and simple, the theft of the presidency. And by definition, the perpetrators of this crime *have* to be denominated criminals."

Recriminations over the election's outcome and the way it was ended was not limited to outrage over actions of the networks and the Supreme Court.★ State and local governments, particularly of Florida, had also disgraced themselves by demonstrating, first, their unwillingness to invest in modern equipment to ensure speedy and

---

★The public's response, it must be noted, was hardly sustained. In general, Americans reacted to the election outcome the way they had to the big stories in the Clinton era—strong immediate response, a sense of disgust, and then . . . nothing, as they moved on to the next news development trumpeted over the tube.

fair tabulation of votes and, second, by failing to institute uniform and fair standards to ensure that every citizen's vote counted. Actions of the leading Florida state election officials, serving under George W. Bush's brother and fellow governor, Jeb Bush, raised serious questions about their impartiality when they rendered a number of rulings that clearly favored the Bush presidency. These immeasurably aided the Bush team and its political strategy of delaying, and hopefully ending, vote recounts by a series of time-consuming court challenges that ultimately ran out the clock on the time remaining before presidential electors had to certify the results.

At the center of the controversy was the state's number one election official, Secretary of State Katherine Harris. Not only was she politically close to Jeb Bush, but she also served as Republican co-chairman of the Bush presidential campaign in Florida. Harris, quickly adopted by the media as the latest instant celebrity and subjected to massive TV coverage—she was dubbed the socialite "citrus heiress," and called Cruella de Vil after the Disney wicked witch–type cartoon character—was widely, and credibly, believed to be seeking an ambassadorship in a future George Bush presidential administration.

American democracy was diminished by these extraordinary events. As the days passed, more and more evidence of voting irregularities and inequities surfaced.

Most egregious was the documentation that blacks, who had died demonstrating for the right to vote just a generation before in the segregated Deep South, remained the most disenfranchised citizens. The most impoverished precincts where blacks voted received the least electoral financial resources, forcing blacks to contend with the least reliable—and often faulty—voting equipment. Much-less-accurate punch-card ballot machines, used in Florida counties with a heavily minority population, were subsequently shown to result in more than three times as many faulty ballots that wound up not being counted as those cast in wealthier, largely white areas that employed more modern and accurate optical ballot-scanning systems. After the election, a computer analysis of the most heavily white precincts of Jacksonville by the *Washington Post* documented this inequity. There, the *Post* found, one of every fourteen ballots was thrown out. At the same time, more than one in every five ballots was thrown out in Jacksonville's largely black precincts. Eight of every ten ballots thrown out statewide in Florida—in all, 180,110 votes were discarded—came in counties using the least-accurate punch-card systems most prevalent in the heavily minority precincts.

Compounding the inequities was another fact: Many of Florida's black precincts did not have computers available to help solve problems when voters found their names missing from registration lists at the polling places. Precinct workers laboring under that handicap were forced to phone central election offices for verification, but, as the *New York Times* reported, "many voters simply gave up and went home because the lines were almost constantly busy."

Blacks were also shown to have suffered from old patterns of discrimination at the polls, including feeling intimidated after being stopped at police checkpoints, not present in similar force at white precincts, and finding their names unjustly stricken from the voting rolls when they attempted to cast their ballots.

The millennial election became the latest, and in some ways the greatest, televised mass spectacle of the Clinton era. All the elements that had typified the excesses of the age were present, including around-the-clock TV coverage complete with specially designed electronic logos and graphics dramatically announcing that *America Waits* the selection of its new president.

Many of the same talking-head commentators and ideologues who first surfaced during O.J., then Monica, then the impeachment, to argue the political merits of those moments reappeared with regularity to offer still more heated polemical arguments over the tube. Some who played less visible public roles in other notable cases of the Nineties became leading players in the election fiasco maneuverings. Notable among them was David Boies who had successfully tried the government's antitrust case against Bill Gates and Microsoft; in what seemed the interminable five-week presidential recount legal process, Boies became the lead and ultimately the losing lawyer for the Gore team's final Supreme Court appeal.

TV techniques that powerfully affected the news business and the nation were reemployed.

In one extraordinary replay of Teletimes past, Americans were treated to the sight of another—and longer—low-speed highway chase as news helicopters hovered for hours over a rental truck carrying disputed presidential election ballots hundreds of miles from southern to northern Florida. O.J. himself, then living in Florida, could not resist the moth-to-light TV instinct of a fading—and infamous—celebrity and made himself available to media interviewers. He commented, neither memorably nor insightfully, on which side had the better legal case.

The entire mass spectacle became a rich object for late-night comedians and ridicule worldwide as problems arose with ill-drawn and hopelessly confusing "butterfly ballots," with "undercounts" and "overcounts," and most memorably with "chads."* Those were the small squares of paper on perforated voting cards that either were or weren't properly pushed through by a citizen to register a vote. Endless wrangling about the efficacy of "hanging, dangling, swinging, dimpled, pregnant" chads entered the public lexicon as primary topics of discussion, and derision.

For Americans, the election debacle was, at minimum, a great embarrassment and a blow to national self-esteem. For America's rivals, it provided a chance to gloat at the all-too-evident humiliating stumbles and ineptitude of the world's colossus. America, the most technologically superior society, couldn't even succeed in providing the technological know-how to count its own votes accurately. America, which prided itself for living up to its stated principles of equality by ending segregation and striking down legal barriers to voting because of race and sex, was shown to be far from perfect in ensuring fairness for all its citizens. America, the richest and most powerful nation in history, was demonstrably unwilling to invest necessary sums to assure that the most precious right of the democracy, the free electoral franchise, was protected, with adequate procedures in place to guarantee it.

---

*Thousands of Palm Beach County voters, many of them Jewish and lifelong Democrats, thought they were voting for Gore. Instead, their votes were counted for Pat Buchanan, the third-party right-wing candidate, because of the ineptly designed and ultimately disastrous "butterfly ballots."

· · ·

When the election year began, a central question loomed: What effect would the scandal era and the cynicism sown by Bill Clinton and his enemies have on voters? Would the dismal voter participation rate, already the worst in the industrial world, continue to decline? Or would voters show that underneath their protective layer of cynicism lay a reservoir of hope and idealism?

At first, in the initial presidential primary contests, voters displayed enthusiasm for candidates who seemed both "authentic" and of unblemished personal character. In those early election-year stages, the candidacies of John McCain, the Republican, and Bill Bradley, the Democrat, produced the highest primary voter turnout in years.

Clearly being expressed by voters then was a hunger for believability, for truth, for change, for renewal. Most of all, they were signaling a desire for reform—*real* reform—and for something better. Although both McCain and Bradley read the country's mood correctly, they did not prevail. It is no disservice to suggest that neither of the eventual presidential nominees, Bush or Gore, stirred the hearts of the people. Nor, as the political process unfolded, did it fulfill hopes for political change; the campaigns were awash in the greatest outpouring of special-interest money ever, with more than $3 billion raised by the candidates and another estimated $2 billion raised and spent on the election. Once more, campaign tactics were aimed at the lowest common denominator. Real extended debate of issues again became a casualty of the political process.★

★Before the election year began, two prospective presidential candidates, McCain and his fellow Vietnam war hero, Democratic senator Bob Kerrey, told me in separate conversations that they had made a private pact. If they won their party's 2000 nomination, they would fulfill the promise that John F. Kennedy and Barry Goldwater had made to each other in the expectation of contending in the 1964 election. That is, they would campaign daily together across the United States engaging in extensive, freewheeling Lincoln-Douglas-type debates. I'm certain the country would have enthusiastically welcomed their approach; but they were not given the opportunity to try it out.

As it became clear that the Bush and Gore candidacies would gain their respective party's nominations, voter participation fell off sharply. This pattern was repeated in the fall election. In the end, barely more than half of those eligible to vote for president chose to do so. Voters who had turned out in such numbers early on and then did not choose to participate in the actual election were not so much disengaged from the political process. They were people who ultimately didn't find what they were looking for.

·   ·   ·

However dismaying or enraging the still-raw election events were, by Inaugural Day 2001 they were all part of history. Now what mattered was the scene being played out on that windswept inaugural stage where the central actors had gathered behind rain-spattered bulletproof glass for the final act of the most tumultuous election in the nation's history.*

By right, the stage belongs to the new president. Attention always focuses squarely on him as the winner and new national leader. Cruelly but inevitably, the erstwhile rival, who happens to occupy that same inaugural stand in a traditional display of the harmonious and peaceful passing of power, is relegated to the sideline and the instant obscurity that awaits history's loser.

Not surprisingly then, the cameras focused little on Al Gore this day, though given the circumstances of his defeat a natural public desire existed to try and fathom the feelings of someone who found himself in such a difficult public position. However stiff and ineffectual a candidate, however insufferably officious a debater, Gore had been a most gracious loser. He had exhibited genuine class in a generous concession speech delivered with ease and a winning, self-deprecating sense of humor. It was by far his best speech of the campaign; had he been able to convey such reassur-

---

*It was not, as often said and written, the *closest* U.S. election. Kennedy-Nixon was decided in 1960 by only 118,550 votes, and eight years later Nixon-Humphrey by 510,314. But Florida's crucial election, of course, was far closer than even the wildest nightmares of ballot projectors and political demographers.

ing personal qualities even occasionally, he might well have won without the wrenching ordeal of all the recounts. Of course, as he knew better than anyone standing there, he *had* won the popular vote. He also had strong grounds for believing he had actually won more than enough of the disputed Florida votes to give him the electoral college victory and the presidency—for believing *he* should be there raising his hand, taking his oath, delivering his address, receiving the tribute of the twenty-one-gun salute that boomed out over the dank, misty Capitol grounds.

However valid the reasons for paying him more than cursory note, Al Gore remained a bit player on that platform.

Though it was George W. Bush's stage and moment, in a striking sense it was the figure of the outgoing president that dominated the scene. This was so not just because of the jaunty, jubilant, smiling manner in which Bill Clinton comported himself as he took his place on that stage for his last presidential bow. Nor was it because the sight of the outgoing first lady—and now incoming new United States senator—beaming cheerfully toward the dignitaries and official guests stirred unusual interest, inevitably raising questions about what *her* next national act would bring.

What focused attention so intently on Bill Clinton was the manner in which he was departing, a manner utterly fitting for his presidency and for the age he singularly represented. Of William Jefferson Clinton it can be said that nothing better exemplified the worst nature of his presidency and its effect on the American people than the way in which he left it.

•   •   •

At the beginning of his second term in 1997, after voters gave him a chance to make a true mark on history by becoming the first Democratic president to serve successive terms since Franklin Roosevelt in 1936 at the apogee of the New Deal, Bill Clinton expressed to his speechwriter Michael Waldman his determination to forge the kind of positive historical legacy that other talented presidents had failed to achieve. For Clinton, the conditions were right: a country at peace, a great boom underway and accelerating, a people optimistic about the future.

Clinton had a plan on how to win that legacy. "All kinds of presidents had significant accomplishments who never got any credit in history because they couldn't control the story line," he told Waldman, as the speechwriter recalled in his memoir, *POTUS Speaks*. Typically Clinton, the gifted improviser, charmer, student of polls and shifting public currents, planned to control that story line. Thus, he would establish his lasting presidential legacy.

That, of course, was before Monica, the impeachment, and all the other Scandal Times diversions that stained his second term and forced him to battle not for a long-term legacy but for the chance merely to survive and not be driven from office in disgrace.

However physically and emotionally draining his fight for survival had been, and however damaging to his public reputation, typically Bill Clinton never forgot his determination to win his legacy by controlling his final presidential story. If anything, his many escapes from disaster only strengthened his resolve. After surviving scandal after scandal, he was determined to avoid making impeachment the first line in his obituary, or in later, more extensive and substantive historical assessments.

His final presidential months were even more frantic than his chaotic opening ones. He was constantly in motion, tirelessly working to demonstrate to the American people and to the world that he was a president who had made a difference. He made eight trips, visiting fourteen countries. He became the first president to visit Vietnam, seeking symbolically to close that tragic chapter in American life. As his days in office dwindled, he redoubled his efforts. He employed all of his charm and skill in an attempt to achieve a breakthrough in the poisonous and again-deteriorating Middle East. He summoned the principal Israeli and Palestinian leaders to the White House, then to Camp David. He dispatched emissaries, hoping to spur an accommodation. In dealing with other great trouble spots—Ireland, North Korea, China, and India and Pakistan—he displayed the same tenacity and energy. Though he did not achieve the historic breakthroughs he sought, the American people credited him with doing the best he could to secure significant easing of tensions and improve the prospects for peace.

On the home front, he worked no less tenaciously to secure

the legacy he craved. He issued a series of executive orders setting aside forever from commercial use the largest unprotected areas of natural wildlife preserves in the United States. He took further actions to protect the environment and issued a blizzard of new regulations governing, among other activities, health and safety working conditions throughout the nation. He began a series of national appearances that in effect represented an extended farewell tour designed to remind voters of what strides had been made during his eight years as president. When his wife was sworn in as New York's new senator, marking a historic first for a first lady and setting the stage for perhaps another Clinton presidency, he was seen glowing with pride at her side. However grave, embittered, and public their marital problems had been, these happy new scenes suggested that Bill and Hillary had successfully put the Monica scandal and its excruciating private and public betrayals and humiliations behind them.

Cynics could—and did—sneer at this flurry of activity and the accompanying feel-good scenes as mere image making, more Clintonesque public manipulation. But the American people, fair as ever, responded positively to them.

As the Inaugural Day approached, Bill Clinton was going out of office with the highest presidential job approval rating of any of his two-term predecessors since pollsters began assessing the chief executive's public standing in the Fifties. Bill Clinton, the bad-boy president—moral leper to many, incorrigible adolescent to others—was leaving with higher approval ratings than the sainted Ike, greater than the beloved Reagan.

The public did not, also fairly, forgive Clinton for his moral lapses; he continued to rank near the bottom of the list of his predecessors for the ethical conduct of his office. Nor did voters signal their approval for many of the sordid actions that had marred his presidency. No greater evidence of underlying public disapproval for those aspects of the Clinton years was shown than in the way voters of Arkansas, his springboard to the presidency, cast their presidential ballots. Al Gore, Clinton's handpicked choice as vice president, couldn't even carry Clinton's home state. Had Gore carried Arkansas—to say nothing of his own home state of Tennessee—he would have won the presidency even without Florida's disputed electoral votes.

Still, Americans were clearly in a tolerant, forgiving, even affectionate mood toward their talented but wayward outgoing president, and Clinton clearly was determined to capitalize on and strengthen those positive feelings.

In the final days of his administration, published assessments of his presidential performance, those journalistic "rough drafts of history" that either do—or don't—set the stage for later historical judgments, were mixed. But all in all, they were more positive than one might have expected from some of the same critics who had chastised him most severely.

Speaking with the weight of its influential editorial voice, the *New York Times,* which had been consistently critical, even unremittingly harsh, in its judgments of Clinton, invoked an editorial cartoon in its final appraisal of Clinton's legacy. The cartoon depicted Clinton lunging to make a diving catch of a huge ball labeled "Greatness." For the *Times,* that image provided "a haunting visual metaphor for the near-miss quality of a presidency that never quite measured up to the potential that every supporter and most critics knew resided within Mr. Clinton. Yet simply stating the fact that Mr. Clinton is not destined to enter the pantheon of great, universally respected presidents does not capture the richness, complexity and drama of those eight years. No citizen is likely to forget them or, for that matter, to quit debating the man's remarkable gifts and the narcissistic indiscipline that diminished them."

Noting, in well-chosen words, that the Clinton years had been marked "by prosperity, rancor, achievements, disappointments and something approaching a national psychodrama involving Mr. Clinton himself," the paper said memorably: "Trying to separate Mr. Clinton from the times he presided over is like trying, in Yeats's phrase, to tell the dancer from the dance."

The *Economist,* which when the Monica scandal broke and impeachment loomed had sternly called for him to resign to save the nation from further embarrassment and the presidency from further diminishment by his disgraceful actions, confessed in the week before the new inauguration that "we have to admit that Mr. Clinton has preserved much more moral authority and effectiveness in office than ever seemed possible. By an irony, he has looked more and more presidential the closer he has come to leaving. In these past weeks, as the focus has shifted to another inexperienced but less clever southern governor, there is a huge sense of talent wasted. With more discipline and less self-indulgence, how good eight years of Bill Clinton could have been."

Time, seeking the views of a number of distinguished citizens, Clinton friends and foes alike, offered assessments that were certainly more positive than they would have been a mere year before. To James MacGregor Burns, the historian, Clinton had been an underrated president, superb at making deals, navigating the political process, dealing with enemies. Burns saw him as an enormously active chief executive who laid out great ideas and dreams and hopes—but who didn't follow through strategically on problems or display the necessary kind of commitment and conviction even to begin to solve them. A surprisingly positive assessment came from one of his strongest Democratic critics, Nebraska senator Bob Kerrey, who told *Time:* "In spite of my sometimes irritation with him when he makes a reversal or doesn't press ahead on important policy areas—like Social Security or Medicare—on the thing that got him where he is today, he has been unwavering: belief in economic growth, and belief that while the economy is growing, you've got to push the circle of freedom out with healthcare and education investments. And he is relentless. Sure, he has altered course from time to time, but he's never backed away from his core set of beliefs. He has redefined what it means to be a Democrat in several important ways. We no longer should be on the defensive about being a tax-and-spend party because we delivered just the opposite. We delivered economic growth. He delivered on law enforcement when the American people used to say Democrats were soft on crime. He hasn't abandoned the core Democratic belief that we should give people an opportunity. Even on defense, he's shown you don't have to have been in the military to be a good Commander in Chief."

Coming from Kerrey, who lost a leg as a Navy Seal in Vietnam and went on to become the only Medal of Honor winner ever to serve in Congress (and who, it later turned out, was a man hiding his own secrets from the bitter Vietnam experience), that last bit of praise for Clinton's civilian leadership of the military was an unexpected and surely welcome bonus for the president. It added to the sense of public goodwill being expressed about his presidential stewardship as he was about leave the White House.

Two days before the inauguration, continuing to work to the end to enhance his presidential legacy by controlling the story line about his presidency, Clinton scheduled a prime-time event to give his formal farewell address from the Oval Office. It was brief— eight minutes—and celebratory and self-congratulatory in word and spirit. He thanked his fellow Americans for their role in transforming the nation during his presidency into "an era of great American renewal." He reminded them of the good times they had enjoyed during the boom and thanked them for their support in helping to build what he would obviously like history to credit him with bequeathing to Americans: "a future of our dreams in a good society with a strong economy, a cleaner environment and a freer, safer, more prosperous world." He called the statistical roll of all the good things his era had produced, among them "22 million new jobs, the lowest unemployment in 30 years, the highest home ownership ever, the longest expansion in history."

It was all a paean to the golden age he saw himself presiding over, and as he neared the end of his remarks he told the American people how he was leaving the presidency "more idealistic, more full of hope, than the day I arrived and more confident than ever that America's best days lie ahead."

In his final words, the outgoing president delivered his ultimate story line: "I will never hold a position higher or a covenant more sacred than that of president of the United States. But there is no title I will wear more proudly than that of citizen."

Though Clinton claimed far too much credit for the benefits of the boom, he nonetheless deserved some credit for it, especially for his early and controversial tax increases in the face of implacable opposition and his subsequent successful efforts to bring down the then exponentially rising national debt and turn it into a surplus. The American people seemed to agree. If not the best of

THE BEST OF TIMES · 541 ·

times, they had certainly been better times for the majority of the people. Even though the glow of the great expansion was now fast fading as the transfer of presidential power approached—dot-coms crashing, profit margins plummeting, markets roiling, layoffs rising, confidence in the economy dropping, specter of recession looming, rolling blackouts hitting western states with the price of electricity soaring—Bill Clinton was exiting the stage with the kind of positive public support he wanted. He had reason to hope, given the warmth of the response to him, that he would still win the kind of historical legacy he believed he deserved.

Scarcely twenty-four hours elapsed before the other sides of the Clinton age were painfully on most public display. Once again, America was forced to confront the very worst characteristics of a president whose final actions made a mockery of his idealistic talk about prideful, idealistic public service and leaving modestly to become Citizen Clinton.

. . .

On Friday night, January 19, in the closing hours of his presidency and on the eve of the inauguration itself, all network newscasts were dominated not by news of the incoming president, but by the actions of the outgoing one.

Even as he had been attempting to add gloss to his public acts as president, it turned out Clinton had secretly been negotiating a deal—in effect, a plea bargain—with the last of the special prosecutors who had plagued him for years and whose pursuit had culminated in his impeachment.

On the morning of the inaugural, readers of the *New York Times* awakened to see an eight-column page-one banner headline. It reported not on George W. Bush, America's about-to-be president, but on the latest installment of Bill Clinton and Scandal Times:

Exiting Job, Clinton Accepts Immunity Deal
Admits Testimony Was False
—Long Legal Fight Ends

Writing from Washington, Neil A. Lewis contributed this perfect characterization of a historic first for the archives of the nation's paper of record: "In a stunning end to the long melodrama and pitched legal battles over President Clinton's relationship with a White House intern, Mr. Clinton today agreed to a settlement in which he will avoid the possibility of indictment in exchange for admitting that he gave false testimony under oath and agreeing to surrender his law license for five years."

Clinton did so to avoid possible criminal charges of perjury, obstruction of justice, and making false sworn statements—all stemming from Monica. He also agreed to pay a fine of $25,000 to the Arkansas Bar Association, which had been considering whether to have him disbarred, and promised not to seek reimbursement for legal fees to which he might have been entitled.

As America once again wallowed in more debris from Scandal Times, on that very day yet another principal of the period emerged to vie for news attention. The plotter, Linda Tripp, still working at the Pentagon and now earning nearly a hundred thousand dollars a year for duties that included sending information to funeral directors about how to conduct military burials, was fired from her job after she refused to fill out resignation forms required of all other political appointees who serve only at the pleasure of a president and do not work under civil service rules.* Such resignations are routinely expected, and made, in order for the incoming president to provide coveted jobs for campaign workers and other loyalists in his new administration. Typically, Tripp refused to do so. Even more typically, through her lawyers she denounced the firing as "vindictive, mean-spirited, and wrong," and announced she would sue to get her job back and serve again in another Bush administration. Tripp even had the chutzpah to attack the Pentagon for supposedly leaking her plans to seek a U.S. government post in Germany.

---

*When she was exiled to the Pentagon from the White House in 1994 for a job she got without a customary interview and for which there was then no opening, she received what the *Washington Post* reported to be "an uncommonly high salary of $69,000" for that position, or 45 percent more than her White House pay. In the next five years it jumped another third, even though she often worked at home, not at the Pentagon.

Nor was Tripp the only Scandal Times figure to share the exit spotlight with the president. Her firing came just after Jesse Jackson, who had reveled in his public role as Clinton's moral adviser and counselor after the Monica story broke, was forced to confess he had fathered a "love child" after an affair with an aide. The child was born at the peak of the Monica scandal during the period Jackson was shown going to the White House to pray with the president, then emerging to offer pious homilies about morality before the waiting TV cameras.

Naturally, these events intruded on Clinton's final hours, producing a flurry of critical commentary and forcing the nation to refocus on more unsavory aspects of his presidency. The criticism, however pointed, was mild compared to the torrent of outraged public reaction that came next.

• • •

Even as the Clintons were leaving the White House for the last time for the inaugural ceremonies at the Capitol, bit by bit news of numerous last-minute presidential pardons and commutations began to seep into the live media coverage.

The list was unusually long—in all, 176 names. Some of the names stirred memories of scandalous actions past: Patty Hearst, the heiress turned bank robber, remembered for brandishing an automatic weapon and wearing a camouflage uniform with black beret back in the Seventies after being kidnapped and brainwashed by a radical group. She had spent two years in jail. Susan McDougal, the Whitewater defendant and former close friend of Bill and Hillary's, who twice refused to testify against the couple and had been seen time and again on TV being led from one judicial proceeding to another wearing leg chains, manacles, and bright orange prison garb like some Public Enemy No. 1. She had served eighteen months for contempt. Roger Clinton, the drug-dealing younger half-brother of the president. He had served a year in prison after pleading guilty to selling cocaine. (It turns out Roger had unsuccessfully tried to get his brother to pardon several friends and associates.)

These were only titillating, not outrageous. But buried in the long list of names was a true outrage.

Just hours before returning to private life, the president granted a pardon to someone named Marc Rich. The name was unfamiliar to most Americans, but not to prosecutors and members of the press who began checking their files to learn more about the identity of those receiving presidential pardons.

Marc Rich had a long and sordid record. He was, as the public quickly learned, a billionaire financier who was a fugitive from justice. Rich had been charged with federal tax violations in the largest tax fraud case in U.S. history, as one of the federal prosecutors termed it, and with trading with Iran during the hostage crisis. He was never tried because he fled the country in the early 1980s, renounced his U.S. citizenship, and took up luxurious haven in Switzerland where he lived in a lakeside mansion surrounded by private security and all the accoutrements of great wealth—private plane, Mercedes and driver, among them. As the *Washington Post* commented editorially, under the deadly and accurate title, UNPARDONABLE, Rich and his financial partner, who also fled the United States to escape prosecution, "while flouting the U.S. justice system in Swiss comfort...have continued to amass enormous wealth, including less-than-savory deals in post-Soviet Russia."

The paper added, scathingly: "With his scandalous present to Mr. Rich, Mr. Clinton has diminished the integrity and grandeur of the pardon power just as surely as he diminished the various privileges he abused by invoking them to defend his tawdry conduct in office."

Making that unpardonable pardon all the worse was the revelation that Rich's former wife, Denise, was a wealthy New York City Democratic fundraiser and major Clinton contributor who gave "an enormous sum of money" to the Clinton presidential library fund, congressional investigators later were told. She also hosted a $3 million event for the president, gave more than a million to Democratic candidates in 2000, and was captured on old TV footage celebrating with both Clintons. It was she who requested that the outgoing president grant the pardon. The Rich pardon reeked of the rankest political quid pro quo, a presidential favor to repay a presidential benefactor.*

Instantly, the tone of published assessments of the president's standing abruptly shifted from sympathetic and positive to negative and critical. Only days after commenting generally favorably about him, under the heading PARDONS ON THE SLY, the *New York Times* observed: "Anger over Bill Clinton's abuse of the pardoning process mounted yesterday, as well it should." Then it said: "A broader look at Mr. Clinton's final pardon list makes clear that the outrage extends well beyond the undeserved leniency for Mr. Rich." The paper cited Clinton's commutation of the sentences for "four Hasidic men from New Square, N.Y., who were in prison for defrauding the government by inventing a fictitious religious school and using it to attract millions in government aid. The commutations were granted after Mr. Clinton and his wife, Senator Hillary Rodham Clinton, met privately in December with supporters of the men, whose politically active sect had overwhelmingly backed Mrs. Clinton in her victorious Senate campaign." Embarrassment for Hillary grew more intense when it was later revealed that her brother, Hugh Rodham, had been paid $400,000 for his help in a pardon and a commutation granted by his brother-in-law for two convicted felons, one a cocaine dealer whose jail sentence was commuted, the other a Miami businessman pardoned after being accused of perjury and mail fraud. Hillary said she was "heartbroken" when she learned of her brother's role; he returned the money after both Clintons insisted he do so.

*The Rich pardon generated news coverage in the months after the inauguration as the U.S. attorney's office for the Southern District of New York pressed ahead in an aggressive investigation before a grand jury and Congress also began its own investigations. In February, Denise Rich invoked her Fifth Amendment protection against self-incrimination in refusing to appear before Congress and was negotiating a limited grant of immunity from New York prosecutors which would permit her to testify before the grand jury. Of the numerous articles about the Rich pardon case, by far the most extensive and revealing was a brilliant article by Maureen Orth, "The Face of Scandal," in the June 2001 issue of *Vanity Fair.*

Outrage over presidential pardons was only part of the latest controversy to swirl around Bill Clinton as he exited the public stage. Another list, also made public in the closing moments of his administration, ignited at least as much, if not more, criticism. This time, it engulfed both Bill and Hillary.

They were leaving the White House, it was revealed, and taking with them an unprecedented $190,027 in gifts—china, silver, oriental rugs, furniture, art—accumulated over the last eight years. As far as could be determined, the *Washington Post* commented in a lead editorial sardonically titled COUNT THE SPOONS, "No previous president appears to have accepted parting gifts of such magnitude.... The list makes it seem as if the Clintons registered for wedding gifts.

"Under Senate rules," the paper noted, "Mrs. Clinton could not have accepted such expensive gifts once sworn in, absent a waiver from the ethics committee. This rule is supposed to keep senators from becoming too beholden to those who would ply them with favors in exchange for their vote." The rules didn't apply to Hillary in this case because gifts accepted before she took the senatorial oath were exempt from the Senate ethics gift-ban rule.

Then, in what amounted to a final plague-on-both-their-houses judgment, the *Post* said witheringly: "The list demonstrates again the Clintons' defining characteristic: They have no capacity for embarrassment. Words like shabby and tawdry come to mind. They don't begin to do it justice."

Many of the gifts were from Hollywood friends and supporters, Democratic donors and associates. They were as expensive as extensive: $21,819 worth of china; $17,966 worth of flatware; $52,021 worth of furniture, $71,650 worth of art; $12,282 worth of carpets. Nor were these kinds of gifts the product of accumulation evenly distributed over their eight years in the White House; most of them came in the last year, and thus could reasonably be seen as departing gifts. In their seven previous White House years, for instance, the most the Clintons reported receiving in gifts was $23,602 in 1999. By comparison, for his last year in the White House in 1992 George H. W. Bush reported $52,853.

Such largesse from Clinton contributors and political backers could not be rationalized as giving financial help to an outgoing first couple facing severe economic hardship. It came on the heels of the Clintons' purchase of two multimillion-dollar homes, one in Northwest Washington, the other in suburban Chappaqua, New York. The gifts would go, obviously, to furnish their new residences. Nor were the Clintons strapped for cash. Hillary had just received an $8 million book advance; Bill could look forward to a similar multimillion-dollar deal for his memoirs.

The same legally required financial disclosure forms that revealed the magnitude and nature of the gifts they were taking with them showed them departing the White House with a variety of significant assets including more than $2 million in personal bank accounts, blind trusts, insurance policies, and common stock. The forms were equally revealing for what they said about some of the price of the scandal years: The Clinton Legal Expense Trust, composed of donations from more donors and backers, had paid out $1.05 million in Clinton legal fees, and the president still owed "between $1 million and $5 million" to two law firms for work they did in defending him in the impeachment proceedings and the other scandal investigations.

Scanning the list of gift givers, two notable names leaped out. One was Denise Rich of New York, Marc Rich's former wife, who had pleaded successfully with the president to pardon her fugitive ex-husband. She had given the Clintons two coffee tables and two chairs. Value: $7,375.

Even more notable for what it said about the inner intrigues and disasters of the Clinton years were the gifts listed from another wealthy benefactor. These included a china cabinet, a travel humidor, a chandelier, and a copy of Abraham Lincoln's 1860 Cooper Union speech in which Lincoln discounted fears that talk of freeing the slaves could trigger a slave insurrection. The donor was Walter Kaye. Yes, believe it or not, the same man who had gotten Monica Lewinsky her job in the White House, who boasted to a Democratic money functionary how he was "an excitement nut" who "would give you a contribution" in return for being shown "some exciting times," who parlayed his fortune into purchasing highest political access, who reveled in telling how he could pick up the phone and call Hillary whenever he needed to reach her, and who, grand jurors were told, paid "for all of Hillary Clinton's parties."

· · ·

To say the Clintons gave their critics a field day is an understatement of historic proportions, and this time there was no question about their being targeted unjustly by ideological enemies. This time, the Clintons had only themselves to blame for their public perception problems. The critics seized their opportunity.* Maureen Dowd, sharp as ever, captured the Inaugural Day scene wonderfully: "Washington is a wet haze of clouds this weekend," she wrote. "Bill Clinton is leaving as he came in—obscured in a Pigpen cloud of dysfunction." She also took note of Clinton's seemingly obsessive need to continue making his case to the very end—and, indeed, beyond. This led him, she added, "to cling to the spotlight until the last microsecond" by giving lengthy self-indulgent and self-celebratory speeches before cheering supporters immediately after the ceremonies, first in Washington then in New York—his "Barbra Streisand–style farewell tour at Andrews and Kennedy," as Dowd nicely put it—in appearances that detracted from TV coverage of the new president.

On the same day, her *New York Times* colleague, Francis X. Clines, in describing how the stain of Monica and scandal followed Clinton to his last hours in office with his acceptance of a plea bargain and then the cascade of controversial presidential pardons and gifts, mused about the impact of this final act on the American public. "And so," he wrote, "in the rain-soaked closing hours of Mr. Clinton's time in power, scandal weary Americans were

---

*So strong was the criticism that within days after leaving the White House the Clintons agreed to reimburse donors for $86,000 worth of gifts received in 2000. Immediately after that, they returned eleven sets of gifts worth $28,000 to the White House when questions arose about whether the gifts were the Clintons' personal property or had been given by donors specifically for the permanent White House collection.

roused for one last surprise act. It was a sign-off scene that, as ever, strained good humor as much as civic patience."

Nor was that the only closing act of Clinton's that strained public patience. Along with the pardons and the gifts, the nation learned their former president sought to lease the top floor of a skyscraper atop Carnegie Hall in midtown Manhattan for his government-paid offices. Annual price to the public: $800,000, a sum so out of scale, so fitting for a potentate rather than a retired civil servant, that the resulting indignant public reaction forced him to seek less lavish—and less expensive—office space in Harlem.

As the *Washington Post* observed, What a way to leave.

His bungling exit produced a denouement to a story that was far from the one the president would have chosen to secure his positive place in history. But it was a story that carried an unintended political consequence. Far from helping himself in history's eye, the circumstances surrounding his last bow managed to make his successor look good by comparison. By closing the Clinton age in such fashion, Bill Clinton bequeathed George W. Bush a splendid gift—a clean slate and a chance, if Bush could seize it, for a fresh start.

Epilogue

*The question for you Americans is: What are you going to
do with all the great things you have?*

—SIR THOMAS HUXLEY

The test of the historian, Henry Adams believed, is to look ahead fifty years and suggest what forces, for good or ill, are at work that will create the critical new conditions society must confront.

By that standard, the events that diverted America during the Age of Clinton—the years of boom and bursting bubbles, scandal and tawdriness, O.J. and Monica, Linda and Lucy, Bill and Hillary, smugness and selfishness, shortsightedness and division, technological advances and scientific and medical breakthroughs—may seem as antiquated within another half century as that flight over Kitty Hawk and those fuzzy electronic pictures of the first Teletimes character, Felix the Cat.

Adams, one of the greatest historians America has produced, attempted to follow his own advice and look far ahead to determine those emerging forces; but in his case, he extended his historical time span to an entire century. He had powerful personal

reasons for doing so. At the turn of the twentieth century, he remembered gazing with both wonder and foreboding at the sights
and sounds exploding around him as his ship steamed into New
York harbor. It seemed to him then that frantic energy and unre-
strained power had been unleashed, hurling great masses of stone
and steam against the sky. Awe-struck, he thought he was witnessing revolutionary forces producing power never before wielded by
man, speed never reached by anything but a meteor.

Forty years had passed since he had last come into that harbor.
On that earlier voyage home after the Civil War, he was returning
from London where his father had served with great distinction as
Lincoln's special emissary to Great Britain, charged with the crucial task of keeping England from entering the war on the side of
the Confederacy. As he entered New York then, Adams remembered thinking that America was about to catapult itself forward
into history. Freed from the divisions of the Civil War, its energies
unleashed, its national ambitions unfulfilled, with a continent yet
to occupy and settle, America was forming into a long caravan
stretching across the plains and beyond the mountains toward the
end of its western frontier. A stupendous acceleration of new forces
propelled the restless young country on its westward march, and as
a young man Adams himself had witnessed the birth of some of
them: the ocean steamer replacing sailing ships; the railway; the
electric telegraph; the daguerreotype. And that was just before petroleum was discovered and energy derived from coal production
began to transform basic industry into the industrial age.

Decades later, an old man returning home again, it was clear to
him that the child born in 1900 was entering a world never known
before. A "new American" who would experience the world of
the next millennium a century away would have to emerge to control forces being unleashed by mankind and by nature that could
tear society apart. His new American—"the child of incalculable
coal-power, chemical power, electrical power, and radiating energy, as well as of new forces yet undetermined—must be a sort of
God compared with any former creation of nature," he thought.
Looking ahead to the next century, he wrote in his *Education:* "At
the rate of progress since 1800, every American who lived into the

year 2000 would know how to control unlimited power. He would think in complexities unimaginable to an earlier mind. He would deal with problems altogether beyond the range of earlier society. To him, the nineteenth century would stand on the same plane with the fourth—equally childlike—and he would only wonder how both of them, knowing so little, and so weak in force, should have done so much."

Farsighted though Adams may have been, the new Americans of our twenty-first century new millennium are hardly God-like, nor have they learned to control unlimited forces. They do live in a world of far greater complexities and face societal problems barely imagined when Adams first began attempting to envision them, but whether they've learned to deal—or even think—of them is another matter. If anything, the evidence of the Clinton age is that instead of developing a habit of thinking and acting for the long-term, these "new Americans" dwell even more in the immediacy of the moment. Witness, as only one sign, their number-one-rated TV show in 2001, the cynical "reality-based" *Survivor: The Australian Outback* that appeals to people's worst instincts to "outwit, outplay, and outlast" by any means necessary their team or "tribe" members, a formula that attracted over 50 million viewers, more than half the number who voted for president of the United States in the 2000 debacle of an election. With similar fare in the offing, such as the vapid and instantly popular *Temptation Island* that celebrates escape and instant gratification, the proverbial American short attention span becomes more so; people are more easily diverted by the many distractions of their electronic culture from contemplating the problems of tomorrow or examining how yesterday's underlying forces created them.

Two of Adams's historical projections, however, stand as constants in the post-Clinton, new millennial era. First, the stupendous acceleration of forces continues to multiply at rates of change almost beyond reckoning with consequences for life on the planet even more unimaginable than in the past. Second, however sobering the new economic concerns sweeping post-Clinton America in the wake of crashing dot-coms and fears of recession, and however serious its unmet problems, the United States still remains the most favored nation. Its people are no less energetic and restless than those Adams saw about ready to launch themselves forward into history on their post–Civil War westward march so long ago. America's physical frontier has long since been closed, but its resources, its multiple strengths and talents, are still the envy of the world. Given its many advantages, and a still-strong public confidence in its future, there should be no reason to keep America from realizing its best times in the decades ahead. None, that is, *if* the lessons of the Clinton years are learned.

* * *

Let it be stipulated that the Clinton age was a time of advances and accomplishments, and also a time of lost opportunities. Not all the blame for failure to take advantage of opportunities arising out of years of peace and prosperity in the Nineties can be laid at the feet of the president. As we have seen, the entire political culture and the attack-and-destroy tactics that animated Clinton's enemies and the press share responsibility with him for that failure. In the more somber aftermath of that period, with economic concerns rising, the American system faces a number of tests that must be met if it is to fulfill its promise for the new millennial future.

Foremost is the self-evident need to reform—*really* reform—the political system. That means much more than reforming the electoral system that broke down so severely in the 2000 presidential year. Even more important, it means reforming the destructive climate that has sullied politics, diminished government, depreciated public service, and induced ever-deepening public cynicism about the functioning of democracy.

In the wake of the 2000 election fiasco, the immediate task is clear: To restore confidence in the fairness of the process by providing the resources, the technological tools, and the educational training necessary to ensure that votes are cast easily and counted accurately and expeditiously. This needs to be accomplished by applying uniform electoral standards nationally.

Solutions to these problems are not that difficult to find. No sooner was the election concluded than two of the distinguished names that have figured in these pages as leaders of the nation's scientific/technological endeavors stepped forward with a promising proposal. Charles Vest and David Baltimore, the presidents respectively of MIT and Caltech, announced a joint effort to design and then produce the most fail-proof computer-age election machinery in order to avoid repeating the voting disasters of the 2000 election.

Better technology alone will not solve all the electoral problems. Clearly needed is reform of the way the national election is conducted. Uniform poll closing times nationally and a ban on the use of TV exit poll projections until all ballots have been cast would ensure that citizens of each time zone get to cast their votes without being influenced by awareness of how other sections are voting. One solution would be to have presidential voting begin on Saturday and conclude twenty-fours later. Election results could then be tabulated, certified and all announced simultaneously the following Monday, a new national holiday. Call it Democracy Day, a celebration of a free people exercising free choice of new leadership.

Another necessary reform involves the electoral college. The electoral college system, that relic of eighteenth-century thinking that offered the franchise only to white men of property and did not envision the concept of equality as requiring true majority rule, has been a disaster waiting to happen, and a disaster that nearly overwhelmed the political system in the last election. Given the advantages that the electoral college confers on the smallest states, each of which receives the same two electoral votes for its Senate seats as those granted the largest states, it is most unlikely the smaller states will forgo their advantage and vote to support a constitutional amendment necessary to abolish the electoral college. But a reform that distributes each state's electoral votes proportionally would preserve the small state advantage and also produce greater fairness in the way votes are actually counted. Instead of the present winner-take-all voting system in which a candidate earns all of the state's electoral votes even if carrying it by the barest fraction over 50 percent, each candidate would receive the percentage of votes cast in each congressional district. That would minimize, if not eliminate, the prospect of a candidate losing the popular vote but winning the presidency by gaining the necessary majority of electoral college ballots.

These reforms, however desirable, will not cure the ills of the system even if they are all adopted—an unlikely prospect. Unless the corrupting money-in-politics question is really addressed and genuine reforms put in place, the system will still inspire public cynicism and deserve the contempt increasing numbers of citizens give it. Yet even if that critically needed reform frees elected officials from having to mortgage themselves by accepting ever-greater special-interest funds, pressuring them to repay their donors through support of special interest legislative agendas, the nation faces still more formidable challenges.

Transcendent is the need to repair the damage inflicted to America's belief in the fairness and effectiveness of its political and governmental system. I speak here not of the obvious distrust sowed by the recent election problems, but of a more debilitating underlying form of disbelief. It is a disbelief resulting not just from the mean-spirited political climate that animated the attack-and-scandal Nineties but from the corrosive events—from Vietnam to Watergate, from assassinations to urban riots—that adversely affected the national state of mind during the decades preceding the Clinton years. As we have seen, the younger the American, the less he or she believes in the role of government as essential to society—and the less likely he or she is to entertain the idea of public service. Young Americans see government and public service as irrelevant to their lives.

At the heart of these concerns lies the oldest question in the American experience: What is the proper role of government in a free society? The question is not about big government or small government or no government. It is about striking the proper balance between public and private interests, about deciding how to allocate national treasure to benefit the entire populace and at the same time ensure the flowering of private initiative and enterprise central to the American system of risk and reward. As usual, Lincoln posed the question best during the agony of the Civil War when he expressed the enduring dilemma of democracy. "It has long been a grave question," he said, "whether any government not too strong for the liberties of its people can be strong enough to maintain its own existence."

That old question about finding the proper balance between an all-powerful government and an ineffectual weak one assumes new dimensions in the world America now faces. It is a question that rises not out of naive notions of idealism, but out of the same practical considerations that led earlier generations to commit the kinds of public resources that led to creation of the Internet, the National Science Foundation, the National Institutes of Health, and the scientific and technological advances that crested in the Nineties and sparked, however fleeting, the great long boom.

Another imperative presents itself, and addressing it may be the most difficult of all for contemporary America.

It requires no insight to state that the millennial America of more than 280 million citizens is more divided than it is united. The disconnect goes far beyond the political divisions that were so sharply exposed during the last presidential election cycle. In truth, Americans don't really know one another; or, at the least, they have relatively little contact with other large segments of their society. They are, in the main, confined to contact within their individual blocs: professionals or working class; inner-city inhabitants or suburban "soccer moms"; black or white; private schools or public; military or civilian; new immigrant or old inhabitant; rich or poor. Absent the crises that forged national unity in the past, and experiencing more, not less, pressures on their time—going to and from work; grappling with the responsibilities of parenting; facing ever-more-intense workplace competition—Americans present a paradox.

They possess more time-and-labor-saving devices than ever and yet, as we also have seen in these pages, they feel increasingly overwhelmed by all the many new demands on their personal and private lives. They are more tolerant socially and take deserved pride, celebrating diversity as a source of national strength. So it is. Yet in the process it's no longer national identity per se that unites and defines Americans but identity within particular gender, ethnic, or racial groups—identity as women, blacks, Latinos (or Hispanics), Asians, Native Americans, gays, lesbians. The motto that Americans adopted to express their uniqueness—*E Pluribus Unum:* Out of Many, One—has been supplanted by *Ex Uno, Plures,* Out of One, Many. The question is whether these changes, reflecting as they do an increasing compartmentalizing of society, lessen a sense of public solidarity and diminish the ability to reach needed consensus on critical public issues, whether indeed these changes erode the special pride in being an American that once inspired national purpose and achievement.

Nor does our educational system adequately provide the kind of historical perspective and emphasis on what used to be called "civics" to provide a broader understanding of the workings of the complex American governmental and political system. Neither does it address the question of how to narrow the gap that exists among the elements of our divided society. In our present all-volunteer military era, I have long believed in the concept of universal service. Not, I add, universal *military* service but compulsory *public* service. By that, I mean as a price of citizenship each person reaching the age of eighteen would be required to spend six months in some form of public service—whether joining Peace Corps, assisting in inner-city or suburban schools, drug education or rehabilitation programs, or environmental cleanup programs or enlisting in the military, if desired. Compulsory public service is no panacea, but at least it would bring elements of our disparate society into greater contact with one another.

George W. Bush's Inaugural address, however overshadowed by the diverting drama of Clinton's ignoble final acts, offered a welcome plea for greater national public purpose. In calling for civility and compassion and working to create a "nation of character," with unity as its principal theme and seeking "a common good" as a central purpose of citizenship, the new president struck a decent, honorable, generous tone. Whether his actions match his words remains to be seen, but it was an impressive beginning toward achieving a more unified society—not a bland, unquestioning society incapable or fearful of vigorous, heated debate over issues, but a society that will need its most talented people to enter public service to meet the challenges and threats of the new millennium.

· · ·

For all the wonders of Technotimes, the most astounding scientific and medical breakthroughs are yet to come. Out at the Jet Propulsion

Laboratory in Pasadena, California, America's leading space scientists speak confidently, indeed almost casually, of the belief that sometime in the next two decades their "origins of life" project will lead to the announcement of the most astonishing discovery in human history—confirmation of the existence of extraterrestrial life. Already, they are in the forefront of the greatest age of exploration ever, with rapidly proceeding discoveries revealing new planets and new solar systems far beyond the infinitesimally small Earth. When scientifically verifiable word comes that we are not alone in the universe, the Jet Propulsion Lab leaders tell me that human beings are going to be forced painfully to reexamine all of our prior assumptions, including our religious assumptions, about our uniqueness.

Less celestial and more mundane, scientific and medical breakthroughs continue at an astounding pace here on our crowded planet. None of these breakthroughs are risk free. All pose new and serious questions for society.

One of the greatest issues of the new millennium involves the rapidly increasing length of people's lives. The longevity curve keeps rising to a point where already centenarians are now the fastest-growing segment of the population; some demographers predict America will include a million people who have reached the age of one hundred by the year 2050. How rapidly America is aging was shown in the 2000 Census data. During the Nineties, the number of Americans over the age of eighty-five rose by 38%. In that same decade, the number of Americans under the age of five increased by only 4 percent. By the year 2000, the number of Americans seventy-five and older had increased by 26% to 16.6 million.

As Dr. Robert Cook-Deegan, director of the National Cancer Policy Board of the Institute of Medicine, tells me: "The question that has to be addressed—and has not been confronted directly—is, how much do we spend to live how much longer? And then, how healthy do we want to be? We could forever and ever and ever pour all the resources in the whole country into trying to deal with these health scourges, but even if we get rid of cancers and Alzheimer's, we're still going to have things like diabetes. We're never going to be free from disease in the whole population. So when do we stop? How much money do we spend to stay healthier for shorter periods of time? That's a serious question for Medicare. It's a serious question for all of health care. We could spend every last dollar in the United States on health care for people in their eighties and nineties if we wanted to. What are we going to balance that against? Those kinds of discussions aren't out in the open at all."

Another distinguished health expert to whom I spoke expanded on those points by raising the profound question of the impact of cloning on society. She is Dr. Ruth R. Faden, one the nation's leading biomedical ethicists, author of numerous books and articles on biomedical ethics and health policy, director of the Bioethics Institute at Johns Hopkins University and also senior research scholar at Georgetown University's Kennedy Institute of Ethics. "I'm not worried about a scientific world where we take a technology and all of a sudden we have a hundred thousand people who look just like you," she says. "I don't think that's going to happen. If the concern is that evil oppressive governments are going to use these means to control us or undermine all the great religions in the world, well, oppressive governments can control people's lives in the most low-tech ways possible. You don't need human cloning to suddenly empower evil states. Go to Afghanistan and ask how many biotechnology experts they have there as they control the lives of their women. Cloning technology will not create evil empires. There won't be a hundred thousand Hitlers. Those are not the real problems.

"There *are* important questions that genetics pose for health, and we're going to have to confront them. Questions about whether to terminate a fetus if genetic evidence indicates the baby will have a disabling disability or will develop breast cancer by the age of forty. These sorts of issues are hot compared to what we would have thought about fifteen years ago. They are troubling and will need to be addressed. We have hard choices to make. We've always had hard choices in medicine, but now we have, I believe, harder choices, and we're still running away from them. We can give everybody decent health care, but we can't give everybody everything. We don't confront that straight on. And if we can't give *everybody* everything they could possibly benefit from, then who is going to decide who gets what and how? Those are huge issues."

Within another decade, she predicts, America will be confronting a health policy crisis of monumental proportions. That is, the rationing of increasingly expensive, life-saving health services. "We have been dancing around rationing in this country for a long time," Dr. Faden says. "That's what we'll be up against." As always, this will pit those with the most capital resources against the those with the least—again, the rich-poor, have–have-not schism already widening in the United States, and even more throughout the world.

Like Faden, the molecular biologist Melvin L. Simon at Caltech is less concerned with cloning than with issues that affect our everyday lives. "Cloning?" he says. "We can clone hordes and hordes of people who all look the same. Why anybody would want to do it, I don't know. It's clear that people are getting healthier and stronger. This is the kind of problem that a creative society needs to learn how to deal with. How do you deal with a brain in an elderly population? We know our population is going to get older. Hopefully these people get ten to twenty years added to their lives and are going to be functional and active people. What are we going to do with them? Put them in front of a television set? That's a *real* issue to deal with. We're a society that's built around beauty and youth and we've got a bunch of people retiring now at sixty with the thought they'll be dead by seventy-five and they're going to find out they'll live to ninety. What the hell are they going to do with the extra fifteen years? We're not preparing for these and the other changes that accompany that."

· · ·

Other concerns confront America's citizens and policy makers. They are international in nature and embrace such things as global warming and population increases. Grappling with these issues demands international cooperation. But at the millennium, the world faces rising tensions between its have and have-not components, growing threats of terrorism accelerated by the dispersal of weapons of mass destruction, and grievously divided fundamentalist and Western industrial societies. The fanaticism of extremist elements in some of those societies was demonstrated shockingly by a mindless act of historical destructiveness in Afghanistan in March 2001. Despite anguished protests world-wide, the ruling Taliban movement there ordered the demolition of priceless ancient art in the form of two magnificent giant Buddha statues carved out of a mountainside that had stood watch over the Bamiyan Valley for fifteen hundred years. This act of cultural barbarism, carried out in the name of Islamic religious fundamentalism but rejected by most Islamic leaders and scholars, was a powerful reminder that no matter how great the promise of the new technological world, human beings continue to face the same kinds of problems that have plagued societies throughout history. It was yet another reminder that the millennial world is far from stable. Indeed, given the evidence of bloody new conflicts growing out of unresolved ancient religious and ethnic hatreds, plus the capacity of smaller groups to sow mass terror through possession of those weapons of mass destruction, the twenty-first century world holds the potential of being even more dangerous than was the twentieth century world during the long Cold War period.

Other problems arise. Take the question of global population.

At the millennium, the population of the world numbered just under 6 billion people. A billion of those people, inhabiting North America, Europe, Japan, a few other places, can be said to be living better materially than ever in history. The other 5 billion are not. All live on a planet with finite resources—oil, water, air—that are being depleted as consumption and population and longevity rates rise and as environmental dangers increase. Another finite resource, land, is also being depleted through destructive farming practices that cause the loss of topsoil and reduce the productive capacity of land presently under cultivation. Add to this mix the projected number of people in the world rising from 6 to 8 to 10 billion within a relatively short time span, along with further depletion of fundamental resources.

At the very least, the *possibility* certainly exists that the planet will not be able to produce adequate food for its inhabitants and clean water for them to drink. If so, that poses the supreme test of Technotimes—part nightmare, part promise—and an even greater test for public institutions and political leaders. For democratic societies who benefit from and depend upon technology to maintain, or even improve, the quality of their lives, the question arising from these possibilities will be how to resolve peacefully issues of vast social, ethical, technological, and political complexity—and do so in ways that permit freedom to flourish even if controls are required to assure adequate and equitable distribution of life-providing resources.

Carry the story forward thirty to fifty years into the future, contemplate global conditions then, and these questions become even more troubling.

For America in particular, this kind of challenge will make the contentious, complex debates over creating the founding charters of democracy seem simple. Exceedingly hard choices existed then as the nation's founders wrestled with how best to strike a fair balance between restricting individual actions of citizens and ensuring the widest latitude possible for a free people. In one way or another, every generation since has wrestled with aspects of those same democratic dilemmas. Americans will continue to do so and the problems promise to become even more tortuous in the future.

Unless, of course, those lulled by the rush of success and the diversions of the Nineties were correct in thinking that the forces of history and the market had been breached forever.

If so, there are no lessons to be drawn from the Age of Clinton except, perhaps, the one Voltaire tried unsuccessfully to teach long ago. History, he reminded us, never repeats itself. Man always does.

Afterword

Of the many misbegotten myths of the Nineties, the greatest was the belief that we were different from all the peoples of the earth. We were immune from the ravages that afflicted other societies, impregnable against distant threats. Everything was in our favor. With the great expansion of Technotimes, we had arrived at a singular moment. Our democratic capitalistic system was the envy of the world. Our military might was unsurpassed. We were No. 1, the world's only superpower, secure and confident in our new riches and new power, unchallenged in our rapid ascension to even greater heights of prosperity and achievement. If threats did emerge, they were out *there*. And we had the power and the capacity to deal with them. It can't happen here.

The shattering of that illusion on 9-11 was the last, and most tragic, of the bubbles that burst during the era of the long boom. In one blinding moment, suddenly and shockingly every American was forced to confront the wrenching reality that we weren't so unique after all. We were as vulnerable as the Israelis or the Palestinians or any others facing threat of sudden attack. We were just like everybody else.

The attacks of 9-11 did more than shred America's overweening sense of complacency about its security and its future. Not only can it happen here; it can happen with an ease and swiftness difficult to fathom given those beliefs in our invulnerability. It was a low-tech attack, one that turned hijacked airliners into missiles, instantly making irrelevant any U.S. edge in technology, advanced weaponry, and military legions. It was superbly planned and executed. It demonstrated the ability of enemies to pierce our borders and strike devastatingly and simultaneously in New York and our nation's capital, aiming a lethal blow at the very symbols of capitalism and U.S. military supremacy. It also taught one of history's oldest lessons, demonstrating once again the power of people willing to die for a cause they believe in.

Everyone witnessing the horrors of that incredible day instantly understood that new realities were brutally being imposed on us. The immediate response from citizens and commentators alike was that 9-11 had "forever changed America." As I shall suggest, despite repeated affirmations of the soundness of that belief, that assumption remains to be proved more than six months later.

But there can be no question that, as Walter Lippmann memorably wrote long ago, the "pictures in our heads" about the world we thought existed and the world that really surrounds us have been irrevocably altered.

In his classic work, *Public Opinion,* written nearly eighty years before 9-11, Lippmann described the impact of the sudden outbreak of World War I on people around the globe. On the eve of that conflict in 1914, he wrote, "All over the world . . . men were making goods that they would not be able to ship, buying goods they would not be able to import, careers were being planned, enterprises contemplated, hopes and expectations entertained, all in the belief that the world as known was the world as it was. Men were writing books describing that world. They trusted the pictures in their heads."

The pictures in their heads were as false as those pictures we held in ours about the world decades later.

For Americans in particular, the pictures implanted on 9-11 will remain indelible, and so will those reflecting the searing events that continued to alarm the public in succeeding weeks. To a degree probably never before felt in the American experience, eclipsing even the shocks of assassinations, sneak attacks on our military bases, and outbreak of civil war on our own soil, the instant changes in daily lives and attitudes wrought by 9-11 were unprecedented.

In the wake of the attacks, for the first time, all commercial air-craft were grounded. People everywhere found themselves react-ing uneasily to the sound of aircraft—U.S. military, as it turned out—in the skies above. For the first time since the depths of the Great Depression, Wall Street was closed for four days, and upon reopening the markets sustained their greatest losses since those grim depression days.

Everywhere, the pattern of daily life was altered. Parents found themselves attempting to deal with the fears of their children, teachers with pupils, family with relatives, husbands with wives, friends and lovers with one another. Travel plans were canceled. Those who did travel, found long lines greeting them at airports where they had to arrive hours before their flight times. There, they encountered security checkpoints that were inadequately managed and poorly staffed. Many people stopped entering the centers of our great cities. Tenants evacuated high-rise buildings and skyscrapers; few new occupants took their places. People ob-sessively speculated over the next source of danger. Fears of bio-logical attacks were rampant. Then it happened. Anthrax-laden letters led to the evacuation and closing of postal and government buildings in the capital and news-center offices in New York. Warn-ings of other anthrax attacks further heightened public anxieties and fears; people found themselves warily viewing unopened mail and even washing it to remove potential anthrax powders.

A succession of well-intended but inept official alerts warned of the prospect of imminent, even greater, attacks by terror cells of faceless assailants poised to strike from within and without our borders. Prospective terrorist targets were cities, offices, bridges, tunnels, water systems, seaports, sporting events, nuclear installa-tions. Inevitably, these alerts sowed further public anxiety. Even less helpful in allaying fears was the fumbling, inconsistent nature of the initial official attempt to reassure the public that, despite the new age of terrorism, essentially all was well. Keep flying, keep spending, keep going to Disney World, keep enjoying yourselves, the government proclaimed—but also keep your eye out for lurking terrorists and be prepared to respond to perceived threats. Other than dialing 911 for emergencies, just how to respond was never adequately articulated.

Above all was the unforgettable memory of the slaughter of thousands of innocent civilians who died horribly in hitherto unimaginable circumstances.

In these and countless other ways, lives and aspirations were altered. Whether they were "forever changed" is another matter. In fact, perhaps the most striking aspect about the aftermath of 9-11 was how quickly America began to slip back into familiar habits and patterns.

F. Scott Fitzgerald once brilliantly described a cast of mind that can apply to a nation as well as to an individual. The test of a first-rate intelligence, Fitzgerald believed, is the ability to hold two contrasting ideas in the mind at the same time and still retain the ability to function. The same can be said of a nation, but even more so: the ability to focus upon and react to totally differing circumstances becomes the hallmark of a successful society.

By that standard, America in the wake of the new world after 9-11 demonstrates an old tendency to rivet its attention on only one challenge at a time, a preoccupation that essentially means either ignoring or failing to deal adequately with other great problems. In the six months after the terrorist attacks shocked and mobilized the nation, those traits were displayed anew.

The first half of that period saw America's attention fixed upon the president's newly declared war on terrorism. The war became such an all-consuming absorption that it underscored the concern expressed in my epilogue that "instead of developing a habit of thinking and acting for the long-term" our Americans of the new millennial world would continue to "dwell even more in the immediacy of the moment."

As the war faded into the recesses of the public mind, the months after 9-11 saw terrorism eclipsed by the sudden emergence of a domestic event that raised questions about the functioning of the American capitalistic system. The monumental collapse of Enron, the Houston-based energy trading company that had become America's seventh-largest corporation widely praised for its digital-age innovation and its golden record of soaring profits that produced a corporate net of 40 percent only months before, represented much more than the largest bankruptcy in American history. It threatened to engender lack of confidence in the integrity of the U.S. financial system.

To some critics Enron became a symbol of the malfunctioning of capital markets produced by a combination of greed, corporate manipulation, corruption, inattentive regulators and auditors, and the purchase of influence through high political connections. Daily revelations about the Enron debacle highlighted Enron's massive attempts to cover up more than a billion dollars in debt through a labyrinthine trail of hidden offshore partnerships and insider deals, permitting those at the top to realize hundreds of millions of dollars in profits by selling their stock while at the same time touting the high value and long-term potential of that stock to their employees and the general public. All of this came at a cost not only to thousands of Enron employees, who lost their life savings, but to state pension plans across the country that had invested in the supposedly soaring Enron stock.

Now Enron dominated the news. It led the network telecasts, filled the front pages, and drove news of the war off the public screen. With the absence of war coverage, also largely removed from public consideration was discussion about whether the ending of combat (or, more likely, the advent of a long guerrilla struggle) would produce a period of stability or further chaos—and the government was not helpful in focusing attention on that critical question. Even more critical is the truism that winning a peace is greatly more difficult than winning a war.

Different though these events at home and abroad were, they were reminders that the United States faces multiple challenges in

the post 9-11 world. One involves the state of its own economic
and political system, another the rising level of resentment of American wealth and power among increasingly impoverished have-not
corners of the world that form the breeding ground for terrorism,
but also rekindled resentment of America among our Western allies.
For these kinds of challenges to be met, the United States will have
to deal with them simultaneously, not one at a time.

Of course, the initial focus on the war was as understandable as
it was inevitable. The reaction to the terrorist attacks whipped
martial feelings to a level unmatched since World War II.

At the same time, the Bush administration's singular focus on
terrorism and the corresponding fixation on the massive military
buildup inverted the Clinton administration's emphasis on domestic events. Throughout the Nineties, public attention was never focused as it should have been on the very real, and very well
known, threat of fanatical Islamic fundamentalism and terrorism.

For the Clinton administration, the operative imperative from
its beginning was domestic: "It's the economy, stupid." With the
Bush administration, the imperative from 9-11 on was foreign:
"Wanted, dead or alive." The focus was on "getting" terrorist
leaders like Osama bin Laden and eliminating terrorism wherever
it existed by destroying those leaders and any regimes that harbored
terrorists within their sovereign borders.

In the Clinton years, top American civilian and defense policymakers were well aware of terrorism and of bin Laden's public
declaration of war against America and his call for a jihad to
destroy all Americans. The World Trade Center, after all, was
bombed by members of bin Laden's Al Qaeda terror network in
1993, and in highly public circumstances four of his lieutenants
were arrested, tried, convicted, and sentenced to life imprisonment
for that crime. Later in the decade, terrorist attacks against American embassies in Africa, military installations in Saudi Arabia, and a
billion-dollar naval vessel, the U.S.S. *Cole,* resulted in the deaths of
hundreds of Americans. While those events were occurring, high-
level blue-ribbon reports offering recommendations on how to
combat the growing terrorist threat were essentially gathering dust
in Washington.

Nor were these the only strong reminders of the threats posed to American interests by terrorists. Since the late 1970s, Americans had been made painfully aware of the dangers of terrorism and the consequences of failing to confront it. The Iranian revolution that deposed the Shah and put in place the harsh fundamentalist regime of Ayatollah Khomeini in 1979 resulted in the storming of our embassy and the capture of Americans who were led away, blindfolded and bound, to Iranian cells. In the 444 days that followed, a daily network news show, "America Held Hostage," televised scenes of screaming Iranian mobs burning American flags and shouting, "Death to America."

In the 1980s, during the Reagan years, hundreds of American Marines stationed in Lebanon were wiped out in a terrorist suicide bomb attack; Shiite terrorists in Beirut murdered a U.S. Navy diver who was a passenger on a hijacked TWA plane; four heavily armed Palestinian terrorists seized four hundred passengers and the crew of the Italian luxury cruise ship, the *Achille Lauro,* off the coast of Egypt, killed an elderly disabled American, and threw his body overboard with his wheelchair; Mideast terrorists planted a bomb aboard a Pan Am plane that detonated over Lockerbie, Scotland, killing all 259 passengers and crew aboard and eleven people on the ground.

To the end of his presidency, Reagan continued to lash out at terrorists, even though he had covert dealings with them, trading arms for Americans held hostage. He called both Iran and Nicaragua part of "a new international version of Murder, Incorporated." Later, he linked those two nations with Libya, North Korea, and Cuba as "outlaw states run by the strangest collection of misfits, Looney Tunes, and squalid criminals since the advent of the Third Reich."

Reagan's successor, the first President Bush, fought and won a war—Desert Storm—against the terrorist regime of Iraq's Saddam Hussein but failed to follow up his military victory with the kind of postwar settlement or occupation that could have brought greater stability to that region. By then, too, no American could have failed to be aware of the shadowy terrorist world that lay beyond our borders, or the threats terrorists posed to us. News reports out of the

Middle East highlighted a seemingly endless cycle of terrorism as group after group there launched increasingly bloody attacks. Hooded, masked terrorists striking suddenly and fearsomely became a staple of network telecasts as well as American films and popular spy thrillers. The names of leading terrorist groups also become familiar to Americans: Hamas, Hezbollah, and Islamic Jihad.

Despite all that history; despite the knowledge that, instead of abating, the threat of terrorism was increasing throughout the Nineties; despite the constant monitoring of terrorist networks like those of bin Laden; despite the tripling of spending and the assignment of additional FBI and CIA personnel to anti-terrorism operations; despite presidentially sanctioned attempts to kill bin Laden during the Clinton years, combating terrorism never assumed the *highest* official priority. Clinton never regarded terrorism as serious enough to marshal support for all-out action. But then neither had any of the presidents during the three previous administrations. Jimmy Carter lost the presidency in large measure because of his handling of the hostage crisis. Ronald Reagan traded arms for hostages, albeit covertly. From Carter to Clinton, terrorism did not get the attention it deserved, and, for that matter, George W. Bush did not treat terrorism as a high priority until 9-11.

There's no mystery about the cause for this official inaction and the monumental failure of public perception. As already suggested, the root failure lay in a collective belief in America's impregnability and superiority—a state of mind enhanced by the special climate and attitudes that typified America in the boom years of the Nineties.

The diverting events of the best of times permitted us to be caught up in a wave of self-indulgence, scandal, illusion, entertainment, and inattention. It didn't matter who was in government or who was president or whether public service was viewed as unimportant or what destructive tactics were being practiced in our politics. It didn't matter what was happening outside our borders. It didn't matter that we and our news media were focused on more frivolous concerns. Now, in our newly altered world, the only question that matters is what lessons America learns from the experience, and what actions it takes in response to it.

I believe the devastating experience of 9-11 requires us to forge a new national partnership between our public and private sectors, one that puts aside destructive, mindless attacks on each other while celebrating the need for vigorous informed political debate. It requires achieving a new unity of purpose of a kind we have not had since World War II. It means putting aside the corrosive belief that government is irrelevant, that public service doesn't matter, that we can privatize everything and allow "the magic of the marketplace" to work for the common good unfettered by the sensible oversight and regulation that ensure the fairness and trustworthiness of the financial system. It means nothing less than changing the way we as a people think and act in the way we educate and inform ourselves as we draw on the ideas and talents of those who can help us see what steps we need to take to preserve and protect ourselves.

After the horror and wreckage of 9-11, no longer can we afford to turn our eyes away from distant lands and peoples. We need to understand the sources of the anger, discontent, and hatred that confront us. We need a new kind of Marshall Plan to address the hopelessness and poverty of the undeveloped world, a new form of Manhattan Project to enlist the best minds in science, technology, and medicine to help forge policies in our, and the world's, best long-term interests—among them, better ways to deal with global warming, with population control, with hunger, with the production and distribution of food, with the eradication of the scourge of such diseases as AIDS, with freeing America from its dependence on Mideast oil. We need to teach history better, to bring into the public policy mix scholars who speak Farsi and understand the cyclonic currents sweeping the Islamic world. We need to expand international exchange programs that help break barriers formed through lack of contact and ignorance. At the least, it would be nice if *all* American pupils studied foreign languages from grade school on.

Admittedly, I recognize such an agenda is easy to proclaim and infinitely harder to achieve. I further realize that offering such a sweeping list for change might seem to border on the naive. But I believe the reaction of the American people in the aftermath of 9-11 gives reason for optimism. *If,* that is, two conditions are met:

That leaders emerge who really lead, and that people will be persuaded to follow them.

As I have said in this book, the real heroes of the bubble years were the American people. No matter how seamy the allegations of scandal leveled against their leaders, no matter how disheartening the actions of those leaders, from the beginning the people retained a steady and mature perspective sadly missing among major elements of the media and the attack-and-destroy political culture. Americans displayed the same traits in the months following 9-11. As always in the past when confronted with a national crisis, they rose to the new and, in many ways, more complex challenges. Despite fears of worse to come, they did not panic. They remained steadfast and supportive of their leaders and their system. The people's response to 9-11 formed an admirable portrait of the resilience of a nation under stress.

A major, comprehensive survey of American attitudes by the Pew Research Center for the People and the Press revealed a reassuring range of national emotions and opinions in the wake of 9-11. Pew's final findings, the result of eight major surveys conducted over three months after the first attacks, documented a people becoming steadily more able to live with the horrors as time put more distance between them and 9-11.

How great an impact 9-11 had on people everywhere was shown in the survey results tabulated between September 21 and 25. Fully 71 percent of those responding then said they felt depressed. Another third were having trouble sleeping. In a matter of days, those negative feelings began to dissipate. Of those surveyed between October 1 and 3, 42 percent said they were depressed and 18 percent were having trouble sleeping. A week later, those figures continued to drop, with 33 percent feeling depressed and 14 percent having trouble sleeping. The next three days, October 12–14, saw apprehension further declining, with 29 percent expressing a feeling of depression and 12 percent having trouble sleeping.

Pervasive anxiety, however, continued. In early October, 73 percent still feared another attack, 37 percent thought a biological or chemical attack was in the offing, 28 percent pronounced themselves "very worried."

Not surprisingly, those same surveys showed virtually the entire populace closely following news reports—this from a people who had been paying less and less attention to news. In mid-September, 96 percent of those surveyed said they were closely following reports about the attacks. Ninety percent were getting their news mainly from TV, while 81 percent said they constantly tuned in, and another 63 percent said they couldn't stop watching. A month later, the percentage of those feeling obsessed by TV war-and-terrorism coverage had dropped from 63 percent to 49 percent.

Other findings were noteworthy for shifts in public attitudes. The status of the news media, an institution that has been viewed increasingly negatively in recent years by larger numbers of the population, registered a sharp upturn in public favor. Early in September, before the attacks, only 35 percent thought the news media usually got its facts straight, while a strong majority, 57 percent, believed the media usually reported inaccurately. Two months later, those beliefs had changed. Forty-six percent thought the media got its facts straight, while only 45 percent believed media reports usually were inaccurate. Not a ringing endorsement of the news media, to be sure, but still a clear improvement in public perceptions about it. An even greater shift in attitudes came in the view of how the media reported on the U.S. In early September, only 43 percent of the people believed the media "stood up for the U.S" and another 36 percent thought the press too critical of the U.S. Two months later, implicit beliefs that the press was anti-America had shifted dramatically. Sixty-nine percent now thought the media stood up for America and only 17 percent thought it too critical.

These findings were rooted in more than impressions. Like the people themselves, in the wake of 9-11, the news media responded splendidly—but it took that crisis to bring out the best in the press. Until those planes crashed into the World Trade Center towers, the scandal-and-celebrity coverage so prevalent during the Nineties had continued to dominate the airwaves. The latest scandal, covered with O.J.–like saturation reporting, involved Congressman Gary Condit and his missing former intern, Chandra Levy. Immediately after 9-11, the media underwent an astonishing

transformation. A press that in large measure had cut back coverage of foreign affairs, closed foreign bureaus, and focused more on
trivia, now produced some of the finest journalism in my lifetime—serious, searching, often eloquent, and filled with the kind
of context and perspective often missing during the scandal years.
The *New York Times,* in a particularly memorable example of journalistic enterprise and excellence, provided a major public service
each day following 9-11 by coming out with an advertising-free
special section of comprehensive coverage titled "A Nation Challenged" that included an extraordinary full page of daily sketches of
those who had died in the attacks. These went far beyond the usual
obituaries. Touchingly, often heartbreakingly, they reminded readers of the range of life and experience of those men and women
who were murdered that day. The sketches were a daily reminder,
too, that people from fifty nations, of all races and religions, had
become the innocent victims of mindless terrorism.

Impressive as the overall news performance was, the greater
problems afflicting a mass media obsessed by the culture of scandal
and celebrity remained. As if to underscore the enduring appeal of
notoriety to mass media merchants seeking to capitalize on it, the
six-month anniversary of 9-11 marked a new low in scandal-and-
celebrity programming. In prime time, viewers of the Fox cable
TV network were able to see a highly promoted encounter pitting
two of the Nineties' trophy names from the era of scandal, Tonya
Harding and Paula Jones, against each other in a boxing ring. The
aptly named show, *Celebrity Boxing,* attracted 15.5 million viewers—a statistic that, as *New York Times* TV critic Bill Carter reported, represented "more viewers than watched almost any other
entertainment show on Fox this year." It also became that week's
No. 1 ranked program among the audience that advertisers covet,
target, and pay most to attract: young males. Nor was this the only
example of the continuing strength of celebrity TV programming.
If anything, the months after 9-11 demonstrated anew that the
news media's conglomerate owners were more driven then ever to
depreciate the value of real news coverage in favor of more entertainment programming. This lesson was delivered shockingly to

the ABC-TV news division by its corporate owners, Disney, who sought to lure the late-night comedian David Letterman away from CBS in a multimillion-dollar deal at the expense of an icon of network TV news: Ted Koppel's widely respected and admired serious news program, *Nightline*. Koppel's program was still making a profit after nearly three decades on the air, and it was attracting more than 5 million viewers nightly—four or five times more than any of the All-Monica shouting-type cable news shows. But corporate executives in Hollywood sneeringly viewed *Nightline* as "irrelevant" and proceeded with plans to kill it without even consulting the top news-division executives of ABC. This came, of course, precisely at a time when serious news could not be more relevant to Americans.

Of all the outraged commentary that Disney's planned move received, none was more telling than that of William Safire, the conservative columnist and former Nixon speechwriter and strategist. Writing from his perch on the op-ed page of the *New York Times,* Safire said:

> If *The Times* were to replace this column with a comic strip, I like to think you would hear a geshrei from me and a keening wail from even critical readers that would rattle corporate rafters. *The Times* would not sacrifice serious opinion-mongering for comic relief because this focused $3 billion outfit understands that its reason for being is primarily to inform. The Disney combine's mission, contrariwise, is to dispense profitable entertainment, and its misbegotten purchase of a news medium allows it to prostitute ABC News's journalistic mission to conform to the parent company's different goal.

Whether the public agreed or not with this attempt to further erode the independence of serious news programming was a question for the future, and so was the question of whether the greed of entertainment conglomerate owners of the news media and their contempt for the public interest would continue unchecked. For now, 9-11 had shocked the news media into behaving admirably and the public had taken notice. The news media wasn't the only quarter to experience change. Elected officials rose to the occasion, laying aside partisan bickering, and the public expressed its approval of Congress and, most notably, the president.

In early September a bare 51 percent of Americans approved of the way George W. Bush was handling his job as president. Three weeks later, 86 percent expressed their approval. Part of this surge in Bush's approval rating was natural; Americans always rally behind their president in a time of crisis, and Bush proved no exception. Part, though, was a deservedly more favorable impression of Bush as a national leader. People credited him, rightly in my view, with exhibiting strong leadership traits in the way he took hold of the government in the midst of the crisis. And many—myself included—who thought he lacked these skills, often embarrassingly so, while possessing an uncurious cast of mind, an appalling lack of knowledge, and an inability to articulate a strong and persuasive message to the people, were relieved and surprised.

Bush struck a necessary tone of bipartisanship, one shared, at least initially, by the Congress, a body that had been bitterly divided along ideological lines during the Clinton years, and a Congress, too, that had been so preoccupied with the hunt for scandal that it turned away from the really serious problems facing the nation. Bush also displayed a welcome capacity for change. From a unilateralist, go-it-alone, American stance he shifted, at least temporarily, to acknowledging the need for international alliances, nation rebuilding, and globalism.

Both Bush and the leaders of his administration also deserved great credit for the way in which they warned Americans against the dangers of stereotyping or attacking those among us who look, dress, act, and speak differently. The message was one of tolerance—and it was a message that the American people for the most part either heeded or figured out for themselves.

A noteworthy finding in that Pew survey of national attitudes came in the surprising, but cheering, way in which Americans viewed fellow Muslim-Americans. When surveyed in March, six months before 9-11, only 45 percent said they felt positively toward their Muslim-American fellow citizens. Two months after 9-11, 59 percent felt positively!

Another impressive shift in attitudes involved the way people had changed their views about U.S. foreign policies pre– and post–9-11. Pew went back and reinterviewed 1,281 Americans whom they had first contacted in late August. The differences in attitudes were striking. Before the attacks, 38 percent of people thought foreign policies should be based mainly on U.S. interests. After 9-11, the number favoring placing America first had dropped to 30 percent, while the number of those placing interests of allies first had jumped from 48 to 59 percent. Similarly, Pew found a tempered, more moderate reaction to the military course the nation should follow against terrorism—one of victory but not vengeance. Despite the public's instinctive desire to strike back at terrorists after 9-11, Pew discovered that "most Americans felt the point of military action should be to prevent future terrorist attacks rather than punish those responsible."

As for domestic concerns, a people already turning pessimistic in the wake of the dot-com debacle had become more so with the advent of the officially proclaimed national recession that March. These feelings worsened after 9-11 with the dramatic drop in stock prices and the closing of Wall Street. The week after the markets reopened saw the Dow Jones industrial average plunge 14.26 percent—which the *Wall Street Journal* tabulated as "the fourth-sharpest weekly plunge for the blue chips in percentage terms since the start of the Twentieth Century." Economic confidence returned as the war effort progressed and the markets recovered much of their earlier losses, but then the collapse of Enron shook the markets, great banking houses, and auditing firms, and left the public wary of being taken by hucksters and hustlers.

Already, Enron was being depicted by many commentators as "an epitaph for the 1990s bubble." The *Washington Post,* in employing that term editorially, captured the essence of the Enron story: "The firm seems to have assembled the various strains of hubris found in different corners of the country: the technological vanity of Silicon Valley mixed with the financial alchemy of Wall Street, the influence-peddling of Washington fused with the ten-gallon brashness of Texas."

However difficult the times had become, the people united

fervently behind their leaders, gathering in emotional mass rallies and engaging in unabashed displays of patriotism. More American flags were unfurled from homes, offices, and cars or affixed to suit lapels than at any time since World War II. Inevitably, this led to an orgy of excess and cheap cashing in on patriotic emotions for commercial purposes. But beneath the often mawkish flag-waving displays was a more tempered patriotism, hard-eyed, not jingoistic, reflecting an America that had suffered and survived the bloodiest single day in its history, having witnessed its greatest loss of life— civilian, this time—since the carnage at Antietam in the Civil War.

Perhaps most reassuring was the sudden shift in perceptions about the role of government and public service. Now, people could see that public service meant heroic police and firemen who willingly went to their deaths attempting to extinguish the inferno that the World Trade Center had become. It meant selfless nurses and emergency room workers, construction crews toiling amid unimaginable scenes of death and destruction. It meant faceless mail carriers who continued their burdensome daily rounds in the face of deadly attacks of anthrax unleashed on postal stations. It meant government workers and bureaucrats reentering closed anthrax-contaminated buildings to carry out the people's business. It meant anonymous government technicians and medical experts toiling to respond to bioterror attacks.

Not all the signs of change were positive. One involved the military. So swiftly were the Taliban forces routed, at such minimal cost to American personnel, that the very ease of the accomplishment engendered a new feeling of American superiority and hubris.

For the military, given the bitter history of public divisiveness over its role in the losing Vietnam War, this new campaign carried with it considerable irony. In the generation since Vietnam, the American military had recovered its lost esteem in the eyes of the public and gone from near the bottom of major institutions that Americans viewed favorably to the top. By 9-11 the credibility of the military among both the public and the press was high.

When the war on terrorism was formally launched three weeks after 9-11, the greatest public concern involving the military was over its fate in operating in the forbidding, rugged landscape of far-off Afghanistan. There, a century before, the British Empire at its peak was brought to its knees with heavy loss of life. A century later Afghan forces, with strong U.S. covert support in advisers and weaponry, dealt a humiliating defeat to Soviet forces stationed there.

Despite that history, the high-tech weaponry employed by the U.S. military, aided by its special forces on the ground, performed superbly. Within two months the lightning swift and, for the most part, deadly accurate air and missile strikes had routed Taliban and Al Qaeda forces, though fighting still continued. Osama bin Laden and his chief deputies were either driven into hiding or killed.

In terms of American casualties, this early triumph was almost cost-free, but the Afghan civilian population paid heavily, suffering unknown numbers of dead and wounded, the "collateral damage" from U.S. strikes. For most of the first six months of fighting, only one U.S. soldier died of combat wounds. One CIA officer died after being shot during a prison revolt. By contrast, during that time ten journalists died while covering the conflict, one of whom, the *Wall Street Journal's* Daniel Pearl, was kidnapped and brutally executed by terrorists who cruelly videotaped themselves slitting his throat and beheading him and then made the grisly video available for public dissemination. Nor could the seeming ease of the U.S. military mission obscure the reality of the hardships in prosecuting this new kind of war against elusive terrorists holed up in caves atop snow-capped mountains, escaping across the borders into neighboring Pakistan and Iran, or melding back into Afghan towns and villages where they were sheltered and aided by local tribesmen. What Pentagon officials billed as a mop-up operation to eliminate the last remaining vestiges of Al Qaeda and Taliban forces in March 2002, quickly showed that far from being finished, the terrorist forces remained a dangerous threat. In a twenty-four-hour period, eight U.S. soldiers were killed in an ambush, many others were wounded, and the Americans had to withdraw under heavy fire. Still, the overall record of the American war to that point had to be regarded as that of a surpassingly easy triumph, but a triumph tempered by awareness that the difficulties and bloodshed were far from over.

For the military, however, this new kind of war engendered a new kind of problem—an obsession with secrecy, with keeping the press and public in the dark about activities that in no way could be construed as jeopardizing military security, and with blatant propaganda activities. The irony was that this reaction threatened to sow new public distrust of the military.

This Big Brother tendency, already strongly held throughout the Bush administration, reached a new peak when it was disclosed that the Pentagon had created a shadowy new unit, funded by millions of taxpayer dollars, and bearing the Orwellian title the Office of Strategic Influence. Its mission was to produce both favorable public reaction among U.S. allies and cultivate disinformation among foes in part by planting false stories in the press. It was yet another example of the failure to learn past lessons, particularly the destructive cast of mind that enveloped the military and the press and poisoned the society during the Vietnam War era. One who did remember that lesson was William Cohen, the former Republican senator who served as Clinton's defense secretary. When news of the strategic influence office broke, Cohen told CNN: "We are talking about deceiving the media and the public in general in foreign countries, and that would be a mistake."

Public reaction to disclosure of the new propaganda office was so strong that the office was disbanded. But damage had been done. This official we-know-best mentality could jeopardize the welcome spirit of unity forged by 9-11 by renewing cynicism about the believability of both the government and the media.

President Bush contributed to the problem of forging a united front against terrorism by his increasingly martial, indeed cocky, rhetoric about "evil doers" and "an axis of evil" arrayed against the United States. The clear message being delivered was that the U.S. could afford to go it alone in its worldwide war against terrorism. To friends and foes alike, the president seemed to say: Toe the American line or get out of our way. America, Rex.

Bush's increasingly bellicose stance created new strains in the American-led alliance against terrorism. Allies in Europe, the Mideast, and Asia reacted uneasily to strong Bush hints that the U.S. might unilaterally strike Iraq's Saddam Hussein—or take action against Iran and North Korea, the two other sovereign states cited by the president as partners in a new "axis of evil." Subsequently, he raised even greater concern with his remarks that use of tactical nuclear weapons was a viable U.S. option. Furthermore, those three nations in no way formed the kind of close alliance as the Axis powers, Germany, Japan, and Italy, did in World War II, and Iran and Iraq had fought each other only years before in a bloody conflict that also saw bioterror weapons employed.

New problems in creating a unified international front against terrorism were matched by vexing new domestic ones. The six months after 9-11 had produced a total turnabout in the long-term outlook for the American economy.

When George W. Bush became president on that wind-swept inaugural stage at the Capitol, the United States had ended the long period of dramatically rising deficits accumulating "as far as the eye can see" that had begun in the 1980s. The cost of this spiraling national debt was borne by every American. It affected all efforts to address dire domestic problems. When the Clinton years began in January 1993, the national debt had multiplied fourfold since the Reagan years, having climbed to $4 trillion. In the year Clinton was running for president, the debt had soared by $373 billion, and was increasing at a rate of $13,000 a second at an annual interest cost to the nation of $292 billion. Three years after his first inaugural, the national debt was still rising, increasing by nearly a trillion dollars in that short span. During that time, an average family of four's share of the burden of debt rose by $7,417 to $62,133—and was still increasing.

Six years later, when Bush took his oath, the boom, and, yes, the bitterly opposed Clinton tax increases in the face of a recession, had transformed the spiral of debt into surpluses for years to come. On his inaugural day in January 2001, the rosy forecast was for a trove of nearly $6 trillion in surpluses that would accumulate between 2002 and 2011.

America was flush; it could do anything it wanted. It could improve education, safeguard and improve the health benefits of an aging society, strengthen its defenses, fight hunger and disease, invest more sums in fundamental research and development for the common good. Just a year later, as the *Washington Post* commented editorially in January 2002, "an astonishing $4 trillion of projected surpluses has vanished from the federal ledger."

Poof, the surplus days were over. The nation was sinking back into the days of rising long-term debt. Already, supposedly sacrosanct and protected Social Security trust funds were being raided to finance daily needs of a government faced with a national security emergency requiring vast new sums for defense and a political system that had approved a reckless $1.3 trillion in tax cuts that will account in future years for 41 percent of that lost surplus money. And all this occurred in the face of an economy that had gone from boom to bust, amid concerns that other major bankruptcies and bookkeeping problems than Enron loomed. Further adding to the sense of public unease was new concern over the supply and price of oil in the face of growing violence and instability in the Middle East; the reemergence of health care as a dominant issue, as costs kept increasing and benefits were being slashed; and new environmental worries about global warming sparked by dramatic photos showing a vast portion of the arctic ice mass breaking off and floating away into the oceans.

· · ·

In spite of these daunting challenges at home and abroad, opportunities for positive change exist. At home, prospects for needed reforms in the economic and political system are clearly better than they were before 9-11. Long stalled, a campaign finance reform bill was passed by a Congress responding to cries for new regulatory financial reform in the wake of the Enron scandal. Reform of the nation's electoral system, seemingly dead in the months after the 2000 presidential election, became more possible as new legislation moved to the forefront in Congress. Measures to require compulsory public service were introduced in the Congress and the president endorsed the idea of enhancing the status of the public servant and public service.

These are praiseworthy, but they do not signal that America is entering a new era of good feelings, politically speaking. The veneer of bipartisanship that formed after 9-11 already is being eroded. Ideological differences between the two main parties are resurfacing over economic policy, judicial appointments, civil rights, and the role of government—debates that inevitably will intensify as the battle for control of the Congress and the White House grows closer amid the prospect raised by some political advisers to the president that he politicize his handling of the war as a tool to win political victories at home. Nor do the improved prospects for re-form add up to the kind of fundamental change necessary for America to safeguard and improve its future prospects. Six months after 9-11, the president had not delivered a call to action that clearly asked—and clearly spelled out—how all citizens should be more engaged in the process of either self-defense or self-protection in the broadest public sense. Despite the renewed spirit of patriotism, Americans for the most part remained as discon-nected from one another as ever. The fighting and dying in far-off lands was waged by a career military force whose common expe-rience was alien to that of the vast majority of citizens.

While the anti-terrorist efforts proceeded abroad, troubling questions loomed about the wisdom and the consequences of the administration's anti-terrorist actions at home. Sweeping new poli-cies aimed at containing suspected terrorists, whether U.S. citizens or not, raised the oldest questions about civil rights and liberties and the protection of the individual against the excesses of the state. Now, as in the past, there are no easy answers to these questions, but the range and scope of activities undertaken by the Bush ad-ministration in the name of the New Security deserve greater de-bate than they have received.

An impressive public airing of these issues came in a lengthy article by Ronald Dworkin, a distinguished law professor and au-thor whom I have cited for his arguments from the liberal side dur-ing the Clinton impeachment debate. Writing in the *New York Review of Books* on February 28, 2002, Professor Dworkin carefully sketched out what he believed to be serious threats to liberty and justice by the government's actions since 9-11. His was no polemic. It was a reasoned exposition of a multitude of new questions spawned by the terrorist attacks. As he said,

> Do we really face such extreme danger from terrorism that we must act unjustly? That is a difficult question. We cannot yet

accurately gauge the actual power of the linked groups of terrorist organizations and cells that apparently aim to kill as many
Americans as possible. Indeed, we scarcely know the identities
and locations of many of these groups.... It is unclear, moreover, how far the administration's various new measures, including military trials [and tribunals], would actually help to prevent
future attacks.

He concluded:

Our government has already gone too far, then, in replacing the
constitutional and legal rights that we have evolved as our own
national standard of fair play in the criminal process. Of course
we are frightened of the power of suicidal terrorists to kill again,
perhaps on an even more massive scale. But what our enemies
mainly hope to achieve through their terror is the destruction of
the values they hate and we cherish.

One can agree or not with Professor Dworkin's arguments,
and many respected conservative scholars and commentators have
made similar points, but he is surely correct in specifying the task
confronting all Americans as they address the age-old conflict between the need for security and the need to protect democratic
rights in a time of war: "We must protect those values as well as we
can, even as we fight the terrorists."

How well the American people and their leaders strike that
balance will go a long way toward answering the question of
whether important lessons of 9-11 have been learned. The maturity and good sense people displayed during the impeachment
trauma of the Nineties and the far more sinister terrorist threats of
9-11 and beyond offer reasons for guarded optimism about the
public's ability to deal effectively with new challenges at home.

· · ·

Even amid the grim uncertainties that affect all foreign relations in
the new age of terrorism, especially in the volatile Middle East,
prospects for positive change can be found.

In our post–9-11 world, the United States possesses an opportunity to take the lead in creating new international alliances, a combination that joins such erstwhile foes as Russia and China in a common front against the mutual threat of terrorism. Similarly, a tantalizing opportunity exists to create new, more realistic alliances in an Islamic world suffering from long decline into despair, paranoia, and hatred.

Clearly, those negative conditions still exist. Indeed, they have worsened notably throughout the Arab world. But 9-11 did not create the Islamic world apocalypse many forecast. Moderates there have not yet been, as feared by many, engulfed by a tide of rising fundamentalism, as zealots preached the heavenly rewards of martyrdom through suicide missions. The Muslim world did not arise in support of bin Laden and the terrorists. Further, the explosive display of American power in Afghanistan in response to terrorism put to rest any illusions among terrorists and fundamentalist mullahs preaching hatred of the West that the U.S. is a paper tiger unwilling to deploy its forces or to pay whatever price may be necessary to prevail.

Now, as in the past, the best hope for long-term change rests with moderates in the Islamic world. For them, self-interest, if not faith, dictates the necessity to deal with the zealots in their midst and to begin the kinds of internal reforms that will reduce over time the desperate conditions that have sent their world descending into greater depths of poverty, ignorance, corruption, and rage. Obviously this cannot be achieved without the strong assistance of a new international commitment to address the underlying problems that afflict the have-not world. Obviously, too, that kind of commitment will be forged only if the United States exercises a strong and ceaseless leadership role internationally.

The same is true, but even more so, when it comes to addressing the single greatest problem inspiring more terrorism and more hatred—failure to achieve peace in the Middle East. To a large degree, the fate of a more stable world lies in the ability finally to reach an accommodation over Israel. There, conditions have deteriorated gravely. The months since 9-11 have brought nothing but more heartbreak, and more bloodshed, to that troubled arena. Both Israelis and Palestinians suffer daily spasms of violence as suicide bombers and full-scale military assaults trigger ever-greater and more deadly military and terrorist escalation. The decision of Israel's Ariel Sharon to send his tanks and troops into Palestinian territory in response to yet more terrorist suicide bombings, many of them from teenage Palestinian girls forming new "martyrs' brigades," led to scenes of destruction and despair rivaling the nightmare ones of World War II and triggered the worst Middle East crisis since the founding of Israel more than half a century before. As the world watched with increasing apprehension while the fighting and dying intensified, Israeli forces battered their way into the headquarters compound where the Palestinian leader, Yasser Arafat, became a virtual prisoner, confined to three rooms without power or water and prevented from leaving by Israeli tanks literally stationed below the stairways to his offices. In the midst of such an explosive situation, dangerously teetering out of control and threatening to spawn more violence if not outright warfare throughout the Arab/Islamic world, President Bush experienced the most severe criticism since taking office. His hands-off policies toward the Israeli-Palestinian grievances, his perceived bias toward Israel, his failure to take an aggressive leadership role to attempt to resolve the conflict, all generated a torrent of criticism not only from within the Muslim world but also among America's European allies. Much more than even the crucial Middle East was at stake; the president's proclaimed war on terrorism itself could well collapse if the United States found itself isolated throughout the world. The new crisis brought home a hard reality: that the Middle East, not the Taliban and not bin Laden, remained central to any prospect of winning a war against terrorism. Yet even in these darkest of times there, a glimmer of hope exists. Again, it stems from moderates on both sides recognizing that self-interest compels them to break the long cycle of hatred and bloodshed. Short of all-out Islamic/Western war, there seems no other solution.

That kind of hope also provides a lesson for America.

Hatred and terrorism will not be eliminated in a lifetime, if ever, nor will age-old problems of human misunderstandings and mistakes. America cannot escape history; we must face the terrors as well as the progress it imposes over the centuries. There is no fail-safe method to guarantee our security. We ignore the old lesson graven on our National Archives Building at our peril: ETER-

NAL VIGILANCE IS THE PRICE OF LIBERTY. That lesson means being prepared to protect ourselves from assaults by enemies who threaten our democratic liberties and freedoms from within, as well as guarding against conventional military or terrorist forces who pose a different kind of threat from without.

As we emerge from the terrors of 9-11, the question is not whether America has changed. The question is whether America can adopt a new realism about its problems and maintain the determination to deal with them—a determination driven not so much by rekindled patriotism as by the more powerful instinct for self-preservation.

Of all the lessons cruelly hammered home to Americans on 9-11, surely the greatest has to be this: Americans in the future can never afford to float through the ether of good times and squander their opportunities to seriously address our long-term problems. We who lived through the Nineties have learned at too high a cost that left unattended such problems will inevitably come back to haunt or even destroy us.

Postscript:
Relics from the Nineties

MARV ALBERT, the sports announcer fired after pleading guilty to a misdemeanor assault charge for biting a lover, found that infamy didn't hurt his career. He was reinstated to full duties at Turner, MSG, and NBC, and married the woman who stood by him during the scandal, Heather Faulkiner, his then-fiancée and an ESPN producer.

MICHAEL BLOOMBERG, founder of the business and media empire Bloomberg L.P., was elected mayor of New York, succeeding Rudolph Giuliani in the daunting task of rebuilding the city after the devastating events of September 11. Bloomberg, one of the richest men in New York, ran on a platform similar to Giuliani's, pledging fiscal conservatism and social liberalism. Giuliani emerged as one of the nation's heroes for his leadership role after 9-11.

SIDNEY BLUMENTHAL, one of many advisers to whom Clinton lied about his relationship with Monica Lewinsky, reportedly received more than half a million dollars to write his memoirs and began producing movies. He dropped his $30 million libel suit against Internet reporter Matt Drudge.

STEVEN BRILL, the mega-media entrepreneur who broke ground with *Court TV,* shut down his magazine *Brill's Content* and unveiled plans to sell the media-news Web site Inside.com to partner Primedia. He subsequently sold *Court TV* to Time-Warner in 1997 for a reported $40 million.

SUSAN CARPENTER-McMILLAN, spokeswoman and strategist for Paula Jones, dropped the "Carpenter" in her unsuccessful bid for the California State Assembly in 2000. She continued her work as a conservative commentator for The Women's Coalition.

MARCIA CLARK, the prosecutor who rose to celebrity status despite losing O. J. Simpson's criminal trial, continued to appear on television as host and commentator, including a stint on the cast of *Power Attorney,* a syndicated courtroom show where high-profile lawyers battle one another on behalf of noncelebrity litigants.

BILL CLINTON began splitting his time between a post-presidential office in Harlem, homes in Chappaqua, New York, and Northwest Washington, D.C., and travel for speeches, which reportedly earned him more than $150,000 apiece. He signed a deal to write his memoirs for $10 million, the highest ever for a nonfiction book. And he took a major step toward shaping his legacy— breaking ground on his presidential library in Little Rock, Arkansas. On the day of the event, however, Clinton stood without his immediate family. Hillary was in Washington and daughter Chelsea was at Oxford. Back at home, Buddy, the chocolate Labrador retriever who was a loyal companion to Clinton during the Starr investigation, was killed by a car near the Clinton's New York home. Later, he considered an NBC offer to become a TV talk show host, which if accepted, would make him the first talking head ex-president—and add another dubious development to Teletimes.

HILLARY CLINTON continued to serve as New York's junior senator. She surprised colleagues with collegiality toward members on both sides of the aisle. At her Washington, D.C., home, she began hosting numerous fundraisers for Democratic colleagues and even threw a baby shower for Texas Republican Kay Bailey Hutchinson. The controversy that Hillary and Bill engendered, however, flared again with release on March 20, 2002, of the special prosecutor's final report on what seemed like the ages-old Whitewater affair that had dogged the Clintons for their years in the White House, divided the nation ideologically, led directly to all the investigations that culminated in impeachment, and, after eight years, cost American taxpayers some $65 million. Like so many of the "Clinton scandals," the final Whitewater report proved to be totally inconclusive, but typically clouded with new controversy. The last prosecutor, Robert W. Ray, a Republican who was seeking the GOP senate seat in New Jersey, concluded the evidence was "inconclusive" as to whether either of the Clintons had committed any crimes. He specifically raised questions about Hillary's handling of billing documents involved in the case. Further, his report accused her of giving "factually incorrect" testimony, but acknowledged that "the evidence [his office compiled about her] was insufficient to obtain and sustain a conviction beyond a reasonable doubt." The grudging nature of his report led the Clintons' attorney, David Kendall, to say caustically of its implied accusation that there might have been evidence of wrongdoing but insufficient evidence to prove it that, "when fairly and carefully read, [it] lends no support to such an innuendo. One might say with equal justification that the Office of Independent Counsel has uncovered no evidence from which a jury might infer beyond a reasonable doubt that the Clintons had pilfered Powerball tickets, trapped fur-bearing mammals out of season, or sold nuclear secrets to Liechtenstein."

JOHNNIE COCHRAN, O. J. Simpson's defense attorney, continued high-profile legal work. Among his endeavors was representing laid-off Enron workers to help them recover lost savings. In late

2001, his sixteen-year-old nephew was killed in an ambush shooting in Los Angeles.

BETTY CURRIE, President Clinton's secretary who was a key witness in the Monica Lewinsky scandal, retired. She continued to take care of the Clintons' former cat, Socks.

THE DIGITAL DIVIDE, the gap that separates America's haves from its have-nots in access to computer technology, threatens to grow wider after the stand taken by Bush's appointee as head of the Federal Communications Commission. As the *Wall Street Journal* reported, Michael Powell, who as FCC chairman is the nation's top telecommunications regulator, said only those "with an unreal understanding of U.S. capitalism would expect the poor, minorities, and rural residents" to have the same access to the Internet as other Americans. Government efforts to bridge the digital divide, he said, veer toward "socialization."

THE DOT-COM BUBBLE burst in 2001, with more than five hundred Internet companies declaring bankruptcy or closing, more than twice the number of the previous year. Even the most famous Internet addresses became defunct, including eToys.com and Drkoop.com.

MATT DRUDGE, who became famous for breaking the Monica Lewinsky story on the Internet, continued reporting on his Web site but gave up his television program in a dispute with Fox News Channel. His radio show "Drudge" reached the 200-affiliate milestone in late 2001 and was heard in all top-ten markets from Los Angeles to New York. A Matt Drudge Defense Fund was set up to counter the various libel suits brought against him.

GENNIFER FLOWERS, whose twelve-year affair with Bill Clinton almost cost him his bid for the presidency, became a saloon owner and torch singer in New Orleans's French Quarter. She said the former president would be welcome with or without his saxophone. The bar's logo is a representation of Flowers's lips.

BILL GATES remained chairman of Microsoft. Despite the recession in 2001, he was reported to be the wealthiest man in America for the eighth straight year, with his fortune down to $54 billion from $63 billion the previous year. The Bill and Melinda Gates Foundation continued to donate large sums to worthy causes.

DAVID GEFFEN remained at DreamWorks SKG and was in the top ten of *Entertainment Week*'s Power List for 2001.

LUCIANNE GOLDBERG, who encouraged Linda Tripp to tape Monica Lewinsky, became a talk-show host for Talk Radio Network and launched an Internet forum, Lucianne.com. Her son, Jonah, a conservative writer and commentator who was also involved in the Tripp tape affair, was hired by CNN.

ELIAN GONZALES continued to live with his father and family in Cárdenas, Cuba.

AL GORE returned to private life and his home in the suburbs of Washington, D.C., but made moves suggesting he might run for the presidency again in 2004. Sporting a beard and more casual clothing, he began appearing at Democratic events. Then he started spending more time in his home state of Tennessee, where he frequently lectured at local universities, and showed off the beard which had become a target of cartoonists and late-night TV comedians. Later, he shaved off the beard and made political speeches.

ALAN GREENSPAN continued as Federal Reserve chairman. As the economy sagged in 2001, he launched a campaign of eleven interest-rate cuts. Greenspan's influence remained strong, with his every remark dissected and analyzed by the financial world.

LANCE ITO, the judge in the O. J. Simpson criminal trial, continued to preside over cases as a Los Angeles Superior Court Judge.

PAULA JONES, who filed the sexual-harassment lawsuit against President Clinton that led to the discovery of his affair with Monica

Lewinsky and his eventual impeachment, married Little Rock neighbor, Steven Mark McFadden, a construction equipment company employee. The wedding took place in Villa Marre, known to many Americans as the house on TV's *Designing Women*. It was Jones's second marriage. The seduction of a further fifteen minutes of fame proved too much for her to resist, however. In the spring of 2002, she donned boxing gloves and padded helmet, the better to protect her new nose job, and battled another half-remembered name from the galaxy of fleeting fame and scandal in the Nineties, Tonya Harding. Their prime-time TV special, *Celebrity Boxing*, wasn't much of a match; by the end of the second round outpunched Paula was trying to hide behind the ref, but the hyped spectacle was a winner for the sponsors and the network. It attracted more than 15 million viewers and so pleased the promoter that he told reporters "Darva Conger is interested if we can find the right person to fight her."

VERNON JORDAN, President Clinton's close friend who helped Monica Lewinsky in her job search, became a senior managing director at New York investment bank Lazard Freres. He maintained an office at the D.C. law firm where he met with Lewinsky during the Clinton years. He was awarded the NAACP's highest honor. In late 2001, he published a memoir entitled *Vernon Can Read!*, which ends just as the Clinton years begin and offers no insight into the Clinton-Jordan-Lewinsky relationship.

KATO KAELIN, the blond witness at the O. J. Simpson trial, continued to cash in on his celebrity. Kaelin estimated his personal income has tripled since the trial. He was a contestant on *The Weakest Link:* "15 Minutes of Fame Edition" along with Tonya Harding and Gennifer Flowers and began work on an autobiography, *O.K. I'm Famous, Now What?*

GARRY KASPAROV, the chess player defeated by the computer Big Blue, officially lost his world chess champion title to Vladimir Kramnik, but continued to be considered by many as the greatest in the game. He seemed to have embraced technology with the launch of his Web site, Kasparovchess.com. In 2002, the chess world planned a rematch of Man vs. Machine, pitting Kramnik against the world's top chess program Deep Fritz in Manama, Bahrain.

MONICA LEWINSKY, the intern whose affair with President Clinton led to his impeachment in 1998, became a handbag designer and studied psychology part-time at Columbia University. She remained a popular subject of cultural interest here and abroad. In March 2002, HBO released a documentary about her. And in Moscow, the Saratov Opera announced plans to present *Monica in the Kremlin,* a new work inspired by the Lewinsky affair.

THE REALITY TV AND QUIZ SHOW genres, which had dominated the airwaves with shows like *Survivor* and *Weakest Link,* started to show signs of weakness. ABC cut back airing of *Who Wants to Be a Millionaire,* the game show hosted by Regis Philbin, to one night a week. At the height of its popularity in 2000, the program was shown up to seven nights a week.

JANET RENO, President Clinton's attorney general, returned to the Kendall, Florida, family home her mother built by hand out of cypress wood and block. She launched a campaign for governor, challenging incumbent Jeb Bush, President George W. Bush's brother.

O. J. SIMPSON continued to tangle with the law. In late 2001, his Miami home was raided in connection with a drug investigation, and he was acquitted in a Florida road-rage trial. He began hosting and appearing at hip-hop shows across the country.

KENNETH STARR returned to his practice at the Washington, D.C., law firm Kirkland & Ellis, taking on issues like school vouchers and weighing in on the Microsoft case.

LINDA TRIPP, whose secret recordings of Monica Lewinsky led to the impeachment of President Clinton, fell on hard times. She was fired from her Defense Department job and narrowly avoided foreclosure of her home outside Washington, D.C. She launched a Web site, Lindatripp.com, where visitors could make credit-card donations to the Linda R. Tripp Legal Defense Fund Trust. In March 2002, she was reported to have been diagnosed with breast cancer.

CRAIG VENTER, the outspoken scientist who was a leader in the effort to map the human genome, stepped down as president of Celera in early 2002. Under Venter's leadership, the company sequenced the human genome in just over two years. But Celera's stock had slumped, and investors were frustrated that the company had not moved more quickly into development of genome-based drugs.

THE VOTE . . . A news media review of Florida's uncounted ballots in the 2000 presidential election found that if either of two limited recounts had been completed, George W. Bush would still have defeated Albert Gore. However, the study also found that Gore could have won if all the state's contested ballots had been counted. Florida changed its voting system after the controversy, doing away with punch-card ballots and mandating a review of over- and undervotes in the event of a recount. Later, Katherine Harris, Florida's Republican Secretary of State who certified Bush's win, remarked: "We can look forward to the next election with confidence." She launched her own bid for United States Congress.

KATHLEEN WILLEY, who went on national television during the height of the Starr investigation to accuse the president of unwanted sexual advances while she was a White House volunteer, sued President Clinton, his wife, and lawyers for releasing personal letters she sent the president. She said they were trying to undermine her credibility. In 2001, she, as with so many other relics of the Nineties, couldn't stay away from the media spotlight. She reappeared on the talk-show circuit to discuss Congressman Gary Condit's relationship with missing staffer Chandra Levy.

Acknowledgments

As always, I owe a greater debt than I can adequately acknowledge to the many friends, colleagues, and professional sources without whose assistance this book could not have been written. To these few, out of so many, I want to express my special thanks: First to Jim Silberman, a great editor and friend, and to Dan Farley, Harcourt's president and publisher, who were enthusiastic and entirely supportive of this project from its beginning in 1997 and more than patient as the twists and turns of the subject matter kept extending the manuscript deadline far beyond what they, or I, ever anticipated. I owe many people at Harcourt special appreciation for their backing, but I particularly want to thank Jen Charat for her splendid editorial assistance and Rachel Myers for a superb job of difficult copy-editing. I'm also grateful to my agent, Bill Leigh, and to my graduate assistants who labored in compiling research material while trying to keep me on track: Kathy Wenner, whose extraordinary efforts on the O. J. Simpson case alone provided extensive and invaluable background matter, and Jennifer Fox and Jennifer Dorroh. I am fortunate in having had the support of my colleagues at the Merrill College of Journalism, University of Maryland, and especially Reese Cleghorn, Tom Kunkel, and Gene Roberts for counsel and friendship. I regret I cannot acknowledge all who took time out of their busy schedules to help me, but I particularly want to thank these among the many: In Hollywood,

Bill Carrick, Beegie Truesdale, and Marcia L. Hale exceeded the bonds of long friendship in opening doors for interviews with such people as Neal Baer, Sean Daniel, David Geffen, Carole King, Ted and Susan Harbert, Norman Lear, Andy Spahn, Gary Ross and Allison Thomas. In Pasadena, at the California Institute of Technology and the Jet Propulsion Laboratory, Roger Bourke was equally accommodating in guiding me to conversations with David Baltimore, David Goodstein, Rob Manning, Mel Simon, Ed Stone, and Rich Terrile. In Silicon Valley, my old friends Larry Stone (and his assistant David Ginsborg) and Regis McKenna once again helped lead me through the ever-changing and ever-more-complex nature of the territory. At Stanford, Adrienne Jamieson, director of the Stanford in Washington program, went out of her way to set up conversations with students and such distinguished administrators and professors as Donald Kennedy and Robert McGuinn, among others. At Pepperdine University, Doug Kmiec offered important insight during a memorable afternoon there. In Los Angeles, Jim Wisely again was of great assistance in suggesting conversations with, among others, Laurie Levenson, Dana Millikin, Anna Soto, Father Modesto, and gang members in South Central L.A. At Microsoft, Mike Kinsley, Merrill Brown, Scott Guthrie, Sam Jadallah, Nathan Myhrvold, Greg Shaw, and Mike Murray were generous with their time. So was Bob Shapiro in Chicago, who took time out of his hectic schedule at Monsanto to talk for several hours. In New York and Washington, D.C., the same was so, among numerous others, with Michael Bloomberg, Steve Brill, Paul Friedman, Dr. Bob Cook-Deegan, Dr. Ruth Faden, Dr. Craig Venter, Sam Dash, and many other political acquaintances who shall be nameless here.

Finally, I want to acknowledge, however inadequately, the special debt I owe two people. Lisa Larragoite, whom I was fortunate to hire from *The NewsHour with Jim Lehrer*, became far more than my assistant. Through the years of reporting, research, and writing she was a full partner in this effort and deserves credit for any strengths the resulting product may have. Its faults are entirely mine. She was, simply, invaluable. At the same time, in ways more numerous than I can express, Kathryn Oberly was indispensable as counselor, critic, friend, and confidante at every stage of this endeavor from beginning to end. It is her book as well as mine.

About the Author

Haynes Johnson has reported on many of the major events affecting America for the last two generations. A native of New York, he earned his master's degree in American history at the University of Wisconsin and then began a newspaper career as a reporter, editor, and columnist at the *Washington Star* and the *Washington Post,* covering events at home and abroad. In 1966 he won the Pulitzer Prize for distinguished national reporting of the civil rights crisis in Selma, Alabama. The award marked the first time a father and son had won prizes for reporting; his father, Malcolm Johnson, won for his *New York Sun* series "Crime on the Waterfront." *The Best of Times* is Haynes Johnson's thirteenth book. Four others became national best-sellers. For many years he was a mainstay on the PBS-TV program, *Washington Week in Review,* and now appears on the "historians' panel" of *The NewsHour with Jim Lehrer.* He has had academic appointments at Princeton, Berkeley, Duke, the Annenberg School at the University of Pennsylvania, George Washington University, and the Brookings Institution, and currently holds the Knight Chair in Journalism at the University of Maryland.

Notes and Sources

As stated earlier, quotations in the manuscript not otherwise identified here or in the text are from transcripts of tape-recorded conversations with people interviewed for this book.

To the Reader

page ix, "...an attempt to tell": Frederick Lewis Allen, preface to *Only Yesterday: An Informal History of the Nineteen-Twenties.* (New York: Harper & Brothers, Publishers, 1931), xiii.

page ix, "...the shortsightedness": Ibid.

Prelude

page 1, "a novel without": William Makepeace Thackeray, *Vanity Fair: A Novel Without a Hero* (New York: The Heritage Press, 1940), xv.

page 1, "not a moral place": Ibid.

page 5, "A wide-spread neurosis": F. Scott Fitzgerald, "Echoes of the Jazz Age," *Scribner's Magazine,* November 1931, reprinted in *The Crack-Up,* ed. by Edmund Wilson (New York: New Directions, 1945), 19.

BOOK ONE: TECHNOTIMES

1. Deep (RS/6000SP) Blue

page 11, "I'm a human being": Bruce Weber, "Swift and Slashing, Computer Topples Kasparov," *New York Times,* May 12, 1997.

page 13, "It had the impact": Ibid.

page 13, "I lost my": Ibid.

page 13, "Garry has a": Ibid.

page 13, "Because it shows": Robert D. McFadden, "Computer in the News: Deep (RS/6000 SP) Blue; "Inscrutable Conqueror," *New York Times,* May 12, 1997.

2. Culture of Success

page 17, "the most transforming": "The Net Imperative," *Economist,* U.S. Edition, June 26, 1999.

Page 19n, "The metaphysical bomb": Henry Adams, *The Education of Henry Adams* (Boston: Houghton Mifflin Company, 1918), 452.

page 20, "fireflies before": "The Real Revolution," *Economist,* U.S. Edition, June 26, 1999.

page 20, "was worth as much": Gary Rivlin, *The Plot to Get Bill Gates; An Irreverent Investigation of the World's Richest Man . . . and the People Who Hate Him* (New York: Random House, Times Business, 1999), 123.

page 21, Ibid, 119n.

page 23, "each home will": John Markoff, "Silicon Valley's Own Work Threatens Its Domination," *New York Times,* July 22, 1999.

page 27, "the booming Internet": Ibid.

page 28, "Chips Nearing": "Science Notebook" *Washington Post,* June 28, 1999. See also Malcolm W. Browne, "Is Incredible Shrinking Chip Nearing End of the Line?" *New York Times,* June 29, 1999.

page 28, "so pervasive": John Markoff, "Tiniest Circuits Hold Prospect of Explosive Computer Speeds," *New York Times,* July 16, 1999. See also "A Renaissance in Computer Science: Chip Designers Seek Life After Silicon," *New York Times,* July 19, 1999.

page 28, "was a major underwriter": John Markoff, "Tiniest Circuits," *New York Times,* July 16, 1999.

page 28, "is just one": Dean Takahashi, "H-P Researchers Shrink Chips to Molecule Size," *Wall Street Journal,* July 16, 1999.

page 29, "the main external influence": In the preface to his *Libraries of the Future* (Cambridge, Mass.: MIT Press, 1965), Licklider credits Bush as "the main external influence" on his ideas and dedicates the book to Bush.

page 29, "the most politically": G. Pascal Zachary, *Endless Frontier: Vannevar Bush, Engineer of the American Century* (Cambridge: MIT Press, 1999), 2.

page 30, "A Memex...is": Vannevar Bush, "As We May Think, " *Atlantic Monthly,* July 1945. The article was reprinted by IBM's "Feed Document" service, July 12, 1999, to "revisit some of the founding documents of the digital age" for Internet users under the title, "Remembering the Memex."

pages 30–31, "turned out to be": "Remembering the Memex." Ibid.

page 31, "a scientist's version": Harry S. Truman, *Memoirs, vol. 1, Year of Decisions* (New York: Doubleday, 1955), 11.

page 31, "Admiral Leahy was": Ibid.

page 31, "This is the biggest fool": Ibid.

page 33, "A nation which": quoted in Bruce L. R. Smith, *American Science Policy Since World War II* (Washington: The Brookings Institution, 1990), 44.

page 36, "The hope is that": J. C. R. Licklider, "Man-Computer Symbiosis," *IRE Transactions on Human Factors in Electronics* I, (March 1960): 4–5. On Aug. 7, 1990, the Licklider article and another one he coauthored with Robert Taylor were reprinted by Digital, Systems Research Center, Palo Alto, Calif., 1990.

page 37, "In a few years": J. C. R. Licklider and Robert Taylor, "The Computer as a Communication Device," *Science and Technology,* April 1968. See Systems Research Center, op. cit, reprint.

page 39, "A network of": Ibid.

page 39, "Do you see the L": from interview, *Sacramento Bee,* May 1, 1996. An excellent, indeed indispensable, overview of the Internet and its leading players, including Kleinrock, can be seen in the 19-page document, "A Brief History of the Internet," by Barry M. Leiner, Vinton G. Cerf, David D. Clark, Robert E. Kahn, Leonard Kleinrock, Daniel C. Lynch, Jon Postel, Larry G. Roberts, Stephen Wolff, published by the Internet Society.

page 41, "For society": Licklider and Taylor, "The Computer as a Communication Device," op. cit.

page 41, "digital divide": Jeri Clausing, "Report Shows Increase in 'Digital Divide,'" *New York Times,* July 8, 1999, and more comprehensive analysis by David E. Sanger, "Big Racial Disparity Persists in Internet Use," *New York Times,* July 9, 1999.

page 46, "The knowledge driving": A text of Vest's testimony before the Joint Economy Committee, June 15, 1999, was made available to me by Vest's MIT office.

page 47, "What is missing": Ibid.

3. Nerd Nirvana

page 58, "In the end": Ken Auletta, "Hard Core," *New Yorker,* August 16, 1999.
page 64, "We knew he": Ibid.

4. Seeding the Future

page 74, "Wow", . . . "This is going to be": Gina Kolata, "In Big Advance, Cloning Creates Dozens of Mice," *New York Times,* July 23, 1998.

page 74, "It's absolutely incredible": Ibid.

page 74, "We have no interest": Ibid.

page 75, "a landmark event": Rick Weiss, "A Crucial Human Cell Is Isolated, Multiplied," *Washington Post,* November 6, 1998.

page 75, "sidestepping the increasing": Nicholas Wade, "Researchers Claim Embryonic Cell Mix of Human and Cow," *New York Times,* November 12, 1998.

page 75, "Should the technique": Nicholas Wade, "Blueprints for People, but How to Read Them?" *New York Times,* December 8, 1998.

page 76, "The question isn't": Michael Waldholz, "Cloning of Humans Grows Increasingly Possible," *Wall Street Journal,* December 18, 1998.

page 76, "the scariest news": Charles Krauthammer, "Of Headless Mice . . . and Men," *Time,* January 19, 1998.

page 80, "how genetic engineering": *Time,* Special Issue, "The Future of Medicine," January 11, 1999.

page 82n, work "any monkey": Dr. James Watson's now-famous congressional testimony was quoted in *Time,* Ibid.

page 87, "the greatest mapping": The Harvard Business School 60-page report, focusing on Venter and Celera, "Gene Research, the Mapping of Life, and the Global Economy," was published October 19, 1998.

page 90, "perhaps the most influential": Gordon Moore, "California Institute of Technology Annual Report, 1996–97," written in Moore's capacity as chair, Caltech's Board of Trustees, p. 2.

page 90, "a central tool": From text of Baltimore inaugural address by Dr. Maxine F. Singer, president of the Carnegie Institution of Washington, March 9, 1998, published in California Institute of Technology's *Engineering & Science* 61, no. 1 (1998):30.

BOOK TWO: TELETIMES

5. *Trial of the Century—Part One*

Direct quotations from network anchors, correspondents, and other people shown on TV and radio newscasts are from transcripts of those broadcasts.

page 107, "This is a special": *ABC News* Transcript # 149–3, June 17, 1994.

page 110, "the bloodiest crime": Nancy Gibbs, "End of the Run," *Time,* June 27, 1994. A brilliant article, and the best, and most comprehensive, of all those I read.

page 113, "You watch this": *CBS Evening News* transcript of broadcast June 20, 1994. Anchor Dan Rather, reporter John Blackstone.

page 115, "one of the top...": *CBS Evening News* Transcript of June 20, 1994 broadcast replaying the Kurtz commentary on O.J. over CNN.

page 116, "We love this": Steve Magagnini, "Angelenos Join Circus of O.J.'s Bizarre Saga," *Sacramento Bee,* June 19, 1994.

page 116, "We were on": Ibid.

page 117, "strangely powerful": Walter Goodman, "Television, Meet Life. Life, Meet TV," *New York Times,* June 19, 1994.

page 121, "first *fighting* gang": *Playboy,* December 1976. This lengthy Q. & A. with O.J. is the most revealing about his early life and the circumstances of his gang involvement.

page 121, "where a couple": Ibid.

page 121, "I only beat up": Gibbs, *Time,* op. cit.

pages 121–122, "One night I": *Playboy,* op. cit.

page 122, "He already is": Gibbs, *Time,* op. cit.

page 122, "That's what happens": Ibid.

page 123, "a lousy student": *Playboy,* op. cit.

pages 123–124, "a whole bunch": Ibid.

page 125, "O. J. Simpson was living": "Last Week in O.J. History; All O.J., All the Time": Week 25," *Boston Globe,* December 11, 1994.

page 126, "This belongs to me": Testimony of Denise Brown, Nicole's sister, O.J. trial.

page 126, "cut their fucking": Among 59 alleged incidents of domestic violence by O. J. Simpson against his wife, Nicole, reviewed before Judge Ito for admissibility in the then-sequestered jury, the most incriminating involved the account given criminal investigators by Eddie Reynoza, who had worked on a movie with O.J. and who told investigators that

Simpson became angry at the mention of Nicole's boyfriends, threatening "to cut their fucking heads off" if he found them driving his cars.

page 127, "It was perplexing": "O.J.'s Fatal Attraction?" *Sports Illustrated,* June 27, 1994. A superb, richly detailed article.

pages 127–128, "911 emergency": Transcripts of tapes broadcast by CNN, June 22, 1994, Transcript # 853–5. Also other police emergency tapes cited later in this chapter are from CNN Transcript # 4124–3.

page 131, "We moved into": *Sports Illustrated,* op. cit.

page 132, "She'd work up": Ibid.

page 138, Immediately, Indianapolis talk shows: Shirely Ragsdale, "Tyson Still Has Much to Learn," *Lafayette (Indiana) Journal and Courier,* June 15, 1994. An especially revealing article by the paper's opinion page editor describing the enraged reaction she hears while driving through Indianapolis after the judge refuses to free Tyson.

page 140, "While television...": Walter Lippmann, column, "Television: Whose Creature, Whose Servant?", October 27, 1959, reprinted in *The Essential Lippmann* ed. by Clinton Rossiter & James Lare (New York: Random House, 1963), 412.

page 145, "So, you think": CNN Transcript # 860–6 from broadcast June 30, 1994.

page 146, "With each case": "CBS Evening News" transcript, anchor Bob Schieffer, reported by Vicki Magay, June 25, 1994.

page 146, "I don't like": quoted in Jill Smolowe, "TV Cameras on Trial," *Time,* July 24, 1995.

page 147, "This book on": From CNN Transcript # 573 of *Showbiz Today,* broadcast June 29, 1994.

page 148, "I'm tired of watching": "We're All Groaning About O.J.," *Publishers Weekly,* August 22, 1994.

page 148, "a less informed": Brad Hayward, "Simpson Recess for Election?" *Sacramento Bee,* August 9, 1994.

page 150, "Al really isn't": "Last Week in O.J.," *Boston Globe,* op. cit.

page 152, "It's been rough": *People,* January 10, 1999.

page 152, "He is probably": "Showbiz wannabe Kato Kaelin may be the Simpson Case's only rising star," *People,* September 12, 1994.

page 153, "The tabloids had...": Elizabeth Gleick, "Leader of the Pack. National Enquirer's Aggressive O. J. Simpson Coverage Raises Legal and Ethical Questions," *Time,* January 9, 1995, is the best account. *People* also detailed a number of other reports about the role of the tabloids. See, for example, its account of the sale of the stiletto to O.J., which was first detailed in the preliminary O.J. court arraignment: *People*, October 10, 1994.

pages 156–157, "It's out of": Harriet Chiang, "Judge Ito, Wife, Take Center Ring in O.J. Trial," *San Francisco Chronicle*, November 21, 1994.

page 157, "a garden variety": Ibid.

page 157, "There's something about": Ibid.

page 160, "All over urban": Michael Wilbon, "A Celebrity Goes Free," *Washington Post*, October 4, 1995.

page 163, nothing since the invention: Daniel J. Boorstin, "TV's Impact on Society," *Life*, September 10, 1971.

page 163, "democratized learning": Boorstin, Ibid.

page 164, "mysterious island-audiences": Ibid.

6. Cult of Celebrity

page 165, "In the future": Andy Warhol's celebrated remark is cited in Leonard Maltin, *Movie Encyclopedia* (New York: Signet, 1994).

page 175, "Television is just": Haynes Johnson, *Sleepwalking Through History: America in the Reagan Years* (New York: W.W. Norton, 1991), 141.

page 179, "I believe that": Ibid.

7. Dream Factories

page 199, "was garish, extravagant": Richard Griffith and Arthur Mayer, *The Movies*, rev. ed. (New York: Fireside/Simon & Schuster, 1970), 231.

BOOK THREE: SCANDAL TIMES

8. Bill's Story

All direct quotations not otherwise attributed are from the voluminous official five-volume Starr report, with appendices, printed as House Document 105–316, "Communication from the Office of the Independent Counsel, Kenneth W. Starr, Referred to the House Committee on the Judiciary." Washington: U.S. Government Printing Office, 1998. These documents contain the sworn grand jury testimony, affidavits, legal proffers, White House logs, depositions, summaries, e-mails, transcripts of telephonic recordings, copies of personal notes and memoranda, search warrants, White House map, litigation history, chronologies of the Paula Jones case history and the Clinton/Lewinsky investigation, citations from pertinent newspaper, magazine, and

TV reports, plus an invaluable table of contents with an index listing all those in any way involved or questioned.

page 247, "By nature and": Lars-Erik Nelson, "Clinton and His Enemies," *New York Review of Books,* January 20, 2000. See also his "The Republicans' War," same publication, February 4, 1999.

page 248, "If ever a nation": Cited in Haynes Johnson, "The American Press and the Crisis of Change," The Goldstein Program in Public Affairs, in *Contemporary Views of American Journalism* (Chestertown, Md.: Literary House Press at Washington College, 1997), 17.

pages 248–249, "If ever there is to be": Ibid.

page 249, "During the course": Ibid, 18.

page 249, "the one great blemish": Charles Dickens, *American Notes for General Circulation and Pictures from Italy* (London: Chapman & Hall, 1913), 203.

page 249, "Any man": Ibid.

page 277, "Bennett knew that": Anthony Lewis, "Nearly a Coup," *New York Review of Books,* April 13, 2000.

page 317, "They were looking": Cited in George Lardner Jr., "The Presidential Scandal's Producer and Publicist: Lucianne Goldberg Frequently Leaked Tidbits to Sustain Clamor," *Washington Post,* November 17, 1998. Lardner's lengthy article is the most meticulously reported, and devastatingly revealing, to appear about Goldberg.

page 318, "I wanted to": Ibid.

page 321, "He was not a bit": "The Roots of Ken Starr's Morality Plays," *Washington Post,* March 2, 1998.

page 344, "almost unhuman sadness": Carl Sandburg, *Abraham Lincoln,* vol. 2, *The War Years* (New York: Harcourt Brace, 1976), 200.

page 344, "it is not all": Ibid, 261.

page 346, "We talked about": Haynes Johnson and David S. Broder, *The System: The American Way of Politics at the Breaking Point* (New York: Little, Brown, 1996), 278.

page 346, "It was the most": Ibid, 279.

page 347, "They've become paranoid": Ibid, 280.

page 362, "we are all captives": Walter Lippmann, *Public Opinion* (London: George Allen & Unwin, 1922), 2.

page 378, "Since the 'explicit nature": Joan Didion, "Clinton Agonistes," *New York Review of Books,* October 22, 1998. In her article she cites the quotations from various publications on the sex scandals, some of which I use here.

page 384, "I am sorry": "President Clinton's Address and the National Prayer Breakfast," *New York Times,* September 12, 1998.

page 389, "It was a surreal": Michiko Kakutani, "Testing of a President," *New York Times,* September 12, 1998.

page 390, "Mr. Starr's report": "What Ken Starr Neglected to Tell Us," Bruce Ackerman, *New York Times,* September 14, 1998.

page 400, "Mr. Starr hoped": Editorial, "No New Evidence," *New York Times,* November 20, 1998.

page 404, "common ground over": Francis X. Clines, "A Dreadful Day Unfolds On and Off House Floor," *New York Times,* December 19, 1998.

9. Trial of the Century—Part Two

page 417, "a well-constituted court": Alexander Hamilton, John Jay, and James Madison, *The Federalist: A Commentary on the Constitution of the United States,* ed. Henry Cabot Lodge (New York: G. P. Putnam's Sons, 1904), 407.

page 419, "The President of": Ibid, 429. Essential for understanding of the Founders' fears of destructive political factionalism is Madison's Federalist paper No. 10, from *New York Packet,* November 23, 1787, in which he brilliantly warns against "the instability, injustice, and confusion introduced into the public councils" by competing partisan factions. A democracy that fails "to break and control the violence of faction," he writes, creates "the mortal diseases under which popular governments have everywhere perished."

page 429, day after: "The End?" *Economist,* February 13, 1999.

page 431, "After decades of": Neal Gabler, "Historic 1974 vs. Histrionic 1998," *New York Times,* December 13, 1998.

pages 434–435, "The very idea": Francis X. Clines, "Over Time the End Was a Mirage: The Scandal Lives On," *New York Times,* February 28, 1999.

page 435, "I no longer": Richard L. Berke, "The Nation: The Far Right Sees the Dawn of the Moral Minority," *New York Times,* February 21, 1999.

page 435, "are complicit in": Ibid.

BOOK FOUR: MILLENNIAL TIMES

10. The People

page 444, "At a crossroads": Douglas A. Blackmon, "Forget the Stereotypes," *Wall Street Journal,* September 17, 1998.

page 445, "changing in far": Daniel Bell, "Into the 21st Century, Bleakly," *New York Times,* July 26, 1992, op-ed page, abbreviated from Bell's longer version in *Wilson Quarterly.*

page 449, "Bedded down usually": Editorial, "They Didn't Live So Long For This," *New York Times,* April 26, 1999.

page 450, "There are no dark": Don McLean et al. "Taste: A Confusion Over Identity," a symposium on "the current state of American culture," *Wall Street Journal,* March 20, 1998.

page 450, "The 90's are the 80's": Maureen Dowd, "Liberties: Vice Takes a Holiday," *New York Times,* December 31, 1997.

page 450, "In our pampered": Dowd, Ibid.

page 451, "When we were": Maureen Dowd, "Liberties: Cowboy Feminism," *New York Times,* April 11, 1999.

page 451, "Rational—and irrational": Frank Rich, "Smash-Mouth 1, Civility 0," *New York Times,* February 17, 2001.

page 451, "more addicted to": Ibid. For Frank Rich's use of "Mediathon," see his "The Age of the Mediathon," *New York Times Magazine,* October 29, 2000. I haven't done justice to Rich's commentaries here, but the "leads" of two of his other articles will perhaps give some of their wonderfully biting flavor. November 20, 1999 "Who Doesn't Want to Be a Millionaire": "When historians look back at America as it rang in the momentous year of 2000, what will they find? Y2K panic? An outbreak of millennial spirituality? 'Prosperity with a purpose'? Guess again. What they are going to see instead is a country drunk on a TV quiz show called 'Who Wants to Be a Millionaire.'" December 18, 1999 "The Future Will Resume in 15 Days": "In the beginning there was sex. Then there was violence. Then there was apocalypse. And finally there was shopping."

page 454, "What was too coarse": Kurt Andersen, "Blunt Trauma," *New Yorker,* March 30, 1998.

page 462, "the number of families": Eric Schmitt, "For First Time, Nuclear Families Drop Below 25% of Households," *New York Times,* May 15, 2001.

page 462, "With more communities": Ibid.

page 462, "We're losing": Nicholas Kulish, "Census 2000: The New Demographics," *Wall Street Journal,* May 15, 2001.

page 463, "Just a few": Patrick Barta, "Looming Need for Public Housing a Big Surprise," *Wall Street Journal,* May 15, 2001.

page 463, "These bare facts": Editorial, "Teen Moms," *Wall Street Journal,* May 8, 1998.

page 464, "at a pace much": Rene Sanchez, "Black Teen Suicide Rate Increases Dramatically," *Washington Post,* March 20, 1998.

page 465, in figures compiled: Shannon McCaffrey, Associated Press, "Gap between rich and poor widens," published, *Arizona Republic,* January 18, 2000.

page 465, "The sobering lesson": Eileen Daspin, "Philanthropy: How to Give More," *Wall Street Journal,* October 2, 1998.

11. The Markets

page 468, "How do we": Greenspan's famous speech was delivered December 5, 1996, at a black-tie dinner before the American Enterprise Institute in Washington, D.C. His "irrational exuberance" phrase became perhaps the most widely quoted remark of the decade. For background and perspective see Bob Woodward, *Maestro: Greenspan's Fed and the American Boom* (New York: Simon & Schuster, 2000), 178–81.

page 469, "one of the oldest": Frederick Lewis Allen, *Only Yesterday,* 303.

page 469, "The poorhouse is": Ibid.

pages 469–470, "History tells us": John Cassidy, "Pricking the Bubble. Annals of Finance," *New Yorker,* August 17, 1998. See also, Richard W. Stevenson, "The Markets: Inflation Remains a Danger, Greenspan Warns," *New York Times,* May 7, 1999.

page 470, "a set of imbalances": Jacob M. Schlesinger and Glenn Burkins, "Greenspan Clarifies View on Stock Market," *Wall Street Journal,* April 6, 2000. See also, E. S. Browning, "Greenspan Warns of Bubble, and Markets Keep Rising," *Wall Street Journal,* January 17, 2000.

page 474, "Two steps up": Frederick Lewis Allen, *Only Yesterday,* 309–10.

page 476, "The wise would": From The Motley Fool Web site home page.

page 476, "Day trading in": Burton G. Malkiel, "Day Trading and Its Dangers," *Wall Street Journal,* August 3, 1999.

page 477, "culture of risk-takers": Bernard Wysocki Jr., "High Rollers: How Life on the Edge Became Mainstream in Today's America," *Wall Street Journal,* August 3, 1999.

page 478n, "When the market": Malkiel, "Day Trading," op. cit.

page 482, "The most-exciting": Greg Ip, "Market on a High Wire," *Wall Street Journal,* January 18, 2000.

page 482, "Boosters have boldly": "Untangling e-conomics," *Economist,* September 23, 2000.

pages 482–483, "The trouble is": Ibid.

page 484, "Wall Street's bull": "A Moment of Economic Suspense," *New York Times,* March 14, 2001.

page 485, "Where is Mary?": Randall Smith and Mylene Mangalindan, "For E-Business Booster Meeker, Fame is E-Phemeral," *Wall Street Journal,* March 8, 2001.

page 485, "It has now": Gretchen Morgenson, "How Did They Value Stock? Count the Absurd Ways. Those Lofty 'New Economy' Measures Fizzle," *New York Times,* March 18, 2001.

page 486, "the magnitude of": Ron Chernow, "Paying for the Potemkin Boon," *New York Times,* March 15, 2001.

page 486, "Think of the": Ibid.

pages 486–487, "We remain confident": Scott Thurm, "Cisco Posts Period Loss of $2.69 Billion," *Wall Street Journal,* May 9, 2001.

page 487, "...this may be": Ibid.

page 487, "We've clearly had": Matt Richtel, "Cisco Takes $3 Billion Charge and Posts Its First Loss," *New York Times,* May 9, 2001.

page 487, "LAID...OFF": *Newsweek,* February 5, 2001.

page 488, "because the extent": "Excerpts from Greenspan's Testimony," *New York Times,* February 14, 2001.

page 488, "It is difficult": Ibid.

12. The Millennials

page 491, "students are increasingly": Ethan Bronner, "College Freshmen Aiming for High Marks in Income," *New York Times,* January 12, 1998.

page 492, "using education more": Ibid.

page 492, "Business executive." Ibid.

page 494, "What remains": Quoted in press release, "Most of the Nation's College Freshmen Embrace the Internet...," accompanying UCLA's 33d annual report, "The American Freshmen: National Norms for Fall 1998," January 25, 1999.

page 496, "Kids who started": Bronner, *New York Times,* op. cit.

page 496, "to learn a": Ibid.

page 497, "a reflection of": "Record Numbers of the Nation's Freshmen

Feel High Degree of Stress, UCLA Study Finds," press release, Oct. 23, 2000, reporting findings of the 1999 annual student survey.

page 498, "These findings": Ibid.

page 516, "If this fall's": Alex P. Kellogg, "Looking Inward, Freshmen Care Less About Politics and More About Money," *Chronicle of Higher Education,* January 26, 2000.

13. The Fiasco

page 519, "Although we may": Dissent of Supreme Court Justice John Paul Stevens in *Bush v. Gore.*

page 521, won the godless: quoted in "One Nation, Fairly Divisible, under God," *The Economist,* January 20, 2001.

pages 523–524n, In a scathing: Daniel J. Wakin, "Report Calls Networks' Election Night Coverage a Disaster," *New York Times,* February 3, 2001.

page 528, "The court's decision": Scott Turow, "A Brand-New Game: No Turning Back from the Dart the Court Has Thrown," Outlook, *Washington Post,* December 17, 2000.

page 528, "It was an act": Ibid.

page 529, "The conservatives stopped": Ronald Dworkin, "A Badly Flawed Election," *New York Review of Books,* January 11, 2001.

page 529, "The stark reality": Vincent Bugliosi, "Five Criminal Justices," *Nation,* February 5, 2001.

page 531, "many voters simply": Editorial, "Race and the Florida Vote," *New York Times,* December 26, 2000.

page 536, "All kinds of": Michael Waldman, *POTUS Speaks: Finding the Words That Defined the Clinton Presidency* (New York: Simon & Schuster, 2000).

page 538, "a haunting visual": "An Appraisal: Bill Clinton's Mixed Legacy," Editorial, *New York Times,* January 14, 2001.

pages 538–539, "we have to admit": "How Was It for You," *Economist,* January 13, 2001.

page 539, "In spite of my": *Time,* November 20, 2000.

page 540, formal farewell address: The president's farewell address was broadcast prime-time, January 18, 2001. Recorded transcript, "I'll Leave the Presidency More Idealistic," published in the *New York Times,* January 19, 2001.

page 542, "In a stunning end": "Transition in Washington: The President," *New York Times,* Jan. 20, 2001.

page 542n, Peter Slevin and Amy Goldstein, "Pentagon Fires Tripp From $98,000 Job," *Washington Post,* January 20, 2001.

page 544, "while flouting the": "Unpardonable," *Washington Post,* January 23, 2001.

page 544, "With his scandalous": Ibid.

page 545, PARDONS ON THE SLY: *New York Times,* Jan. 25, 2001.

page 546, COUNT THE SPOONS: *Washington Post,* January 24, 2001.

page 548, "Washington is a wet haze": Maureen Dowd, "Liberties: 41 on 43's Terrible 37," *New York Times,* January 21, 2001.

pages 548–549, "And so . . . in the rain-soaked": Francis X. Clines, "The Nation: Closing the Book: Changing the Story on the Last Page," *New York Times,* January 21, 2001.

Epilogue

page 550, The test of the historian . . .: Henry Adams, *The Education of Henry Adams* (Boston: Houghton Miflin, 1918), 395.

page 551, he thought he was witnessing: Ibid, 499.

page 551, "the child of incalculable": Ibid, 496.

page 551, "At the rate": Ibid, 496–97.

page 555, "It has long been": speech, Nov. 10, 1864. *Abraham Lincoln: Speeches and Writings,* vol. 2, *1859–1865* (New York: Library of America, 1989), 641.

Bibliography

GENERAL WORKS, HISTORY AND FICTION

Adams, Henry. *The Education of Henry Adams.* Boston: Houghton Mifflin, 1918.

Adams, Samuel Hopkins, *The Incredible Era: The Life and Times of Warren Gamaliel Harding.* Boston: Houghton Mifflin, 1939.

Allen, Frederick Lewis. *Only Yesterday: An Informal History of the Nineteen-Twenties.* New York: Harper & Bros., 1931.

Barzun, Jacques, *From Dawn to Decadence: 500 Years of Western Cultural Life, 1500 to the Present.* New York: HarperCollins, 2000.

Beer, Thomas. *The Mauve Decade: American Life at the End of the Nineteenth Century.* New York: Alfred A. Knopf, 1926.

Bellamy, Edward. *Looking Backward: 2000–1887.* New York: Modern Library, 1951.

Bryce, James. *The American Commonwealth.* 2 vols. New York: Macmillan & Co., 1895.

Carson, Rachel. *Silent Spring.* Boston: Houghton Mifflin, 1962.

Chernow, Ron. *Titan: The Life of John D. Rockefeller, Sr.* New York: Random House, 1998.

Dickens, Charles. *American Notes for General Circulation and Pictures from Italy.* London: Chapman & Hall, 1913.

———. *A Tale of Two Cities.* New York: Grossett & Dunlop, 1948.

Fitzgerald, F. Scott. *The Great Gatsby.* New York: Charles Scribner's Sons, 1925.

———. *The Crack-Up.* Edited by Edmund Wilson. New York: New Directions, 1945.

Frum, David. *How We Got Here: The 70's, The Decade that Brought You Modern Life (for Better or Worse).* New York: Basic Books, 2000.

Furnas, J. C. *Great Times: An Informal Social History of the United States, 1914–1929.* New York: G. P. Putnam's Sons, 1974.

Galbraith, John Kenneth. *The Great Crash, 1929.* Boston: Houghton Mifflin, 1955.

Goodwin, Doris Kearns, *No Ordinary Time: Franklin and Eleanor Roosevelt, The Home Front in World War II.* New York: Touchstone, 1995.

Griffith, Richard, and Arthur Mayer. *The Movies.* Rev. ed. New York: Fireside/Simon & Schuster, 1970.

Hamilton, Alexander, John Jay, and James Madison. *The Federalist: A Commentary on the Constitution of the United States.* Edited by Henry Cabot Lodge. New York: G. P. Putnam's Sons, 1904.

Hutchens, John K., editor. *The American Twenties: A Literary Panorama.* Philadelphia: J. B. Lippincott, 1952.

Huxley, Aldous. *Brave New World.* New York: Harper & Row, 1946.

Johnson, Haynes. *Sleepwalking Through History: America in the Reagan Years.* New York: W. W. Norton, 1991.

Kennedy, David M. *Freedom From Fear: The American People in Depression and War, 1929–1945.* New York: Oxford University Press, 1999.

Kidder, Tracy. *The Soul of a New Machine.* Boston: Little, Brown, 1981.

Leighton, Isabel, editor, *The Aspirin Age. 1919–1941.* New York: Simon & Schuster, 1949.

Lincoln, Abraham. *Speeches and Writings, vol. 2, 1859–1865.* New York: The Library of America, 1989.

Lippmann, Walter. *Public Opinion.* London: George Allen & Unwin, 1922.

Manchester, William. *The Glory and the Dream: A Narrative History of America, 1932–1972.* 2 vols. Boston: Little, Brown, 1973, 1974.

Myrdal, Gunnar. *An American Dilemma: The Negro Problem and Modern Democracy.* New York: Harper & Bros., 1944.

Parrington, Vernon Louis. *Main Currents in American Thought: An Interpretation of American Literature from the Beginning to 1920.* New York: Harcourt, Brace, 1927.

Sandburg, Carl. *Abraham Lincoln.* 4 vols. New York: Harcourt, Brace, 1939.

Schlesinger, Arthur M. Jr., *The Age of Jackson.* Boston: Little, Brown, 1972.

Schulberg, Budd. *What Makes Sammy Run?* New York: Random House, 1941.

Sullivan, Mark. *Our Times: The United States, 1900–1925.* 6 vols. New York: Charles Scribner's Sons, 1927–1935.

Thackeray, William Makepeace. *Vanity Fair: A Novel Without a Hero.* New York: Heritage Press, 1940.

Tocqueville, Alexis de. *Democracy in America.* 2 vols. Translated by Henry Reeve. New York: Colonial Press, 1899.

Truman, Harry S. *Memoirs, vol. 1. Year of Decisions, vol. 2, Years of Trial and Hope.* New York: Doubleday, 1955, 1956.

Tuchman, Barbara. *The Guns of August.* New York: Macmillan, 1962.

————. *The Proud Tower: A Portrait of the World Before the War, 1890–1914.* New York: Macmillan, 1966.

————. *A Distant Mirror: The Calamitous Fourteenth Century.* New York: Alfred A. Knopf, 1978.

Turner, Frederick Jackson. *The Frontier in American History.* New York: Henry Holt, 1920.

Twain, Mark, and Charles Dudley Warren. *The Gilded Age.* 2 vols. New York: Harper & Bros., 1899.

West, Nathanael, *Novels and Other Writings.* New York: Library of America, 1997. (Includes *Miss Lonelyhearts* and *The Day of the Locust.*)

Wolfe, Tom. *The Bonfire of the Vanities.* New York: Farrar, Straus, Giroux, 1987.

————. *A Man In Full.* New York: Farrar, Straus, Giroux, 1998.

Woodward, Bob, and Carl Bernstein. *All the President's Men.* New York: Simon & Schuster, 1974.

TECHNOTIMES

Abbate, Janet. *Inventing the Internet.* Cambridge, Mass.: MIT Press, 1999.

Andersen, Kurt. *Turn of the Century.* New York: Random House, 1999.

Auletta, Ken. *World War 3.0: Microsoft and Its Enemies*. New York: Random House, 2001.

Berners-Lee, Tim, with Mark Fischetti. *Weaving the Web: The Original Design and Ultimate Destiny of the World Wide Web by Its Inventor*. New York: HarperCollins, 2000.

Bronson, Po. *The Nudist on the Late Shift*. New York: Random House, 1999.

Brooks, David. *Bobos in Paradise: The New Upper Class and How They Got There*. New York: Simon & Schuster, 2000.

Ceruzzi, Paul E. *A History of Modern Computing*. Cambridge, Mass.: MIT Press, 1998.

Cook-Deegan, Robert. *The Gene Wars: Science, Politics, and the Human Genome*. New York: W. W. Norton, 1994.

Davis, Stanley M., and Christopher Meyer. *Blur: The Speed of Change in the Connected Economy*. Reading, Mass.: Addison-Wesley, 1998.

Friedman, Thomas L. *The Lexus and the Olive Tree: Understanding Globalization*. New York: Farrar, Straus, Giroux, 1999.

Fukuyama, Francis. *The End of History and the Last Man*. New York: Free Press, 1992.

Gardner, David, and Tom Gardner. *The Motley Fool Investment Guide*. New York: Simon & Schuster, 1996.

Goldstein, Judith R., *Millikan's School: A History of the California Institute of Technology*. New York: W. W. Norton, 1991.

Goodstein, David L., and Judith R. Goodstein. *Feynman's Lost Lecture: The Motion of Planets around the Sun*. New York: W. W. Norton, 1996.

Himmelfarb, Gertrude. *One Nation, Two Cultures*. New York: Alfred A. Knopf, 1999.

Kovach, Bill, and Tom Rosenstiel. *Warp Speed: America in the Age of Mixed Media*. New York: Century Foundation Press, 1999.

Lewis, Michael. *The New New Thing: A Silicon Valley Story*. New York: W. W. Norton, 2000.

Licklider, J. C. R. *Libraries of the Future*. Cambridge, Mass.: MIT Press, 1965.

McCartney, Scott. *ENIAC: The Triumphs and Tragedies of the World's First Computer*. New York: Walker, 1999.

McKenna, Regis. *Who's Afraid of Big Blue? How Companies Are Challenging IBM—and Winning*. Reading, Mass.: Addison-Wesley, 1989.

———. *Real Time: Preparing for the Age of the Never Satisfied Customer*. Boston: Harvard Business School Press, 1997.

Rivlin, Gary. *The Plot to Get Bill Gates: An Irreverent Investigation of the World's Richest Man . . . and the People Who Hate Him.* New York: Random House/Times Business, 1999.

Segaller, Stephen. *Nerds 2.0.1: A Brief History of the Internet.* New York: TV Books, 1998.

Smith, Bruce L. R. *American Science Policy Since World War II.* Washington, D.C.: Brookings Institution, 1989.

Swisher, Kara. *AOL.COM: How Steve Case Beat Bill Gates, Nailed the Net-heads, and Made Millions in the War for the Web.* New York: Random House/Times Business, 1998.

Zachary, G. Pascal. *Endless Frontier: Vannevar Bush, Engineer of the American Century.* Cambridge, Mass.: MIT Press, 1999.

THE CLINTON AGE

Baker, Peter. *The Breach: Inside the Impeachment and Trial of William Jefferson Clinton.* New York: Charles Scribner's Sons, 2000.

Conason, Joe, and Gene Lyons. *The Hunting of the President: The Ten-Year Campaign to Destroy Bill and Hillary Clinton.* New York: St. Martin's Press, 2000.

Drew, Elizabeth, *On the Edge: The Clinton Presidency.* New York: Simon & Schuster, 1994.

Dionne, E. J., Jr., and William Kristol, editors, *Bush v. Gore: The Court Cases and the Commentary.* Washington: Brookings Institution Press, 2001.

Johnson, Haynes. *Divided We Fall: Gambling With History in the Nineties.* New York: W. W. Norton, 1994.

——— and David S. Broder. *The System: The American Way of Politics at the Breaking Point.* New York: Little, Brown, 1996.

Isikoff, Michael. *Uncovering Clinton: A Reporter's Story.* New York: Crown, 1999.

Maraniss, David. *First in His Class: A Biography of Bill Clinton.* New York: Simon & Schuster, 1995.

———. *The Clinton Enigma: A Four-and-a-Half-Minute Speech Reveals This President's Entire Life.* New York: Simon & Schuster, 1998.

New York Times Staff, *36 Days: The Complete Chronicle of the 2000 Presidential Election Crisis.* New York: Times Book/Holt paper, 2001.

Schmidt, Susan, and Michael Weisskopf. *Truth at Any Cost: Ken Starr and the Unmaking of Bill Clinton*. New York: HarperCollins, 2000.

Toobin, Jeffrey. *A Vast Conspiracy: The Real Story of the Sex Scandal That Nearly Brought Down a President*. New York: Random House, 1999.

Waldman, Michael. *POTUS Speaks: Finding the Words That Defined the Clinton Presidency*. New York: Simon & Schuster, 2000.

Washington Post Political Staff. *Deadlock: The Inside Story of America's Closest Election*. New York: Public Affairs, 2001.

Woodward, Bob. *The Agenda: Inside the Clinton White House*. New York: Simon & Schuster, 1994.

———. *Maestro: Greenspan's Fed and the American Boom*. New York: Simon & Schuster, 2000.

Index